Video

Digital Communication & Production

Second Edition

By

Jim Stinson

Photos and graphics by the author, unless otherwise credited.

Publisher
The Goodheart-Willcox Company, Inc.
Tinley Park, Illinois
www.g-w.com

Manufactured in the United States of America.

Library of Congress Catalog Card Number

ISBN-13: 978-1-59070-767-8
ISBN-10: 1-59070-767-2

2 3 4 5 6 7 8 9 – 08 – 12 11 10 09 08 07

The Goodheart-Willcox Company, Inc. Brand Disclaimer: Brand names, company names, and illustrations for products and services included in this text are provided for educational purposes only and do not represent or imply endorsement or recommendation by the author or the publisher.

The Goodheart-Willcox Company, Inc. Safety Notice: The reader is expressly advised to carefully read, understand, and apply all safety precautions and warnings described in this book or that might also be indicated in undertaking the activities and exercises described herein to minimize risk of personal injury or injury to others. Common sense and good judgment should also be exercised and applied to help avoid all potential hazards. The reader should always refer to the appropriate manufacturer's technical information, directions, and recommendations; then proceed with care to follow specific equipment operating instructions. The reader should understand these notices and cautions are not exhaustive. The publisher makes no warranty or representation whatsoever, either expressed or implied, including but not limited to equipment, procedures, and applications described or referred to herein, their quality, performance, merchantability, or fitness for a particular purpose. The publisher assumes no responsibility for any changes, errors, or omissions in this book. The publisher specifically disclaims any liability whatsoever, including any direct, indirect, incidental, consequential, special, or exemplary damages resulting, in whole or in part, from the reader's use or reliance upon the information, instructions, procedures, warnings, cautions, applications, or other matter contained in this book. The publisher assumes no responsibility for the activities of the reader.

Library of Congress Cataloging-in-Publication Data

Stinson, Jim
Video: digital communication and production / by Stinson, Jim
p. cm.
ISBN-13: 978-1-59070-767-8
ISBN-10: 1-59070-767-2
1. Video recording. 2. Video recordings-Production and direction. I. Title. II. Title: Video: digital communication and production

TR850.S7375 2007
778.59--dc22 2006048694

About the Author

After graduating from Harvard, Jim Stinson studied theater history at the Yale Graduate School and directing at the Yale School of Drama before transferring to the UCLA film school, where he earned the degree of Master of Fine Arts. Although he has worked on filmed commercials, TV series, and feature films, he has spent most of his career as a writer, producer, director, videographer, and/or editor of educational and corporate video programs.

In the classroom, he has taught film production at Art Center College of Design, film studies at California State University, Los Angeles, and video production at La Cañada High School, La Cañada, CA. Recently, he has been presenting twelve or more seminars each year on video production.

About this Book

Video: Digital Communication and Production fulfills the promise of its title by covering both the ways in which video communicates with viewers and the methods by which it does so. Communication is featured because production by itself has no purpose. If communication were excluded, this book would be like a carpentry manual that covered sawing, drilling, and nailing without ever explaining how to build anything.

This book treats video as a mature and independent medium, rather than merely a variant of television or a recording alternative to film. Video has become fully empowered by the digital revolution that is transforming so many aspects of twenty-first century life.

The topics in this book have been selected and organized with two groups of readers in mind: students preparing for careers in communications media and creators of personal programs who expect to make videos of professional caliber. Though the text does not pretend to include all there is to know about video, it does cover all you need to get started.

To organize this sprawling subject, ***Video: Digital Communication and Production*** is presented in six major sections:

- Chapters 1 and 2 help you start making videos immediately.
- Chapters 3 through 8 cover video communication: the concepts and principles behind the hardware and production techniques.
- Chapters 9 and 10 present the crucial process of preproduction: preparing to make successful programs.
- Chapters 11 through 16 introduce all major aspects of videography, lighting, and audio.
- Chapters 17 and 18 survey the art of directing—both the camera and the people it records.
- Chapters 19 through 24 explain the basics of postproduction. This edition provides greatly expanded coverage of digital editing processes and techniques.

This organization may be termed "semi-random-access:" On the one hand, it is possible to read only the chapters desired, in any order. On the other hand, individual chapters will generally be more useful in conjunction with the others in their sections. In most cases, larger subjects have been distributed among multiple chapters for simplicity of presentation. For this reason, expect to find occasional duplication of material, since the same concepts and techniques may apply to procedures covered in different chapters.

About This New Edition

The word "Digital" has been added to the title of this book for this edition, reflecting the fact that video production is now entirely digital, from the camcorder at one end to the DVD at the other. The computer has also taken over film-based production so completely that, outside of certain technicalities of lighting and cinematography, a person trained in video can work in film with equal confidence.

To reflect this trend, the original four chapters on postproduction have been largely replaced by six new or heavily revised units that treat digital editing in greater depth, including professional-level DVD authoring.

Other major revisions reflect requests from readers:

- The chapters on lighting have been reorganized and expanded from two to three, to clarify procedures and include more solutions for practical lighting problems.
- The chapter on visual composition has been doubled in length, to provide more examples and to address the challenge of visual design for the wide-screen video format that is starting to replace the traditional TV proportions.

- The chapter on program development has been expanded to deal more thoroughly with scriptwriting for different video genres.
- The five chapters on video communication have been augmented by a sixth unit that brings together the separate concepts of video space, time, composition, language, and sound, to present a unified esthetic of expression in this medium.

Finally, the text throughout has been reviewed and revised as needed, to clarify information, update technical details, and add useful information. Hundreds of illustrations have been added, revised, or replaced to expand content and/or improve presentation.

Overall, the author and publisher hope that we have increased the attractiveness of this book in the process of enhancing its usefulness.

The author and publisher cordially solicit corrections and suggestions from readers. Please help us improve the book by e-mailing us at **www.g-w.com**.

Finally, as you use this book, remember that video production is both a demanding discipline and a source of great satisfaction. Like pro golfers, video professionals are generally pleased to be paid for doing things that they would happily do for free. If ***Video: Digital Communication and Production*** enhances your pleasure as it increases your skills, then the book will have fulfilled its purpose.

Jim Stinson
Portland, Oregon
www.jimstinson.com

A note on "tape:" This new edition somewhat reduces the use of words like "taping" and "footage" as alternative recording formats become more popular. However, classic terms are highly resistant to change. After all, people still speak of "filming" with a camcorder and actual film is often called "celluloid," which disappeared over 60 years ago!

JS

Contents

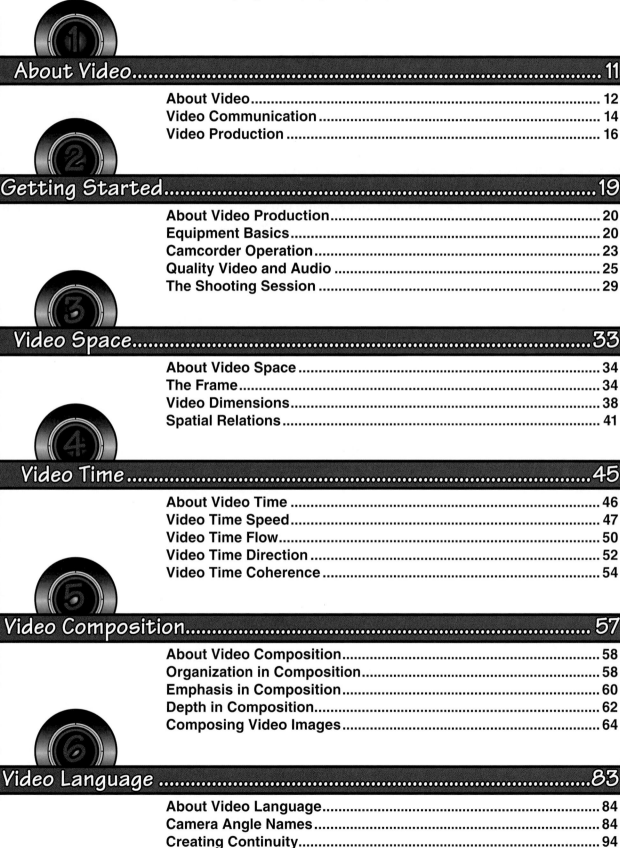

Acknowledgments

Dedication: For Sue, still more than ever.

The following organizations supplied photographs for the book:

ACD Systems
Adobe
Alpenlite/Western Recreational Vehicles
Anton Bauer
Azden
Barber Technologies
Bogen Imaging
Canon USA
Casablanca by MacroSystem
Cool-Lux
Corel
J. L. Fisher, Inc.
Hollywood Lite
JVC Company of America
Lowel-Light Manufacturing, Inc.
Mamiya America Corporation
Microdolly Hollywood
Minnesota Public Radio
Movie Magic Budgeting/Creative Planet
NASA
Neumann
Panasonic Broadcast
PBS
Photoflex Lighting
Pinnacle Systems
Sennheiser
Shotmaster/The Badham Company
Sony Electronics, Inc.
Sound Forge
The Tiffen Company LLC
Ulead Systems, Inc.
Videoguys
Videonics

Special thanks to Sue Stinson for the many photos she contributed to the book.

Many photos feature the stars of The McKinley Dramatic Arts and Vintage Tractor Society, especially, Jake Bennion, Ileana Butu, Claudia Conroy, James Conroy, Mark Conroy, Garrett Nasalroad, Mauro V. Escobar, Jerry Ferree, Louanne Ferree, Tommie L. McAuley, Jessica Kidd Millar, Sam Millar, Dalan Musson, Annegret Ogden, Dunbar Ogden, Jennifer Schalf, J. D. Schall, Louanne Silveira, Marge Simas, Norm Simas, Alex Stinson, Anji Stinson, Lillie Stinson, Sue Stinson, Vincent Vaughn, and Alicia Wilkinson.

The fictional town of McKinley was created from shots of people, places, and events in the Humboldt County, California, cities of Arcata, Eureka, Ferndale, Fortuna, McKinleyville, and Trinidad. The annual three-day Kinetic Sculpture Race begins in Arcata and ends, more or less, in Ferndale.

1

About Video

Objectives

After studying this chapter, you will be able to:

- Explain the meaning of "Video Communication."
- Explain why it is important to understand the nature of the video world.
- Describe the three major phases of video production.

About Video

Video is a pursuit that attracts people with a wide variety of talents, so you are almost certain to find a place in it for your particular abilities and interests. Because making videos is so interesting, you may want to pick up a camcorder right now and get started. But if you first take the time to review this chapter and the next, you will have a better idea of what video is all about, and you will be able to make better video programs from the start.

What Is Video?

Exactly what is "video?" The answer is not as obvious as it seems. Until recently, video—as we treat it in this book—did not exist. Instead, there were just two main audiovisual media: *film* and *television.*

Film was the medium used for creating most audiovisual programs, from movies to TV commercials. Film was (and is) an excellent production medium for several reasons:

- Film equipment is relatively portable, so location filming is practical.
- Film's ability to reproduce quality images in black and white or color is highly refined.
- Film picture and sound tracks are usually recorded on separate strips of film (or audio tape), so sophisticated editing is possible.

Television was the medium used for broadcasting studio programs as they happened (**"live"**), and other programs previously produced on (or copied onto) film, **Figure 1-1**.

Originally, television was not an ideal production medium for several reasons:

- Its equipment was heavy, complex, and tied down by cables to its control systems.
- Its image quality was markedly lower than that of film, and its ability to render shades of gray from black to white was limited. (Color could not be reproduced at all, except in experimental setups.)
- It could not be recorded for later editing, except by copying the live signal to film and then treating it as if it were a filmed program. These "kinescope" films degraded picture quality even further.

As the popularity of television grew over the years, equipment manufacturers gradually solved most of its problems. They miniaturized hardware until a broadcast-quality camera and its recorder could be combined in a package smaller than the size of a film camera plus its attendant sound tape recorder. This combination ***camcorder***, **Figure 1-2**, also considerably reduced the tangle of studio cables. At the same time, engineers greatly improved picture sharpness and gray scale range and developed high-quality color. They perfected videotape recording, so the television signal could be electronically copied and edited.

Today you can obtain high quality video and audio from camcorders that weigh less than three pounds.

In short, television technology eventually improved until the medium rivaled the abilities of film. Today you can produce professional quality programs on either film or videotape with comparable ease and practicality.

Figure 1-1.
Comedians Dean Martin and Jerry Lewis in a 1950s TV studio. (Gene Lester/Archive Photos)

Figure 1-2.
A compact professional camcorder. (Panasonic Broadcast)

Nevertheless, the two media remain quite different in many essential respects. Though film has continued to evolve over the years, it is still essentially the same medium. By contrast, the electronic production technology that began as "television" has changed so drastically that it needs a name of its own.

That is why we now call it ***video***.

Video versus Film

Some film makers think that video is less desirable than their own medium in several ways:

- Video picture clarity is comparatively coarse because its resolution is relatively low, **Figure 1-3**.
- Video color sometimes lacks a certain richness and "snap" that is hard to describe but easy to see in film. That is why some TV commercials are shot on film, even though they will be seen only on video.
- Until recently, video has offered less flexibility in sound editing. It is common for even simple films to mix as many as eight or more audio tracks. Sound editing has been more cumbersome in video.

In other ways, however, film is less desirable than video:

- Film is much more expensive to shoot and process to a final composite positive print. While small format video can cost as little as two dollars an hour, finished film costs hundreds of times as much.
- Film is less tolerant of different light levels, so it must be supplied in several grades of light sensitivity, to suit different conditions.
- Film sound is more cumbersome, since it is almost always recorded on a separate tape recorder.
- Film color balancing is time-consuming and expensive.
- Film titles and effects, such as dissolves and double exposures, cannot be added in real time and evaluated on the edited working copy. Instead, they must be created separately in the film laboratory.
- Film editing requires negative cutting, a tedious and expensive extra step. Since the original camera film is rarely used to create an edited program, a negative cutter must match it, frame-for-frame to the program's completed "work print," before viewable "release prints" can be made.

Figure 1-3.
At close range, the low resolution of video is obvious. (PBS)

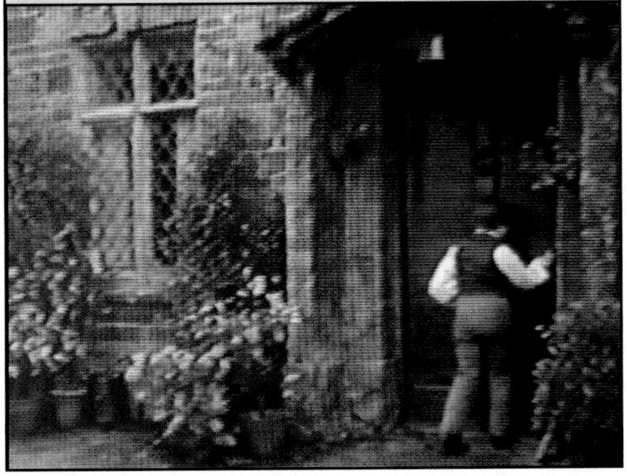

Many films now avoid these problems by digitizing the original footage to create a "digital intermediate" (DI), completing postproduction in pure video mode, and then converting the finished production back to film for release.

Converging Technologies

Today, however, the arguments for and against film or video are growing outdated as the two media grow ever closer together. High-definition video, for example, is close to the visual quality of film; and modern film stocks have wider exposure ranges.

At the same time, visual hybrids are being created. For example, commercials that are shot on film are then transferred immediately to tape. The rest of the production process is pure video. In other cases, videotape work prints are made, editing is completed, and then the film negative is cut to match the tape. Finally, in many theatrical films, the special effects are created electronically and then transferred to film.

Working in video mode involves *digitizing* the film: scanning it frame-by-frame and converting it to a pattern of tiny dots. Since a large computer can convert a frame of film to a matrix of about 4,000 by 3,000 pixels (dots), the digitized images remain film-sharp.

Digital postproduction helps video as well as film, especially in sound editing. Since computerized editing software can handle an almost limitless number of audio channels, multilayer sound tracks are now common.

Although film and video use different production techniques, they speak the same audiovisual "language." If you master this language, you can use it to communicate in either medium or in any hybrid of both.

Types of Video Production

Today, people make a wide variety of video programs ranging from five-second commercials to thirteen-hour miniseries. These are distributed by broadcast, cable, satellite TV and the Internet, and on discs played in schools, businesses, and homes. Some videos are produced to entertain, to persuade, or to teach. Others capture a family holiday, inventory a stamp collection, or document a vacation.

Beyond the kinds of videos we think of as "programs" lie still other types, such as the specialized videography employed in areas as diverse as medicine, industry, science, and law enforcement. In fact, professional uses for video are expanding in much the same way as computer applications. It is probably safe to say that all but a few 21st Century careers will involve video in some way.

Video Talents and Jobs

If you are considering a career in video, you have a wide range of specialties to choose from.

If you like the story side of production, try writing, directing, or editing. Though these crafts involve different skills and techniques, they are really all part of the same process. If you have graphic talents, the specialties of art direction, set and costume design, and makeup are vital to sophisticated video production. You can also create postproduction video graphics and titles.

If you are intrigued by the nuts and bolts of production, then camera operation, lighting, and audio recording are skills that command respect.

If you have technical aptitude, audio and video engineering are challenging occupations. If management is your aim, video producing and production management require well-developed skills in organization, personnel, and finance.

Every type of business expertise is employed in running a video production unit or company. To function effectively, corporate managers need to know the fundamentals of video as thoroughly as their colleagues on the production side. In particular, they need to understand how the video medium communicates its message.

Not for Professionals Only

This book treats the art and craft of video on a professional level. Making video programs is usually a collaborative process carried on by personnel working in organizations — that is, by professionals.

However, you may not anticipate a career in video production. Instead, you may want to master this medium purely for personal expression, **Figure 1-4**. If so, you need not be concerned that video is discussed here in a professional context.

Figure 1-4.
Consumer camcorders offer professional features.

Like music or painting or photography or cabinetmaking, video is an art in which "amateur" practitioners can and should develop exactly the same skills that full-time professionals use to make their living in this medium.

Video Communication

You may become the best camera person on your block, the finest sound recordist, or the most talented makeup artist. But you still will not know how to make a video unless you master the art of video communication.

This book includes several chapters on video communication, covering three broad topics: the nature of the video world, the language of video expression, and the construction of video programs.

Visual Literacy

Even if your career never involves producing video, you spend considerable time consuming it — several hours a day if you are a typical TV viewer. Video is the most persuasive and powerful system ever invented for delivering facts, ideas, and opinions. You may not think video affects you all that much, but it does:

- It persuades you about what to buy and how to vote.
- It explains who is important in the world and why.
- It teaches you how your community, your nation, and your planet are working — and how they should work if the world were an ideal place.
- It shows you what is fashionable, what is desirable, and what you are supposed to want in life.
- It models ways to love, fight, work, worship, and dream.

Video does all this with a vividness and immediacy that make you feel that you are experiencing the actual sights and sounds of the real thing.

In fact, you are not seeing the real thing. Instead, you are presented with a sophisticated and carefully contrived imitation of reality that the video makers want you to accept. They may present the "reality" that soap A is better then soap B. They may "prove" that political candidate X is superior to candidate Y. They may "demonstrate" that wolves are essential to sound ecology (or, alternatively, that wolves are livestock killers that must be exterminated).

Even when they are not trying to sell you a product, a person, or an idea, makers of video programs do not present reality; rather, they offer their own versions of it. They cannot help doing this, because even the most objective program simply cannot be made without selecting and condensing reality. This essential process imposes a certain point of view on every video program, because the selection and presentation of material reflect a set of standards adopted by the program's makers.

What does that mean to you as a viewer? If you understand the techniques of video communication, you can separate the information you are watching from the methods used to organize and present it. At best, you will be able to get past the artificial reality on your screen to search for the actual reality beyond it.

At least, you will not be deceived by a medium that looks real *always* but is real *never.*

This ability to understand how media work is often called "visual literacy," and the skills involved are the subject of much of this book.

The Nature of the Video World

When you watch TV, you may think you are looking at a picture of the actual world, but you are not. The TV screen is a window that looks out on a completely different universe — a strange cosmos in which normal laws of space and time and gravity do not work at all. In the video universe:

- An actor can open a door in London and walk through it to Los Angeles.
- A car can turn a corner and jump forward a week in time.
- An actor can fall ten stories onto concrete and walk away unhurt.

Figure 1-5.
The outside of this door is a location.

The inside of the door is on a set.

Except during spectacular special effects, most of the strange behavior of the video universe is quite invisible to the audience. Writers, directors, designers, and editors understand how to use the laws of the video world to fool the viewer, **Figure 1-5**. These laws are so powerful that unless you understand them, you cannot effectively tape a single scene or competently edit two shots together.

Chapters 3 and 4 explain the rules and regulations of the video universe and show how to take advantage of them to make effective programs.

The Language of Video Expression

Video communication uses a visual language — a language with rules much like those of a written language such as English or Chinese.

For example:

- An image is much like a single word.
- A shot is like a complete sentence.
- A scene is like a paragraph.
- A sequence is like a chapter.

Unlike their written languages, the video languages of England and China (and almost all other nations) are the same. Video is a powerful social force, partly because almost everyone on the planet understands the universal language of film and video.

At the basic level, video has a grammar, with the equivalents of subjects and verbs and tenses. On a more sophisticated level, video has its own rhetoric — a wealth of techniques for creating distinctive styles of expression. These techniques include the management of composition and camera movement, the creation of visual continuity, and the control of program rhythm and pace.

Like literary styles, powerful visual styles are recognizable. Just as a knowledgeable reader can tell Dickens from Hemingway, an informed viewer can identify a film by Fellini or Bergman or Scorcese.

The Construction of Video Programs

Video communication is like written communication in yet another way: it is not enough to write a grammatically correct sentence or compose a short paragraph. You must also be able to organize and develop a coherent story or essay — or even a whole book. To make a professional video program of any length, you have to design and construct it as carefully as you would a piece of professional writing.

Effective nonfiction programs, such as training films and documentaries, depend on logical subject organization, clear presentation, and energetic pacing. Fiction, music, and performance videos also demand the skills of the artist and storyteller.

The shortest videos can be as hard to construct as the longest. A 30-, 10- or even 5-second commercial must organize and present its material with rigorous economy, so that it creates its intended effect.

Video Production

Video production falls into three broad topics: preproduction, production, and postproduction.

Preproduction

The preproduction phase includes everything you do before actual shooting begins. Preproduction includes scripting (or storyboarding) the video, **Figure 1-6**, as well as scouting locations, gathering cast and crew, and planning production equipment and other requirements.

Preproduction may not seem as exciting as shooting or as creative as editing, but experienced professionals know that it is a mistake to shortchange the preproduction phase. Video programs

Figure 1-6.
A storyboard is a common alternative to a script.

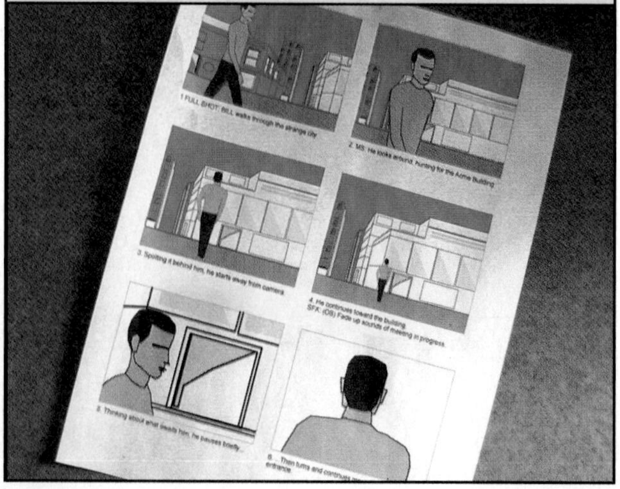

Figure 1-7.
The storyboard is revised during preproduction.

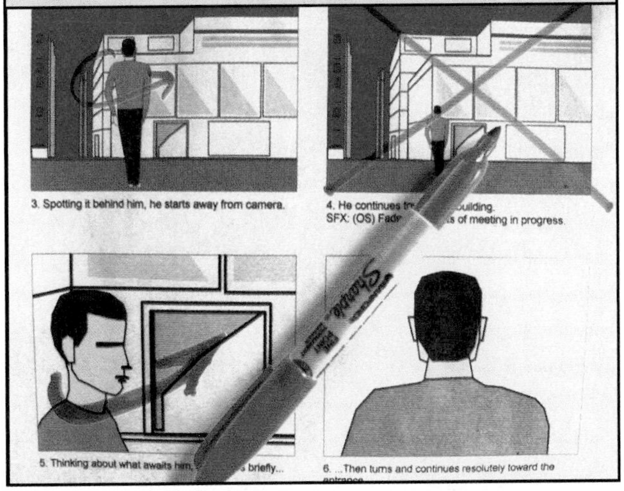

are generally made in small, disconnected pieces. Advance planning is essential to determine how these pieces should fit together, **Figure 1-7**. Also, in professional video production, time really is money, so the more thoroughly you have planned your ***shoot***, the faster you can complete it. In creating professional programs, the preproduction phase is often longer and more complicated than the production phase.

Production

The production phase covers the actual shooting of the material that will become the video program. In most productions, responsibility for the visual character, or "look," of the program is shared by the director, videographer, and production designer. The director also guides the performers through their roles and ensures that the program content is recorded from appropriate points of view.

The director is usually supported by a production management staff, while the videographer is assisted by a lighting director, sound recordist, and key grip. (Grips are staff members who practice many of the technical crafts associated with production.)

Working in close collaboration, the production people, technicians, and performers stage and record the video footage that will become the basis of the finished program, **Figure 1-8**.

Postproduction

At the end of the production phase, many video newcomers are essentially finished. "Shoot it and show it" might be their motto. But for more experienced video makers, the postproduction phase is just as important as production, and just as enjoyable.

When you finish shooting, you do not yet have a video program, but only a collection of footage. In postproduction you select the shots you want to include, assemble them in order, add music and sound effects to the audio, and create titles and visual effects. Collectively, this process is called editing, **Figure 1-9**. When practiced skillfully, editing is so creatively satisfying that some video makers think of the production

Figure 1-8.
Shooting a documentary program. (Sue Stinson)

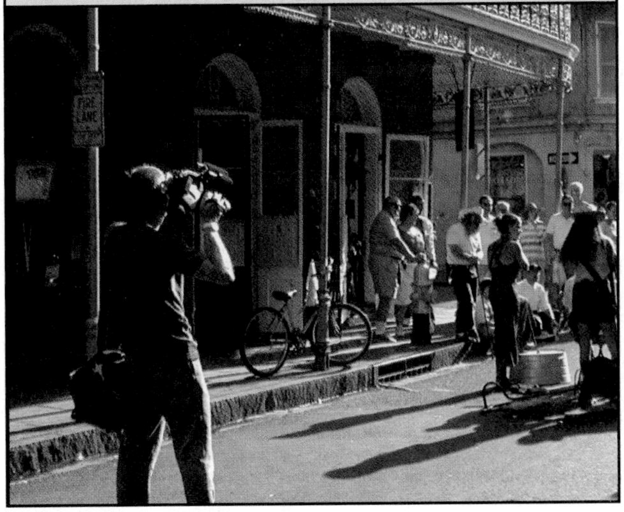

Figure 1-9.
Most of the postproduction phase is spent editing the program. (JVC)

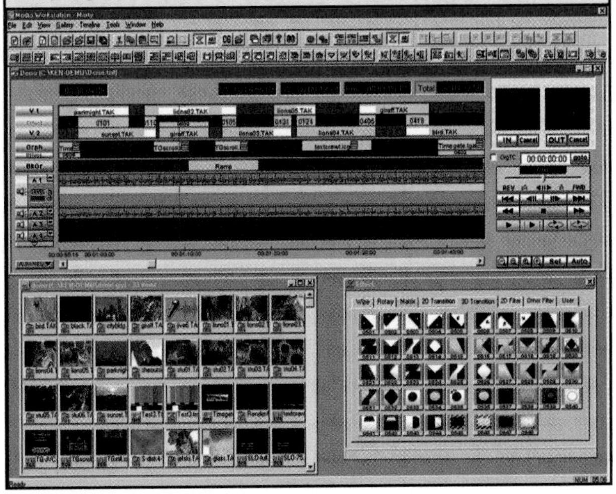

phase as primarily a source of raw material for postproduction.

The director Alfred Hitchcock said he did not find the production phase especially interesting because by the time it started, he had already finished making the entire movie in his head!

The Tip of the Iceberg

This introductory chapter has provided a very brief overview of video communication and production. The rest of this book is devoted to the details of this rich and fascinating subject.

You do not have to master every chapter, however, before you begin to shoot videos. Chapter 2 contains the essentials you need to make short, simple programs. When you have completed that chapter, you should be able to achieve satisfying results in beginning video production.

Development of the camcorder made it possible to blend film and television techniques into a new kind of medium known as "video."

Chapter Review

Answer the following questions on a separate piece of paper. Do not write in this book.

1. The TV screen is a window that looks out on a completely different universe —the ______ universe.
2. A ______ combines the functions of a video camera and a sound recorder.
3. Film and video speak the same audiovisual "______."
4. *True or False?* Amateur video makers do not need to develop the same skills as professionals.
5. Video is a medium that looks real always, but is real ______.

Technical Terms

Camcorder: An appliance that both captures moving images (camera) and stores them on tape or other media (recorder).

Digitize: To record images and sounds as numerical data, either directly in a camcorder or in the process of importing them to a computer.

Film: An audiovisual medium that records images on transparent plastic strips by means of photosensitive chemicals.

Live: Recorded and (usually) transmitted for display continuously and in real time.

Shoot: To record film or video; also, "a shoot" is an informal term for the production phase of a film or video project.

Television: Studio-based, multicamera video that is often produced and transmitted "live."

Video: An audiovisual medium that records on magnetic tape or other media by electronic means; also, single-camera taped program creation in the manner of film production rather than studio television.

Visual literacy: The ability to evaluate the content of visual media through an understanding of the way in which it is recorded and presented.

2 Getting Started

Objectives

After studying this chapter, you will be able to:

- Operate basic video equipment.
- Videotape a simple program.
- Avoid common shooting mistakes.
- Conduct a safe and courteous shoot.

JVC

About Video Production

This chapter is designed to get you ready for successful video making as quickly as possible.

Of course, you could skip this chapter and begin right away. Many new camcorder owners unpack their hardware, drop in a tape, and start shooting immediately. But the results are often unsatisfying, because even the simplest video project demands that you know at least a few fundamental ideas and techniques.

To create a good program, you need to know how to operate the equipment, how to record good-quality video and audio, and how to tape footage that can be edited effectively. By the time you have finished this chapter, you should be able to do these things comfortably. We will begin with a quick run-through of your camcorder controls and other equipment.

Equipment Basics

Camcorders are unusual because they are both simple and complicated at the same time. On the one hand, they are so simple that small children can use them without instruction. On the other hand, they are so complicated that many owners never do figure out how to use all the buttons and menus on their hardware.

Basic Camcorder Controls

To operate a camcorder successfully, you need to understand four essential controls: power, record, zoom, and white balance.

The ***power switch*** turns the camcorder on and off. Many power switches include a small light to remind you whenever the unit is running. This information is important because a running camcorder uses battery power even when it is not recording.

Technically, a video *camcorder* includes both a camera and a videotape recorder in a single unit, while a video *camera* is connected by cable to a separate recorder. In practice, however, camcorders are often referred to as "cameras."

Some cameras include power-saving circuits that put the unit into "standby" or "sleep" mode when it has not been recording for a certain period. This will happen often when you are shooting, so you need to learn how to return the camera to full power. Different units are reactivated in different ways.

The ***record switch*** starts and stops the actual recording process. Some camcorders have two record switches, one in a convenient location on the handgrip and the other on the camera body.

Many camcorders also have a record switch on a wireless remote control.

Note that you do not have to hold down the record switch continuously while taping. Simply press it once to begin and then press it again to stop.

The ***zoom control*** may be a pair of buttons or a bar that rocks back and forth, **Figure 2-1**. In either system, you press one button (or bar end) to zoom in and fill the screen with a narrower part of the scene and the other button (or bar end) to zoom out and fill the screen with a wider part of the scene. On some camcorders you can zoom the lens at different speeds, depending on how firmly you press the zoom controls.

Figure 2-1.
Most zoom controls are rocker bars. (JVC)

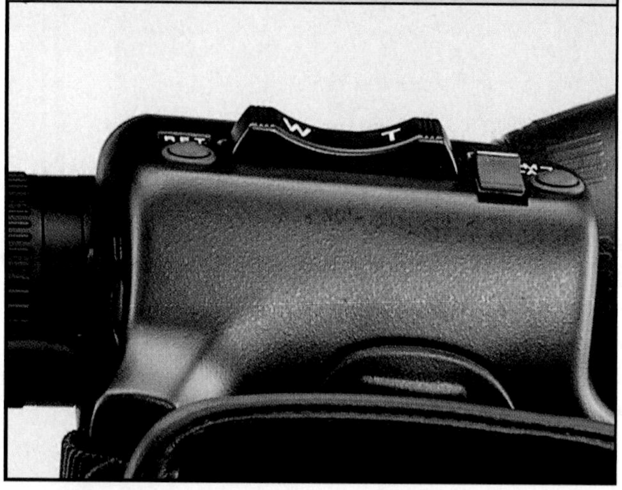

The ***white balance*** switch matches your camcorder to the overall color quality of the light in which you are shooting. All camcorders have an "automatic" setting for white balance. For now, use this setting to let the camcorder read the incoming light color and adjust itself automatically.

In some camcorders, the white balance and many of the other controls are controlled via menus displayed in the viewfinder or on a special LCD panel.

Automatic Camera Controls

White balance is not the only control that can be set to work automatically. The camcorder can also automate lens focus and exposure control.

The ***autofocus control*** keeps the picture sharp and clear. (Like your eyes, the camcorder lens must be set one way to record distant objects and another way to record close ones.) On the automatic setting, your camcorder estimates the distance to the central subject in the viewfinder. Then the camcorder sets the lens for that estimated distance. For now, leave the autofocus enabled, even though this may present occasional focus problems. Later chapters discuss how and when to use autofocus to best advantage.

Throughout this book, *enabled* means that a function is turned on (activated) and *disabled* means that it is turned off (deactivated).

The ***autoexposure control*** regulates the amount of light admitted through the camcorder lens. Whether shooting a candle-lit birthday or a day on bright ski slopes, the camcorder must have exactly the same amount of light to form a good picture. The auto exposure system delivers that precise amount. Like autofocus, the autoexposure system can be fooled sometimes. Leave it enabled for now.

Default Settings

A ***default*** is an action or condition that is automatically chosen by the equipment, unless you actively select a different one. In most camcorders, the autofocus and autoexposure functions are default settings, enabled automatically when the power is turned on.

In some camcorders, the white balance does not default to the automatic setting. Instead, it remains on the setting selected before the unit was last turned off. If the light color is different when you turn it on again, you may record footage with an unpleasant color cast, **Figure 2-2**. For this reason, always check the white balance before you shoot. For now, set it to automatic.

Videotape

Videotape, of course, is the medium on which the camcorder records both picture and sound. All camera videotape is supplied in closed plastic cassettes. Later, you will learn how videotape works, but for now, you need to know just the basics of handling, loading, and running it.

Some digital camcorders record on disks or other media.

Handling tape is simple because it is well protected in its plastic armor. Store a tape in its sleeve or case when not in use, and keep it out of direct sunlight and hot car interiors. Never open the cassette's protective front flap and expose the tape inside. There is no reason to open this flap except, on rare occasions, to inspect a damaged tape. Touching the delicate tape can degrade its recording ability.

Loading videotape can be easy or difficult, depending on the design of the camcorder. Some

Figure 2-2.
White balance. A—Used outdoors, the indoor white balance makes everything look blue. B—Outdoor white balance shows colors accurately.

units will accept a tape, however carelessly you insert it. Others demand that you position the cassette precisely and install it with care. No camcorder, however, will permit you to insert a tape upside-down or backward. Most cameras accept the tape with the cassette face toward you and the tape flap on the lower (or inner) edge. One caution: miniaturized camcorders are somewhat delicate. You can break them by forcing a tape. If the tape does not slide in at once, make sure that it is properly oriented.

Preparing the tape

Once the tape is inside the camcorder, it must be prepared for use. If the tape is brand-new, roll it forward about 30 seconds. Do this by pressing the record button with the cap still covering the lens. This is done because the first few inches of tape will gradually stretch as the cassette is repeatedly rewound to its beginning, eventually ruining any material recorded there. By starting your recording 30 seconds in, you can avoid this problem.

If you have used some of the tape before, you must prepare it by ***rolling down to raw stock.*** This means previewing the tape to find the end of your previous footage, recording a few seconds of black (again, with the cap on the lens), then leaving the tape at that position, **Figure 2-3**.

To save precious battery power, it is best to use the camera's AC adapter. To record black, simply press RECORD.

Tape preparation is more important than it may seem, because video makers usually review their footage after returning from a shoot, and they sometimes forget to roll past their final shot in preparation for their next session. Too often, valuable footage is taped over and erased by the next day's work.

Figure 2-3.
Preparing the tape. A—The existence of a picture indicates previously recorded tape. B—"Snow" or solid blue indicates a blank section. C—Black color indicates recording with the lens cap on or black color from a computer.

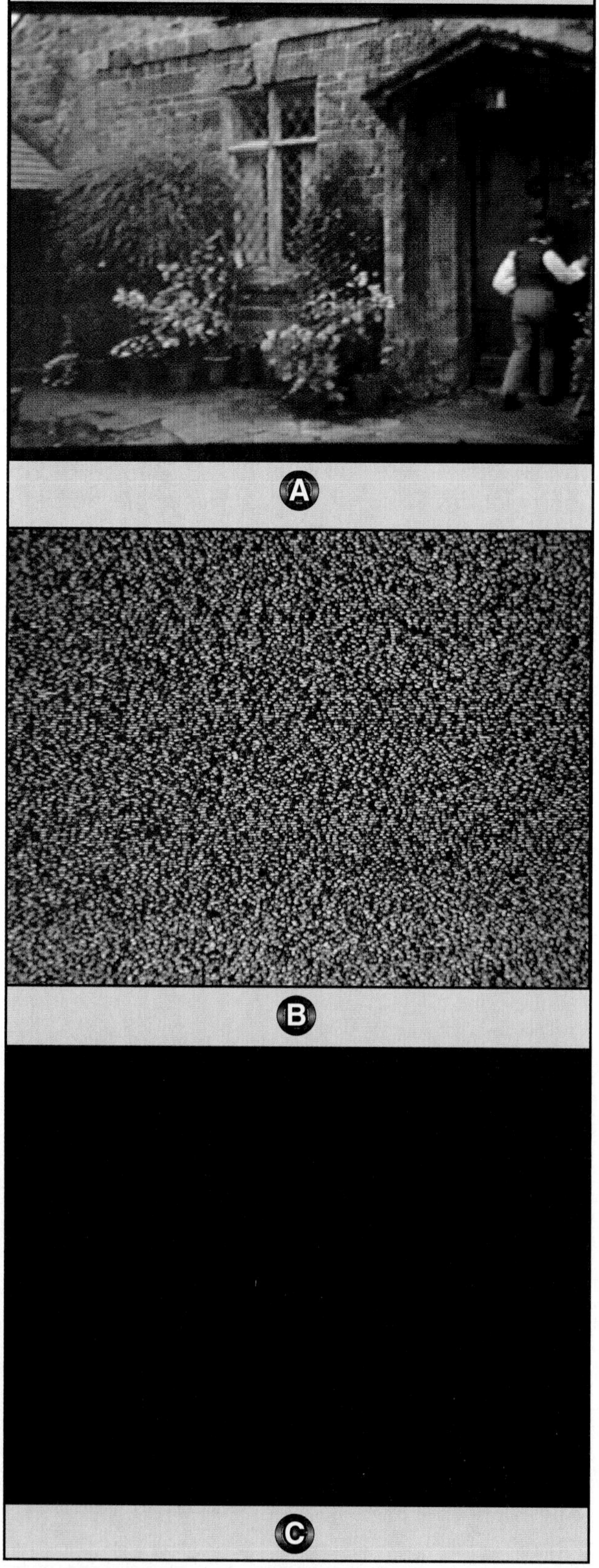

Batteries

Camcorders run on battery power, though they can use household current when available. Batteries are covered in detail in Chapter 11. For now, you need to know how to charge them, handle them, and ensure an adequate supply of camera power.

Though batteries appear simple and tough, they do require care. Dropping them can split their cases. Leaving them outdoors in cold weather can reduce the amount of power they deliver. Most importantly, be careful not to short-circuit a battery by bridging its two contact terminals. This will produce high heat levels and eventually ruin the battery.

A camcorder battery can be charged in two ways: by leaving it in the camera and plugging in the camera's power/charging cord, or by attaching it directly to the camcorder's charging unit. Whichever method you choose, always charge a battery immediately after use, then unplug the unit for storage. Just prior to your next use, recharge the battery again.

Finally, never go out shooting without at least one fully-charged spare battery. Though camcorder batteries have advertised capacities of up to two hours, they seldom last that long. Older batteries, in particular, may stop working after 30 minutes or less, so a spare is indispensable.

Tripod

Another essential accessory is a camcorder tripod. Although a skilled camera operator can shoot good quality footage by hand-holding, it is difficult to obtain steady images without the support of a tripod. Shaky pictures are the most obvious signs of amateurish production.

Many beginning video makers avoid tripods because they can be clumsy and a nuisance to manage. But if you routinely use a tripod from the start, you will come to find it a natural part of the shooting process.

You can adjust the stiffness with which tripods move sideways and up and down. Locate

Figure 2-4.
A typical tripod head.

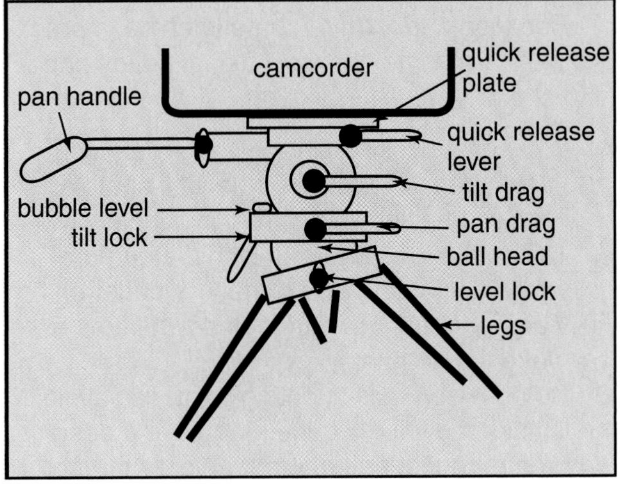

Preshoot Checklist

- Battery charged and inserted in camcorder.
- Spare battery charged and ready for use.
- New videotape positioned at the end of 30 seconds of blank recording.
- Previously used videocassette rolled forward to blank tape.
- White balance control set to automatic.
- Tripod included in shooting kit.

Also make sure that the date is not visible in the viewfinder. If it is, disable the "date" function on the camcorder so that the date does not appear on the footage you will shoot.

these controls, **Figure 2-4**, on your particular unit and set them so you can move the camcorder smoothly. Be careful to set the vertical tension tightly enough to hold the camera level even when it is unattended.

Some tripods include a quick-release mechanism that makes it easier to attach and remove the camcorder. Before using one of these tripods, make sure you understand how the release mechanism works.

Now that you have located the camcorder controls and learned about tape, batteries, and tripods, you are ready to start shooting. But before you do so, use the accompanying checklist to ensure that you are ready to go.

Camcorder Operation

Camcorders are so simple to operate that anybody can obtain pictures of some sort. But if you follow just a few professional practices, you can improve your footage dramatically. The following suggestions assume that you will edit your first projects in the camera; that is, you will shoot each shot in order and make each one last as long as it will in the completed program. (Some of these tips do not apply when you are shooting footage to edit later.)

Checking the Viewfinder

All camcorder viewfinders display several types of information, **Figure 2-5**.

The problem is that many people look right through this information and pay no attention to

Figure 2-5.
Different viewfinders display different kinds of information.

it. By checking the viewfinder data before making each shot, you can avoid many common shooting problems.

Different camcorders display different types of information, but most of them will show you the following, in one form or another:

- *Battery charge:* shows roughly how much power is left in the camcorder battery. When it gets too low, change batteries, so you do not lose power in the middle of a shot.
- *Tape loaded*: shows whether or not a tape cassette is in the camcorder.
- *Tape counter*: shows how much footage has been shot. Digital camcorder displays read in hours, minutes, seconds, and frames.
- *Record symbol*: shows whether you are actually recording or only previewing the image in the viewfinder.
- *White balance*: shows whether the camcorder is set for outdoor or indoor color, or whether the control is set to the automatic position.
- *Date*: shows whether or not a date stamp will be printed over the shot.
- *Zoom*: shows the image size, compared to the extreme wide angle setting.

In addition to these common displays, some camcorders show other kinds of information in the viewfinder. To interpret them, consult the instruction manual for your particular model.

Checking Camera Settings

No matter how carefully you set up your camera at your base of operations, check it again when you are ready to shoot. In particular:

- Make sure the white balance switch is set to automatic.
- Verify that the autofocus and autoexposure controls are enabled. Though they default to automatic, they can be accidentally changed after the camcorder is turned on.
- Just before making each shot, ensure that the camera has not gone into standby (sleep) mode.

Using the Tripod

Here are some tips for mounting the camera on the tripod and using the tripod effectively.

- Make sure the tripod head is level by adjusting the lengths of the legs or using the ball head.
- Screw the camcorder onto the head firmly enough so it does not wobble, but not tight enough to make it difficult to remove.
- If the tripod has a center column, do not raise it. This will make the unit less stable.
- Point one leg at the subject to be videotaped. Doing this will let you stand close behind the camera, in the space between the other two legs.
- When you pan the camera (pivot it from side to side), stand facing the center of the move, **Figure 2-6**. To make the shot, twist your upper body to frame the start of the shot, then follow the camera, twisting your body the opposite way until you frame the end of the shot. By doing this, you will avoid tangling yourself in the tripod.

Hand-holding the Camera

Because a tripod will sometimes be impractical, here are some tips for making steady handheld shots:

- Whenever possible, brace yourself on something. Lean your elbows on a wall or table. Prop yourself up with a tree, light pole, or wall.
- To shoot a low angle, sit with the camera firmly in your lap, swing the viewfinder or viewing screen up and look down into it to make the shot.
- Do not walk while shooting if you can avoid it.
- Unless the shot is quite long, take a deep breath and let half of it out before starting, then hold your breath as you shoot.

Figure 2-6.
Start a pan with your feet aimed at the center of the pan and twist your body as you rotate the camera.

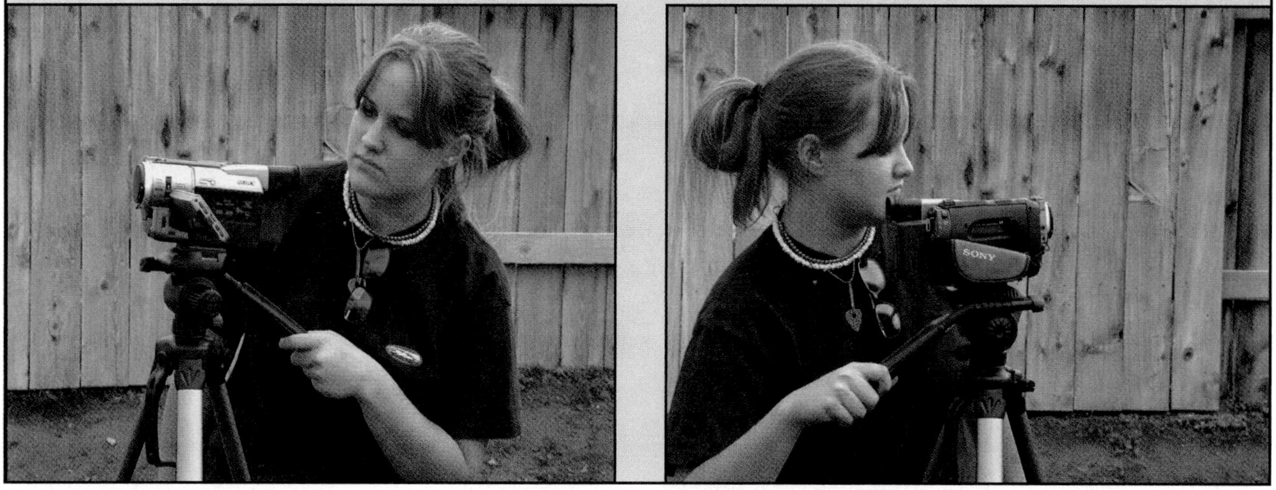

- Use the widest-angle lens setting. The wider the angle, the less obvious any camera shake will be.
- If your camcorder has a lens stabilization feature, make sure that it is enabled.

Avoiding Camera Problems

The do's and don'ts of good camera work are covered more completely in later chapters. At this stage, just remember four tips:

- Avoid swinging the camera around to center one subject, and then another, and then another. Instead, get a good-looking picture of each subject and shoot it as a separate shot.
- Do not make shots too brief for the viewer to look at. Three to five seconds is generally a good minimum length for this kind of project.
- Do not pose subjects against the sky, white walls, or other backgrounds that are lighter than their faces. This will deceive the auto exposure system and result in poorly-lit pictures.
- Avoid zooming while shooting. If you want to change from a wider to a closer view, for example, shoot each angle as a separate shot.

As you will see later on, professional video makers seldom use the zoom except in studios or when covering live activities like sports.

If you follow these four tips, your footage will look technically competent, but it still may not be very pleasing. To record good-looking video and clearly audible sound, you need to remember just a few more basic ideas.

Quality Video and Audio

Your program will be judged on how it looks and sounds, so here are some tips for improving both video and audio quality.

Good Quality Video

You can make better-looking images by using four key ideas: head room, look room, lead room, and the rule of thirds.

Allowing proper ***head room*** means positioning subjects at a pleasing distance from the top of the picture. When videotaping people, most beginners center the subjects' heads in the frame because that is how we look at people with our own eyes. The problem is that the results look awkward on a video screen, **Figure 2-7**. For better composition, place the subject's eyes in the top third of the frame, except in wide shots.

Look room is similar to head room. When watching people, we tend to center them left-to-right as well as top-to-bottom, **Figure 2-8**. As you can see, this appears pleasing only when the subject is looking straight at the camera. If the subject is looking to one side, the composition seems constricted on that side. The remedy is to shift the subject away from the direction of the look.

Lead room is look room with the subject moving, **Figure 2-9**. If you center moving subjects left-to-right, they seem about to run into the edge of the picture. By positioning subjects away from the frame edge toward which they are moving, you achieve more satisfying results.

Figure 2-7.
Head room. A—Centering the eyes in the frame results in too much head room. B—Try to keep the eyes in the upper third of the frame.

Figure 2-8.
Look room. A—Looking forward, the subject may be centered in the frame. B—When looking to one side, a centered subject crowds the frame edge. C—Allow somewhat more space on the side toward which the subject is looking.

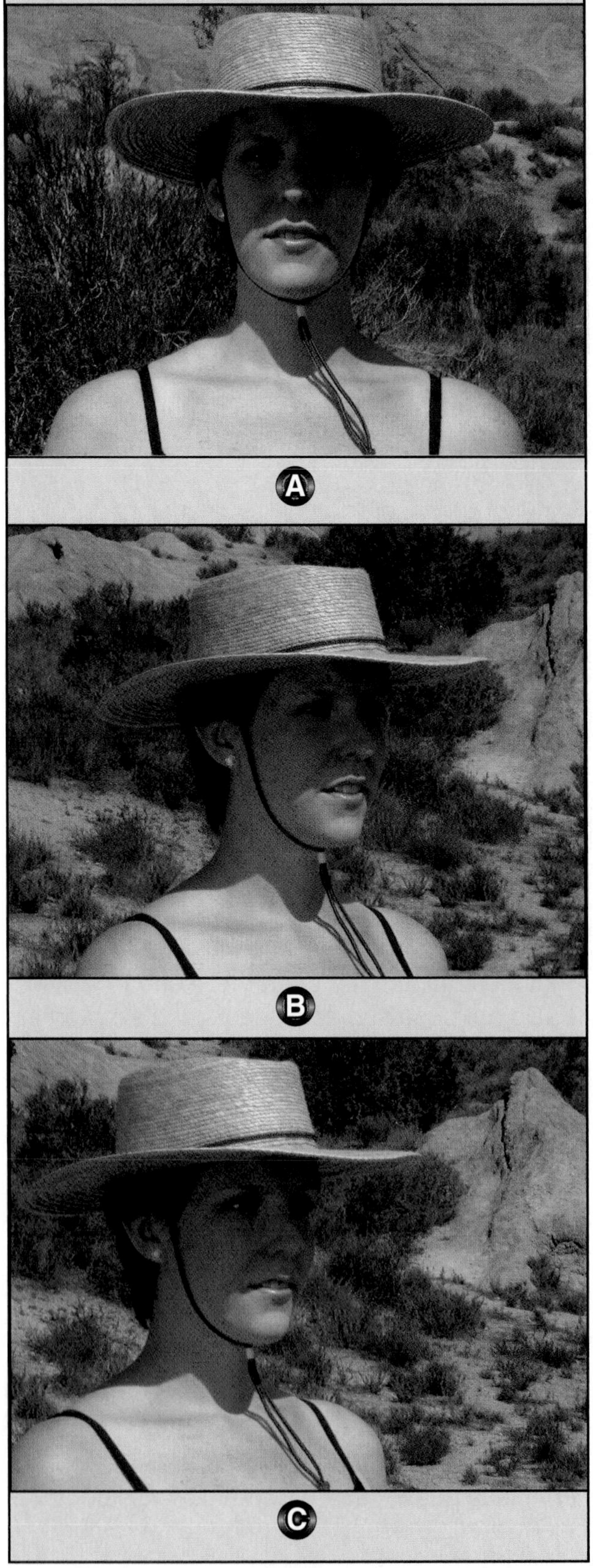

The ***rule of thirds*** offers the simplest way to achieve good pictorial composition. People tend to center subjects in their pictures. A tree is photographed dividing the frame vertically. The horizon is placed so it divides the image horizontally. The resulting picture looks balanced and rather dull. You might call this kind of composition "the rule of halves" because the frame is divided in half on both axes.

If you imagine a tic-tac-toe grid in front of your picture, you can divide the image into thirds instead of halves, **Figure 2-10**. The resulting compositions will be much more interesting.

An "axis" is the same in video as in graphed algebra equations. The X axis is horizontal (left-to-right) and the Y axis is vertical (top-to-bottom).

Figure 2-9.
Lead room. A—Do not allow a moving subject to crowd the frame edge. B—Allow extra room ahead of a moving subject.

Figure 2-10.
Rule of thirds. A—Centered compositions may look stiff. B—This composition is organized on a grid of thirds.

Good Quality Audio

For your first video projects, you will probably use the camcorder's built-in microphone. Here are three simple tips for recording better-sounding audio.

Stay close to the subject

Place the camera, with its microphone (mike), as close as possible to the subject. The farther away the microphone is, the more it picks up interfering background noise. Instead of setting the lens at full telephoto and shooting from, say, twenty feet away, set it at wide angle to obtain a similar composition from a five-foot distance, **Figure 2-11**.

Minimize background noise

Set up your shoot so that the camcorder is aimed away from major noise sources. Do not place subjects in front of a busy street because the mike will be pointed directly at the traffic noise behind them. Instead, position your subjects so that the camcorder mike points away from the traffic.

Direct silently

Do not give verbal instructions from the camera position while shooting. Camcorder mikes pick up sounds to the sides as well as in front. You can ruin sound takes because your own voice is mixed with the production sound.

Figure 2-11.
Moving the camcorder/microphone closer to the subject captures better quality sound.

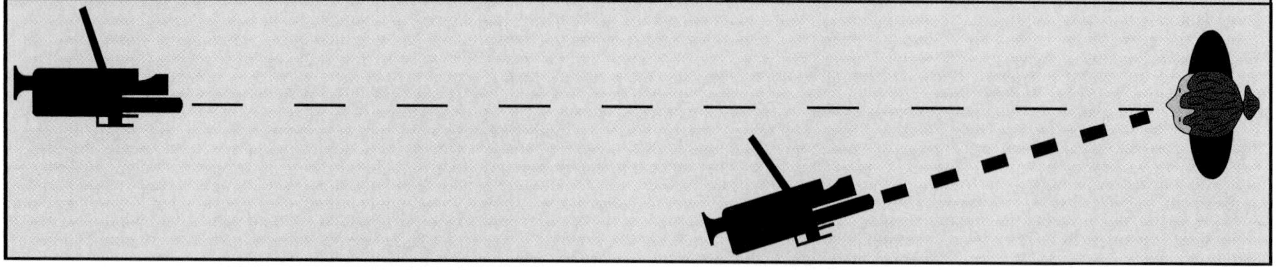

Camera Angles

When people begin making videos, they often set up shots that look very similar. The three frames shown simply document a person walking. Because the person is nearly the same size in every shot and the camera is in nearly the same position, the resulting shots are not very interesting.

As you can see, the sequence is more interesting when the shots are more varied.

Too-similar shots make uninteresting video.

Varying subject size and camera position makes more interesting shots.

The Shooting Session

Speaking of directing, we need to mention a few things about guiding both the performers in front of the camcorder and the crew behind it.

Directing the shoot

On more advanced video projects, you will be shooting to obtain raw footage for later editing. On beginning projects, however, you will probably be editing in the camera to create a finished video in the process of shooting it.

To edit in the camera successfully, you need to observe a few simple rules:

- Record every shot in the order in which it will appear in the finished program.
- If you are taping continuous action, such as somebody walking, try to leave the screen

Notice that the shots vary in both size and camera position. Changing only the image size makes the camera appear to jump toward or away from the subject.

On the other hand, changing only the camera position produces shots that are too similar.

Only subject size changes.

Changing only the horizontal camera angle also makes a jump cut.

empty between shots. For example, let people walk off-screen at the end of one shot and then show them walking again at the start of the next shot. See **Figure 2-12**.

The rule works as well in reverse: show people walking at the end of the outgoing shot; then have them walk into an empty screen at the start of the incoming shot. By doing this, you will not have to match the actions in the two shots.

- Try to make each shot look different from the preceding one. Varied images will make your video much more interesting to watch. (See the sidebar *Camera Angles* for tips on varying image look.)

Finally, it is worth repeating that the crew should keep quiet during each shot, because the camcorder mike will pick up all the sounds of the crew as well as those of the cast.

Managing the Shooting Session

Shooting video can sometimes become such an intense activity that perfectly responsible people get carried away. A few reminders about video security, safety, and courtesy may help avoid this problem.

Security refers to the care of your expensive and somewhat delicate video equipment. To keep your hardware safe:

- Designate one person to take responsibility for the camcorder and tripod at all times. That person is never to leave the immediate area of the equipment.

Figure 2-12.
The first two shots (A and B) make a jump cut. Allowing the subject to leave the frame (C and D) hides the mismatched action.

CORRECTING COMMON MISTAKES

Symptom	Cause	Cure
Viewfinder blank	Power not on Battery dead Camera on standby	Turn on power Replace battery Reactivate camera
Camera won't record	No tape inserted VCR mode enabled	Insert tape Enable camcorder mode
Colors too blue or too orange	White balance set incorrectly	Switch white balance to automatic setting
Shaky pictures	Hand-held camera	Use tripod
Too much head room	Subject head centered in frame	Place subject's eyes in top 1/3 of frame

- If you are hand-holding the camcorder, never set it down on anything except its tripod. It is easy to grab a camera from a table or a sidewalk and run. It is much harder and more conspicuous to snatch both camera and tripod. Also, the tripod makes the camera more visible, so it is less likely to be bumped or knocked to the ground.
- Detach the camcorder from the tripod when transporting them. If the camcorder has a handle, never use it to lift both camera and tripod together. The handle is not strong enough to carry the weight of both pieces.
- Always protect the camcorder from the weather. Keep it out of hot sun except when actually shooting. Do not allow it to get wet, particularly the delicate glass of the zoom lens.

Security for cast and crew is important. Many video projects involve action sequences such as fights and car chases. To keep everyone safe:

- Do not ask people to perform feats for the camera that they would not normally attempt. Professional stunt people are highly trained, very experienced, and well paid for taking risks. Your cast and crew are not.
- Do not put the crew at risk, (say, by hanging out over balconies or climbing on high roofs) in pursuit of interesting camera angles.
- Remember that the videographer is concentrating intently on the viewfinder. It is easy for that person to run into objects or stumble on stairs. An assistant should guide the videographer during moving shots.

Finally, practice simple courtesy. You have a right to shoot outdoors on public property as long as you observe the rules of both law and good manners. Do not shoot inside (in stores or offices, for example) without permission. Though passers-by may walk through the background of your shots, do not use them as actual subjects without their permission. Some people feel uncomfortable or irritated when suddenly confronted with a camcorder. Respect their right to privacy.

Getting Started

By following the suggestions in this chapter, you can make simple but satisfying videos. Soon, however, you will want to undertake more ambitious projects and create programs of fully professional caliber. To do this, you will use the more sophisticated production techniques covered in the rest of this book. All of these techniques have essentially the same purpose: to communicate effectively with viewers in the medium of video. For this reason, Chapters 3 through 7 cover the principles of video communication.

Chapter Review

Answer the following questions on a separate piece of paper. Do not write in this book.

1. *True or False?* Default settings on the camcorder cannot be changed by the user.
2. The ______ control matches the camcorder to the overall light quality of the scene.
3. The "rule of ______" is a guide to good composition.
   A. quarters
   B. thirds
   C. halves
4. *True or False?* The terms "look room" and "lead room" are *not* interchangeable.
5. For steady video images, use a ______ instead of hand-holding the camcorder.

Technical Terms

Autoexposure: The camera system that delivers the correct amount of illumination to the recording mechanism, regardless of the light level of the shooting environment.

Autofocus: The camera system that ensures that the subject in the image appears clear and sharp.

Camcorder: An appliance that both captures moving images (**cam**era) and stores them on tape (re**corder**).

Default: A setting that is selected automatically unless the user changes it manually.

Head room: The distance between the top of a subject's head and the upper edge of the frame.

Lead room: The distance between the subject and the edge of the frame toward which it is moving.

Look room: The distance between the subject and the edge of the frame toward which it is looking.

Pan: To pivot the camcorder horizontally on its support.

Roll to raw stock: To advance a videotape through previously recorded sections to blank tape, in preparation for additional recording.

Rule of thirds: An aid to composition in the form of an imaginary tic-tac-toe grid superimposed on the image. Important picture components are aligned with the lines and intersections of the grid.

White balance: A camera system that neutralizes the color tints of different light sources, such as sunshine and halogen lamps.

Zoom: To magnify or reduce the size of a subject by changing the focal length of the lens.

3

Video Space

Objectives

After studying this chapter, you will be able to:

- Use the video frame to control what viewers see.
- Compose images and action for the two-dimensional screen.
- Create the illusion of depth in the image.
- Exploit the elements of scale, distance, position, and relationship.

About Video Space

This chapter explains the very different laws that govern space in the video universe and shows how you can turn these odd-seeming laws to your advantage in communicating information and feelings to your audience.

Though we use the term "video" here, as usual, the rules of the film universe are identical. It was the pioneer film makers, of course, who discovered these rules to begin with.

When viewers watch video, they unconsciously think they are looking through a sort of window at the events on the other side. In real life, both sides of a window belong to the same world and observe the same laws of space and time. For this reason, viewers assume the same about the world on the other side of the screen "window."

However, the screen is not a window and it has no "other side." Video images are not *behind* the screen but *on* it, and that creates two fundamental differences from the real world.

First, video information is limited by the borders of the image, **Figure 3-1**. By opening a real-world window and leaning through it, you can see more of the scene outside. But no matter how close you get to a screen, you can never see anything beyond its borders, and of course you cannot thrust your head through it.

Second, the screen itself is flat, or nearly so, which means it has only two dimensions instead of three.There is no true depth in a movie image. No matter how dramatically that spaceship seems to zoom away from you, it is not actually moving farther back from the screen surface, but merely growing smaller on it.

In short, space (and time as well) work very differently in the special world on the screen. It is these crucial differences that make video possible. If you master the laws of video space, you will be able to use some of the major tools of video communication.

The Laws of Video Space

This chapter discusses the four laws of video space. Here are those laws in summary form:

- What is outside the frame does not exist, unless that existence is implied.
- Height and breadth are determined solely by the frame, and depth is only an illusion.
- Size, position, distance, relationship, and movement are not fixed.
- Direction is determined solely by the frame.

The Frame

The frame is not just a passive border around the picture. It is a powerful video communication tool. First, the frame defines and controls what the viewer sees. Even the most honest videos show only selections from reality, and those selections are made by including them within the frame. Second, screen composition and movement are defined by reference to the frame around the image.

In video, the word *frame* has two completely different meanings: 1) the border around the screen and 2) a single film or video picture, as in "Video is displayed at 30 frames per second." See Figure 3-2.

Figure 3-1.
Screen proportions (aspect ratios). A—The traditional video frame, proportioned 4 to 3 (1.33 to1). B—The wide screen video frame, proportioned 16 to 9 (1.78 to 1). (Sue Stinson)

Figure 3-2.
A "frame" is both the border around the screen and a single video image in a series.

The Frame Controls Content

The first law of video space is perhaps the most important: *what is outside the frame does not exist* in the video world, no matter how real it is in the actual world. This is the law that makes media production possible. The sound stage or location may be packed with people, cameras, lights, and sound equipment, but because the camera operators exclude these extraneous elements from the frame, they are completely invisible to the audience. Because the audience cannot see something, it does not exist for them.

Video makers use the frame to alter reality in countless ways. Here are a few examples:

- By having a small number of spectators sit close together and framing off the many empty seats around them, you can make an audience seem bigger than it really is. Since all the empty seats are outside the frame, they do not exist.

Note that to *frame off* something is to exclude it completely from the image, while to *frame* something is not only to include it but to feature it. The command, "Frame off that tree" means keep it out of the picture, while "Frame that tree" means make it the center of interest in the picture.

- By framing off the many well-kept homes in a neighborhood and framing only the few run-down houses, you can make a prosperous area look like a slum. In the video world, the attractive homes outside the frame do not exist.
- By showing the skyscraper tops behind your actor but framing off her feet and the flat roof on which she is standing, you can make her seem dangerously close to the edge of the building. On the screen, the roof outside the frame does not exist.

Using the frame to include and exclude information is perhaps the most universally exploited technique in video communication.

Although real world elements outside the frame do not exist in the video world, the converse is also true: elements that do *not* exist in the real world *can* exist in the video world if the video maker implies that they lie somewhere outside the frame.

For example, suppose viewers see three shots: a rodeo grandstand, a spectator, and a shot of the rodeo action. Crowd noises and the announcer on the public address system fill the sound track across all three shots, **Figure 3-3**. Imagine, too, that the director has sent other people walking in front of the spectator.

In fact, the spectator never was at a rodeo, as you can see from the wider view of her in **Figure 3-4**. The passing extras and the sound effects laid under her closeup *imply* that the rest of

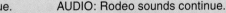

Figure 3-3.
AUDIO: Rodeo background sounds AUDIO: Rodeo sounds continue. AUDIO: Rodeo sounds continue.

The Power of the Frame

The frame that bounds the video world is a versatile tool. The frame can conceal production.

The frame isolates the subjects from the production around them. (Photoflex)

The frame can alter reality.

Though the subject seems to be high in the air, he is actually standing on firm ground.

The frame can create a visual composition.

The frame isolates a composition within the random elements of the scene. (Corel)

The frame can define movement.

Against a featureless background, the car's movement is in relation to the frame.

Figure 3-4.
The "spectator" as actually videotaped.

the scene extends far beyond what viewers can see on the screen. Because the frame prevents them from seeing the reality of the desert background, they believe that the spectator is at a rodeo.

To include this ability to suggest things that are not really there, we must amend our law of video space to add this idea: what is outside the frame does not exist, *unless that existence is implied*. By exploiting both parts of this law, video makers use the frame to determine what information viewers receive and how they interpret it.

The Frame Determines Composition and Movement

Video makers also use the frame to control how information is presented. The frame is the viewers' only fixed reference (though they are almost never consciously aware of it). They depend on the frame to locate and organize the parts of the pictures they are watching, and to judge the speed with which things move. That is, viewers use the frame to perceive composition and movement.

Composition is the organization of the elements (parts) of a picture. These elements are arranged in relation to one another. More importantly, they are also arranged by their relationship to the frame that surrounds them. Without the defining border supplied by the frame, composition would be difficult or even impossible. (Composition is covered more fully in Chapter 5.)

Movement refers to motion on the screen, and viewers perceive that motion by two sets of clues:

- How rapidly the background changes in relation to the subject in the foreground (for example, scenery flying past a car window).
- How quickly the subject moves in relation to the frame.

Since this discussion concerns the frame, we are more interested in that second clue to movement: motion in relation to the frame.

Think of the video images you have seen of satellites in orbit, as recorded from a space shuttle. They do not move on the screen at all, but seem to float in space as quietly as a boat on a pond, **Figure 3-5**. In truth, of course, those satellites are whirling along their orbits at thousands of miles per hour. But since the cameras recording them are moving at exactly the same speed, the images of the satellites do not move *in relation to the frame*.

Now think of the spaceships that zoom across movie and TV screens at warp speeds, spanning whole galaxies in minutes. In reality, these spaceships are often miniature models that do not move at all. They are fixed in place, and the illusion of rapid motion is created by moving the camera past them, **Figure 3-6**. Making the camera fly past the stationary ship models causes them to move violently *in relation to the frame*.

Figure 3-5.
A fast-moving satellite does not move in the frame. (NASA)

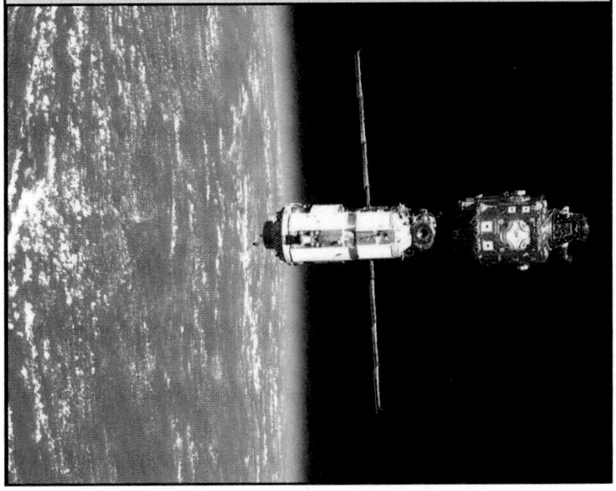

Figure 3-6.
Making a spaceship move. A—Plan of the shooting setup: the spaceship remains fixed, while the camera moves right to left. B, C —On screen, the spaceship seems to enter the frame from the left and leave it on the right.

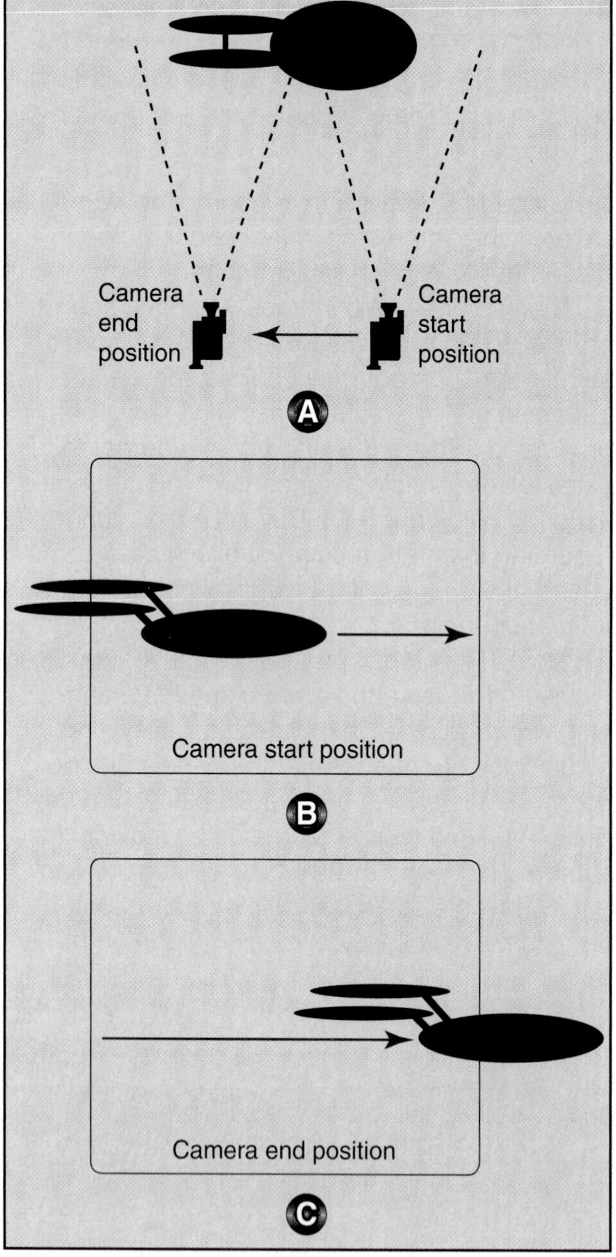

Today's spaceships are often computer animations rather than models; but for big-screen science fiction movies, the large scale and fine details of models are still sometimes preferred.

In the actual world, the satellite in space is traveling at thousands of miles per hour, while the model spaceship is not moving at all. But in the video world, the exact opposite is true: the satellite is not changing position with respect to the frame, while the model is.

It is fair to say that the frame around the screen world is the video maker's most powerful single tool. It determines what does or does not exist in that world, how objects in that world relate to one another, and how things move in that world. Every decision made by the director, the videographer, the lighting director, the production designer, and even the sound recordist, is determined first and foremost by the characteristics of the frame.

Video Dimensions

While the frame around the screen forms the boundaries of the video world, it is the screen itself that displays that world as a strange, flat universe in which space acts differently from real-world space. In the video world, the dimension called "height" is not always vertical and the dimension called "breadth" is not always horizontal. The dimension called "depth" does not even exist. What viewers perceive as that third dimension is merely an illusion.

This brings us to the second law of the video world: *Height and breadth are determined solely by the frame, and depth is only an illusion.*

If you understand how this second law operates, you can make it do all kinds of tricks. So let us see how video height and breadth differ from those dimensions in the actual world.

Horizontal and Vertical

In the real world (at least on Earth), height and breadth are defined by gravity: height is

Figure 3-7.
In the real world, horizontal and vertical are defined by gravity.

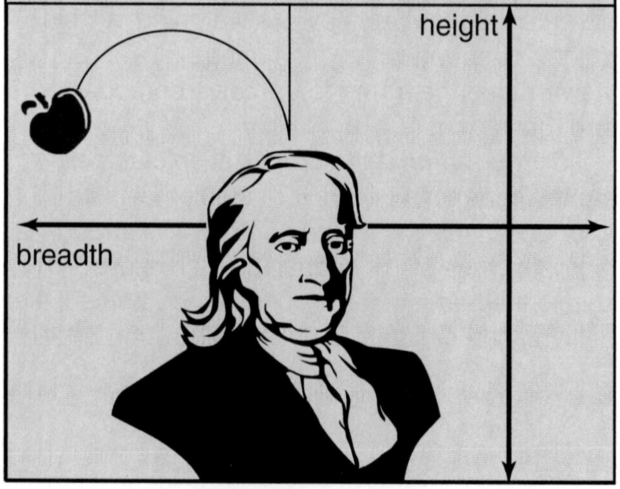

the dimension parallel to the direction in which things fall; breadth is the dimension at right angles to that direction. Newton's apple always falls in a vertical path to the horizontal ground because the orientations of height and breadth are unchangeable, **Figure 3-7**.

In the video world, however, height and breadth are not defined by gravity but by the frame around the image. Height is the dimension parallel to the sides of the frame. Breadth is the dimension parallel to the top and bottom.

The difference is important, because it means video height and breadth are determined entirely by the video maker. By tilting the camera, you can make "vertical" and "horizontal" whatever directions you choose.

Naming Screen Dimensions

In technical discussions of screen-world space, the dimensions are sometimes given the names of the axes of a mathematical graph because such a graph, like a video screen, occupies a two-dimensional space.

The X, Y, and Z axes of the video world.

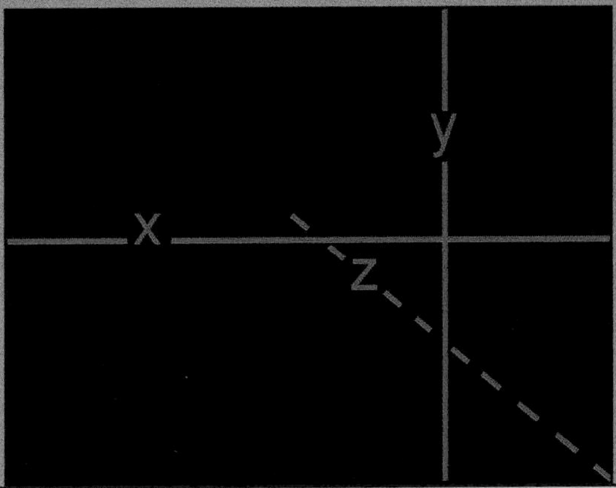

The horizontal dimension is called the "X axis" and the vertical dimension is called the "Y axis." The apparent third dimension, at right angles to both the others, is called the "Z axis." It is indicated by a dashed line because in video, depth is only an illusion.

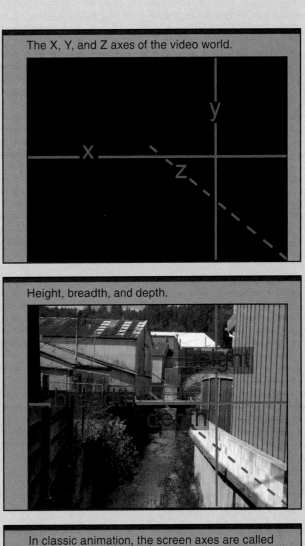

Height, breadth, and depth.

Artists working in animation have long used a different system. In calling for the animation camera to move up, down, left, or right on a drawing they use compass directions: North, South, East, and West.

Though all three sets of names have advantages, this book uses the familiar terms, height, breadth, and depth.

In classic animation, the screen axes are called North/South and East/West.

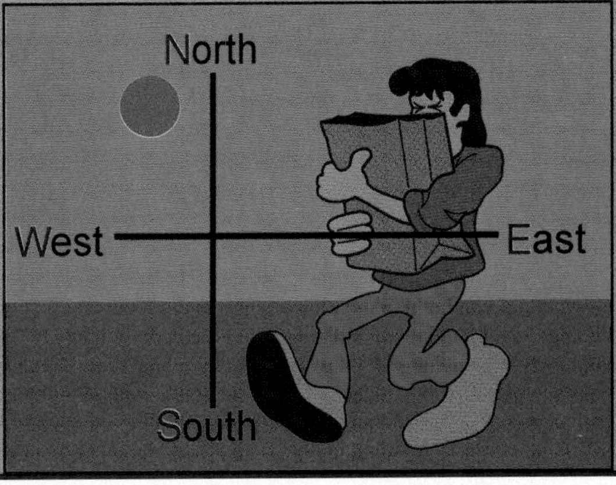

Playing Tricks with Video Space

Since height and breadth are not fixed in the video world, you can play all kinds of tricks with them. For example, by turning the camera 90 degrees, you can reverse the two dimensions. In this example, an actor standing on a walkway appears to be scaling a vertical concrete wall.

In the next example, the actor appears to be able to do pushups with just one finger. (In a moment, he will remove his finger from the "floor" and float in mid-air!) Note how the cup affixed to the wall helps sell the gag that the wall is actually a floor. (In movie jargon, a gag is any trick or stunt, and to sell an illusion is to include information that makes it more believable.)

The subject appears to be scaling a wall, but the camcorder is actually turned sideways.

Because the camcorder is turned 90 degrees, the subject seems to be floating above a concrete floor.

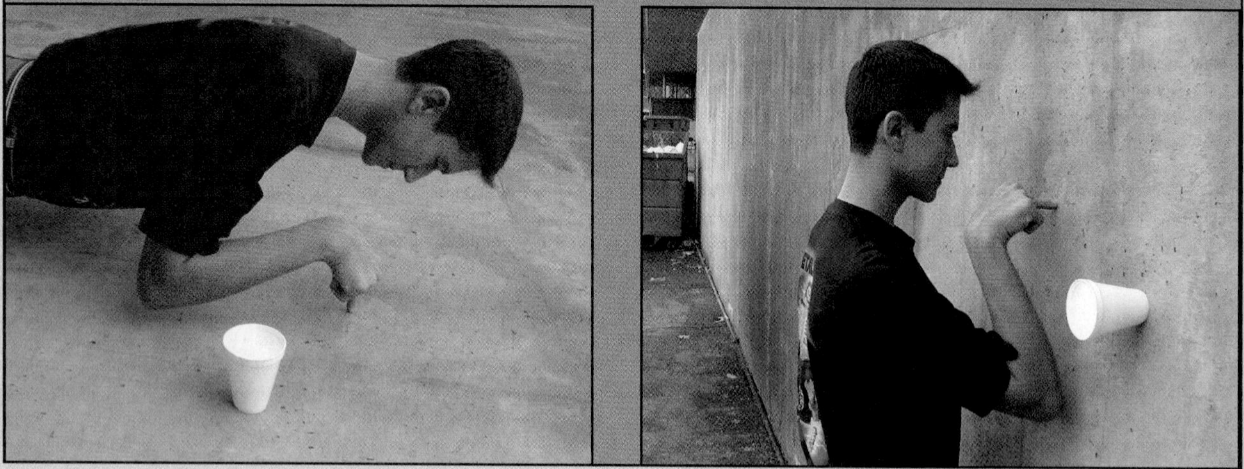

The Illusion of Depth

Though screen height and breadth are different from their real-world counterparts, they do exist in some form, at least, because the screen itself has height and breadth. However, the flat screen surface has no depth, and so the image on it cannot have depth either. In video, depth is purely an illusion.

In both the real and video worlds, the farther away objects are, the smaller they appear. But though the appearance is the same in both worlds, the realities are exact opposites. In the actual world, receding objects really do get farther away but they only seem to shrink in size. In the video world, receding objects only seem to get farther away (they remain on the surface of the screen) but they really do shrink in size, **Figure 3-8**.

The lack of true depth in the video world is both an advantage and a disadvantage. On the

Figure 3-8.
Though the spaceship seems to be traveling away from the viewer, its image has simply shrunk on the screen surface.

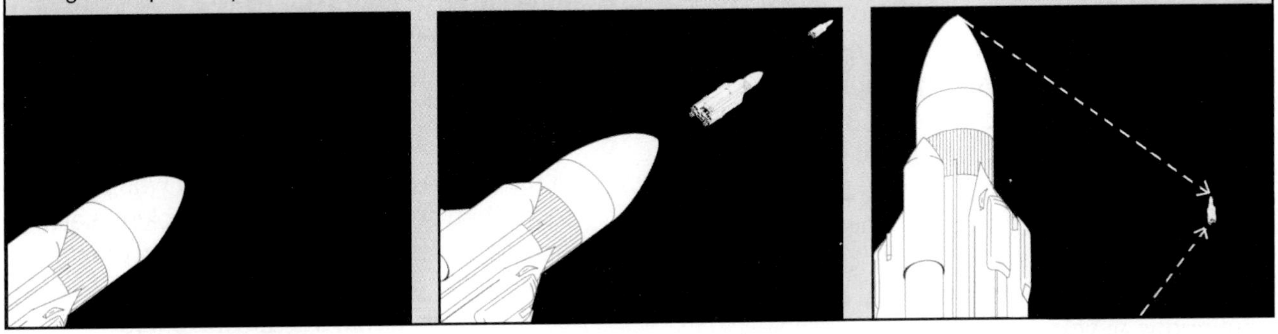

one hand, since depth is just a trick, video makers can manipulate it like any other illusion. On the other hand, since the screen is really flat, you must use a variety of compositional techniques to create the illusion of depth upon it.

Spatial Relations

The second law of the video world leads directly to the third law: *size, position, distance, relationship, and movement are not fixed.* Video space is completely fluid and is controlled by the video maker. In the actual world objects keep the same size and position; but in video space nothing is fixed. You can use this fact to your advantage.

Size, Position, and Distance

In the actual world, an inch is always an inch and a meter is always a meter. That means the size of an object is always the same. The object's position also remains the same (at least until it actually moves). Finally, the distance between objects stays the same, too. None of this is true in the video world. For example,

- Two actors may escape from a full-size automobile just before it explodes, but the car that actually blows up is only a foot-long model. An object's size can be altered as long as it looks unchanged to the viewer.
- An actor standing a foot away from a wall in a long shot may be ***"cheated"*** several feet farther from it for a closeup (to allow more room for camera and lights). Though the actor's position has changed, the shift is not visible to the audience.
- Two aircraft that actually pass each other a safe distance apart seem, on screen, to be a hair's breadth from colliding, **Figure 3-9**. The trick is done with a telephoto lens, which greatly reduces the apparent distance between the two objects.

As these examples suggest, movie makers routinely change the space in the screen world to fit the needs of the moment. Changes in size, position, and distance are all forms of changes in scale. In the video world, the apparent length of an inch or a meter or a mile can vary tremendously from one shot to another.

Figure 3-9.
As photographed by a telephoto lens, the planes nearly collide. In reality, they are a safe distance apart. (Corel)

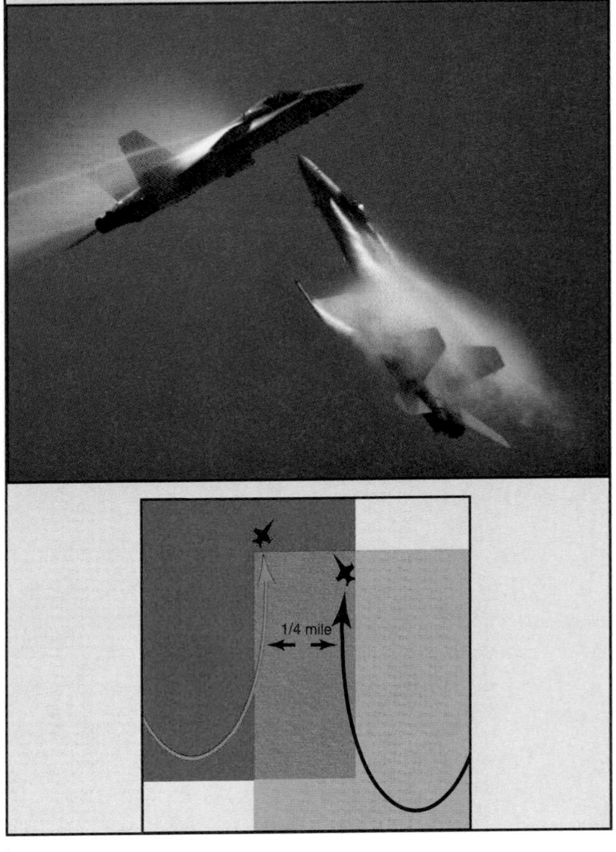

Relationships among Objects

You can control more than just the scale of screen world space. You can also control the relationships among different objects.

Controlling spatial relationships usually means putting things together in the screen world that are far apart in the real world. This is so common in movie making that professionals take the process for granted. Typically, spatial relationships are altered by editing two or more separate shots together into what seems like a continuous action. For example:

- On screen, a character walks down a street, turns a corner, and continues up a side street. In the actual world, the main street is a location and the side street is a construction on a studio lot.
- On screen, a character climbs six flights of stairs. In actual shooting, a single flight was used repeatedly, with different floor number signs and camera angles.
- On screen, two characters carry on a conversation. In actuality, each actor was photographed separately in similar-looking sets, playing the scene with an off-screen crew member. In postproduction, the editor replaced crew members with the absent actors.
- On screen, a subject crosses a courtyard in fifteen steps. The real-world location is 50 steps wide and obstructed by a large fountain, but the director and editor have

Figure 3-10.
Controlling spatial relationships. A—The audience sees the subject start to cross the courtyard. B—The audience sees the subject complete the cross. C—The whole courtyard, showing the central portion omitted from the two shots.

condensed the space by means of clever camera setups and editing, **Figure 3-10**.

A *setup* is a single camera position.

In short, screen-world relationships between unconnected places and people can be created by recording them separately and editing the different shots together.

Direction of Movement

In addition to controlling the relationships between screen-world places, video makers can control the way in which on-screen people get to those places. That is, they also control direction in the video world. To state this fact in the form of the fourth law of video space, *direction is determined solely by the frame.*

In the real world we express directions in compass terms: North, South, East, West, and those directions are fixed. If the street on which you live runs north and south, it *always* runs north and south. In the screen world, by contrast, compass directions are quite meaningless. Instead, direction is perceived with reference to the borders of the screen. For instance, successive shots may show an actor walking south, east, and then north; but as long as she is oriented in approximately the same way *on the screen*, the audience perceives her to be continuing in the same direction, **Figure 3-11**.

Figure 3-11.
In the real world (A), the subject walks south, turns east, then turns north. On the screen, however (B and C), she always walks left to right.

Figure 3-12.
Conventional maps show the U.S. on the left and Europe on the right. Flying from Europe to the U.S., a plane is conventionally shown heading screen left. (Corel)

Directions in the video world are subject to certain conventions. For instance, a plane traveling from England to the U.S. is usually shown moving from screen right to screen left, **Figure 3-12.** That is because conventional maps show North America on the left side of the Atlantic Ocean and Europe on the right.

If the plane flying west were photographed from the north side, it would move from screen left to screen right instead. Though the resulting video would be a perfectly accurate record of the plane's movement in the actual world, it might confuse audiences watching it on the screen.

Back to the Frame

In discussing direction on the screen, we come full circle to the idea of the frame —the powerful border that lets video makers dictate what is in the screen world, where it is located, how big it is, and how it relates to other things in that world.

At first, the laws governing screen-world space may seem like abstract theories, and in a way, they are. But remember that professional video makers understand these fundamental rules of screen-world operations and use them constantly in practical ways to make their programs.

Chapter Review

Answer the following questions on a separate piece of paper. Do not write in this book.

1. The video world has ______ dimensions and is limited by the ______ of the image.
2. The act of excluding something from the image is referred to as ______.
   A. masking out
   B. cheating
   C. framing off
3. *True or False?* In video, what is outside the frame does not exist.
4. Because the video screen is flat, the dimension of ______ is an illusion.
5. By convention, an airplane that is flying toward the ______ is shown moving across the screen from right to left.

Technical Terms

Cheat: To move a subject from its original place (to facilitate another shot) in a way that is undetectable to the viewer.

Composition: The purposeful arrangement of the parts of an image.

Frame: The four edges that make up the border of the screen. Also, a single video image.

To frame: To feature a subject in a composition.

To frame off: To exclude an object completely from the screen.

Gag: Any effect, trick, or stunt in a movie.

Screen direction: The orientation of on-screen movement with respect to the left and right edges of the frame.

Sell: To add details to increase the believability of a screen illusion.

Setup: A single camera position, usually including lights and microphone placements as well.

4 Video Time

Objectives

After studying this chapter, you will be able to:

- Explain the differences between actual time and video time.
- Describe the characteristics of video time.
- Exploit the characteristics of video time in making programs.

Meanwhile, back at the ranch...

About Video Time

This chapter examines the nature of video time, just as Chapter 3 explores video space. Like video space, video time is quite different from real-world time, and operates by its own set of rules. If you master these rules, you can alter video time in your programs by cutting, pasting, stretching, squeezing, and otherwise manipulating it to suit your own purposes. You can time-travel backward, forward, and sideways—or operate in multiple time flows at once. In short, you can control video time with an ease and power that any science fiction writer might envy.

Space and time are as closely connected in the video world as they are in the real one, but we are considering them separately to simplify the discussion.

Not only *can* you control video time, you *must* control it. Otherwise, you could not possibly fit the actions of your program into the brief time allotted to it. And if you did present a program in real-world time—say, a two-hour hike that actually lasted two hours on screen—your audience would be profoundly bored long before your literal video ended. A wise observer once noted that movies are life with the boring parts cut out. You remove those boring parts by managing video time.

The film 2001 A Space Odyssey begins in the prehuman past and concludes in the 21st century, a real-world time period of perhaps a million years

To manage video time you have to understand the laws by which it operates. An easy way to do that is by comparing the nature of video time with the laws of actual time, as we perceive it in the real world.

Real-World Time vs. Video-World Time

The most important characteristic of real-world time is its constancy. Time always behaves in exactly the same way with respect to its speed, flow, direction, and coherence.

We are talking about time as humans experience it. How time may be viewed by philosophers or scientists lies outside our subject area.

Speed

Time always passes at exactly the same speed. A second is always one second long and a year (as the calendar counts it) always lasts one year—no more, no less. Despite our subjective feelings that time sometimes drags and sometimes flies, it actually proceeds at a rigidly fixed pace, **Figure 4-1**.

Figure 4-1.
Real-world time always flows at the same rate.

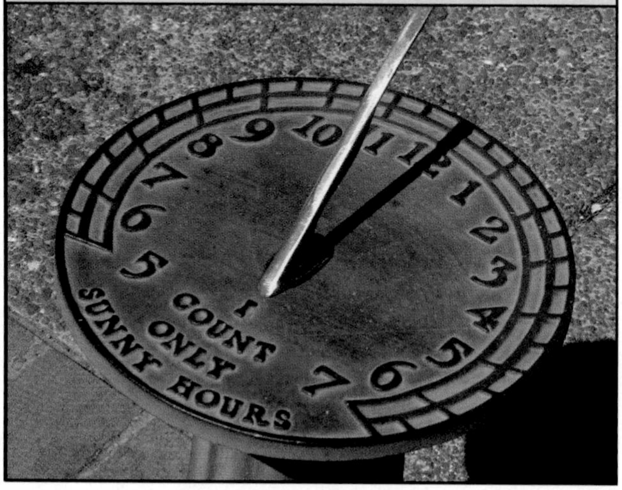

Flow

That pace is never interrupted, because real-world time is always continuous. Though we may sometimes feel that time seems to stop, it actually rolls along without even the briefest pause.

Direction

Real time flows one way only: the future becomes the present and the present becomes the past, like a river that never passes the same spot twice. In the actual world we can revisit the past only in memory.

Coherence

In our world, time affects everything identically, so everything moves together. Here, time is not like a river, in which some water drops flow along faster than others. In the world as we perceive it, everything is always at exactly the same "point in time."

At any one moment, people in different parts of the world may be at different hours of the day or even days of the week; but in the flow of time, all of us are always at exactly the same point.

Time in the video world is the opposite of real-world time because nothing about it is constant. Its speed, flow, direction, and coherence are always controlled and continually changed by the director and editor.

Video Time Speed

You can control the speed at which time passes in your program, expanding an eye blink or compressing a hundred-yard stroll so that each event fills exactly the same amount of screen time.

Screen time is the length of real-world time in which a video program (or a designated piece of it) is actually displayed on-screen.

There are two ways to change time speed: you can either alter the *actual* speed within individual shots or else alter the *apparent* speed of the action in editing different shots together. The two methods of varying time speed have markedly different effects on the viewer.

Altering speed within a shot

Altering speed within a shot is achieved by recording or playing the shot at a slower or faster rate than normal. The result is an obvious change in the speed of time on the screen. For example:

- In slow motion, the lovers float lazily toward each other across a flowery field; the whirring wings of a hummingbird seem to flap as deliberately as a buzzard's, **Figure 4-2**.

Figure 4-2.
Slow motion can make a hummingbird's wings visible. (Sue Stinson)

- In fast motion, the clowns skitter around the screen like doodlebugs; the flower's night-long opening unfolds in two brief seconds.

As these examples suggest, different degrees of slow or fast motion serve different purposes.

Moderate slow motion imparts a dreamlike effect. The feeling may be pleasant or romantic, as with the loping lovers; it may be nightmarish, as when a victim falls sedately toward certain death—and slowly falls and falls and falls. The mood may be nostalgic, as when a former football star recalls the 80-yard touchdown run that was the highlight of his career.

More pronounced slow motion has a different effect: it allows us to see otherwise invisible actions. It can divide a horseshoe pitch into many separate parts for analysis or reveal the complex movement of a hummingbird's wings. In its most extreme form, slow motion lets us watch as a bullet explodes a lightbulb.

With fast motion, the differences are similar, though of course the effect is the opposite. Moderate fast motion alters the mood, usually making an ordinary action look funny. Silent comedy chases were often filmed in moderate fast motion to heighten the comic effect.

More pronounced fast motion condenses action to make its overall shape or pattern easier to see. It can show us a sky full of small clouds boiling up into a full-fledged thunderhead right before our eyes. Very fast motion goes further by shortening actions that are so long we cannot normally see them at all, like the passage of the moon across the sky, **Figure 4-3**, or the blooming of a flower. In its most extreme form, fast motion can compress whole months into a few seconds, as when the audience sees an office building go from vacant lot to finished structure in what seems like a single shot.

Popular editing software includes stop-motion commands that allow you to create time lapse photography by automatically selecting, say, one frame for every minute of source footage, speeding up the motion by 1800 percent.

In short, slow motion expands the screen time during which an action happens and fast motion contracts it. In both cases the effect is obvious to the audience. When you watch a swinging baseball bat float majestically across the screen or the moon spring up in the sky, you realize that time has been stretched or

Figure 4-3.
Very fast motion can condense hours into seconds. (Sue Stinson)

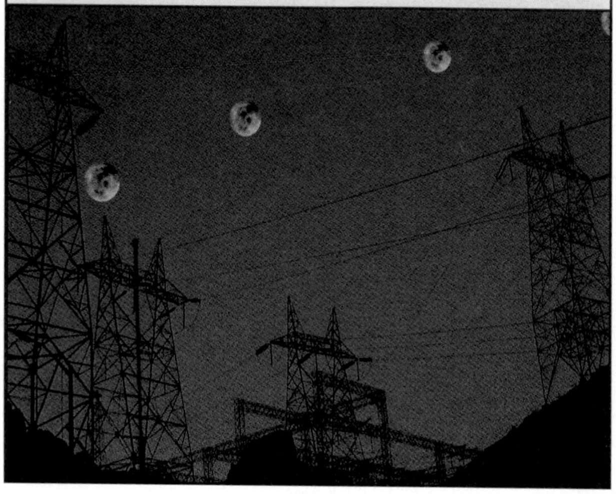

condensed and you know that the appearance of the action is not realistic.

In the film medium, model shots such as toy-size trains exploding are generally recorded in slow motion to make the action look more realistic, while car chases are often shot in mild fast motion to increase the apparent speed. Today, some camcorders can also shoot at variable speeds.

Slow and fast motion are used extensively in educational video programs, where their ability to present actions for analysis is valuable. In fiction programs, however, the movie makers want you to believe that the story is real. Since slow or fast motion shots remind you that you are watching a video or film, these effects are usually used sparingly in fiction programs. Fiction and nonfiction programs alike rely mainly on a second method of controlling video time. This method is *editing*.

Altering speed in editing

Slow and fast motion techniques are always visible because they vary time speed within each shot. In controlling time through editing, however, the action within every shot unrolls at a perfectly normal, realistic pace. It is *between* shots that editors expand or condense the passage of time. Because the manipulations occur between shots instead of within them, they are usually invisible to the viewer. Editors control video time speed in two ways: they slow time down by overlapping action and speed it up by omitting action.

Overlapping and Omitting. To slow an action down, editors show parts of it more than once. For example, suppose you want to dramatize the feelings of a character who is wearily descending a flight of stairs. To do this, the director first covers the action from above, and then again, from below, **Figure 4-4**.

Suppose that the staircase has 15 steps. To extend the time required to descend it, the editor shows the actor walking down steps 1 through 10 from the rear, and then cuts to the front angle. Here is the trick: even though the first shot ends on step 10, the second shot starts back up on step 5. Because the shots have such different points of view, the audience does not notice the overlap and the editor has extended the action by five steps.

To speed an action up, editors use the opposite technique: they cut out repetitive or otherwise uninteresting action, **Figure 4-5**. Imagine the same staircase scene, only this time the character is jogging briskly down the steps. To enhance the impetuous feeling of the scene, the director uses exactly the same camera placements as before; but this time the editor ends the first shot with the actor on step 5 and starts the second shot on step 10. This shortens the time required to descend the stairs and, as before, the audience does not notice that the character has mysteriously skipped steps 5 through 9.

In both speeding the action up and slowing it down, the key is invisibility. The editor is able to compress or expand video time by adjusting it *between shots*, where the audience cannot see what is happening.

Altering Video Time To Edit Performance

Editors routinely adjust video time in order to fine-tune the performances of actors. To see how this works, imagine a romantic comedy in which a young woman is taking her boyfriend home to meet her parents, who believe she is marrying a wealthy man. In successive shots she says, "Don't forget, you're worth 50 million dollars," and he replies, "I'm not worth 50 dollars."

In the first version, his line follows hers immediately. In the second version, the editor begins the boyfriend's shot about 1.5 seconds before he speaks. Though the actor is simply looking at his partner, the added time makes him appear to be thinking about the hazards of this lie.

Figure 4-4.
Overlapping action through the use of a reverse angle.

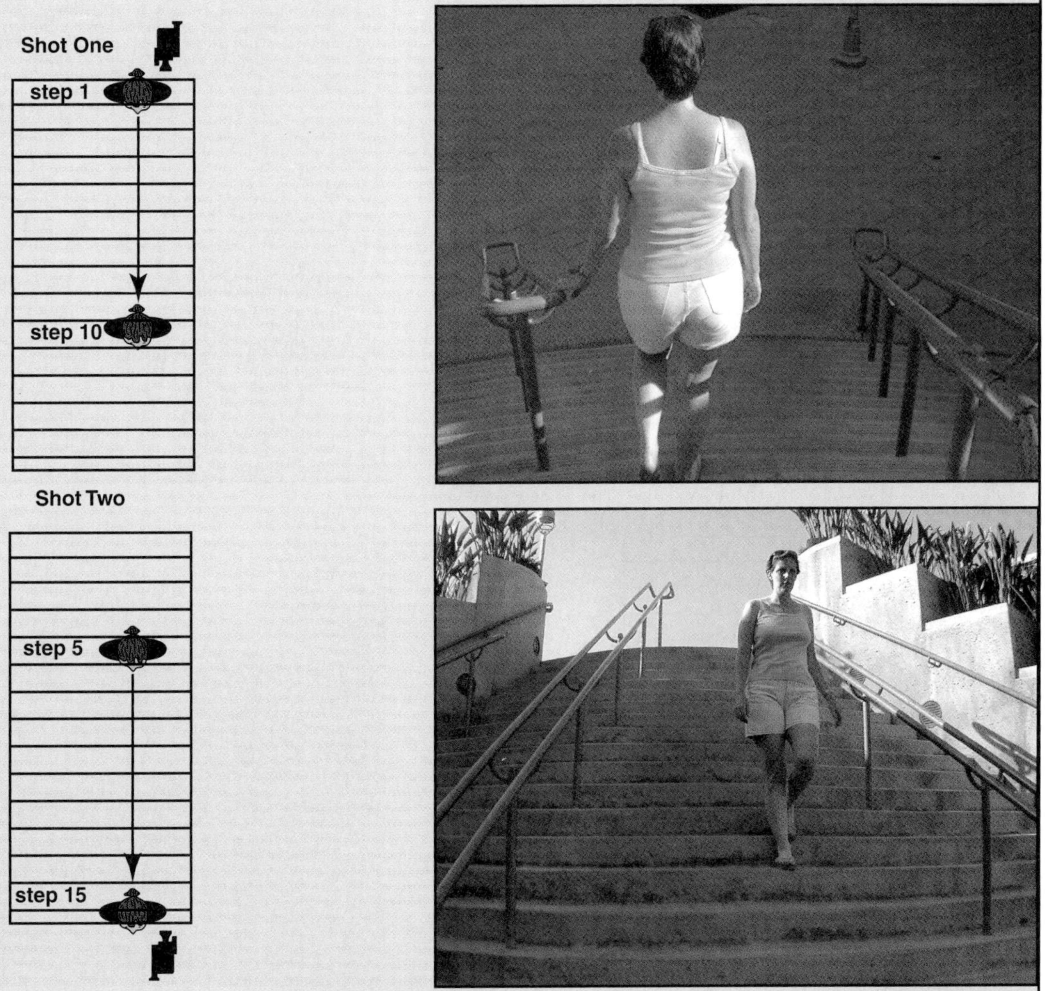
Shot One
step 1
step 10
Shot Two
step 5
step 15

Overlapping Action

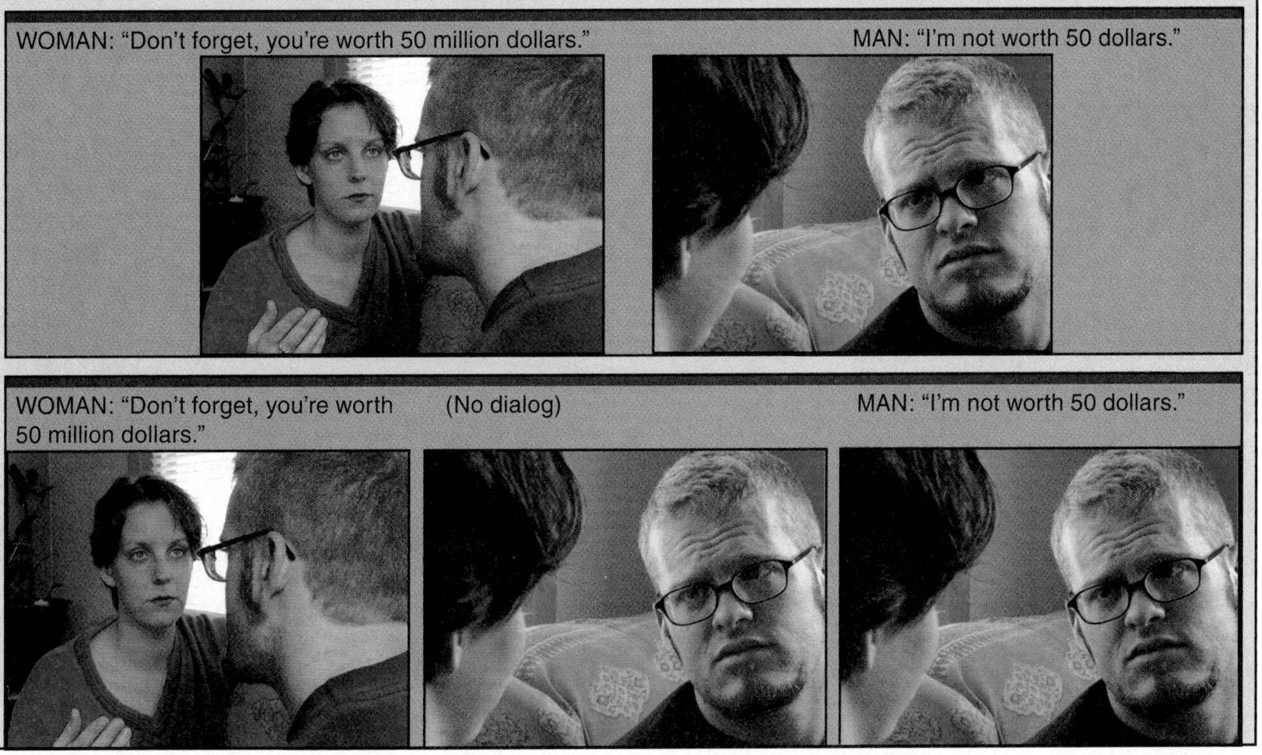
WOMAN: "Don't forget, you're worth 50 million dollars."
MAN: "I'm not worth 50 dollars."
WOMAN: "Don't forget, you're worth 50 million dollars."
(No dialog)
MAN: "I'm not worth 50 dollars."

Figure 4-5.
Omitting action through the use of reverse angles.

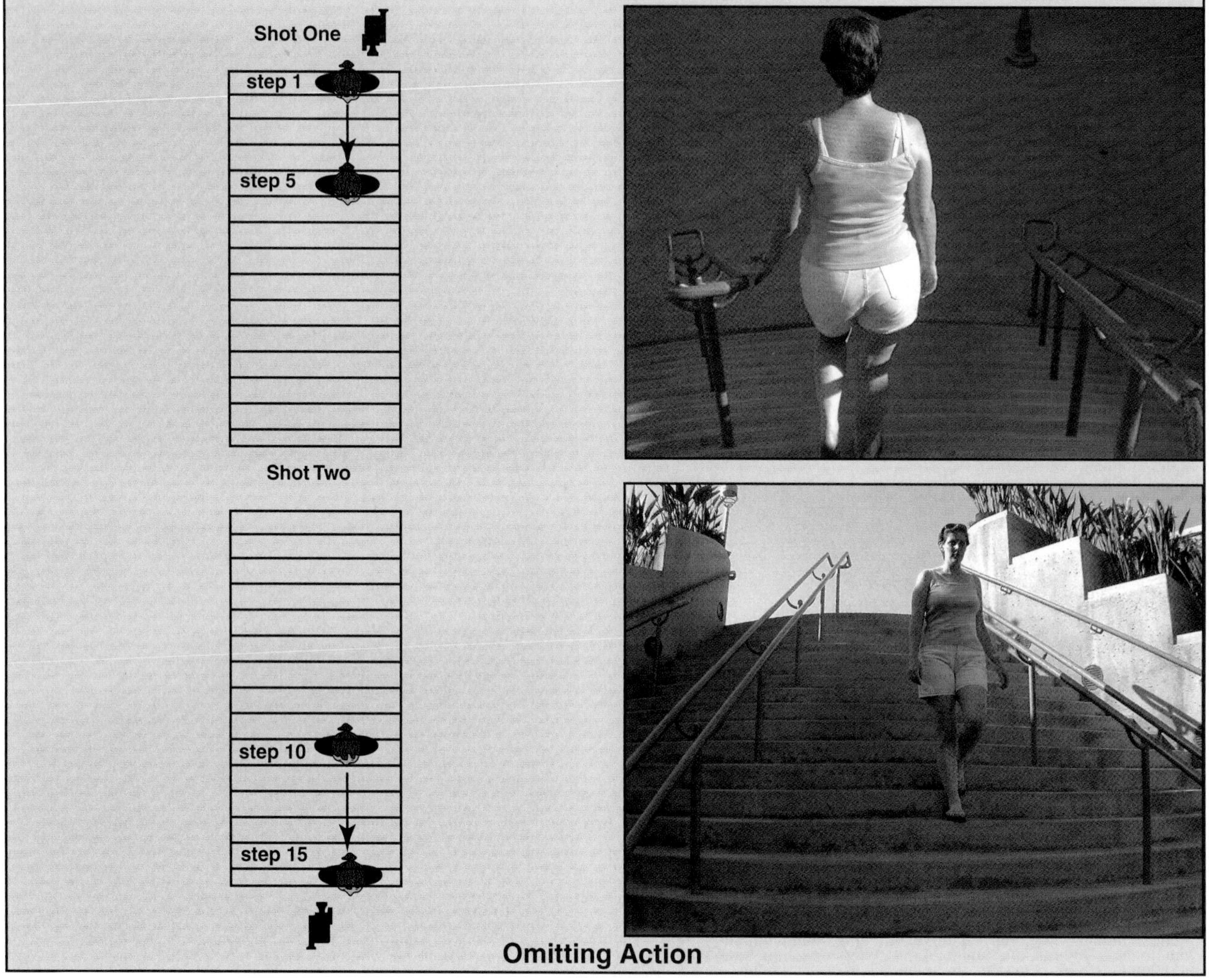

Video Time Flow

Omitting or overlapping action changes not only the speed of video time, but its flow as well. Real-world time flows continuously, without so much as a moment repeated or cut. But as we have seen, time flow in the very different world of video is constantly interrupted to control its speed. We adjust video time flow for other purposes, too. To see how this works, we need to consider the three main forms of video time progression: *serial*, *parallel*, and *disjointed*.

Serial Time Flow

Up to this point, we have been discussing time flow in *serial* form: a stream of events moving forward in a single sequence. For example, a character first showers, then gets dressed, then eats breakfast, then leaves for work. This serial flow of actions is the most common depiction of time in video because that is how we perceive time in real life.

Single-stream time flow can be compressed to an amazing degree. A story can span three or four generations in two hours. A documentary video may cover hundreds of years of history. Despite its many advantages, however, a strictly serial time flow is limiting because you can show only one set of related actions. When you want to show two or more sequences of events at a time, you need to establish multiple streams of video time, flowing *parallel* to one another.

Parallel Time Flow

In the video world we often see two or more sequences of action that are happening at the same time. To use a melodramatic cliché as an

Figure 4-6.
Two sequences take place simultaneously: the hero hangs from the railing, as the heroine races to save him.

example, we might see the hero hanging from a railing around the roof of a building while the heroine races up an endless stairwell to rescue him, **Figure 4-6**.

The editor switches back and forth among these parallel sequences in what is called ***cross cutting***.

Cross cutting is so widespread in movie storytelling that the silent movie title, "Meanwhile, back at the ranch..." has become a common phrase.

The audience never notices that in watching the movie, they are seeing the same moments of time twice-over (though in two different locations).

Here, video time is flowing in just two parallel streams; but skillful directors and editors can keep three, four, or even more parallel time-streams going at once.

The record for parallel cutting may be held by the film, *The Longest Day*, which follows several people at once through the D-day invasion of Normandy in World War II.

Disjointed Time Flow

If serial and parallel time flow are alternate methods of presenting sequences on the screen, *disjointed* time flow describes the way those sequences are usually recorded in production. Most programs are shot out of chronological order and then reorganized into proper sequence during the editing process.

To save time and money, a production will typically shoot all the scenes that take place at one location (or on one set) at the same time, regardless of their positions in the edited program.

For example, imagine that we see the parents of a naive young woman putting her on a train for the big city. Four years later, in story time, they return to the same station to welcome back their now successful and sophisticated daughter. Typically, those scenes will be shot back-to-back in real-world time, often on the same day, to avoid the expense of a second trip back to the station location.

On the other hand, many sequences that seem continuous on the screen are shot in separate pieces days, weeks, or even months apart. The most common example involves moving from exteriors to interiors and back. Often, the outside of the door is a real door on location and the inside is part of a set, **Figure 4-7**. Though the actor's passage from outside to inside seems continuous in screen time, the two shots of that passage were made at very different points in real time.

In fact, *every* seemingly continuous video sequence is actually built from individual shots made at different points in real-world time. For example, a dialogue sequence that plays for two minutes on the screen may take a whole production day to shoot. (This is not true of multi-camera TV programs like comedies and talk shows, in which all of the camera angles are recorded simultaneously by multiple cameras.) Different points in real-world time can even be combined in single video shots, as when actors play multiple roles via double exposure or are matted into backgrounds photographed at other times.

Figure 4-7.
The two sides of a movie door are often not the same door. A—Exterior. B—Interior in a studio.

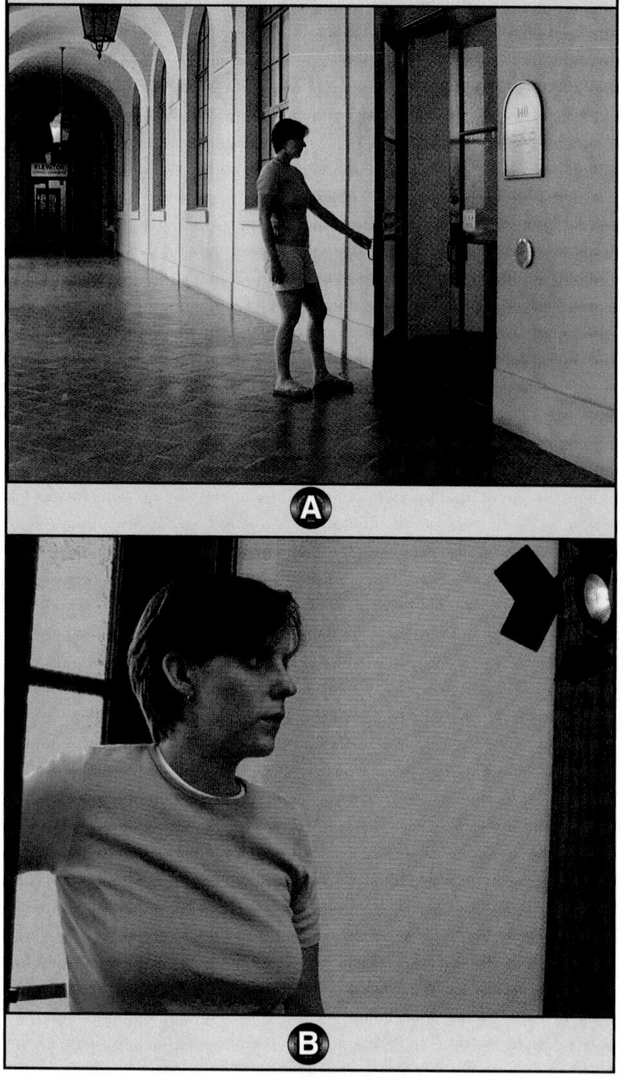

As long ago as 1921, in *The Playhouse*, Buster Keaton populated an entire theater with copies of himself, at one point playing nine different actors in the same shot!

But whether video time is manipulated to adjust pacing, to show parallel actions simultaneously, or to combine multiple real-world moments into single video time sequences, its flow is routinely stopped, started, and moved around in a way that is impossible in the actual world.

Video Time Direction

In the real world, we perceive time as moving always in one direction, from future to past. Or perhaps we think of ourselves as moving *forward*, so that the future is always becoming the present as the present becomes the past. Though events in the video world usually move in the same way, the rules of this different universe also allow us to reverse time direction and to jump forward and backward in time. The simplest way to reverse video time is by playing a recording backward, either for comedy or to achieve a special effect. We have all seen shots that show things like pie fragments leaving an actor's face and reassembling themselves into a complete pie, or milk rising like a small water spout to re-enter a milk bottle. Like fast motion, reverse motion within a shot often has a comic effect.

Time Running Backward

Reverse-motion can also be used to simulate actions that would otherwise be impossible, **Figure 4-8**. For example, imagine two actors making an impossible leap from the ground to a wall six feet high. To accomplish this feat, they are recorded jumping *backward* from the top of the wall to the ground. The editor then reverses the shot.

Like slow and fast motion, reverse motion shots tend to be used sparingly because when they are noticeable, they look obviously unnatural and contrived.

Flashing Backward (and Forward)

Though audiences do not really believe that time itself can run backward, they readily accept the idea that the video program can jump backward in time and then resume its forward direction. A time jump like this is called a *flashback*. The flashback is such an old and familiar technique that two examples will illustrate it sufficiently.

Figure 4-8.

A—The actors approach the wall and crouch to jump.

B—They leap upward.

C—They appear to jump to the top of the wall and…

D—land there.

In shot C as originally recorded, they jump backward off the top of the wall…and drop toward the ground. The original ending of shot C is discarded.

- At a high school class' 25th reunion, the now-sedate participants remember the crazy stunts they pulled as teenagers; and, as each story unfolds, we jump back 25 years to watch it happen.
- As she takes the oath of office, the new governor thinks of her long climb to get to this moment; and we begin reliving that struggle with her.

In the first example, we jump back to the past in order to *contrast it* with the present. In the second, we explore the past in order to *explain* the present.

In both cases, the moviemakers flash us backward in time—a leap that is impossible in the actual world.

It is not a flashback when a character in a program actually travels through time (as people often do in science fiction stories). In a flashback, it is the story, rather than the character, that changes position in time.

Of course, we can also flash *forward*, jumping into the future for a look, before returning to the main narrative in the present. In this case, time changes direction not when we make the leap, but when we jump backward again to return to the present.

Figure 4-9.
WOMAN: I can get there in an hour.

WOMAN (offscreen) I've got time to spare!

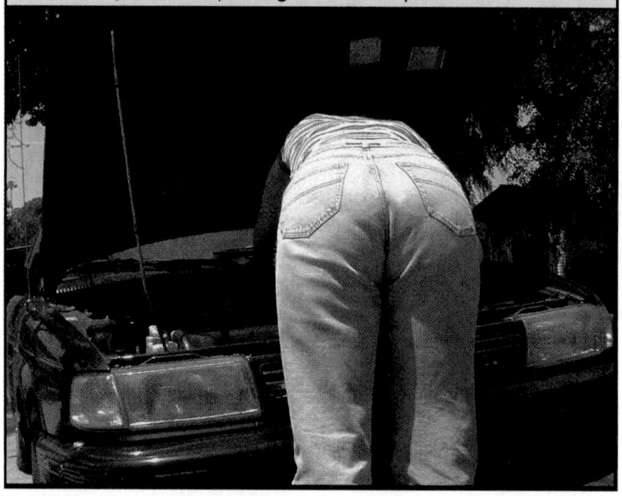

Video Time Coherence

Video makers manage time in one additional way. They pull moments in time apart by splitting their video and audio components and mixing them with those of other moments. A common application of this technique is *voiceover narration*. When a boy on the screen delivers newspapers from his bicycle as an adult voice on the sound track says, "Little did I know what would happen on that Spring day so long ago...," the picture is unfolding at one point in time, while the sound is at a point years later.

Split Edits

An equally useful time-splitting technique is called, appropriately, a *split edit*: a transition from one shot to another in which the video and audio portions do not change simultaneously. Split edits take two forms: *video* changes first or *audio* changes first.

To illustrate both forms, imagine a young woman planning a trip. In the first scene, she estimates her driving time and in the second, she tries to fix her broken-down car.

Video leads. When video changes first, the old audio continues offscreen, over the new video, **Figure 4-9**.

Audio leads. When audio changes first, the sound from the new scene begins over the preceding video, **Figure 4-10**.

Whichever way the edit is split, the time of one element is overlapped with the time of the other.

About Time

It is evident that video time behaves very differently from actual time because nothing about it is

Figure 4-10.
WOMAN: I can get there in an hour. I've got time to spare!
SOUND EFFECTS: Clanking of tools.

WOMAN: SOUND EFFECTS: Clanking of tools continues.

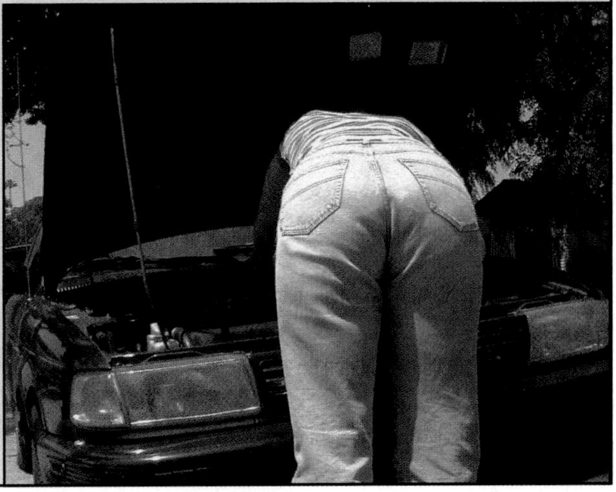

fixed: not its speed, flow, direction, or coherence. The skillful video maker uses the changeability of video time, manipulating it to condense the action, unfold the narrative, and create special effects.

Our discussion of time has been, necessarily, somewhat abstract and theoretical. In actual production most video makers deal with space and time instinctively, and so can you. However, by taking the trouble to understand the very different laws that rule time in the video world, you will be able to use them to greater advantage in your own programs.

Now that we have considered video time in this chapter and video space in Chapter 3, we are ready to look at video and audio: the two components of every program. Chapters 5 and 6 concern communicating with video images: through composition in Chapter 5 and through video grammar in Chapter 6. Communicating with audio is discussed in Chapter 7.

Chapter Review

Answer the following questions on a separate piece of paper. Do not write in this book.

1. Video time is variable; real-world time is ______.
2. *True or False?* A video editor can shorten time, but cannot lengthen it
3. To produce a dreamlike effect, a shot can be recorded in moderate ______.
4. A technique called ______ is used to convey parallel time flow.
5. *True or False?* "Flashback" is the term used to describe a character's travel through time.

Technical Terms

Cross cutting: Also called "intercutting", cross cutting is showing two actions at once by alternating back and forth between them. It can also be used to show three or even more parallel actions.

Flashback: A sequence that takes place earlier in the story than the sequence that precedes it.

Screen time: The length of real-world time during which a sequence is displayed on the screen (in contrast to the length of video world time that apparently passes during that sequence).

Split edit: An edit in which the audio and video of the new shot do not begin simultaneously. When "video leads," the sound from the preceding shot continues over the visual of the new shot. When "audio leads," the sound from the new shot begins over the end of the preceding visual.

Space, Time, and Persuasion

This chapter and Chapter 3 explain how video makers manipulate video space and time and why they do it. They do it to change reality to suit their purposes.

What are those purposes? Some are unavoidable, such as condensing time and space to cut out the boring parts. Some are constructive, such as focusing viewers on the essentials of a subject and presenting it coherently. Some are harmless and dramatically useful, such as turning two different doors into one or adjusting the pace of an actor's performance.

However, one very common purpose is not always so innocent: some videos falsify reality to encourage viewers to buy a product, vote for a candidate, or support a social agenda. That is, they seek to persuade viewers by distorting the truth.

Of course, false or misleading rhetoric is not new. People have used it for many centuries in speeches and writings. The problem with video is that the persuasive powers of the medium are inherently greater than any found in other forms of communication. This means that viewers are more likely to believe the manipulated "reality" in video than to accept false facts and arguments in the older media of speech and writing.

What is it about video that makes it so compelling, so good at selling false realities? There are several answers, but perhaps the central one was articulated by the French critic Andre Bazin in an essay called "The Ontology of the Photographic Image." (Ontology is the formal study of existence, of being.) The essay was translated and published in What is Cinema? University of California Press, 1967.

Though Bazin's essay is concerned with the medium of film, his ideas are equally applicable to video; and though his reasoning is extensive and subtle, it is based on a simple idea. Bazin suggests that photographs compel belief because they appear to prove the existence of their subjects.

His argument is something like this: the original photos that most people see are snapshots made with simple cameras and without manipulation by the photographer. Therefore, if something appears in a snapshot, it "must" be real, or, at least, it must have been real when that photo was taken. After all, you cannot take a snapshot of something that does not exist; so when you look at a photo of someone on a ferry boat with the Statue of Liberty behind her, you see apparent evidence of three things. At the instant when that photograph was snapped:

- The person existed.
- The statue existed.
- The person stood in front of the statue.

This belief in the literal truth of photographs has been reinforced by over 150 years of photography, most of it news pictures, portraits, and snapshots—all records of actual people and places. Consequently, an unconscious assumption has developed that moving photographs too (cinema and video) do not lie, just as mathematical figures do not lie.

But remember the old cliché: "Figures don't lie, but liars figure." You could say with equal truth, "Photographs don't lie but liars photograph."

A snapshot seems to prove that the person was at the Statue of Liberty

But the subject...

was composited into the original image.

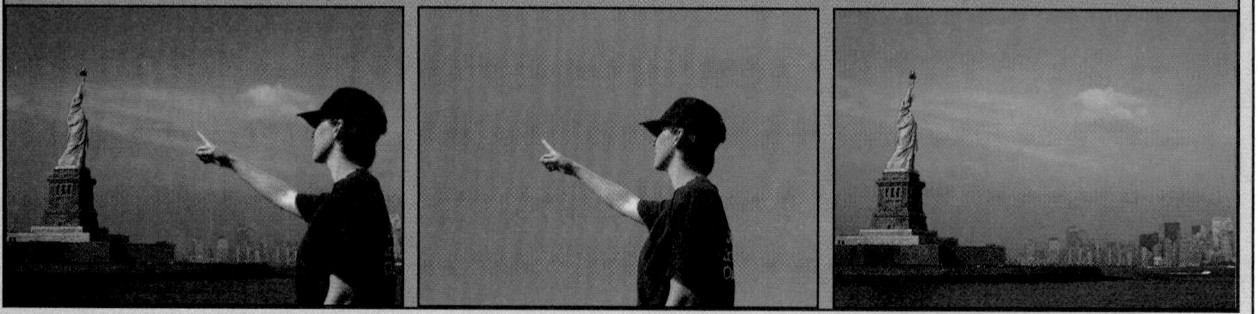

5 Video Composition

Objectives

After studying this chapter, you will be able to:

- List the key principles of video composition.
- Explain how contrast is used for emphasis.
- Describe techniques for creating depth in compositions.

(Sue Stinson)

About Video Composition

The word *composition* (literally, "putting in the same place") has many different, though related, meanings. In video, composition is the purposeful arrangement of the components of a visual image. These components, often called pictorial elements, may be objects such as people, buildings, or trees. They may also be abstract elements such as lines, shapes, degrees of brightness, or colors.

What is the purpose of composition? One easy answer is that it somehow renders a picture more "pleasing" or more "artistic." But that reply makes composition seem rather unimportant: nice to have, but not absolutely necessary. In fact, however, composition is absolutely essential in video because it helps images communicate with viewers more quickly, efficiently, and powerfully. Composition does this by:

- Organizing pictorial elements so that viewers can quickly sort them out and identify them.
- Adding emphasis to direct attention to the most important elements on the screen.
- Creating an illusory third dimension in a two-dimensional picture.

Overall, composition is such a powerful communication tool that every video maker needs to master its fundamentals. So let us see how composition affects visual organization, emphasis, and depth.

Figure 5-1.
Composition helps viewers "read" the image. A—Disorganized. B—Organized.

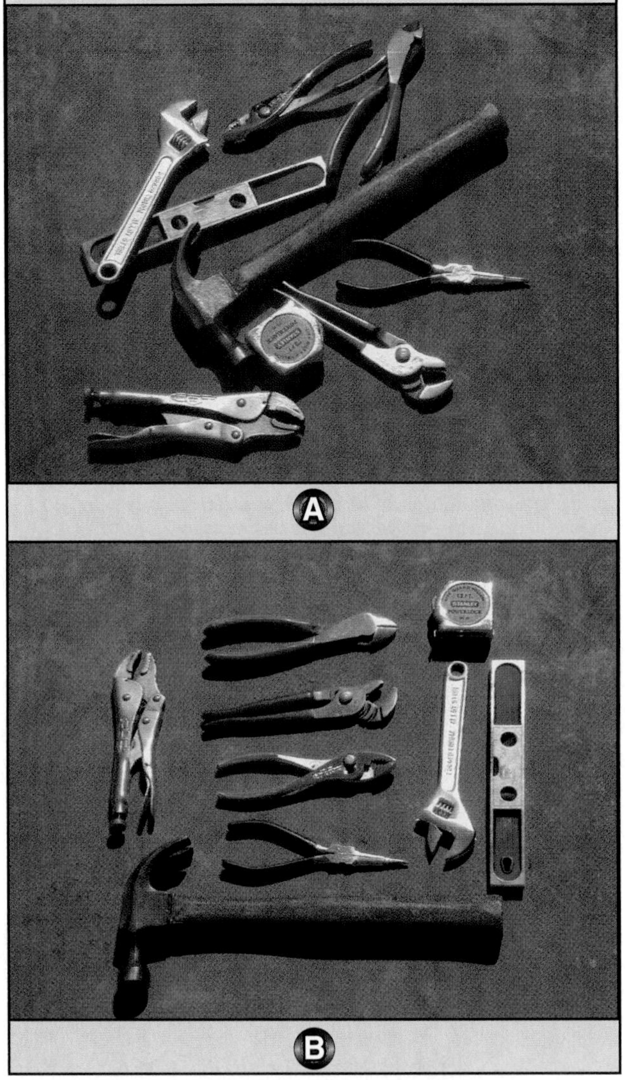

Organization in Composition

Composition organizes the elements within an image, to help viewers decode it.

Decoding an image means recognizing the elements in it, assigning meaning to each one, and identifying a relationship among them.

For example, the first image in **Figure 5-1** shows a number of small tools. Dropped in a disorganized heap, they are hard to identify. In the second image, the individual tools are easy to spot because they have been arranged in a composition. Viewers need composition to help them decode pictures, because they are unable to organize screen images as easily as they organize visual information from the real world.

Simplicity, Order, and Balance

In organizing the pictorial elements in images, composition follows the key principles of simplicity, order, and balance, **Figure 5-2**.

Simplicity means eliminating unnecessary objects from the image, so that the viewer has fewer things to identify. Notice that the first picture is already easier to decode, merely because the number of tools in it has been reduced.

Order means arranging objects in the image. In the center photo, the tools are even easier to recognize because they have been lined up neatly.

Balance means distributing the objects in a way that gives about equal visual "weight" to each section of the image. Also, notice that

Figure 5-2.
Principles of composition.

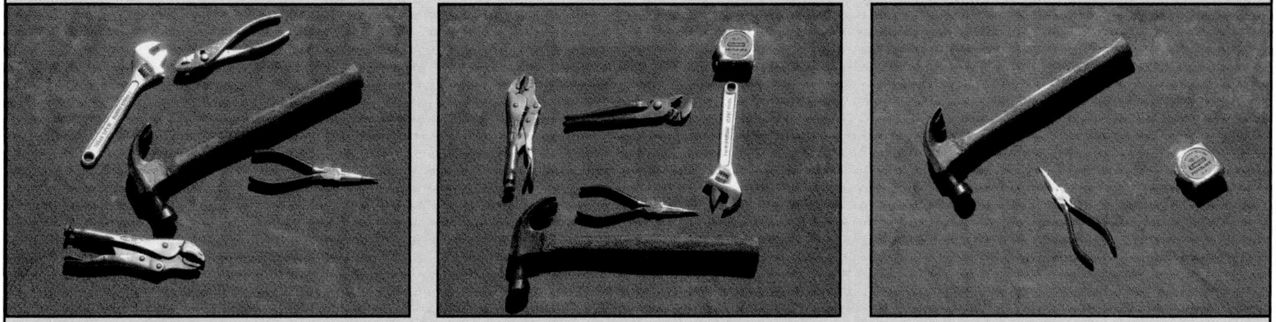

Simplicity Order Balance

Composition and the Frame

We have noted that the **frame** around the image forms the border of the video world. At the same time, the frame acts as the most powerful component of visual composition. In fact, it is the frame that makes composition possible. A composition is the visual organization of a defined area and that definition is provided by the frame.

To see how this works, look at Photo A. Though the edges of this picture do form a frame of sorts, the image as a whole lacks an overall sense of composition. However, as you can see from Photos B, C, and D, we can create several different compositions, according to how we place frames around different components of the original picture.

Different compositions within a frame. A—Frame that is poorly composed. B—Composition 1. C—Composition 2. D—Composition 3. (Corel)

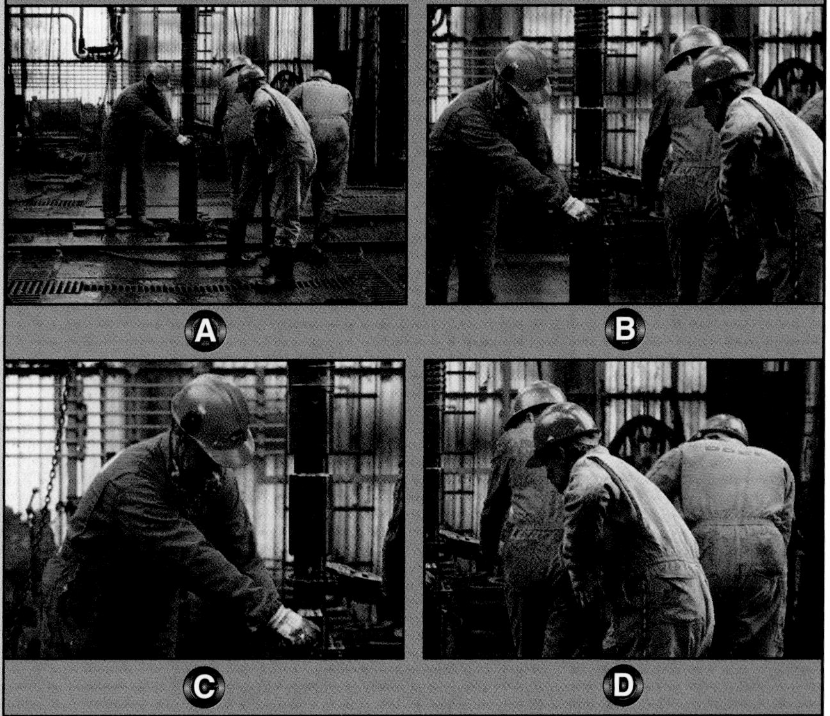

A B C D

In addition to defining an area, the frame tells you what is **horizontal** (parallel to the horizon) and what is **vertical** (perpendicular to the horizon). In both the following images, the person is positioned upright. In the second one, however, she seems to be tilted sideways. In fact, it is the frame and not the person that has been tilted in the second shot. But because we derive our sense of vertical and horizontal from the frame, the picture continues to appear level, while the person within it seems tilted.

A portrait with a level horizon. A portrait with an off-level horizon.

arranging the large hammer so it balances the group of smaller tools makes it stand out in the composition.

Emphasis in Composition

Organization helps viewers make sense of the elements in an image, but by itself, it does not confer special importance on any of them. That is the function of emphasis. *Emphasis* is any technique that attracts the viewer's attention to one part of a composition. Techniques of emphasis include position, relationship, significance, and contrast.

Position

You can emphasize a compositional element by the way in which you position it within the frame. In any composition, the upper left quadrant attracts the eye first, because that is where we automatically look to start reading a new page of text.

Obviously, not everyone begins reading at the upper left. Readers of Hebrew, for example, begin the page at a different point.

Psychological experiments have confirmed that in a composition, the upper left quadrant is most attractive, the upper right is next, the lower left is third, and the lower right is the least attractive. See **Figure 5-3**.

Figure 5-3.
In Western cultures, the upper-left quadrant is the most powerful part of the image.

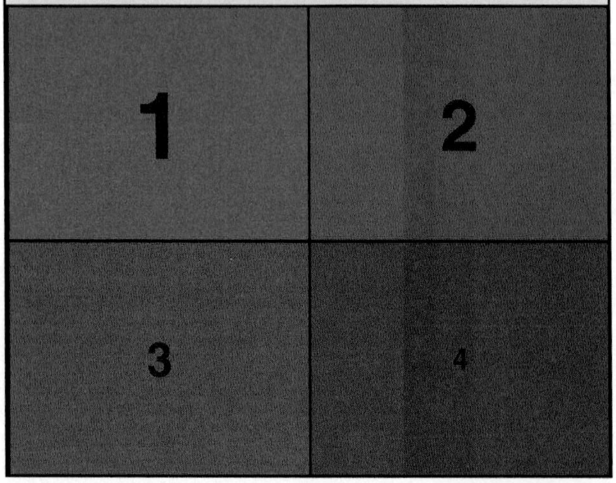

Figure 5-4.
Leading lines, such as the diagonal lines of this stairway, lead the viewer's eye to the center of interest. (Sue Stinson)

Relationship

You can emphasize an element by placing it in relationship to other elements in the composition. These other components lead the eye to the center of interest. You can see this in **Figure 5-4**. Since the elements of video images are usually moving, a good director can exploit their ever-changing relationships.

Significance

Some compositional elements attract attention simply because they mean something to the audience. For instance, when you look at a photo of a celebrity, **Figure 5-5**, your eye goes immediately to the famous person because you recognize the

Figure 5-5.
President Clinton stands out in this group photo.

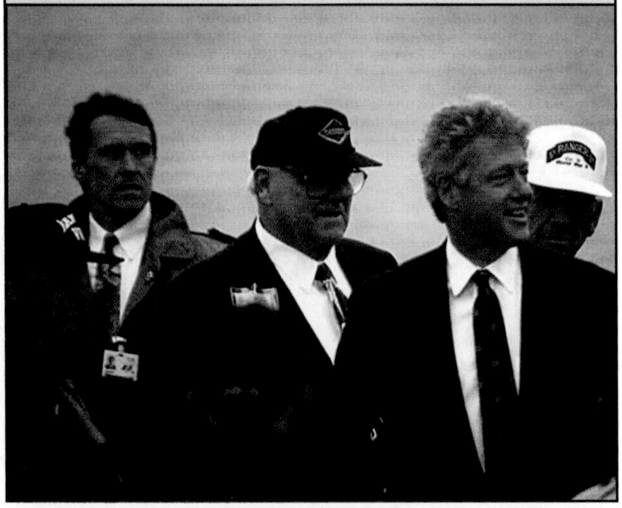

face. It means something to you. Significance in a composition is widely used in Hollywood movies because the highly recognizable stars automatically attract viewers' attention.

Emphasis through Contrast

Contrast is the technique of emphasizing one element in a picture by making it look different from the other elements. An element can be contrasted by adjusting its traits—traits that include size, shape, brightness, color, and focus.

In videography, the word *contrast* is also used in the more technical sense of difference in luminance (*brightness*) between the lightest and darkest picture elements.

Contrasting size

The bigger something is, the more it calls attention to itself, **Figure 5-6**. It is simple to make one object appear bigger than other objects of a similar size: just place it closer to the camera. Not surprisingly, important action is typically staged in the foreground, where the actors are largest.

Contrasting shape

The rectangle in **Figure 5-7** attracts attention because its shape contrasts with the circles around it.

Contrasting brightness

Light elements attract the eye more than dark ones, **Figure 5-8**. Much of the craft of video lighting involves emphasizing important elements through careful control of brightness.

Contrasting color

Perhaps the simplest form of contrast is color. The viewer's eye is attracted to elements colored differently from the prevailing hues of the overall image. In **Figure 5-9**, the red circle stands out.

Figure 5-6.
Using contrast in size to emphasize one element in a picture.

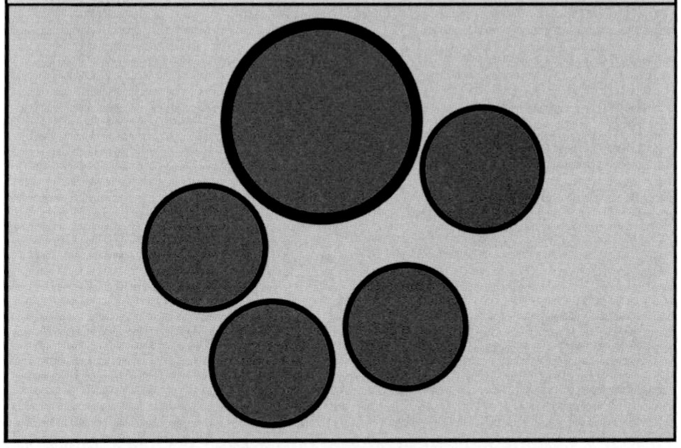

Figure 5-7.
Using contrast in shape to emphasize one element in a picture.

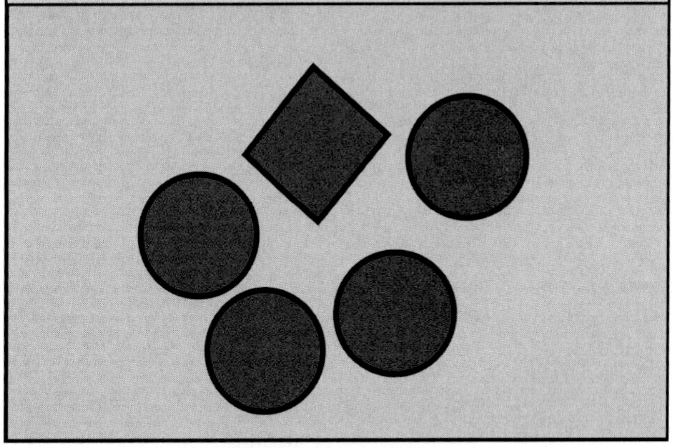

Figure 5-8.
Using contrast in brightness to emphasize one element in a picture.

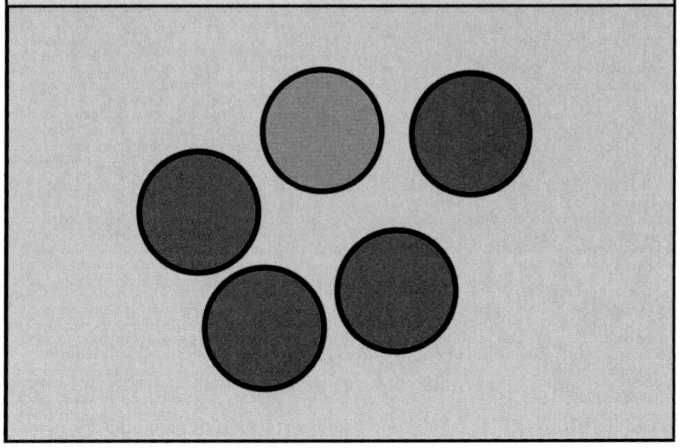

Figure 5-9.
Using contrast in color to emphasize one element in a picture.

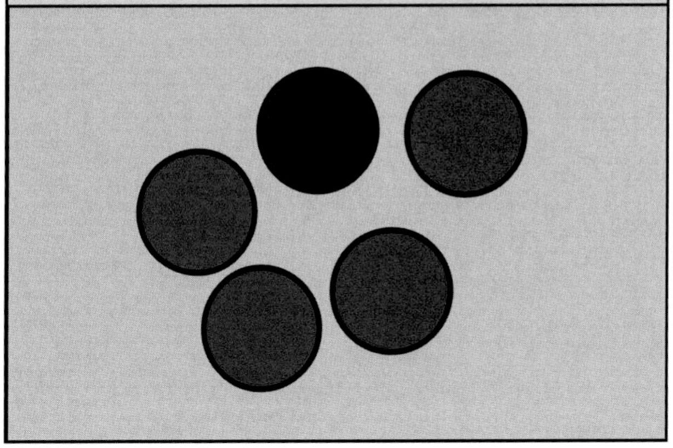

Contrasting focus

Focus refers to the visual sharpness (clarity) of a pictorial element. Sharp, well-detailed elements attract viewers' attention away from soft, out-of-focus elements. Because video images are rarely equally crisp throughout their depth, videographers set the focus on the elements they want to emphasize. Although the pink and yellow flowers in the background of **Figure 5-10** are bright and colorful, they have been de-emphasized by throwing them out of focus. Sometimes you will see the focus change in the middle of a shot, to shift attention from one element to another.

Figure 5-10.
Using contrast in focus to emphasize one element in a picture. (Sue Stinson)

Great painters like Renoir often achieved the effect of selective focus by painting important picture elements in fine detail and leaving less important ones sketchy and less distinct.

Depth in Composition

As we noted in Chapter 3, the video world has only two true dimensions: height and breadth. On the essentially flat screen, what appears to be depth is only an illusion. Since most video programs are intended to mirror the actual world, depth-enhancing techniques play an important role in pictorial composition. Collectively, these are called perspective.

Perspective

Perspective is a group of techniques used to suggest the presence of depth on a two-dimensional surface—a surface such as a painting, a photograph, or a video screen. Over many centuries, artists have evolved several ways to represent three dimensions on two-dimensional surfaces. These effects will appear in your compositions whether or not you employ them intentionally. But if you purposely apply them, you can greatly enhance the sense of depth in your images. Perspective techniques include *size, overlap, convergence, vertical position, sharpness,* and *color intensity.*

Size

In the actual world, the closer an object is, the larger it appears, **Figure 5-11**. So in a two-dimensional image, apparent size is a powerful clue to distance.

Figure 5-11.
Using apparent size as a perspective technique.

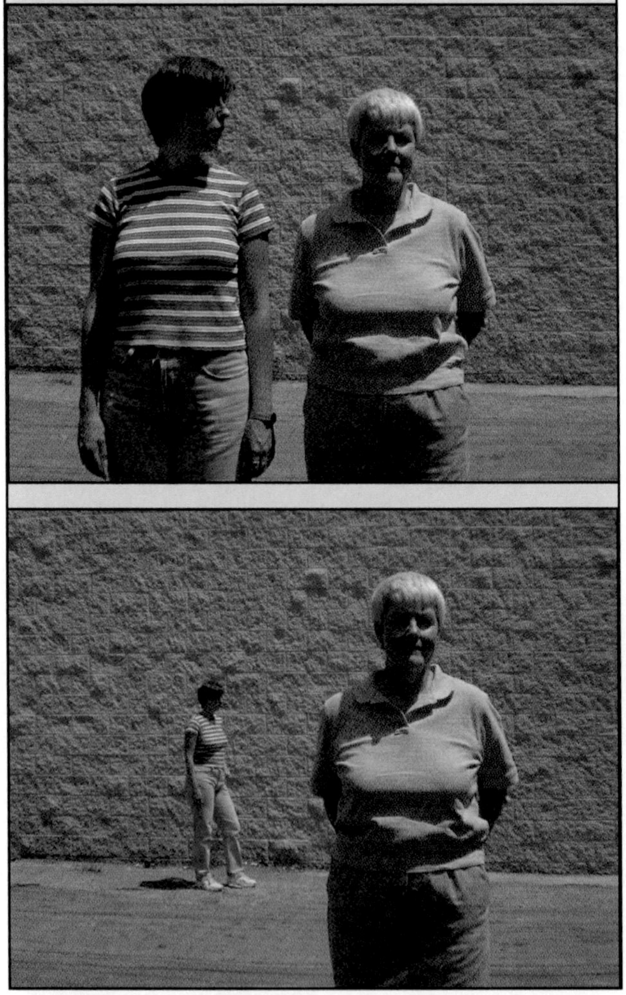

Overlap

In the actual world, one object will mask part of another if it is in front of it. So in two-dimensional compositions, an object that overlaps another is perceived to be closer, **Figure 5-12**.

Figure 5-12.
Using overlap as a perspective technique.

Figure 5-14.
Using vertical position as a perspective technique.

Convergence

In the actual world, parallel lines such as railroad tracks or the edges of roads seem to gradually come together as they recede into the distance. In two-dimensional compositions, converging lines convey a powerful sense of depth, **Figure 5-13**.

Vertical position

In the actual world, the farther away objects are, the higher they usually appear in the field of view. In **Figure 5-14**, notice that the subjects in the background are higher on the screen than the foreground objects. Also, the fence lines and horse heads gradually rise up the picture plane as they recede.

Sharpness

The farther away objects are, the more indistinct they appear, for two reasons. First, their fine details are smaller and harder to distinguish. Second, distant objects are softened by the amount of air between them and the viewer. Because the effect is created by air, this is often called *atmospheric perspective.* This type of perspective is evident in a picture of foothills with mountains behind them, **Figure 5-15**. Each ridge appears softer and less distinct than the one in front of it.

Color intensity

The same atmosphere that softens the appearance of objects also reduces the

Figure 5-13.
Using converging lines as a perspective technique. (Sue Stinson)

Figure 5-15.
Using decreasing sharpness as a perspective technique.

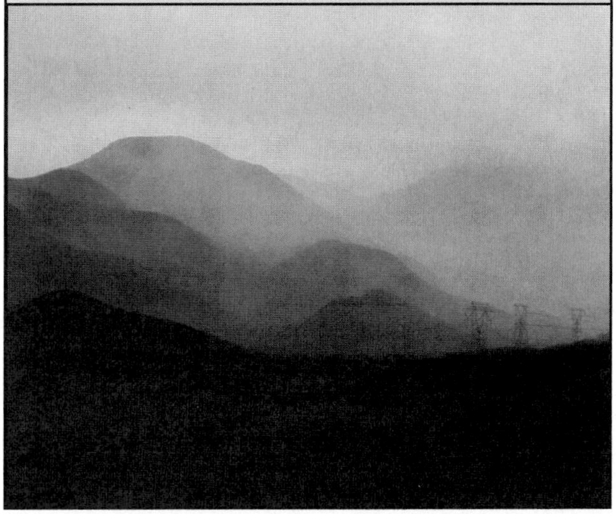

Figure 5-16.
Using color intensity as a perspective technique.

intensity of their colors. The farther away something is, the paler or more pastel its colors appear. In **Figure 5-16**, the orange color of the runner's shirt is much more intense in the closer view at right than in the long shot at left.

Composing Video Images

Once you understand the ideas behind pictorial composition, you can make them work for you by creating compositions, directing the viewer's eye, and controlling apparent depth in your images. The resulting images will be both better-looking and more effective.

Compositional Schemes

As you gain experience in framing video shots, you will instinctively use the principles of simplicity, order, balance, emphasis, and contrast to design your images. You do this to enhance their visual coherence by applying one of several compositional schemes.

The American Heritage Dictionary defines coherence as "...a logical, orderly, and aesthetically consistent arrangement of parts."

Asymmetrical balance

The most common of these schemes is called "asymmetrical balance," in which visual

Figure 5-17.
A—Symmetrical balance is static. B—Asymmetrical balance can make a more interesting composition. (Sue Stinson)

A B

Figure 5-18.
The small figure balances the door.
(Sue Stinson)

Figure 5-19.
A large, dark mass can be balanced against a smaller and brighter object.

elements are not equally opposed, like two children on a level seesaw, but distributed less formally to give an overall impression of balance. Generally speaking, compositions are more pleasing when their elements appear to have the same visual "weight." But when visual elements are balanced symmetrically, the result is often static and dull. It is usually better to organize them so that their characteristics are different and their positions are not evenly opposed. To see the difference between symmetrical and asymmetrical compositions, look at **Figure 5-17**.

In the first image, the church is planted squarely in the middle of the frame, so that the resulting composition is not very interesting. In the second version, the church is smaller and placed far down in the lower left corner, creating a more striking composition. (Notice that it is also given emphasis by its contrasting brightness, color, and position.)

In **Figure 5-18**, the composition is strongly asymmetrical, with the archway and the bright sky leading the eye strongly to the right half of the frame. Notice how the eye is then drawn to the large door on the left side, as if someone were about to come through it.

In **Figure 5-19**, the smaller, brighter subject on the right is balanced by the dark mass of the column on the left.

The rule of thirds

Perhaps the easiest way to design asymmetrical compositions is by using the so-called "rule of thirds." To do this, mentally fill the camcorder viewfinder with a tic-tac-toe grid that divides the frame into three equal vertical and horizontal sections. Then, create compositions in which the important pictorial elements match lines and intersections on the grid. It is not necessary to

Figure 5-20.
A—The scene to be composed. B—A grid imposed on the image. C—The image reframed to align better with the grid. (Corel)

Figure 5-21.
Notice that most compositions do not align visual elements with every line or intersection on the grid.
(Sue Stinson, Corel)

Figure 5-22.
None of these compositions follows the rule of thirds. The canoe builder is close, but note how the tea kettle in the foreground and the large cylinder on the right side do not fit anywhere on a grid. (Sue Stinson, Corel)

match *all* parts of the grid. **Figures 5-20** and **5-21** illustrate the use of the rule of thirds.

The rule of thirds can be applied to most kinds of subjects. It is a useful guideline, especially when you are composing images quickly to record events as they happen.

Other compositional methods

When you have time to arrange your subjects as well as your frame, it is often useful to ignore the rule of thirds, as you can see from **Figure 5-22**. Though each composition is balanced, it does not align on a tic-tac-toe grid.

Composition for Widescreen Video

Traditional video images are proportioned 1.33 to 1, meaning that the width of the frame is one and one-third as long as its height. The proportion is also expressed as 12 to 9. All HDTV images, many DVD programs, and some TV shows and commercials are now recorded in a wider format, proportioned 1.78 to 1—or, as it is commonly known, 16 to 9.

For more direct comparison, we will express the two screen shapes as 12 to 9 and 16 to 9, **Figure 5-23**.

Widescreen advantages

The 16 to 9 frame works very well whenever the image is strongly horizontal. This is usually the case with action scenes and scenic panoramas, **Figure 5-24**.

Figure 5-24.
A—Widescreen lends itself to action. B—It also works well for panoramic vistas.

Widescreen drawbacks

Wide frames are not without problems, however. For one thing, they make closeups more difficult, as you can see from **Figure 5-25**.

Figure 5-23.
A—The traditional video frame is proportioned 12 to 9. B—Widescreen video is proportioned 16 to 9.

Figure 5-25.
A—Framing a "closeup" leaves a lot of empty space. B—Framing to utilize the screen area cuts off part of the head.

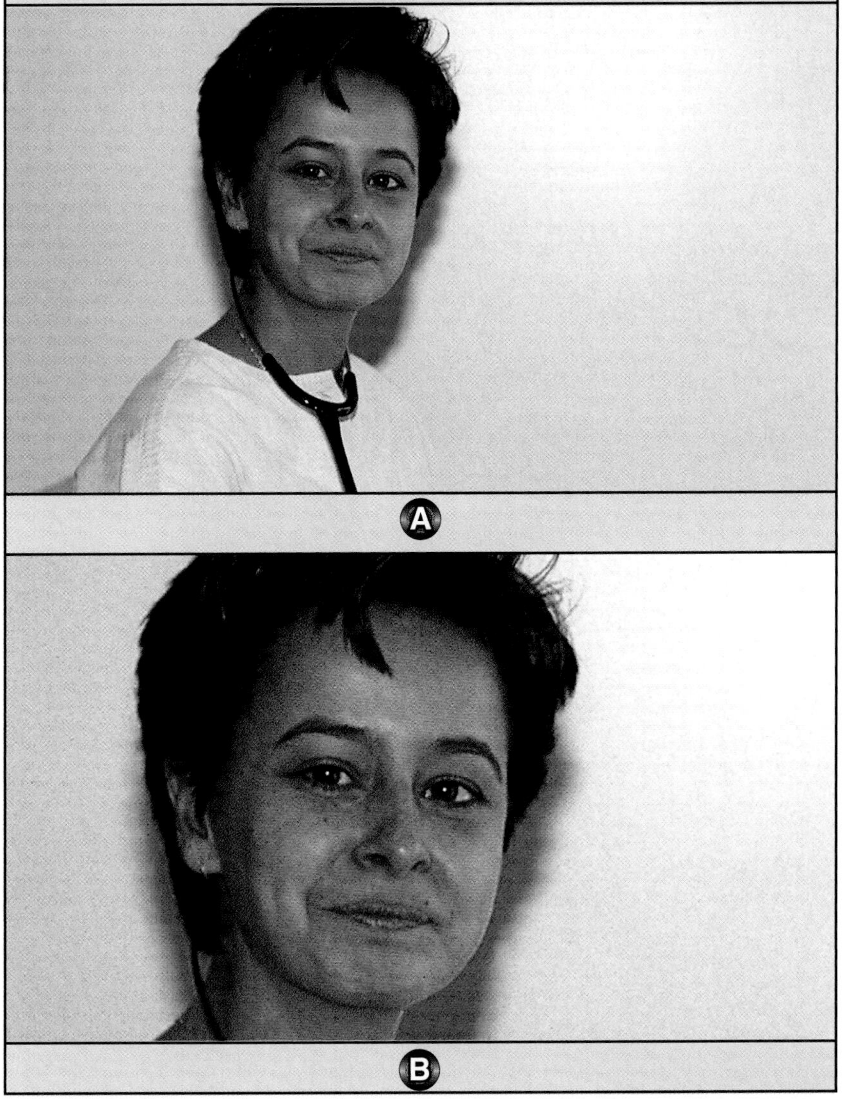

Also, widescreen images can be difficult to display on traditional TV screens. If the program is to be shown on a 12 to 9 screen, the edges of the composition will be clipped, **Figure 5-26**.

Widescreen images are often displayed "letterboxed" on regular screens, with black bands above and below the picture.

When this happens, "pan-and-scan" copying techniques are sometimes used to recompose the image for rerecording in 12 to 9 proportions, **Figure 5-27**.

Composing Widescreen Images

If you keep a few basic principles in mind, 16 to 9 frames can yield effective compositions.

Exploiting horizontals

Since widescreen is a horizontal format, look for camera angles that can exploit a wide view. Obviously, horizontal subjects can be easily framed, **Figure 5-28**.

Figure 5-26.
A 16 to 9 composition loses quality when clipped to 12 to 9 proportions.

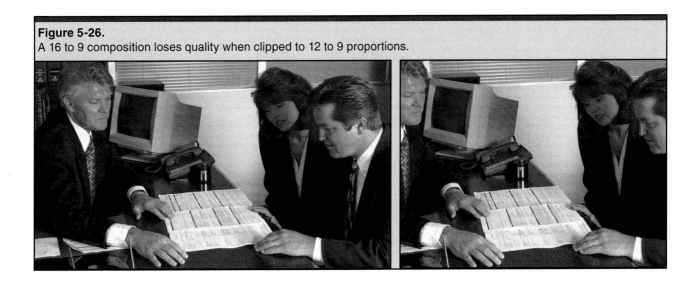

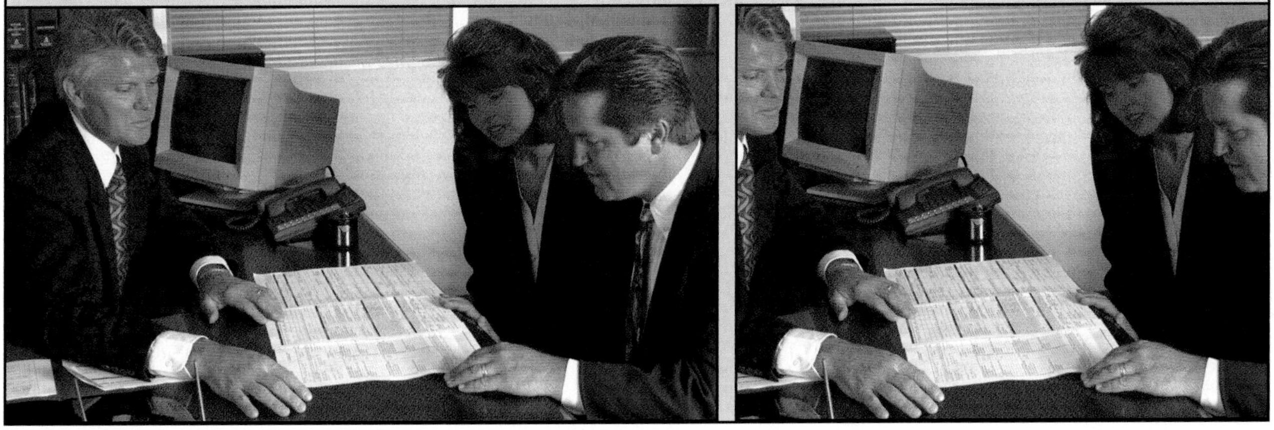

Figure 5-27.
In pan-and-scan copying, the original widescreen image is re-framed as a pair of close-ups (usually alternating with the dialogue). The image of the deer can be successfully cropped to match the 12 to 9 format.

Figure 5-28.
A horizontal subject. (Corel)

Figure 5-29.
Low angles can be effective in widescreen. (Corel)

Low angles and wide angle lenses can produce strong feelings of space and speed, **Figure 5-29**.

Using the rule of thirds

The rule of thirds works as well with widescreen compositions as it does with regular ones, **Figure 5-30**.

Composing at the frame borders

When the rule of thirds will not yield a good composition, try the opposite approach by placing important elements near the edges of the frame, **Figure 5-31**.

Figure 5-31.
Placing important picture elements near the edge of the screen can be an effective compositional tool. (Corel)

Framing subjects

Widescreen images can be framed like traditional images, with a few notable exceptions. Obtaining large close-ups requires cutting off part(s) of the subject's head. In these cases, framing off the chin produces an unattractive

Figure 5-30.
The rule of thirds applies to widescreen composition, as well. (Corel)

Figure 5-32.
Widescreen close-ups. A—With her chin excluded, the subject looks cramped. B—Framing the chin (and losing a bit more hair) solves the problem. (Corel)

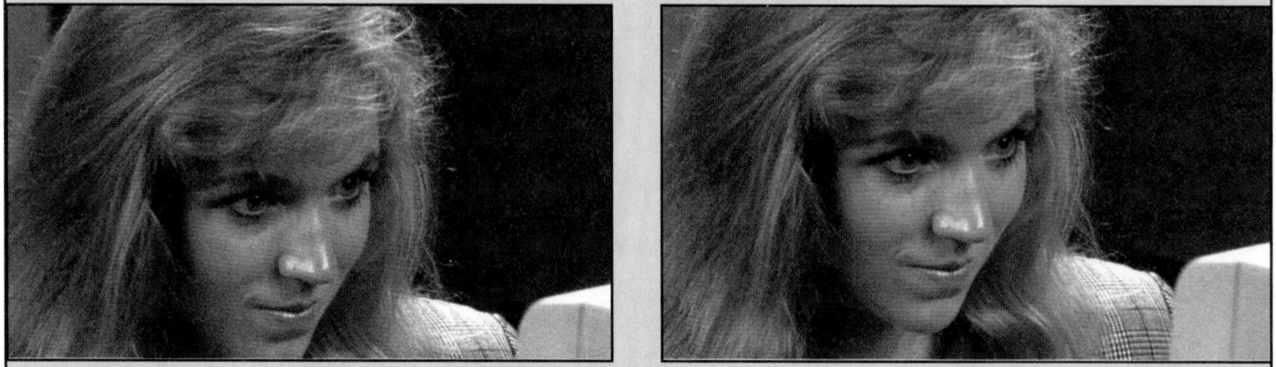

A B

effect. It is better to exclude more of the top of the head instead. Leave extra room below the chin so that it remains in the frame when the subject is speaking, **Figure 5-32**.

The horizontal format of wide screen allows more "lead room" in front of moving subjects. In **Figure 5-33**, notice that the plane formation has ample sky in front of it.

Figure 5-33.
Allow ample lead room for moving subjects. (Corel)

Directing the Viewer's Eye

The second job of composition, in addition to creating a coherent image, is placing the attention of viewers where you want it. To do this, you use emphasis and leading lines.

The most powerful way to attract the eye is through movement within the image, but that lies outside the subject of composition.

Emphasis

Emphasis, as noted earlier, can be added by adjusting subject position, relationship, or significance, and—above all—contrasting size,

Figure 5-34.
Using distinctive shapes to lead the eye. A—Though tiny in the frame, the human shapes are distinctive. B—The sand dollar draws the eye. (Adobe)

A

B

Figure 5-35.
Using multiple composition techniques to direct the eye. A—Size, position, and color. (Corel) B—Size and shape. (Adobe)

A B

shape, brightness, color, and/or focus. **Figure 5-34** shows two compositions that lead the eye to distinctive shapes.

Many compositions add emphasis using several techniques at once. In **Figure 5-35**, the first image directs the eye with a combination of size, position, and color. The second depends on size and shape.

Leading lines

The most straightforward way to emphasize a subject is simply to point to it by using one or more of the lines in the composition. **Figure 5-36** provides some examples.

Leading lines can be used more indirectly, **Figure 5-37**.

Controlling the Third Dimension

Composition plays a key role in creating the illusion of a third dimension on the two-dimensional video screen. In most cases, the goal is to enhance apparent depth, using several different methods.

Figure 5-36.
Leading lines. A—The subject is pointed to by no less than four leading lines. B—Here, the subject is so small that he would be unnoticed without the leading lines of the fence rails.

A B

Figure 5-37.
More leading lines. A—The skywriter's smoke trail leads to the small biplane. (Corel) B—The dark cow is emphasized by both horizontal and vertical lines, as well as contrasting color. (Sue Stinson)

A B

Staging in depth

First, try to place subjects and camcorder setups to suggest depth in the image. In the second image in **Figure 5-38**, several tricks are used to enhance apparent depth. The revised camera placement makes the left subject larger than the right one, the building windows diminish in size, and the clapboard plank lines converge. Notice also that the nearer subject "breaks the frame,"—that is, parts of her head and left side extend beyond the image border.

Nineteenth century French artists learned this technique from Japanese prints and the then-new art of photography.

Figure 5-39 shows two other ways to stage action in depth. In the first image, the long street curves away out of sight, suggesting that it continues for some distance. The second image exploits apparent scale, placing the large boulder in the left foreground and the subjects so far away that they appear tiny.

Figure 5-38.
The first staging appears flat, while the second emphasizes depth.

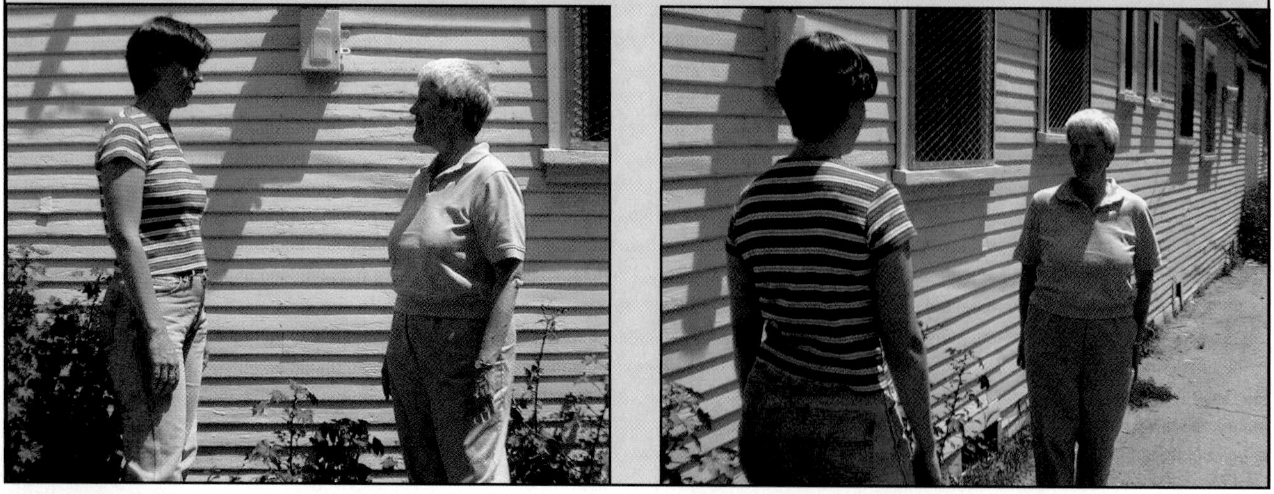

Figure 5-39.
Staging in depth. A—Street continuing out of the frame. B—Using apparent scale. (Sue Stinson)

Exploiting perspective

Staging in depth uses tricks of perspective, each of which creates the illusion of a third dimension. See **Figures 5-40** and **5-41**.

Also, as Figure 5-41 indicates, wide angle lens settings and low camera angles enhance depth, especially with prominent foreground figures.

Figure 5-40.
Creating the illusion of a third dimension. A—Size. B—Overlap. C—Convergence. D—Vertical position. E—Sharpness. F—Color intensity. (Sue Stinson)

Figure 5-41.
Enhancing depth using a wide angle lens and a low camera angle. (Adobe, Sue Stinson)

Framing the image

To add extra depth to an image, it often helps to frame it with foreground elements. The most common framing devices include tree branches and archways, **Figure 5-42**.

Architectural elements such as overpasses and roofs also work well, **Figure 5-43**.

Windows make excellent framing elements, **Figure 5-44**. Try zooming in until the window is excluded and the subject fills the frame, as shown by the "frame" in Figure 5-44B. Also, by editing the shot together with an exterior shot of a building, you can establish that the view is from that building—even if the two real-world locations are entirely different.

Working on the picture plane

Though you usually try to enhance the feeling of depth, there are times when you want the opposite effect: an image composed on the flat plane of the screen. (Commercials and travel videos frequently use this technique.) Extreme telephoto lenses are often used to create this flat appearance, as shown in **Figure 5-45**.

You can achieve the same effect without long lenses by carefully omitting all the perspective "clues" that you would normally include if you were aiming to enhance the feeling of depth, **Figure 5-46**.

Figure 5-42.
Framing with foreground elements. A—Tree branches. (Sue Stinson) B—Archway.

A B

Figure 5-43.
Framing with architectural elements. A—Overpass. B—Roof overhang.

Figure 5-44.
Windows as frames. A—Window framing nearby scene. B—Window with more distant scene. The black frame indicates how a subject can be extracted from the larger shot.

Figure 5-45.
Flattening perspective. A—The telephoto lens compresses the scene, placing all three racers on almost the same plane. (Corel) B—The same flattening effect is used in this scene of stone walls and a farm structure.

Figure 5-46.
Omitting clues to perspective. A—Lack of a horizon and the relative sizes of the gulls and boats mislead the viewer, providing no sense of perspective. B—This semi-silhouette provides little help in determining perspective.

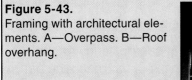

Composition and Movement

Professional-looking camera moves should be planned so that they both start and end with strong compositions.

Panning and tilting

Without planning, moving shots often end with weak compositions. **Figure 5-47** shows the courtyard of a palace with groups of tourists on the right and left sides.

If the shot is composed for the right-side group and then panned to frame the left group, **Figure 5-48**, the movement will end unsatisfactorily because the size and shape of the left group is different from the size and shape of the right group.

If you have time to rehearse the shot, it is better to set the ending composition first, then set the starting composition and make the pan, zooming and reframing during the move to create the ending composition. See **Figure 5-49**.

When tilting the camcorder up or down, the procedure is exactly the same, **Figure 5-50**.

The moving camera

The same concept applies when making a moving shot—whether hand-held or on a camera dolly: rehearse both the ending and starting

Figure 5-47.
A palace courtyard. (Sue Stinson)

Figure 5-48.
A—The start of the pan. B—The end of the pan. C—The beginning and ending frames.

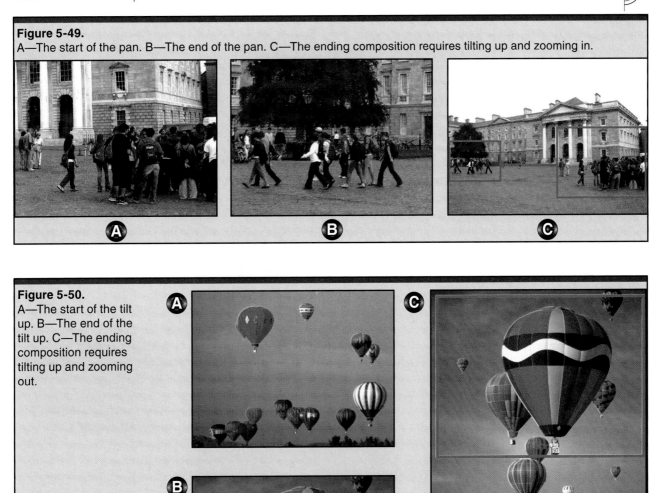

Figure 5-49.
A—The start of the pan. B—The end of the pan. C—The ending composition requires tilting up and zooming in.

Figure 5-50.
A—The start of the tilt up. B—The end of the tilt up. C—The ending composition requires tilting up and zooming out.

compositions, then make the move, adjusting camera angle and lens focal length setting as needed.

Trust Yourself

Although the craft of creating coherent, effective images appears complicated and technical, you can rely upon your own taste and instincts to guide you to good compositions. Creating a pictorial composition is like driving a car. When you are just learning, you have to remember every part of a complex process; but before you know it, driving has become so instinctive that you do it without conscious thought. The same thing happens as you make video programs. You gradually internalize the principles of composition until you can use them naturally and instinctively.

From Image to Program

This chapter on composition has focused on the design of individual images. A video program consists of many thousands of these single frames, organized into shots, scenes, and sequences. Chapter 6, *Video Language,* presents the principles behind this organization.

Perspective through Movement

Of course, the camera adds to the feeling of depth when it moves forward or backward, into or out of a composition. But it also enhances depth when moving sideways, parallel to the picture plane. That occurs because we notice even tiny changes in the relationships between objects and use those changes to calculate depth.

In Photo A, the two pilings overlap and only the tip of the distant island is visible at right edge of the frame. In Photo B, the camera has moved only about ten feet to the left, while framing the same composition. Now the pilings have separated and much more of the island has appeared. Our brains use these changes as clues to distance. That is why people who have lost the use of one eye can still perceive depth, even though they lack stereo vision.

A—Starting from its first position, B—the camera dollies to the left.

A

B

In Photo C, the bridge structure flashing by adds a strong sense of movement. In addition, the brain can notice and interpret even the tiny movement of the distant dock (Photo D), enhancing the feeling of depth.

C—The bridge structure adds a feeling of movement. D—Note the changed position of the dock marked by the red lines. The camera dollies to the right. (Sue Stinson)

C

D

Chapter Review

Answer the following questions on a separate piece of paper. Do not write in this book.

1. Composition _______ the elements within an image, to help viewers decode it.
2. When visual elements are balanced _______, the result is often static and dull.
3. Which of the following is *not* a technique used to create perspective?
   A. convergence
   B. rule of thirds
   C. sharpness
   D. vertical position
4. *True or False?* Wide screen images have a 16 to 9 ratio.
5. "Staging in depth" uses tricks of _______ to create the illusion of a third dimension.

Technical Terms

Asymmetrical balance: A composition in which dissimilar elements have equal "visual weight."

Brightness: The position of a pictorial element on a scale from black to white.

Composition: The purposeful arrangement of elements in an image.

Contrast: The difference between one pictorial element and others; also, the ratio of the brightest part of an image to the darkest.

Decoding: Identifying and understanding the elements in a composition.

Emphasis: The process of calling attention to a pictorial element.

Focus: Photographically, the part of the image (measured from near to far) that appears sharp and clear; also, generally, the object of a viewer's attention.

Leading lines: Lines on the picture plane that emphasize an element by pointing to it.

Leading the eye: Using compositional techniques to direct the viewer's attention.

Letterboxed image: A wide screen image displayed in the center of a regular TV screen, with black bands above and below it filling the frame.

Panning: Moving the camera horizontally.

Perspective: The simulation of depth in a two-dimensional image.

Picture plane: The actual two-dimensional image.

Rule of thirds: A method of composition that aligns important visual elements with the lines and intersections of a tic-tac-toe grid.

Staging in depth: Positioning subjects and camcorder to exploit perspective in the image.

Symmetrical balance: A composition in which visual elements are evenly placed and opposed.

Tilting: Rotating the camcorder vertically.

Widescreen video: Video using a screen proportioned 16 to 9, in contrast to the traditional TV screen's 12 to 9 proportion.

The elements of good composition must be carefully considered before shooting begins.

6 Video Language

Objectives

After studying this chapter, you will be able to:

- Explain the concept of camera angles.
- Name the principal types of camera angles.
- Vary shot types effectively.
- Create continuity of action.
- Select and use scene transitions.

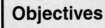

(Sue Stinson)

About Video Language

No matter how seamless videos may appear, they are really long assemblies of individual pieces, placed one after another in strings as long as one thousand units or more. These individual pieces are ca lled shots.

A ***shot*** is a single, uninterrupted visual recording, a length of tape during which the camera has operated continuously. Of course, shots are not just strung together in random order. On the contrary, they are organized according to a set of rules. Collectively, these rules may be called the language of video.

A "take" is one attempt to record a shot. Slating (labeling) a single recording as 37A3 would translate to "Scene 37, shot A, take 3."

Video Language Terms

Though the analogy between video and a language like English is not perfect, it does provide a useful comparison. For example:

A ***frame*** is like a unit of sound (a "phoneme"), an essential building block that is usually too brief to deliver meaning by itself.

An ***image*** (a more or less static segment of a shot that is perceived as a picture) is like a word.

A ***shot*** (a continuously recorded stream of evolving images) is like a sentence. It is the shortest assemblage that conveys a complete piece of information.

A ***scene*** is like a paragraph. Composed of several shots (visual sentences), it conveys meaning about a single topic.

Though the terms *shot* and *scene* have separate meanings in this book, they are sometimes used interchangeably.

A ***sequence*** is like a chapter. It assembles a number of scenes into a longer action that is also devoted to a single (though larger) part of the narrative. The analogy between video and verbal language could be carried still further. In dramatic programs for TV, a major assembly of video sequences is called an act (just like an act in a stage play) and a complete video program could be compared to a book or play.

Studying Video Language

Mastering the formal language of video can be quite intriguing. Individual shots are like the building blocks in a fascinating construction set that is very rewarding to experiment with. This chapter presents the basic rules for using that video construction set. Since we cannot discuss video components without identifying them, our construction guide begins with the naming of camera angles.

Camera Angle Names

A ***camera angle*** is a distinctive, identifiable way of framing subjects from a particular position at a particular image size. Most camera angles have specific names, **Figure 6-1**, because they are relatively standardized setups that are used again and again. This is not because video directors lack originality but because a century of film and video production has developed a certain number of angles that convey information effectively and look attractive to viewers.

Technically, a *shot* is a single piece of video recording, an *angle* is a particular view of the subject, and a *setup* is a distinct camera position. In actual video production, however, these words are sometimes used interchangeably.

Usually, angles are named for one of several different sets of characteristics: subject distance, horizontal camera position, vertical camera position, lens perspective, shot purpose, and shot population.

Angle names are not standardized, so the labels used in the following discussion are only typical. For more on angle name variations, see the sidebar, *What's in a name?*

Figure 6-1.
A "worm's-eye" angle of a subject.

Subject Distance

The most common angle names describe the apparent distance between the camera and a standing adult human, as you can see from **Figures 6-2** through **6-11**. (These names can be confusing because, strictly speaking, they *actually* refer to the subject size in relation to the frame.) These angles are:

Extreme long shot: The figure is tiny and indistinct in a very large area.

Long shot: The figure is small in the frame (half the frame height or less) and slightly indistinct.

Medium long shot: The standing human is distinct and somewhat closer, but with considerable head and foot room.

Full shot: The standing figure fills the screen from top to bottom, often with just a small amount of head and foot room.

Three-quarter shot: The shot shows the subject from about the knees to the top of the head.

Medium shot: The shot shows the subject from the belt line to the top of the head.

Medium closeup: The shot shows the subject from about the solar plexus to the top of the head.

Closeup: The shot shows the subject from the shoulders to the top of the head.

Big closeup: The shot shows the subject from below the chin to the forehead or hairline.

Extreme closeup: The shot shows the subject from the base of the nose to the eyebrows.

Although these names all use a standing human for reference, they are also employed

Figure 6-3.
Long shot.

Figure 6-4.
Medium long shot.

Figure 6-2.
Extreme long shot.

Figure 6-5.
Full shot.

Figure 6-6.
Three-quarter shot.

Figure 6-7.
Medium shot.

Figure 6-8.
Medium closeup.

Figure 6-9.
Closeup.

Figure 6-10.
Big closeup.

Figure 6-11.
Extreme closeup.

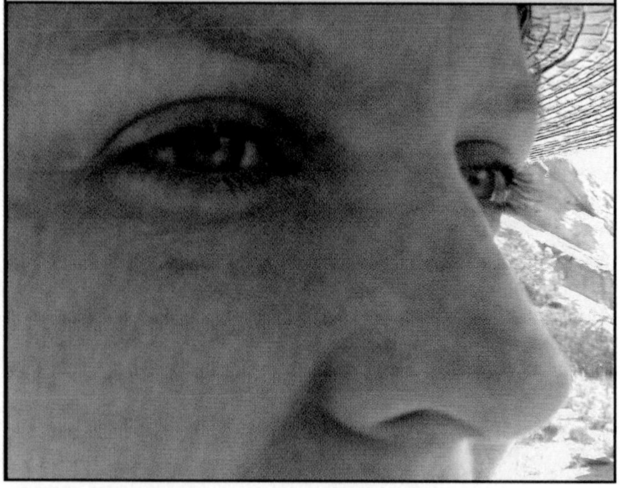

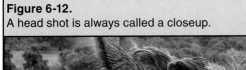

Figure 6-12.
A head shot is always called a closeup.

with both smaller and larger objects. For example, though a horse's head is obviously larger than a human's, a shot framing it would still be called a closeup, **Figure 6-12**.

Horizontal Position

After subject distance, the most frequently used angle names are based on horizontal camera positions. In the angles named according to subject distance, as shown in Figures 6-2 through 6-11, we always see the person from the front; but the camera can view its subject from other directions as well, and the more common horizontal positions have specific names. These are shown in **Figures 6-13** through **6-17**.

Figure 6-13.
Front angle.

Figure 6-14.
Three-quarter angle.

Figure 6-15.
Profile angle.

Figure 6-16.
Three-quarter rear angle.

Figure 6-17.
Rear angle.

Front angle: The camera faces the front of the subject.

Three-quarter angle: The camera is placed between 15 and 45 degrees around toward one side of the subject.

Profile angle: The camera is at a right angle to the original front shot.

Three-quarter rear angle: The camera is another 45 degrees around, so that the subject is now facing away.

Rear angle: The camera is directly opposite its front position and fully behind the subject.

Taken together, horizontal camera angle and subject distance provide the most typical shot descriptions; for example, "three-quarter closeup." However, an equally important component of every angle is the camera's height.

Vertical Angle

Shots may also be labeled by the vertical angle from which the camera views the action. The easiest way to describe standard camera heights is by pretending that the camera is at the end of a clock hand and pointed at the center of the dial, as you can see from **Figure 6-18**.

Using the clock for reference, here are some common camera height names**:**

Bird's-eye angle: An extremely high camera position (around 1 o'clock) that simulates the view from a plane or high building, **Figure 6-19**.

What's in a Name?

If you consult various books on videography and cinematography, you will discover that there is no standard set of angle definitions—let alone the shot names that describe them. The reasons for this are simple. First, movie production developed independently in the United States, Canada, the United Kingdom, Australia, and New Zealand, and each country evolved its own terminology. Second, video, which began as broadcast television, developed its own vocabulary. It borrowed some terms from the film medium and invented others of its own. To illustrate this confusing situation, the table at right lists just some of the alternative names for common shot lengths. (The terms used in this book are in bold type.)

Of these alternative names, the most common variants are extreme wide shot, wide shot, and medium wide shot instead of extreme long shot, long shot, and medium long shot.

Notice that the same terms are sometimes applied to different image sizes. For instance, some sources label a medium closeup as a closeup and a tight closeup as an extreme closeup. One British source also includes something called a "mid shot," reaching from the mid-hip level to the top of the head.

The cutoff levels of all these shots are based on two very different theories. One theory advises videographers to avoid cutting subjects at body joints because the results look unattractive. The opposite theory (endorsed by this book) says that joints and other natural bending places like the waist are appropriate levels for framing off the subject.

While you are reading this book, you need to understand its shot definitions and terminology. Then, as you actually make videos, you will pick up the terms used in your particular production environment.

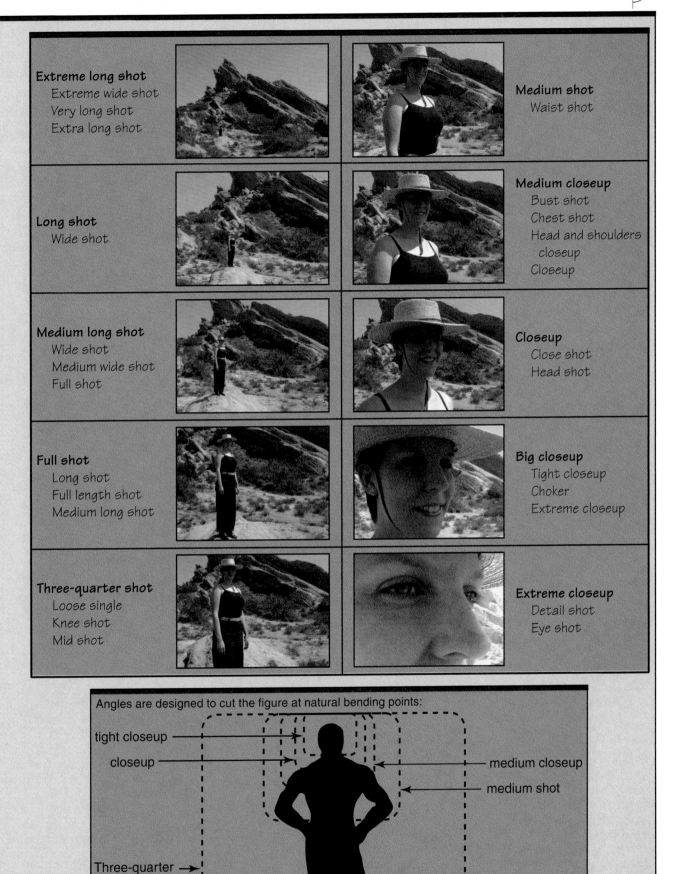

Extreme long shot
Extreme wide shot
Very long shot
Extra long shot

Medium shot
Waist shot

Long shot
Wide shot

Medium closeup
Bust shot
Chest shot
Head and shoulders closeup
Closeup

Medium long shot
Wide shot
Medium wide shot
Full shot

Closeup
Close shot
Head shot

Full shot
Long shot
Full length shot
Medium long shot

Big closeup
Tight closeup
Choker
Extreme closeup

Three-quarter shot
Loose single
Knee shot
Mid shot

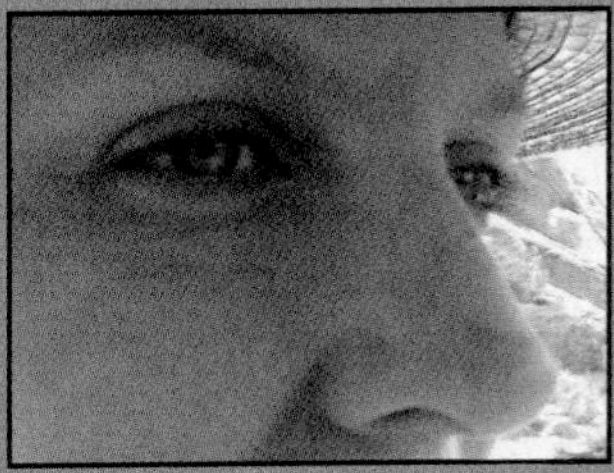

Extreme closeup
Detail shot
Eye shot

Angles are designed to cut the figure at natural bending points:

Figure 6-18.
Vertical angle shot names.

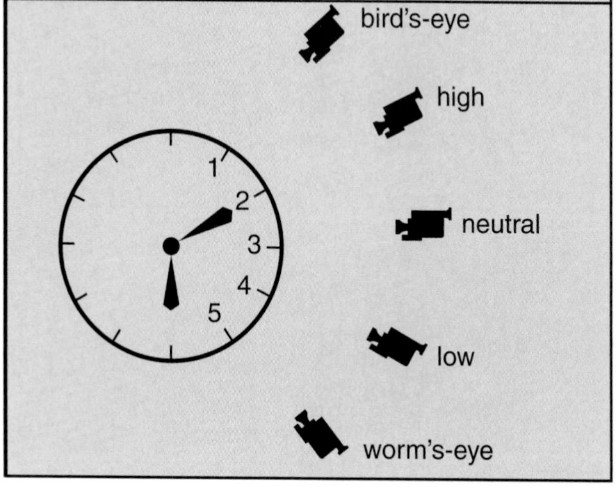

Figure 6-19.
Bird's-eye angle.

Figure 6-20.
High angle.

Figure 6-21.
Neutral angle.

Figure 6-22.
Low angle.

Figure 6-23.
Worm's-eye angle.

High angle: A shot in which the camera is evidently higher than the eye level of a human subject (between 1:30 and 2:30), **Figure 6-20**.

Neutral angle: A shot in which the camera is more or less at the subject's eye level (3:00), **Figure 6-21**.

Low angle: A shot in which the camera is evidently below eye level (between 3:30 and 4:30), **Figure 6-22**.

Worm's-eye angle: An extremely low camera position, looking dramatically upward (5:00), **Figure 6-23**.

The height of an angle names the position of the camera, not the subject. For instance, the position of a dog may be very close to the ground, but in taping the pet from a standing position, you are shooting downward from a high angle.

Lens Perspective

Another type of shot name is based on the appearance created by a particular camera lens. Directors will often call for a shot by the type of lens to be used when they want a stylized rendering of perspective. See **Figures 6-24** and **6-25**.

Wide-angle: A wide-angle lens exaggerates apparent depth and dramatizes movement toward and away from the camera. Directors often employ a wide-angle shot in taping a chase, a fight, or some other sequence full of dynamic action.

Normal: A normal-angle lens renders perspective approximately the way human vision perceives it, neither increasing nor reducing the apparent depth.

Telephoto: A telephoto lens compresses apparent depth and de-emphasizes movement toward and away from the camera. Telephoto lenses are used to dramatize congestion and to intensify composition on the two-dimensional screen.

Strictly speaking, the opposite of a "wide-angle" lens ought to be called a "narrow-angle" lens. But in real-world production work, the term "telephoto" is used instead.

Shot Purpose

In addition to camera position, camera height, and lens type, shots are often named for the purpose they serve in the program. Here are some common examples:

Master shot: The purpose of a master shot, **Figure 6-26**, is to record all or most of a scene in a full shot or even wider, capturing the action from beginning to end. After the master has been taped, matching closer shots of the same action are made to replace parts of the master in the final edited version. Today, shooting complete master shots is often considered old-fashioned.

Establishing shot: The purpose of an establishing shot is to orient viewers to the general scene and the performers in it. Though beginning a sequence with an establishing shot is also considered old-fashioned, it is often helpful to introduce one early in a sequence.

Since an establishing shot shows all the important parts and players in a scene, it is sometimes the beginning section of a master shot.

Figure 6-24.
Scene viewed through a wide-angle lens.

Figure 6-25.
Scene viewed through a telephoto lens.

Figure 6-26.
A master shot typically records a scene from beginning to end.

Figure 6-28.
An over-the-shoulder shot.

Reverse shot: The purpose of a reverse angle is to show the action from a point of view nearly opposite that of the main camera position, **Figure 6-27**. The position is *nearly,* but not fully, the opposite point of view because the camera usually stays on the same side of an imaginary action line to preserve screen direction.

Over-the-shoulder shot: The purpose of an over-the-shoulder shot, **Figure 6-28**, is to include part of one performer in the foreground while focusing on another performer. In addition to controlling emphasis, an over-the-shoulder shot enhances the feeling of depth.

Cutaway shot: The purpose of a cutaway shot is to show the audience something outside the principal action, or to reveal something from an on-screen person's point of view. For example, if we see several angles of a man mowing his lawn and then a shot of a dog lying in the uncut grass, the dog is a cutaway shot, **Figure 6-29**.

Insert shot: The purpose of an insert is to show a small detail of the action, often from the point of view of a person on the screen. The middle frame in **Figure 6-30** is an insert because it shows the steering wheel from the driver's point of view. Because an insert often serves the same functions as a cutaway, it is sometimes called a "cut-in."

POV shot: The purpose of a POV (point-of-view) shot is to show the audience what someone on the screen is seeing. The insert shown in **Figure 6-30** is from the driver's POV.

Figure 6-27.
Reverse shot. Left, the original angle. Right, the reverse angle.

Figure 6-29. Cutaway.
The third frame in this sequence is a cutaway shot.

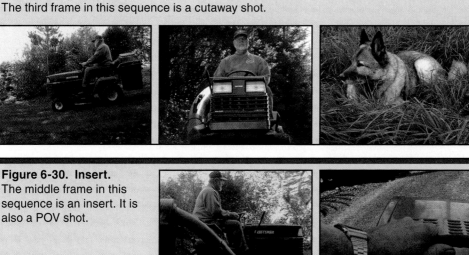

Figure 6-30. Insert.
The middle frame in this sequence is an insert. It is also a POV shot.

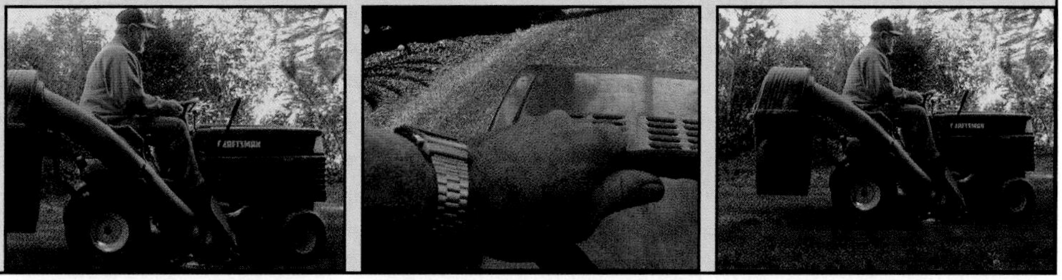

Figure 6-31. Glance-object pair.
A—The subject checks his watch. B—He then looks off-screen at ... C—the "object" of his glance.

Glance-object pair: Often, the cutaway, insert, and POV shots are used with other angles to make pairs of shots that some critics call "glance-object," **Figure 6-31**. This term means just what its name implies. In the first (glance) shot, a performer looks at something off screen. In the second (object) shot, we see the person, place, or object the performer was looking at.

Shot Population

Finally, shots may be labeled according to how many performers are in them:

Single: A shot showing one person.

Two-shot: A shot showing two people.

Three-shot: A shot showing three people. See **Figure 6-32**.

By custom, a one-person angle is not called a "one-shot," but simply a "single."

Note that these terms number only the principal performers. For example, a shot of two people having dinner with other restaurant diners in the background could still be called a two-shot.

Shot Name Confusions

Shot names can be confusing at first, for several reasons. Many shot names are synonymous with the camera angles from which

Figure 6-32.
A single, a two-shot, and a three-shot.

they are made ("low angle")—but others are not ("reverse"). Shots are often given multiple names. For instance, a director might call for a "neutral over-the-shoulder two-shot," or a "high-angle POV closeup." Shots may be identified by different names. For example, if you have just made an "over-the-shoulder two-shot," asking for a "reverse" indicates that you want *another* over-the-shoulder two-shot, from the opposite point of view. Finally, shots often have different names in different countries (U.S., U.K.), in different media (video, film), in different production centers (New York, Hollywood), or and even in different studios in the same area.

In practice, however, you will soon pick up the shot names favored in the environment where you work.

To add to the confusion of names, some people use "scene" interchangeably with "shot." In this book, a scene is a short, single piece of action, usually conveyed through a number of consecutive shots.

Creating Continuity

If the "words" in the language of video are shots, then video "sentences" are scenes made of several successive shots, **Figure 6-33**. Why break the action of a scene into separate shots? Why not simply record everything from start to finish in one continuous piece of video? The most important reason is to show each individual part of the action from the best possible position. This would be impossible to do if the camera were limited to one continuous take. Separate shots also enhance the interest of the material by providing visual variety.

In most video programs, you wish to conceal the fact that you are constantly changing camera angles. Every time viewers notice a new angle they are briefly distracted from the program content; so the goal is to make the action appear to happen in a single, continuous flow. The process of doing this is called ***continuity***. In creating video continuity, you stage each shot so that it can be combined unobtrusively with the shots that precede and follow it. The key requirement for doing this is *matching action.* In matching action, you make the incoming shot appear to begin at precisely the point where the outgoing shot ends.

Of course, the action is never truly continuous (unless you are editing shots from two cameras recording simultaneously). In fact, apparently uninterrupted action is an illusion created by a collaboration between the director and the editor.

The continuity described here is the classic Hollywood style that has become the worldwide standard for commercial production.

Varying Shots

It might seem that the best way to conceal a change in shots would be to make the new angle as much as possible like the old one, as in **Figure 6-34**. But as you can see, the opposite is true. Instead of two shots that cut together invisibly, the result looks like a single shot with a piece chopped out of its middle. This is called a ***jump cut***, because the subject seems to jump abruptly on the screen.

Editors use the term "cut together" to describe how successfully one shot follows another.

To make a smooth edit, the technique is to match the action closely, while decisively altering the camera angle. See **Figure 6-35**.

Generally, this means changing at least two of the angle's three major characteristics: camera position, camera height, and subject size. Unlike Figure 6-34, both **Figure 6-36** and **Figure 6-37** *do* change one trait out of three. However, altering only a single trait still results in an undesirable jump cut that calls attention to the edit. To achieve a smooth edit, a new camera setup should change *two* angle traits, most often the image size plus either the camera position or height.

Figure 6-33.
On screen, what appears to be continuous action is actually constructed of multiple shots.

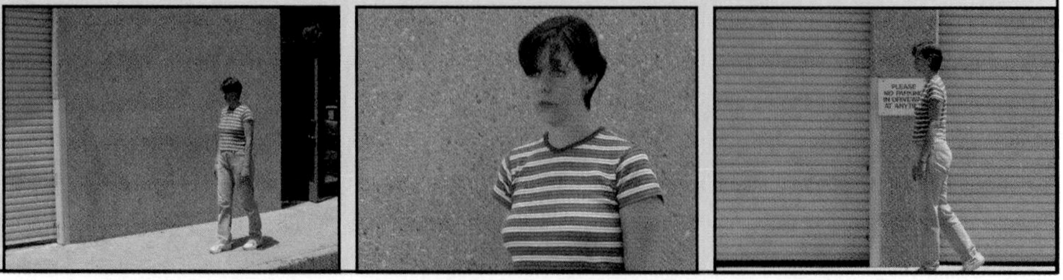

Figure 6-34.
A too-similar angle makes an obvious cut.

Figure 6-35.
A change in angle conceals the cut.

Figure 6-36.
Changing only the vertical angle creates a jump cut.

Figure 6-37.
Changing only the image size also creates a jump cut.

Figure 6-38.
Changing to a new image size and horizontal angle.

Figure 6-39.
Changing to a new image size and vertical angle.

*A **setup** is a completely new camera location, including the lighting and the microphone position.*

Examples of effective old/new angle combinations are shown in **Figure 6-38** and **Figure 6-39**.

Matching Action

While the angle should change decisively from one shot to the next, the *action* should be closely matched, to make the second shot seem to begin exactly where the first shot ends. There are two basic ways to match action. The first is to have the performer repeat part of the action from the first shot in the second shot. In **Figure 6-40**, for example, the subject pours milk into a mixing bowl, repeating this action in each of two shots. In the establishing shot, she raises the milk carton over the bowl. In the second shot, she pours the milk. In the third shot, the subject tilts the carton upright and sets it down. If the edits are made correctly, the three different shots will appear to display a single continuous action.

A second (and easier) way to synchronize action is by concealing the match point.

*A **match point** in two shots that record the same action is any point at which the editor can cut from one shot to the other, continuing the action without an apparent break.*

In **Figure 6-41**, the performer exits the frame, leaving the shot empty. The next shot starts with an empty frame, and then the performer enters. Though the shots may have been made in completely different locations, the action appears continuous. The performer does not have to both exit the first shot and enter the second one. Typically,

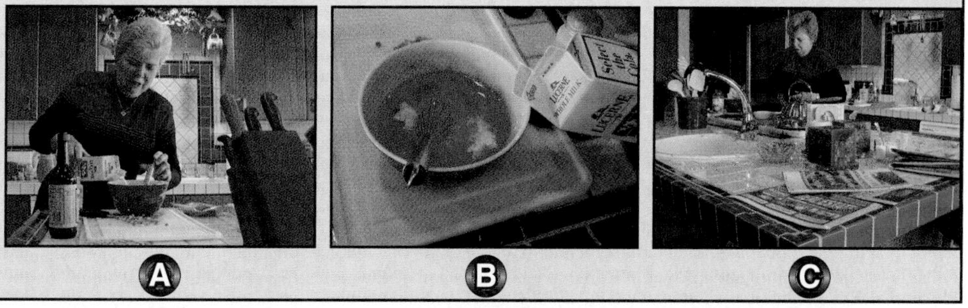

Figure 6-40.
A—The subject begins pouring...
B—finishes pouring...
C—and sets the carton down.

Figure 6-41.
In the first shot, the subject leaves an empty frame, then enters an empty frame in the second shot.

an actor will leave the first shot and then be shown again as the second shot begins — or vice versa.

Making Transitions

Up to this point, we have discussed the language of video at the level of "sentences" (shots) and "paragraphs" (scenes). Now we need to add a few rules for making transitions between "chapters." In video language, a chapter is a sequence: (a program component assembled from several related scenes.) The conventions for transitions from one sequence to another were established long ago and were strictly observed in classic Hollywood movies. Today, however, these rules are broken as often as they are observed.

Classic Hollywood Transitions

With few exceptions, traditional movies have marked transitions with two effects, fades and dissolves.

Silent films also used an effect called an iris out, in which a black screen gradually revealed an image in an ever-expanding circle. An iris in reversed the effect.

A ***fade-in***, **Figure 6-42**, begins with a black screen, which lightens gradually until the image reaches full brightness. A ***fade-out*** is the exact

Figure 6-42.
A fade-in and a fade-out are exact opposites. (Sue Stinson)

Video Program Components

We have identified shots, scenes, sequences, and acts as components of a video program; and we have established that a shot is a single, uninterrupted recording. But what do the other terms mean? To obtain a feel for these different program components, imagine a dramatic hour-long TV show about a daring museum robbery.

Program

The program includes the entire story: planning the robbery, committing the deed, fleeing with the loot, and finally being captured and jailed.

Act

Typically around 10-15 minutes in length, an act is devoted to each of the four main parts of the story outlined above. In Act Two, for example, we see the entire museum robbery, from approach to escape. An act contains several — even many — separate sequences.

Sequence

A sequence is a major component of an act and has its own beginning, middle, and end. One sequence in the museum robbery act might be breaking in through the roof. First the robbers cut a hole in the roof, then they anchor a climbing rope to a chimney and drop the rope through the hole. Two men climb down the rope while a third robber remains on the roof as lookout. Notice that this sequence is composed of several different actions, each of which is a scene.

Sequences are often linked by transitions like this wipe.

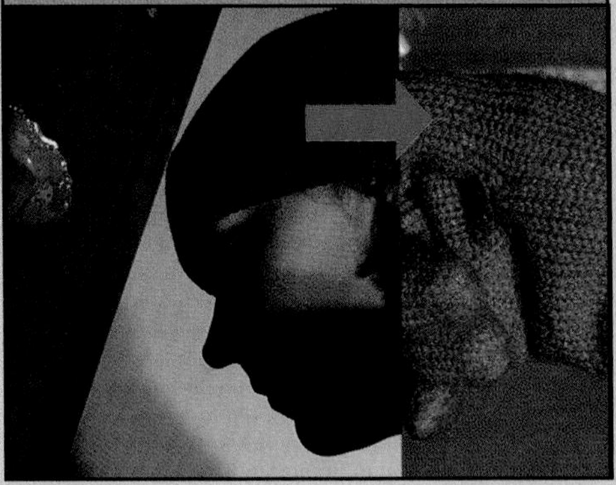

Scene

In the sequence of breaking in through the roof, one scene shows the anchoring of the climbing rope. This scene is built up from several individual shots.

Shot

A shot is a single uninterrupted recording. The rope-anchoring scene might require three separate shots. In a medium shot, a robber loops the rope around the chimney. In a closeup insert, he ties a knot in the rope. In a full shot, he tosses the other end of the rope out of frame. The camera pans to follow the rope, revealing a second robber dropping its end through the roof hole. (Remember: a "take" is not a program component, but only a single attempt to record a shot.)

Image

The last shot in the rope-tying scene contains two distinct pictures: first, the man throwing the rope and then, the second man lowering it down the hole. Since the shot includes two quite different pieces of visual information, we may say that it contains two images. A long moving shot may consist of a great many identifiably individual images.

Frame

For completeness, we should include the still picture called the frame. Each second of video consists of 30 frames (in the North American NTSC format). However, a frame is too brief to be used as a distinct video component, except in very special cases.

Figure 6-43.
A dissolve gradually replaces one image with another.

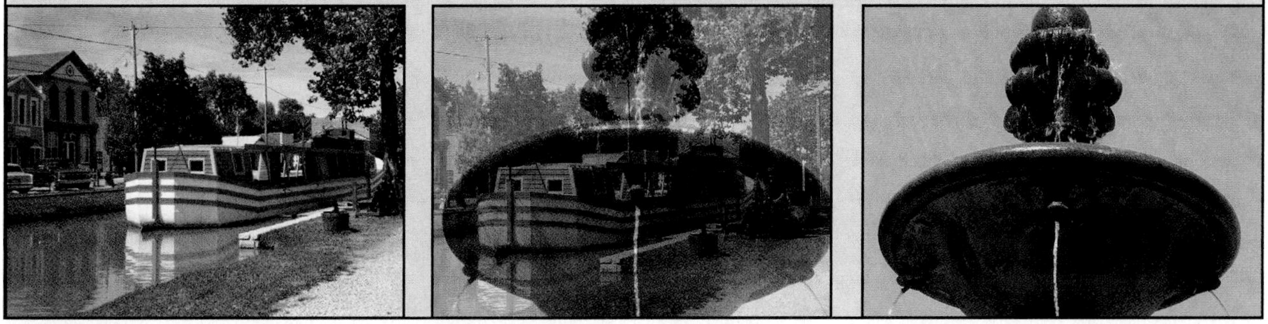

Figure 6-44.
A wipe is a traditional form of transition. (Sue Stinson)

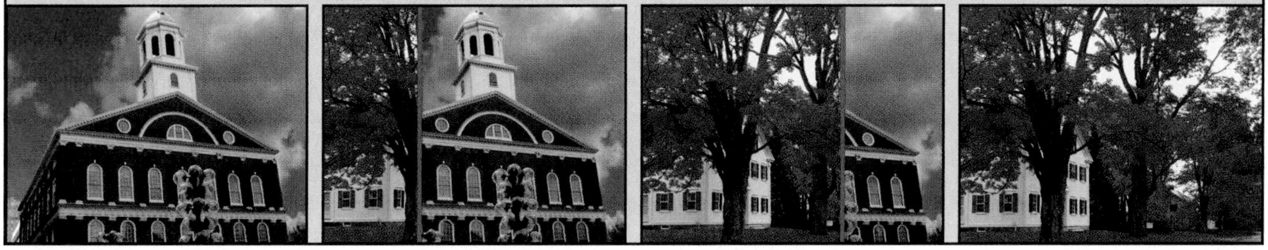

opposite. Fades mark the beginnings and endings of major program segments. For example, most commercial TV programs fade out before each commercial break and then resume by fading in again.

A ***dissolve*** (often called a "mix" in video) is an effect in which one image is gradually replaced by another, **Figure 6-43**. A dissolve, in fact, is a fade-in superimposed over a fade-out. A dissolve is used instead of a fade to indicate a shorter break in the action or a less decisive transition in the story.

A ***wipe*** is an effect in which a line moves steadily across the screen, progressively replacing the old image ahead of it with the new one behind it, **Figure 6-44**. Though a wipe may move along any axis, it is typically a vertical line that progresses from left to right. After falling out of fashion for some years, wipes are once again quite popular.

Modern Transitions

Today's audiences are so well-trained to follow story continuity that fades and dissolves are used less often, though fades almost always begin and end a program. Instead, editors use straight cuts or, sometimes, digital video effects.

Today, fades and dissolves are also created digitally.

A ***digital video effect (DVE)*** is any computer-generated transition from one shot to another, **Figures 6-45** and **6-46**. Modern software offers many different types, including:

- ***Flips***, in which the screen appears to revolve, as if to show its other side.
- ***Fly-ins***, in which the incoming picture swoops to the center of the screen, growing from a tiny dot until it fills the frame completely.
- ***Rotations***, in which geometric forms with the outgoing shot on one face revolve to reveal the incoming shot on another face.

In addition, there are hundreds of other DVEs available — far more than an editor might ever use.

In this respect, DVEs are like the thousands of type faces available for word processing and desktop publishing. Typically, people settle on a few favorites, reserving the others for occasional use in special situations.

Digital video effects defy the main rule of editing, that transitions should be unobtrusive. On the contrary, DVEs purposely emphasize transitions by dramatizing them. Used judiciously,

Figure 6-45.
A digital rotation. (Sue Stinson)

Figure 6-46.
A digital page return. (Sue Stinson)

these arresting visuals add excitement and punch to a program. If they are overused, however, they soon grow tiresome and annoying.

DVEs are lavishly employed in TV commercials and music videos, however. Commercials are very brief, sometimes as short as 5 seconds. They need to get viewers' attention quickly, and DVEs are effective attention-grabbers. Because commercials are so short, their rapid-fire effects don't last long enough to wear viewers out.

As for music videos, these programs often use the music as an excuse to explore the possibilities of the video medium. This means that they don't always play by the rules of video grammar. At their best, music videos and other experimental programs may be compared to poems. Just as poets take "poetic license" with the rules of verbal language, so video artists sometimes use camera angles, editing continuity, and transition effects without regard to the normal rules of the video medium.

Playing by the Rules

You can disregard the rules of video language when you make a program like a music video, or, for that matter, when you make any other type of video. Nevertheless, you need to learn and practice the language of video until you are thoroughly at home with it. Remember that even the most experimental art is grounded in a mastery of classical technique.

Though the artist Picasso usually created "unrealistic" images, he was a master draftsman who could execute highly realistic works when he chose to do so.

Video Language in Action

This simple sequence demonstrates the wide variety of shot names commonly used in making videos—to include every name would require three times as many images.

Establishing shot

Two-shot

Over-the-shoulder shot

Reverse

Insert

Profile shot

Medium closeup

Single

High-angle

Glance

Object (cutaway)

Big closeup

Chapter Review

Answer the following questions on a separate piece of paper. Do not write in this book.

1. *True or False?* "Take" is the label used to identify the best version of a shot.
2. A ______ shows the subject from the belt line to the top of the head.
3. "POV" is an abbreviation for
   A. point of view
   B. perspective optical vector
   C. perfect outdoor venue
4. Fade-in, dissolve, and wipe are examples of traditional ______.
5. Digital video effects (DVEs) are extensively used in music videos and ______.

Technical Terms

Act: A major section (usually between 10 and 45 minutes) of a longer program.

Camera angle: The position from which a shot is taken, described by horizontal position, vertical angle, and subject distance.

Continuity: The organization of video material into a coherent presentation.

Cut together: A term with two meanings: 1) to follow one shot with another; 2) to select edit points on outgoing and incoming shots so that an edit is not apparent.

Digital video effect (DVE): Informally, any digitally-created transitional device other than a fade or dissolve. (Technically, fades and dissolves are digital, too).

Dissolve: A fade-in that coincides with a fade-out, so that the incoming shot gradually replaces the outgoing shot. Typically used as a transition between sequences that are fairly closely related.

Fade-in: A transition in which the image begins as pure black and gradually lightens to full brightness. Used to signal the start of a major section such as an act or an entire program.

Fade-out: A transition in which the image begins at full brightness and gradually darkens to pure black. Used to signal the end of a major section such as an act or an entire program.

Frame: A single still picture, 30 of which make a second of NTSC video. (Also, the border around the image.)

Image: A single unit of visual information. An image may last for many frames, until the subject, the camera, or both, creates a new image by moving. Most shots contain several identifiable images.

Jump cut: An edit in which the incoming shot is too similar visually to the outgoing shot.

Match point: The places, in two shots, where they can be cut together to make the action appear continuous.

Point of view (POV): A vantage point from which the camera records a shot. Unlike a camera *angle*, a point of view is not described by subject distance ("closeup," etc.) and unlike a *setup*, a point of view is not concerned with production equipment.

Program: Any complete video presentation, from a five-second commercial to a movie two or more hours long.

Scene: A short segment of program content, usually made up of several related shots.

Sequence: A longer segment of program content, usually consisting of several related scenes.

Setup: An arrangement of production equipment (typically camera, microphone, and lighting) placed to record shots from a certain point of view.

Shot: A single continuous recording. In editing, shot A may be interrupted by shot B before resuming. When this happens, the second part of the original shot A may be considered a new shot.

Telephoto: A lens or a setting on a zoom lens that magnifies subjects and minimizes apparent depth by filling the frame with a narrow angle of view.

Wide-angle: A lens or a setting on a zoom lens that minimizes subjects and magnifies apparent depth by filling the frame with a wide angle of view.

Wipe: A transition between sequences in which a line moves across the screen, progressively covering the outgoing shot before it with the incoming shot behind it.

7 Video Sound

Objectives

After studying this chapter, you will be able to:

- List the most important functions of audio (video sound).
- Explain the major contributions that sound makes to video programs.
- Listen analytically to video sound tracks.
- Describe the component parts of a typical composite sound track.

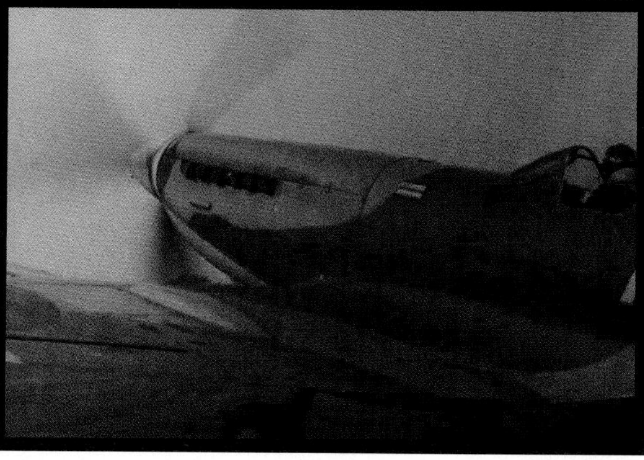

(Corel)

About Video Sound

Sound is vitally important in the medium of video. It communicates so powerfully that you can create successful programs of audio without video (which, of course, is called *radio*), but not video without audio. Even so-called silent films were not truly soundless. Because stories unfolding in absolute silence conveyed an eerie feeling, even the smallest silent movie theaters showed their programs with musical accompaniment.

And yet, sound is often undervalued in video production. In amateur programs especially, the finished audio track may be nothing more than whatever the camcorder's microphone happened to pick up.

Sound is what you record and hear, while *audio* is the audible half of a program (as opposed to "video"). In practice, the terms *sound* and *audio* are often used interchangeably.

Quality audio depends on both a mastery of the technical processes and an understanding of how sound contributes to the effect of a video program. The technical processes are covered in later chapters on sound recording and editing. In this chapter we focus on how to use sound as a tool to communicate information and stir emotion. Along the way, we will explain the principal components of a typical sound track, consider the role of music in video, and look briefly at the history of audio recording. We will begin with sound's most important role: *delivering information.*

Delivering Information

The biggest task that sound performs is delivering information, most often through dialogue spoken by on-screen performers or the words of an unseen narrator. The audio track also provides information through sound effects. As we will see, these humble noises laid down by sound editors form a remarkably eloquent (and economical) method of communicating.

To "lay down a track" means to record one or more audio components. To "lay in" an individual piece of audio means to add it to an audio track in postproduction.

Dialogue

Video is essentially a visual medium, but part of good scriptwriting lies in knowing when it is more efficient to tell something than to show it. Since reading lengthy on-screen titles is tiresome, "telling" typically involves some form of audible speech, usually ***dialogue***, which is the on-screen conversation between characters, **Figure 7-1**.

Figure 7-1.
Dialogue is on-screen speech.

Though "dialogue" properly refers to conversations between two or more performers, it is commonly used to label all on-screen speech.

Dialogue often delivers information explicitly. "I'm going to the zoo" tells us, in a straightforward way, where the speaker is headed. Dialogue may also contain implicit information. "I'm going back to the zoo," also tells us that the speaker has come from the zoo, while, "I'm going to the zoo again" indicates that the speaker has been there at some time in the past. If a second character says, "You're not going to the zoo again?" we learn three things: that the first character is going to the zoo, that he or she has been there often, and that the second character disapproves of these repeated visits. This dialogue has delivered three pieces of information in seven words and two seconds, working far more economically than visuals could in this case.

In a well-directed story video, dialogue also provides information about emotions and character. If "You're not going to the zoo again?" is said with fond amusement, it conveys certain things about the speaker's feelings. But if the line is delivered with weary resignation instead, or perhaps in a resentful whine, the speaker's feelings appear quite different.

Dialogue is also important in many nonfiction videos, where it often takes the form of interviews. Some interviews are presented as conversations between two people on-screen. (This approach is common in news and discussion programs.) In other cases, the interview is conducted and edited so that the interviewer is invisible and the questions are unheard, **Figure 7-2**. The interviewee seems to be speaking only to the viewer.

Figure 7-2.
In the contemporary interview style, only the subject appears on screen. (Corel)

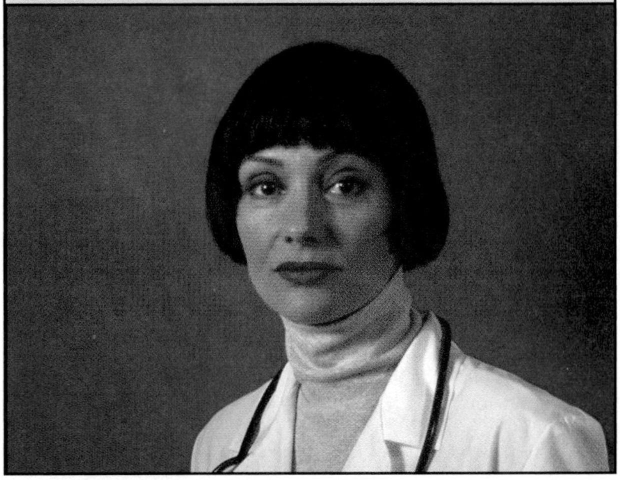

Narration

Often, a person being interviewed begins on screen, but after a few moments the visual component of the interview is replaced by footage that illustrates what the person is talking about, while his or her voice continues on the sound track. This is called ***voiceover*** (usually abbreviated "V.O." in scripts).

The terms "offscreen," "off camera," and "V.O." are essentially interchangeable.

Voice-over interviews are widely used in documentary programs, where they may alternate with fully scripted narration. Spoken by professional voice-over actors, narration is by far the most efficient way of delivering verbal information. A good script can pack a great deal into a short space, as you can see by the brief excerpt shown in **Figure 7-3**.

Figure 7-3.
An excerpt from a script.

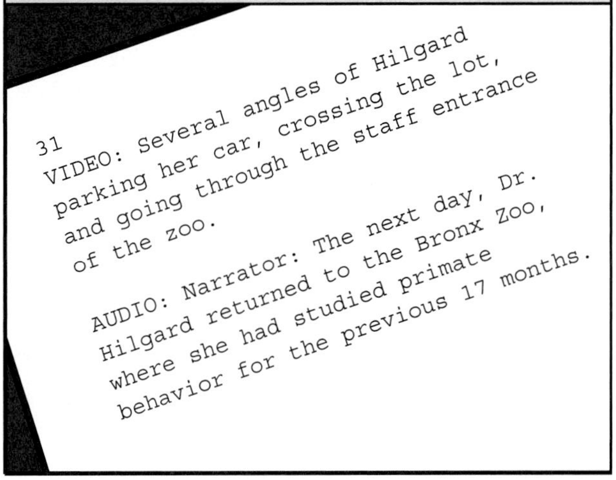
31
VIDEO: Several angles of Hilgard
parking her car, crossing the lot,
and going through the staff entrance
of the zoo.
AUDIO: Narrator: The next day, Dr.
Hilgard returned to the Bronx Zoo,
where she had studied primate
behavior for the previous 17 months.

If you take a moment to study the narrator's sentence, you will discover that just 22 words deliver ten pieces of information:

1. The subject is female.
2. Her name is Hilgard.
3. She has a doctorate.
4. She went to a zoo.
5. It was the Bronx Zoo.
6. She had been there previously.
7. She had studied there.
8. Her study was in primate behavior.
9. Her study lasted 17 months.
10. The study period was quite recent.

The visuals, on the other hand, are essentially meaningless filler included to occupy the screen while the narration is spoken.

Filler visuals, often called "wallpaper," are covered more thoroughly in the chapter on Program Development.

Another reason for narration's power lies in the skill of the professional narrator. Using pace, inflection, and emphasis, a narrator organizes and highlights text in the process of speaking it. This makes the narration easier to understand and more interesting to listen to. If you have ever winced at the sound of an amateur reading a text in a droning monotone, you will appreciate the complex techniques involved in speaking narration with professional skill.

Sound Effects

Sound effects are the noises that come from the world on the screen. Some are

Figure 7-4.
Waveform displayed by a digital sound editing program. (Sound Forge)

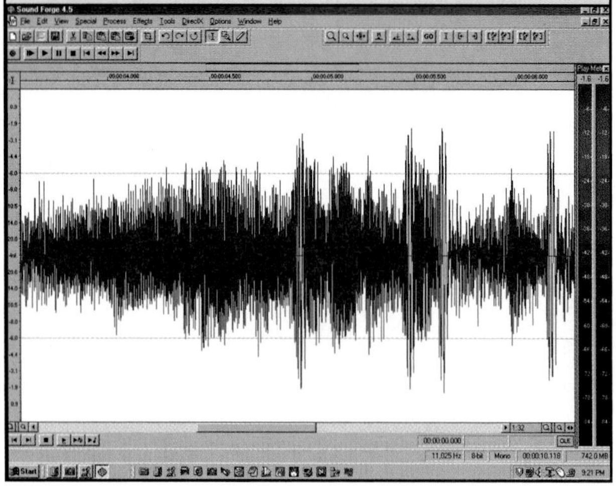

recorded with the video and consist of the actual sounds produced by the on-screen action. Many others are laid in by the sound editor in postproduction, **Figure 7-4**.

Typically, the sound effects synchronized with on-screen action are designed to heighten the sense of reality. To see how this works, try muting your TV sound during a sequence of vigorous action such as a car chase or a fight. Notice how the on-screen activities suddenly grow remote and uninvolving.

Although sound effects are typically used to heighten the realism of visual action, they can also deliver information independently of the video. For example, imagine a car chase during which our hero's brakes suddenly fail. The director communicates this information with an insert, a close shot of a foot repeatedly punching a brake pedal. The problem with this is that the foot movement looks no different from that of heavy braking, so the shot does not convey the essential information that the car's brakes have failed.

To supply this information, the sound editor lays in repeated sound effects of the pedal clanking on the floorboard of the car. It is that sound, not the picture, that tells the audience the brakes are useless**, Figure 7-5**.

Conveying Implications

Although sound is often used to supplement the video, its most important informational role is to tell you about things that are not on screen—things that exist only by implication. Remember the first law of video space: what is outside the frame does not exist, *unless that existence is implied.* If you provide the audience with evidence that something they cannot see is nonetheless there, they will believe it.

Sound Implies Existence

By far the easiest way to imply the presence of things off screen is through sound. In **Figure 7-6**, for example, a mother says goodbye to her daughter and watches her ride away.

But what, exactly, is the daughter riding in or on? We can't tell from the visuals because her transportation remains outside the frame.

Now suppose we add sound effects to the sequence, as in **Figure 7-7**.

Figure 7-5.
By itself, the video communicates only "braking." With an added sound effect, it communicates "brakes failed!"

Figure 7-6.
The mother and daughter embrace... the daughter leaves the frame... ...and her mother watches her ride away.

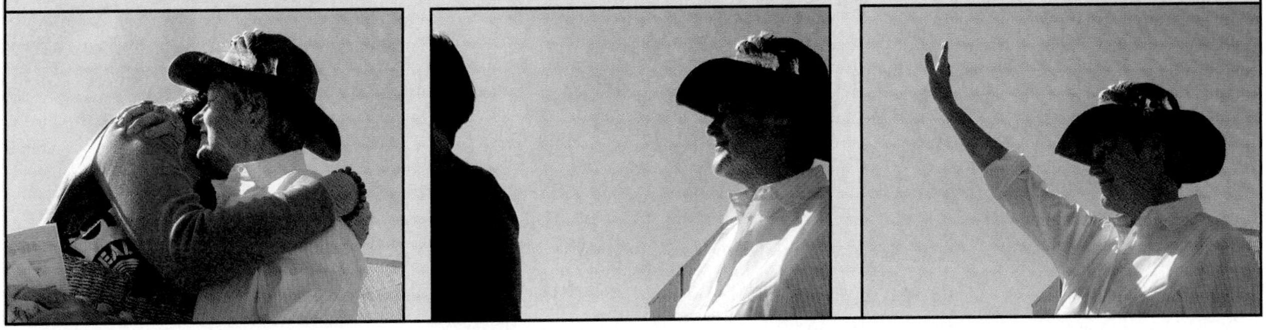

Figure 7-7.
SFX: A bus horn honks impatiently.
DAUGHTER: G'bye, Momma
MOTHER: Honey, you take care, y'hear?... SFX: the hiss of bus doors opening... SFX: ...and then closing; followed by the diesel rumble of a bus engine, fading in the distance.

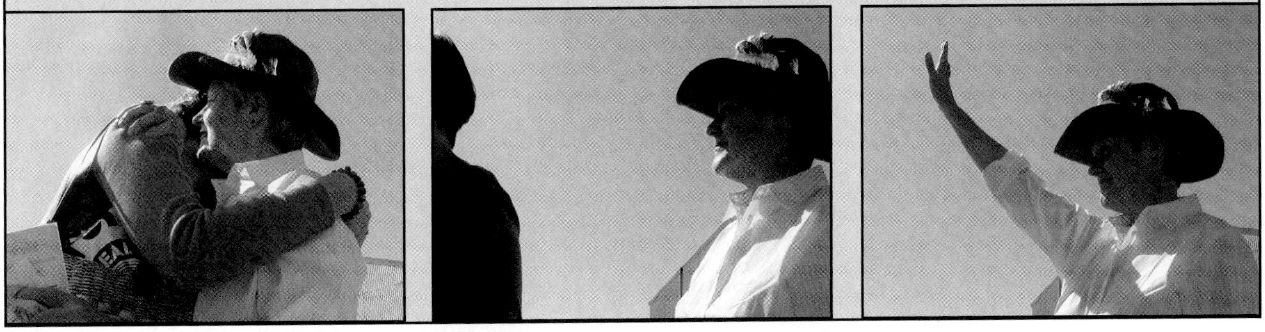

In video scripts, "SFX" is the standard abbreviation for "sound effect."

With only a honk, a hiss, and an engine roar, we have constructed a complete bus, **Figure 7-8**, and added it to the scene; a bus whose existence is only implied, and implied exclusively by sounds.

Figure 7-8.
A phantom bus has been created entirely by sound effects. (Sue Stinson)

We could change the young woman's transportation just by substituting different effects:

- A brass bell chimes over an idling speedboat motor. The motor revs up and then gradually fades, leaving the sound of gulls and water lapping around a pier.
- A conductor calls, *"Boooooard!"* and we hear the first slow chuffs of a big steam locomotive. The puffing grows louder and faster, then diminishes in the distance, with the wail of a whistle.
- The motor of a single-engine plane idles smoothly. After a moment, the door *chunks* shut, the engine revs, and the sound moves farther away as the plane starts to taxi down the runway.

By using sound effects alone, we could supply the daughter with four completely different modes of transportation.

Sound Implies Locale

With the bus and the airplane we can only assume that the scene is set on a road or an airstrip. In the case of the train, however, the

The Perception of Sound

Sound does not really exist in nature. What we think of as "a sound" is actually a disturbance of the air, moving in a pattern of waves. These waves apply rapidly shifting air pressures to our ears, which create nerve impulses, which our brains organize and interpret as "sound."

Sound in the actual world

Human hearing is marvelously acute and subtle, because our brains are skilled at organizing, processing, and interpreting the audio babble that continuously assaults our ears. We can regulate overall volume, adjusting our tolerance for loudness to suit the environment. For instance, imagine driving at 65 miles per hour with the car radio playing. When you slow to 30 mph, the radio suddenly sounds uncomfortably loud, even though the volume has not increased. The noisy radio did not seem too loud at 65 mph, because your brain can adjust for the higher overall sound levels of high-speed driving.

We can also listen selectively, processing some parts of the overall sound but ignoring others. For example, most people quickly filter out the drone of an air conditioner until they don't notice it at all. On the other hand, parents can often hear the faint cries of a baby two rooms away, even through the sound of a TV program. With the air conditioning, they suppress and ignore the sound; but with the baby, they isolate and recognize the sound, despite the competition from other noise. Even though people hear everything, they listen selectively.

The listener hears the TV sound and the baby's cry, but filters out the air conditioner noise.

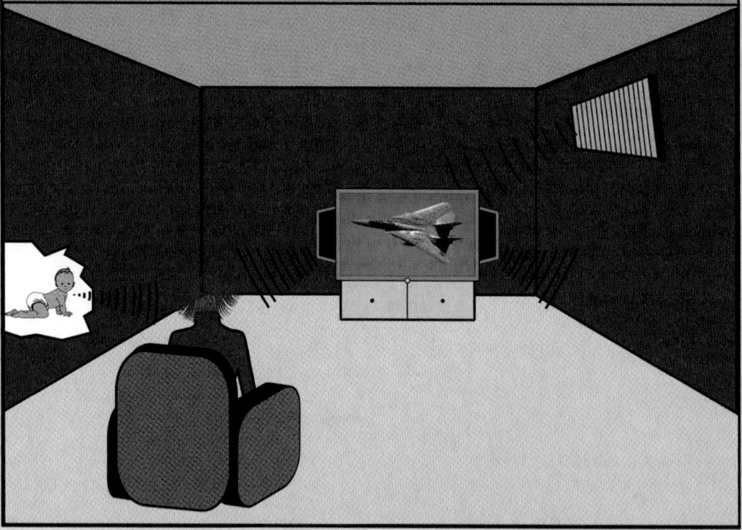

Sound from video programs

Unfortunately, we do not seem able to listen to an audio track as selectively as we listen to real-world sound. This creates both a problem and an opportunity for video makers. The problem is that the sound editor must do the job that our brains do with real-world sound: regulate volume and filter out unwanted sounds, while isolating and emphasizing wanted ones. The opportunity lies in the brain's ability to recognize and process many sounds at once. This allows editors to build audio programs out of many layers of audio. In complex, big-budget productions, individual sound track elements can number in the dozens.

Until recently, complex audio was uncommon in video programs because the hardware required to play back, synchronize, and mix audio tracks on videotape was very expensive. The alternative of **mixing** a few tracks at a time and then remixing the subassemblies resulted in loss of audio quality. (Larger studios could build up multiple tracks on one- or two-inch audiotape.) With the establishment of digital recording, however, these difficulties have disappeared, and videotape-based audio can now match the sophistication of audio for film.

Figure 7-9.
The sounds of a steel mill are as overwhelming as its images. (Corel)

conductor's call supplies the train station for us. In the same way, the gulls and lapping water inform us that we are on a boat dock.

You can easily think of the locations that are established by sounds like the crackle and roar of a forest fire, the hiss of rain and rattle of thunder, or the deafening roar of a steel mill, **Figure 7-9**. We could multiply examples but the point is clear: sound is psychologically powerful in evoking a sense of place.

Sound Creates Relationships

Sound is very useful in bringing together and unifying program components that were originally shot separately. For example, look carefully at the sound elements supporting the visuals in **Figure 7-10**.

The exterior shots of the truck were probably made on location with stunt doubles doing the driving. The interiors with Bob and Bill were shot separately, with their pickup towed safely behind a camera truck. To meld these different times and places into a single sequence, the sound editors used three common techniques:

- They laid some of the actors' dialogue in as off-screen voiceovers accompanying the exteriors.
- They continued the car sound effects over shots of the auto's interior.
- They varied the volume, perspective, and quality of the dialogue and sound effects from shot to shot to match the different viewpoints of the audience.

Figure 7-10.
BOB: Yer goin' too fast, Bill!
SFX: (muted) engine roar.

BILL: (O.S., faint) Not fast enough!
SFX: (loud) engine roar, gravel spraying.

BILL: (muffled by window) Hey, there's a shortcut.
SFX: (loud) engine continues.

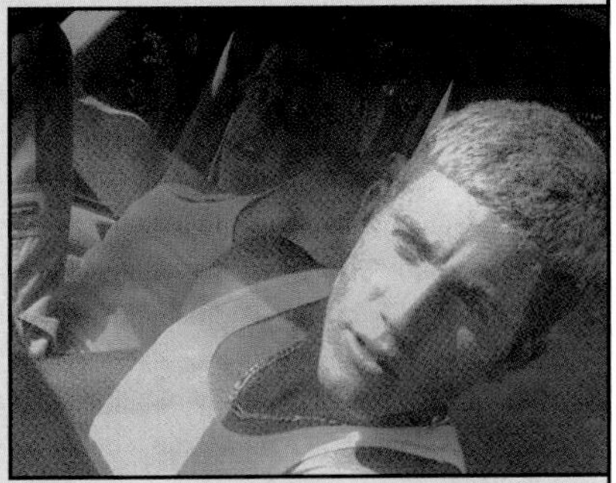

BOB: (O.S., faint).That's a field!
SFX: (loud) engine, gravel, continue.

BOB: (muffled by windshield) Now ya done it!
SFX: (loud) engine, springs creaking, rocks hitting metal.

BOB: (O.S., faint) We're stuck.
SFX: (loud, close) engine, hiss of broken radiator.

Layers of Sound

The audio part of a video program is usually built up from several different sound tracks. These individual components can vary considerably in both their level of reality and their function in the overall sound track.

Levels of reality

Some sounds are absolutely authentic, some are completely artificial, and some fall in between these two extremes. Listed in order of reality, we find:

- **Live sound.** This audio is recorded simultaneously with the video and is often called the production track. Live sound includes spoken dialogue, the noises created by the action, and the ambient sounds of the shooting location.
- **Sweetened sound.** This is live audio that has been processed in postproduction to improve its quality. In some cases, a process called ***equalization*** strengthens certain sound frequencies (pitches) while relatively weakening others. Often, extra live sound is added to conceal the audio differences between different shots.
- **Foley sound.** This is a track of sound effects recorded live in a special studio while the video is displayed, so that Foley technicians can synchronize the sounds they are creating with the program in real time. For example, a Foley studio may have several sections of floor covered variously with gravel, asphalt, concrete, or tile, for use in recording footstep sound effects synchronized with the footfalls of on-screen actors walking on similar material.
- **Library sound.** This is audio that is not created specifically for a particular program, but is stored on discs for use where needed. Some sound libraries contain thousands of different effects. General background sounds such as waves or forest noises or city traffic are often recorded on digital "loops."
- **Synthesized sound.** This is digital audio that is created on a computer instead of being recorded acoustically.

Audio functions

Audio tracks can also be classified by their functions. Here are some of the most common uses for sound:

- **Production sound.** As noted, this is sound that is recorded with the video.
- **Presence.** Often called room tone, presence is recorded at a shooting location while the cast and crew remain perfectly still. The result is a recording of the location's faint background sounds, such as ventilation systems or equipment hum. Presence is used to fill gaps in the production track left when unwanted sounds are cut out.
- **Background sound.** This is an ongoing sound effect like the chatter of diners and the tinkle of china and silver in a restaurant. When the

A Foley setup for recording footsteps on different surfaces.

Thousands of sound effects are available on CD-ROM disks.

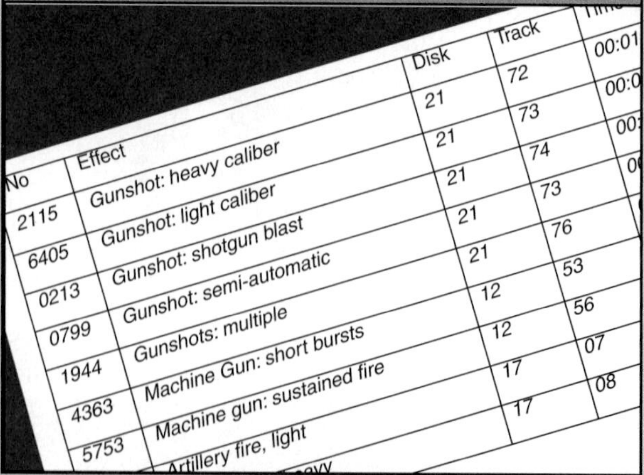
No
Effect
Disk
Track
2115
Gunshot: heavy caliber
6405
Gunshot: light caliber
0213
Gunshot: shotgun blast
0799
Gunshot: semi-automatic
1944
Gunshots: multiple
4363
Machine Gun: short bursts
5753
Machine gun: sustained fire
Artillery fire, light

background track is mixed with the live production tracks, shot-to-shot changes in sound quality are smoothed over. Like presence, background sound smoothes out differences in the production track. But it is also intended to supply sound that was not present during production recording.

- **Sound effects.** Sound effects (often abbreviated "SFX") are specific sounds such as walking, gunfire, or car engines that are added in postproduction, whether from Foley recording or from a sound effects library.
- **Dialogue replacement.** This is a process by which actors rerecord dialogue in a studio, synchronizing it to the lip movements of performers as played back on a monitor or screen. The procedure is still often called "looping" because in predigital days it was accomplished by splicing the film containing a few lines of dialogue into an endless, head-to-tail loop, which could be projected continuously.
- **Music.** Music is added to the sound track to enhance continuity and communicate feeling. Music composed for a specific program may be timed right down to the frame. Library music (precomposed and stored selections) may be selected by length, or else faded in, faded out, or both, in order to time it to the picture. Digital music loops contain embedded information that allows them to be adjusted precisely to fit any length required.

Typical uses for Foley sound effects:

Microphone too far away to record actual sound.

Actual sound doesn't sound the way it should.

Exaggerated sound desired for emotional effect.

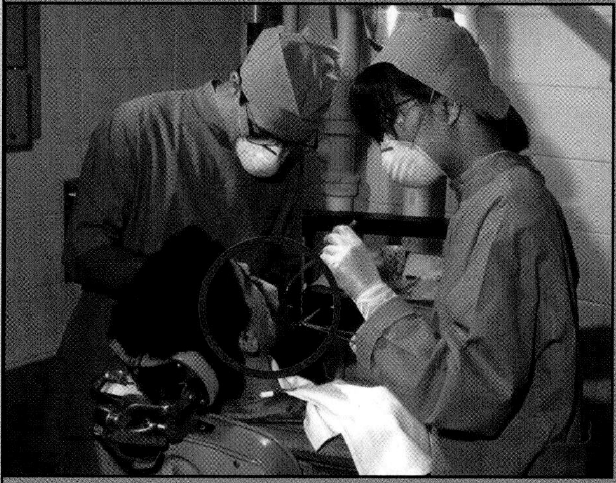

Actual sound obscured by background noise.

The combination of audio styling and matched-action cutting reinforces the illusion of a single, continuous sequence.

Strengthening Continuity

Sound is a great smoother of rough connections. In the example of Bill and Bob, **Figure 7-10**, those connections are between shots in a sequence. Sound is also used to tie separate sequences together through a technique called a split edit.

Splits edits are also discussed in Chapter 4, Video Time.

The Split Edit

A ***split edit*** is a transition from one shot to another in which video and audio do not change simultaneously. Either the new video is accompanied by a few extra seconds of the old audio or, conversely, the new audio is laid over the last part of the old video. Split edits are usually, though not always, used during transitions between two sequences. To see how split edits work, we return to Bob and Bill.

Since the next four trios of pictures are identical, pay close attention to their captions, which demonstrate the different options for split edits.

Straight cut version

For reference, **Figure 7-11** shows a straight cut (without a split edit). Notice that, in the middle shot, Bill's picture and sound conclude together. In this straight cut version, the joke is plain but heavy-handed.

Split edit, video leading

In **Figure 7-12**, the video changes to the new sequence ahead of the audio. We cut to Bob sitting in the darkness (last shot) as Bill's line from the preceding scene is completed in voiceover. Placing the old line over the new shot adds an ironic comment.

Split edit, audio leading

Figure 7-13 shows the opposite form of split edit. In this version, we hear the obvious sounds of night creatures as Bill completes his line in the middle shot. Instead of conveying a sense

Figure 7-11. Straight Cut

BOB: (O.S., faint) We're stuck!
SFX: (loud, close) Engine, hiss of broken radiator.

BILL: Aw, simmer down. Someone'll come along any minute.

SFX: Night frog/cricket sounds: ribbet, ribbet, ribbet.

Figure 7-12. Split Edit; Video Leads

BOB: (O.S., faint) We're stuck.
SFX: (loud, close) Engine, hiss of broken radiator.

BILL: Aw, simmer down.

BILL: (O.S) Someone'll come along any minute.
SFX: Night frog/cricket sounds: ribbet, ribbet, ribbet.

Figure 7-13. Split Edit; Audio Leads
BOB: (O.S., faint) We're stuck.
SFX: (loud, close) Engine, hiss of broken radiator.
BILL: Aw, simmer down.
SFX: Night frog/cricket sounds: ribbet, ribbet, ribbet.
BILL: Someone'll come along any minute.
SFX: night sounds continue.

Figure 7-14. Two-way Split Editing
BOB: (O.S., faint) We're stuck.
SFX: (loud, close) Engine, hiss of broken radiator.
BILL: Aw, simmer down.
SFX: Night frog/cricket sounds: ribbet, ribbet, ribbet.
BILL: (O.S.) Someone'll come along any minute.
SFX: night sounds continue.

of irony, like Figure 7-12, this opposite approach allows the audience to anticipate the joke.

Two-way split edit

Figure 7-14 shows the audio split both forward and backward. The sound effects for the incoming (night) sequence begin offscreen over the outgoing (day) sequence, then the day-sequence dialogue continues, offscreen, over the start of the night sequence.

Which version is best? It depends on the dramatic needs of the program.

Evoking Feelings

In addition to performing all its other tasks, sound is very useful for conveying and enhancing feelings, both momentary responses and overall moods. Music is the most powerful mood setter, as we will see, but sound effects and background noises can also stimulate viewers' emotions.

Sound and Mood

Certain background sounds are useful for setting the overall mood or tone of a video sequence. For example:

- Gentle surf and steady rain both sound peaceful and soothing.
- Howling wind sends a message of desolate emptiness.
- Steady mechanical vibration imparts a mood of tenseness. (That is why the Death Star megaship in the original *Star Wars* film is always accompanied by an ominous, almost subsonic rumble.)
- The sounds of busy city streets convey feelings of energy.
- Jet engine background sounds deliver a feeling of power, **Figure 7-15**.
- The crackling of logs in an open fire suggests cozy safety.

These sound effects are mood setting, partly because they are more or less continuous, running in the background of the audio track.

Figure 7-15.
Jet engine sound effects convey a sense of power. (Corel)

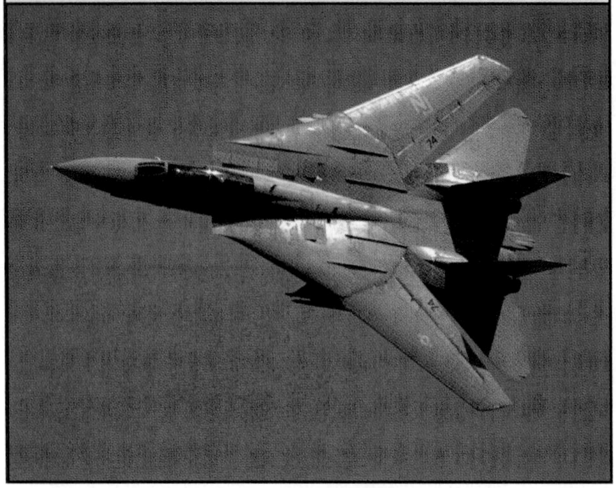

Sound and Emotional Response

Other sounds can evoke more specific emotional responses. This is often because they are associated with particular actions or objects. For instance:

- Echoing footsteps on concrete create suspense.
- The crackle of crunching bones and the pulpy thud of smashed flesh can arouse disgust and even pain.
- The metallic click when a rifle hammer is cocked can provoke a sense of dread.
- Fireworks explosions, **Figure 7-16**, can signal triumph.

Figure 7-16.
Triumphant fireworks are enhanced by sound explosions.

When any sound is weirdly distorted it can impart a feeling of disorientation and even insanity. Listed on the printed page, these evocative sounds may seem like clichés; but when editors lay them skillfully to appropriate visuals, they can be remarkably effective.

Unrealistic Sounds

Some sound effects are frankly unrealistic, especially in comic programs. For example, as a running character skids around a corner, we hear the sound effect of squealing tires and when he rushes past the camera, he is accompanied by an engine roar.

Unrealistic effects can also be used in serious programs. For instance, in a fine nature documentary titled *The Year of the Jaguar*, we hear the ominous rattle of a shell or bamboo wind chime whenever a jaguar spots some offscreen prey. Note the care with which this effect was selected. The sharp tinkling of a glass wind chime would sound out of place in this Central American rain forest. But the more organic sound of shells or bamboo is acceptable, even though it is equally unrealistic.

Perhaps the most unrealistic effects of all are the sounds made by the spaceships in many science fiction programs. We all know that sound cannot exist in the vacuum of space, but we accept it anyway, as an evocation of the power of these mighty vessels.

The Role of Music

The most powerful sound tool for evoking emotion is music, and music has been an integral part of movies since their very beginning. Joy, sorrow, excitement, dread, and almost every other human feeling have been summoned in music. Building upon the innovations of opera composers, the creators of film and video music have evolved a wide variety of techniques for supporting and enhancing the visual track, **Figure 7-17**. Though music's emotional effect is evident in fiction programs, it is equally important in other types of videos.

Commercials and promotional videos rely heavily on emotions to sell their ideas or products. Even documentary and training programs utilize music to set an overall tone, create transitions between program sections, and supply a sense of closure at the end. Like production sound, sound effects, and background sound, music is treated by sound editors as a separate audio component to be blended with the other sound elements to create a final composite sound track.

Figure 7-17.
Library music compositions are offered in several versions.

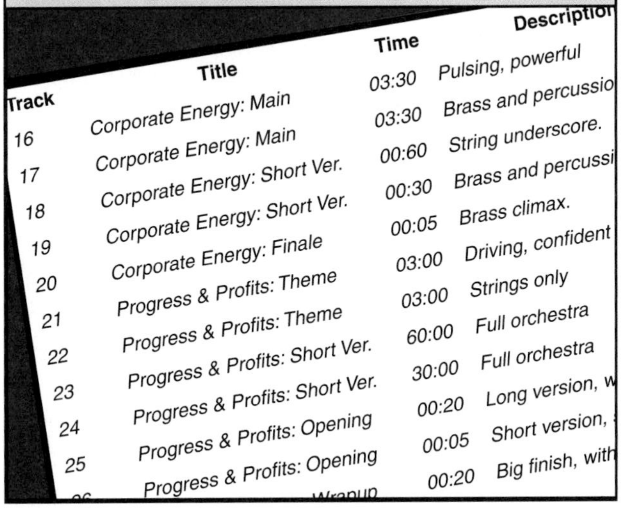

Track	Title	Time	Description
16	Corporate Energy: Main	03:30	Pulsing, powerful
17	Corporate Energy: Main	03:30	Brass and percussio
18	Corporate Energy: Short Ver.	00:60	String underscore.
19	Corporate Energy: Short Ver.	00:30	Brass and percussi
20	Corporate Energy: Finale	00:05	Brass climax.
21	Progress & Profits: Theme	03:00	Driving, confident
22	Progress & Profits: Theme	03:00	Strings only
23	Progress & Profits: Short Ver.	60:00	Full orchestra
24	Progress & Profits: Short Ver.	30:00	Full orchestra
25	Progress & Profits: Opening	00:20	Long version, w
	Progress & Profits: Opening	00:05	Short version,
		00:20	Big finish, with

Communicating with Sound

As you can see, sound is a powerful resource in creating effective video programs. It can deliver abstract information far more efficiently than visuals could. It can suggest places, things, and even people that don't appear on screen. Sound helps bind the many separate shots that make up the visuals into a single, flowing whole. Sound has the power to arouse all kinds of emotions in viewers. And now, digital editing offers the ability to weave subtle audio textures, creating composite tracks as rich and complex as the sound in any theatrical film.

Chapter Review

Answer the following questions on a separate piece of paper. Do not write in this book.

1. ______ is on-screen speech.
2. *True or False?* Voice-over (V.O.) is a technique used only for narration.
3. When synchronized with on-screen action, ______ heighten the sense of reality.
4. *True or False?* In a "split edit, audio leading," the sound changes before the picture changes.
5. Audio components used by editors include production sound, background sound, sound effects, and ______.

Technical Terms

Audio: Collectively, the sound components of an audiovisual program.

Dialogue: Speech by performers on-screen.

Equalization: The adjustment of the volume levels of various sound frequencies to balance the overall mixture of sounds.

Mixing: The blending together of separate audio tracks, either in a computer or through a sound mixing board.

Production track: The "live" sound recorded with the video.

Sound: The noises recorded as audio.

Sound effects: Specific noises added to a sound track.

Split edit: A cut in which audio and video do not change together. When "audio leads," the sound with the incoming video is heard over the end of the outgoing video; when "video leads," the sound with the outgoing video continues over the incoming video.

Straight cut: An edit in which audio and video change simultaneously; also, an edit that does not include an effect such as a fade or dissolve.

Voiceover: Narration or dialogue recorded independently and then paired with related video.

The History of Video Sound

Radio

The first medium to use audio was radio, and radio (of course) was sound only. It was here in this "theater of the mind" that sound technicians invented ways to suggest places and events through sound effects. Background noises such as theater audiences or rushing rivers were recorded on phonograph records.

Sound effects teams provided "live" sounds, operating miniature doors, "walking" shoes across different surfaces, and firing blank shots in real time, while the story unfolded. Once sound movies arrived, these live sound effects operations became the Foley recording studios of movies.

Sound effects technician Tom Keith rehearses for the Prairie Home Companion radio program. (Minnesota Public Radio photo by Frederic Petters)

Sound films

At first, movie audio was necessarily simple: a live production track "sweetened" with some effects and underscored with music.

After World War II, movie sound advanced with the introduction of magnetic recording on both tape and film. (Previously, sound had been recorded optically by photographing the changing light outputs of photoelectric cells.) Magnetic recording offered improved quality, flexibility, and economy, so that the sound track elements used to make a composite audio track could multiply from fewer than a dozen to as many as fifty or more.

Magnetic recording brought another benefit, too: it allowed sound technicians to lay down control tracks that made it easier to synchronize sound tracks with one another (and with the picture) without a mechanical connection. As the techniques of film and video gradually merged, film sound engineers began using the time code method, in which the length of a recording is measured in hours, minutes, seconds, and frames, and the time "address" of every single frame is recorded with it.

Television

In the early days of TV, music and background effects were usually prerecorded on audiotape, but individual sound effects were often produced in real time. The arrival of videotape permitted more sophisticated sound editing and the development of time code made it practical to lay in individual synchronized sound effects.

Video sound today

Video sound came of age with the development of computer-based editing, in which audio components are digitized and all editing is performed on the computer. Digital sound editing allows almost instant access to all sound components, and the computer can handle dozens of tracks at once. As a result, the possibilities for creating video sound tracks are almost limitless.

8 Video Communication

Objectives

After studying this chapter, you will be able to:

- Explain how the characteristics of video fit and work together.
- Build programs by assembling individual components.
- Distinguish video “truth” from real-world truth.

A NASA probe approaches the third moon of Tralfamador.

The Moon Is Not Green Cheese—It is a Tortilla.

The shot on the opening page of this chapter is a digital composite. First, a tortilla was shot against a green screen. Next, the sun and stars were created in a paint program. The sun and stars were then composited with the tortilla.

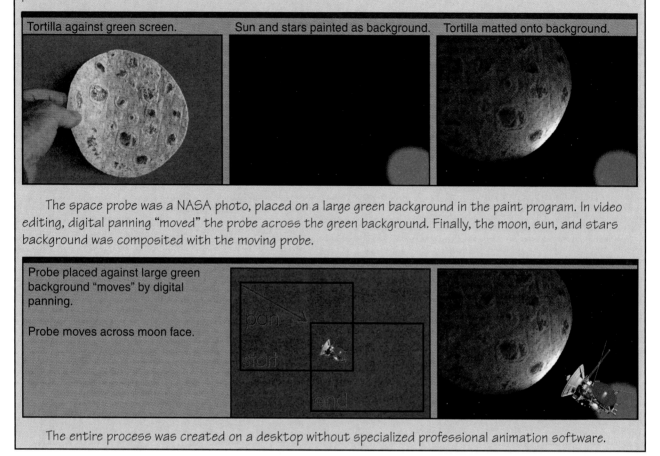

Tortilla against green screen. Sun and stars painted as background. Tortilla matted onto background.

The space probe was a NASA photo, placed on a large green background in the paint program. In video editing, digital panning "moved" the probe across the green background. Finally, the moon, sun, and stars background was composited with the moving probe.

Probe placed against large green background "moves" by digital panning.

Probe moves across moon face.

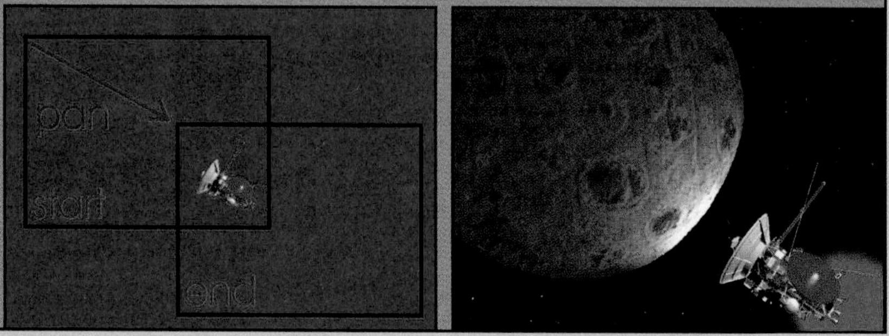

The entire process was created on a desktop without specialized professional animation software.

About Video Communication

In Chapters 3 through 7, we examined video *space, time*, *composition, language*, and *sound*. For easier discussion in those chapters, the topics were treated separately; when you actually make videos, however, you must work with all of them at the same time. This chapter brings the components of the video world together, showing how you manage them through the craft of *video communication*.

To do this, the following section briefly summarizes the characteristics of the video world. The next section shows how video professionals build programs from pieces of raw material. The chapter concludes with a discussion of "truth" in the video world and honesty in presenting it.

Throughout, the most important idea to keep in mind is that all video—even the simplest home movie—is not a recording of the *real* world. Instead, it shows a self-contained *video* world, specially constructed to the design of its maker.

You could say that from fade-in to fade-out, every video is simply an assembly of special effects.

The Video World

Because the components of the video world operate together, affecting and reinforcing one-another, **Figure 8-1**, it is useful to review them as a single system made up of space, time, composition, and sound.

This section condenses and reorganizes the contents of Chapters 3 through 7.

Figure 8-1.
Space, time, composition, and sound are the components of the video world. (Corel)

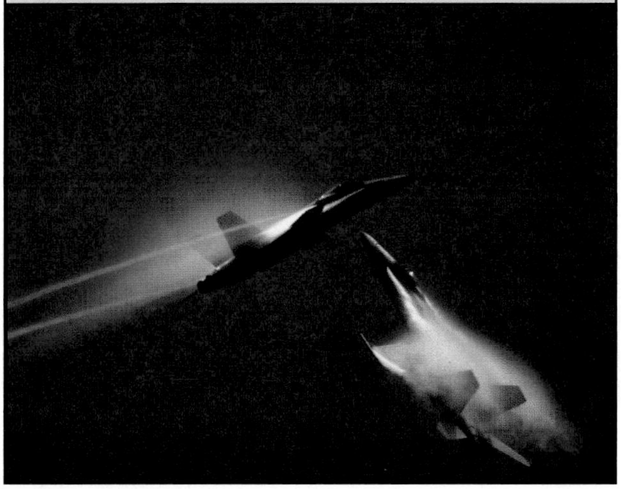

Video Space

The laws of video space are aspects of two basic concepts:

- the power of the frame.
- the illusion of size and distance.

By understanding these ideas, you can manage video space with confidence.

The power of the frame

The frame around the image is the most powerful shaper of the video world, because it determines what exists in that world and what does not. To use it, you:

- ***Frame*** (include) everything you want to exist, and
- ***Frame off*** (exclude) everything you do not want.

This means, of course, that all your production crew and equipment "do not exist," because they are outside the frame. It also means that your shooting locations do not have to match the places you wish to show the audience, so long as you include the parts that do fit and exclude the parts that do not.

Suppose, for example, you need to tape a sequence in the courtyard of a Spanish royal palace. By framing the parts you want and framing off the parts you do not want, you can conceal the fact that you are really in the garden of an American municipal building. So, when hunting locations, do not look for the places you need, but for the places in which to *create* what you need by selective framing, **Figure 8-2**.

The Laws of Video Space

Here are the laws of video space, as presented in Chapter 3:

- What is outside the frame does not exist, unless that existence is implied.
- Height and breadth are determined solely by the frame; depth is only an illusion.
- Size, position, distance, relationship, and movement are not fixed.
- Direction is determined solely by the frame.

Figure 8-2.
Creating a location with selective framing. A—A modern city hall built in the Spanish style. B—The frame excludes the modern offices, signs, and other features elsewhere in the courtyard.

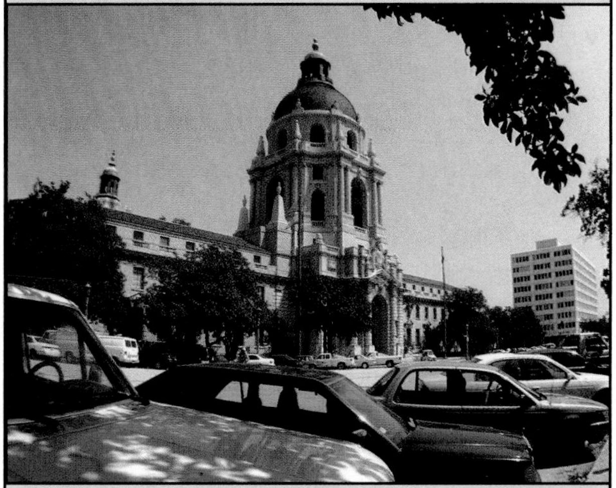

A

B

Figure 8-3.
Suggesting a location. A—The climber could be thousands of feet up. B—Instead, he is only four feet from the ground. (Sue Stinson)

Many spaceship interiors in the film Alien were shot in the working parts of an ocean-going freighter.

The frame also allows you to suggest that some things exist outside the borders of the image. It can do this in several ways. If something in the frame, such as a rock face, continues past the frame border, then viewers tend to think that it goes on for some distance, when, in fact, it may be very small. See **Figure 8-3**.

Things outside the frame can also be suggested by actions. In **Figure 8-4**, the runner is poised on the starting line. When the starter's gun goes off, he sprints out of frame, suggesting that he is in a stadium, rather than merely on a small pad painted blue with white lines. (Like the rock face, the lines and blue paint also suggest that the track continues outside the frame.)

The most powerful suggestive tool is sound. Anything can be made to exist outside the frame if its characteristic sounds are laid on the audio track, **Figure 8-5**.

Finally, you can combine these techniques to reinforce their suggestive power. For instance, you could enhance your city hall courtyard sequence by starting with stock footage of a real Spanish palace, **Figure 8-6**.

In the same way, a stock establishing shot can make your climbers appear to ascending the Rock of Gibraltar! (Be sure to add dramatic wind blowing on the audio track.) See **Figure 8-7**.

Figure 8-4.
The rest of the "track" is suggested by the action of running out of frame. (Adobe)

Figure 8-5.
The phantom bus described in Chapter 7 is created outside the frame by sound.

Figure 8-6.
A stock shot of an actual Spanish palace...

...establishes the courtyard "inside." (Sue Stinson)

Figure 8-7.
The Rock of Gibraltar establishes the location...

...of the rock climber.

Figure 8-8.
Stock shot of a New York street.

Figure 8-9.
The "wall climber" described in Chapter 3 was shot with the camera turned on its side, converting the horizontal plane to the vertical.

Figure 8-10.
Conventional directions. A— "West-bound" planes, as shot by Camera 1. (Corel) B— "East-bound" planes, as shot by Camera 2. (Corel) C— The plan view shows that both groups of planes are actually flying in the same direction.

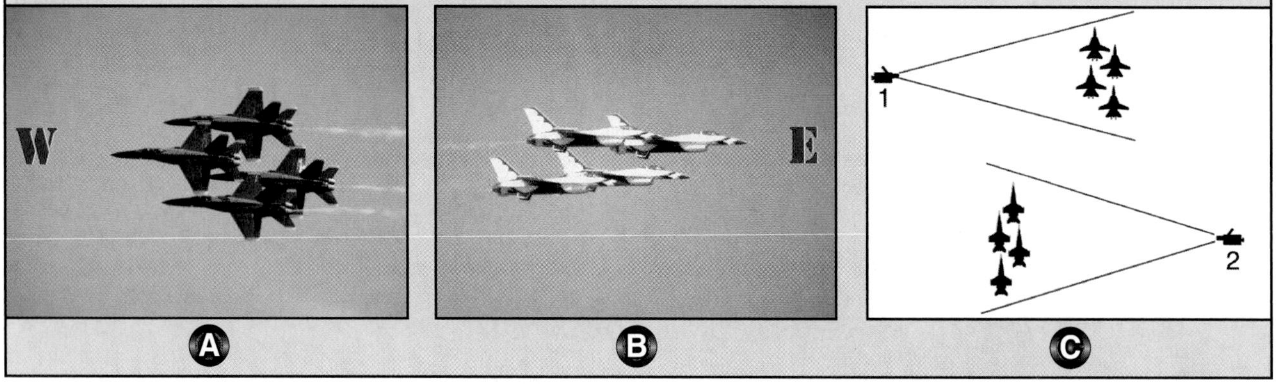

A "stock shot" is a piece of video obtained from a commercial library of commonly used material. For example, stock shots of New York exteriors, Figure 8-8, are often inserted into TV programs shot in Los Angeles studios.

In addition to deciding what exists, the frame also determines what is up and down. *Vertical* is always parallel to the sides of the frame while *horizontal* matches the top and bottom. By rotating the camera onto its side, you can reverse the horizontal and vertical dimensions, **Figure 8-9**. (If you turn the camera completely upside down, you can create human fly or zero gravity effects.)

By convention, the frame can determine the compass direction, with left-pointing subjects moving "west," and right-pointing ones moving "east." See **Figure 8-10**.

The frame can also affect apparent subject motion. Allowing the subject to cross the frame and leave it implies strong movement, while a subject that remains stationary with respect to the frame can seem motionless, **Figure 8-11**.

The control of scale

The video world has no actual depth, because everything is displayed on the flat surface of the screen. Subjects only *seem* to move closer or farther away as they grow or shrink in size on that surface.

In the real world, of course, things are just the opposite: subjects only *seem* to grow or shrink as they move forward or away, while actually remaining the same size. We learn to judge distances by the apparent sizes of people and things, because we know how large they look at

Figure 8-11.
Apparent subject motion. A—The racing boat flashes across the frame. B—The spacecraft appears motionless. (NASA)

Figure 8-12.
Tricking the eye. A—The "leprechaun" is 50 feet from the camera; the hand, only three feet. B—A toy locomotive can look full-size in a close-up.

A

B

close range. We use this ability when watching videos, judging subjects that are smaller *on the screen* to be farther away and bigger ones to be closer.

Because viewers apply real-world depth perception to screen images, all kinds of deceptions are possible, **Figure 8-12**. A man 50 feet away can look like a leprechaun, because he is aligned with a hand three feet away. A toy train can seem to be full size because it is so close to the camera.

On movie sets, children or "little people" extras are sometimes used in the background to make rooms, and especially hallways, appear deeper than they are.

Figure 8-13.
The hand reveals the true size of the engine.

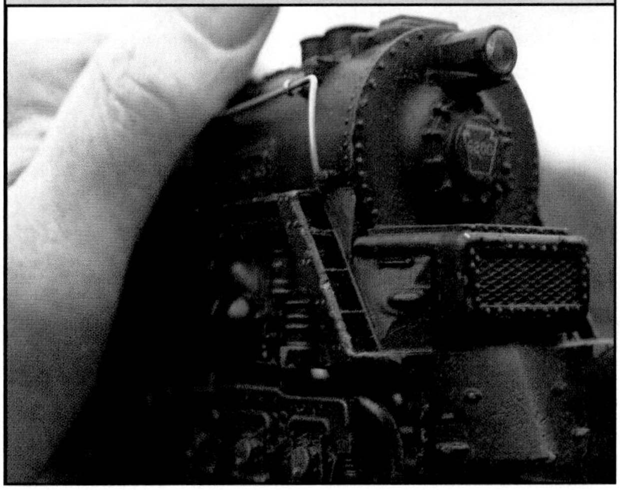

The leprechaun and the locomotive can fool viewers because of one vital clue: *apparent scale*. The leprechaun looks tiny compared to the hand, while the locomotive looks large until compared to a similar hand, **Figure 8-13**. Without the sense of ***scale*** (comparison) provided by the contrasting figures, viewers would not be able to tell their sizes.

By providing different clues about scale, you can make subjects and objects seem to be any size you choose.

Video Time

In making videos, you are constantly dealing with time as well as space. Unlike real-world time, which never changes its characteristics, video time can be shaped in any way you choose.

Speed

You can change the speed of time within a shot by slowing it with slow motion or speeding it up with fast motion.

Flow

You can control the way video time flows. In addition to the customary single stream of events, you can present multiple streams, either by placing one after another, or by cutting back and forth among them. These streams may run parallel in time to the main story, or they may represent the past, the future, or both.

Direction

Usually, videos move forward in time, from the beginning of events to their ending. Time can also move backward when you reverse a shot in editing.

Coherence

In the video world, you can separate the video and audio, moving them around in time independently. The most common way to do this is the split edit, in which either the video moves forward in time before the audio does, or vice versa. Another common use is ***voice-over*** narration, **Figure 8-14**, where the off-camera speaker is commenting on visuals that happened in the past (or will happen in the future).

Figure 8-14.
NARRATOR (V.O.) Some tombs have remained standing for thousands of years… (Sue Stinson)

A Christmas Story is a well-known film in which the narrator is in the present while his story, on the screen, is in the past.

Space, Time, and Editing

Though space and time can be managed in individual shots—space through the control of framing, and time through slow, fast, or reverse motion—the most powerful tool for shaping the video world is *editing*. In placing one shot after another, you literally create video worlds as you build sequences.

A sequence is an assembly of shots organized to present a single action, process, or idea.

Through editing, you can compress or expand both space and time—usually both at once. By switching camera angles, you can cut out unwanted parts of actions (such as walking long distances). In the courtyard example, **Figure 8-15**, notice that editing eliminates both the undesirable *space* in the courtyard and the unwanted *time* required to cross it.

Editing can create single locations out of two or more places that may be far apart in the real world. This happens each time a subject starts through a door at a location and finishes this entrance in a different location, or even a studio set. For example, suppose you need a scene in an historic mansion, but are not permitted to shoot inside it. Instead, you can record establishing shots outside the building, and then use a matching set for the interior, **Figure 8-16**.

Controlling time flow, direction, and coherence is largely a function of editing. Notice that the reversed shot in this jumping sequence, **Figure 8-17**, would be meaningless without the shots placed before and after it by the editor.

Parallel time streams are dependent on editing to show first a piece of one, then the other, then the first again. Sometimes, two

Figure 8-15.
Condensing time and space. A—The wide courtyard can be condensed by cutting between two camera angles. B—The first shot shows the subject entering the courtyard. C—In the second shot, the subject has completed the crossing.

A B C

Figure 8-16.
An exterior shot of this Victorian mansion...

...is matched with an interior shot in a studio.

Figure 8-17.
The appearance of actors jumping *onto* a high wall is achieved by having them jump *off* backward, then reversing the video in editing.

Figure 8-18.
Parallel time streams may be shown sequentially... or simultaneously on a split screen.

streams are shown simultaneously, either in different parts of a multiple image or superimposed for a "double exposure" effect. See **Figure 8-18**.

The split edits that break up time coherence are made in postproduction.

Figure 8-19.
Lens focal length effects. A—Wide-angle lens increases apparent depth. B—Telephoto lens decreases apparent depth.

The Role of Composition

Chapter 5 explains that ***composition*** in video is the purposeful arrangement of the components of a visual image, using the principles of simplicity, order, balance, and emphasis. Good composition enhances the overall quality of the image and leads the viewer's attention where the director wants it to go.

Directing the eye

Composition directs viewers' attention to certain elements in the image by managing these elements:

- position in the frame,
- relationship to other elements,
- significance (importance for their own sake),
- contrast (difference from other elements in the image).

The most powerful tool for emphasizing a pictorial element is movement, which involves controlling composition continuously throughout a shot.

Managing depth

Most importantly for creating the video world, composition controls the illusion of depth — the false third dimension on the flat screen. The techniques for suggesting depth in an image are termed ***perspective***.

In the video world, perspective is every bit as changeable as space and time. The most common tool for managing perspective is lens focal length, **Figure 8-19**. A ***wide-angle*** setting increases apparent depth, rendering distant subjects smaller than "normal" and making backgrounds seem

Figure 8-20.
Wide-angle lens perspective.

Figure 8-21.
Telephoto lens perspective. (Corel)

more distant than they really are. A ***telephoto*** setting does the opposite: enlarging distant subjects and making the background seem closer. For these reasons, wide-angle images have greater than average "depth" and telephoto images have less.

In making videos, you use these characteristics constantly. You may select a wide-angle lens setting to make a small space look bigger, to de-emphasize a distant background, or to make movement along the "Z axis," (apparent depth) more dynamic, **Figure 8-20**.

You may pick a telephoto lens setting to make subjects appear closer together than they are, to make an environment feel close and congested, or to bring a distant background closer, **Figure 8-21**.

You can also use perspective to fool the eye by changing the apparent size and/or distance

The Principle Tools of Perspective

Visual tools used to create perspective:

A—Size.

B—Overlap.

C—Convergence.

D—Vertical position.

E—Sharpness.

F—Color intensity.

of pictorial elements, like the leprechaun and the locomotive examples shown earlier in this chapter.

The Power of Sound

Because all the parts of the video world work together, we have already noted the role of sound. Most often, audio adds to the information presented by the visuals, but it also plays important roles in creating the video world.

Implication

As mentioned earlier, sound can imply the presence of something that is outside the frame (such as the "phantom bus").

Reinforcement

Sound can also make a created environment feel more realistic. For example, you can make a single table against a blank wall appear to be in a restaurant by adding running sound effects of silverware, glassware, and conversations.

Blending

Finally, sound can pull together visuals recorded at different times and/or places by running continuously under the edited shots.

In summarizing the roles of video space, time, composition, sound and editing, we see how they create each video world.

It is "each" video world because the world of every program is different—custom-made for it during production and postproduction.

Video Organization

Like a masonry wall, a video world is constructed of many individual pieces, each one obeying the rules of video space, time, composition, and sound. Despite these hundreds and often thousands of separate parts, the resulting programs *appear* to be seamless and unified. That is because each separate element has been individually created, selected, modified, and sequenced to give the illusion of continuity.

Why not simply record and present the real world unmodified, the way videos such as reality programs, news reports and documentaries do it? The fact is that these programs do *not* present the real world at all. Even the most "neutral" news report or "unbiased" documentary is not a recording of reality but someone's *version* of it, **Figure 8-22**.

Figure 8-22.
A "standup" television news report.

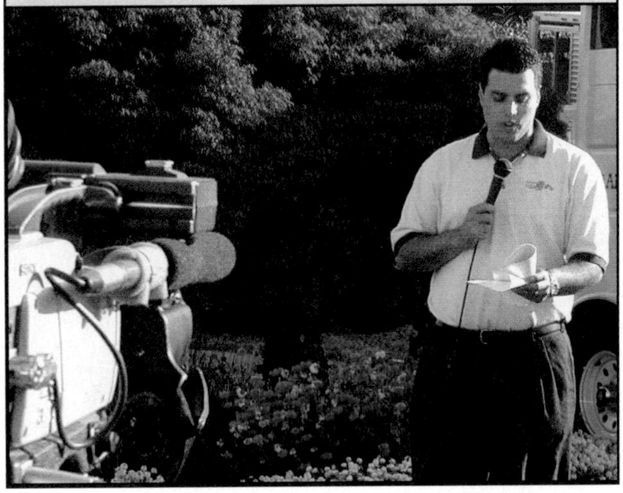

No matter how hard people may try, it is impossible to present a truly complete or unbiased "reality" in a video because the techniques required to make it cannot avoid creating a *video world,* rather than capturing the real one. This is unavoidable because,

- The video cannot show everything, only a few pieces.
- To make sense to viewers, those pieces must be created, selected, and sequenced according to a plan (usually in the form of a script).
- A plan is a way of organizing material, and any organization requires a point of view—a particular way of seeing the subject matter of the program.

By the time the raw material has been fully organized for presentation, it has been shaped by *recording, selecting, modifying, sequencing,* and *reinforcing.*

To see how this happens, we will refer to the lighthouse example in the sidebar.

The following topics are covered more fully in the chapters on editing. They are summarized here to show how video is manipulated.

The Lighthouse Example

In a recent film, a scene supposedly set in a lighthouse, **A** was actually staged in an old lighthouse top that had been moved miles inland to decorate the entrance of a county fairgrounds, **B**.

To recreate that scene, you would start with a slightly low angle that excludes most of the background, **C**, then frame off the fairground gate supporting the light, **D**.

To make both the performers and the light visible, you would shoot the sequence near sunset. In postproduction, you would change the warm sunset light to a bluish tint that suggests twilight, **E**.

And lay in an audio background of waves and gulls to enhance the effect, **F**.

Finally, you would blend it with an establishing shot taken at a completely different lighthouse, **G**.

Edited together, the establishing shot makes at the actual lighthouse and the following shots of the fairground light, **H**, seem to exist at the same location.

Why not shoot the whole sequence at the real lighthouse? Camera placement is much easier at the fairground light, which is only 20 feet off the ground, instead high in the air.

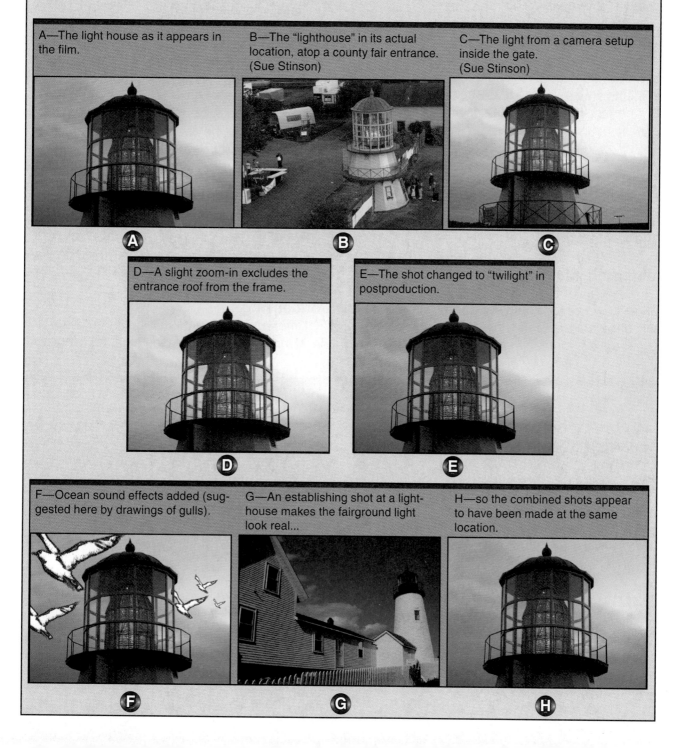

A—The light house as it appears in the film.

A

B—The "lighthouse" in its actual location, atop a county fair entrance. (Sue Stinson)

B

C—The light from a camera setup inside the gate. (Sue Stinson)

C

D—A slight zoom-in excludes the entrance roof from the frame.

D

E—The shot changed to "twilight" in postproduction.

E

F—Ocean sound effects added (suggested here by drawings of gulls).

F

G—An establishing shot at a lighthouse makes the fairground light look real...

G

H—so the combined shots appear to have been made at the same location.

H

Recording

During production, every single frame of video footage is shaped by two essential decisions: what subjects to frame (and what to frame off), and when to start recording (and when to stop).

Framing the subject

In the lighthouse example, the fairground light is framed by the camera, while its base and background are framed off, as shown in illustration D. As a result, the fairgrounds do not exist in the video world, while the rest of the "lighthouse" below the frame is implied.

Timing the recording

The sequence is shot at sunset, the only hour at which both the lighthouse lamp and the subjects are easily seen, as shown in illustration E. The rest of the day, with unsuitable lighting conditions, is omitted.

Selecting

As postproduction begins, the material selected for inclusion in the program is identified and processed.

Choosing shots

Usually, the action is covered in more camera angles than needed, and each angle may have two or more takes. Everything that is not selected is excluded from the program, **Figure 8-23**.

A *shot* is a continuous recording, from camera start to stop. A *take* is one attempt to record a shot.

Trimming shots

Each shot is cut to length (called ***"trimming"***), and all the material before the start point and after the end point is discarded.

Modifying

Digital postproduction offers powerful tools for changing the color, speed, screen direction, and framing of each shot.

Altering shots

In our lighthouse example, both shots G and H were darkened and tinted to simulate night.

Compositing and computer graphics (CG)

Elements can be added to the original footage by ***compositing***. To intensify the seaside effect in our example, flying gulls might be superimposed on the shots of the fairgrounds light.

Sequencing

Changing and shaping the real world continues as individual shots are edited together.

Condensing

Different angles are cut together to eliminate unnecessary space and time. Leaving

Figure 8-23
Shot selection. A—The alternate establishing shot is discarded. B—The selected establishing shot.

Figure 8-24.
Editing different angles together allows the removal of most of the walk.

our lighthouse example, here is the courtyard sequence mentioned previously, **Figure 8-24**.

Repositioning

Shots are rarely placed in the order in which they were recorded, and sometimes they can be used in ways different from the original intention. For example, a performer's reaction shot made to be used at one point in a conversation can often be used (or even re-used) at another point.

Relating

Editing can bring together elements that have no real connection. So-called "glance-object pairs" are edits in which the subject in the first shot looks off screen and then another subject is shown in the second shot. The result makes it appear that the first subject is looking at the second. In **Figure 8-25**, the man and the sundial are actually in two completely different locations. The man's actions are parts of a single shot, with the sundial edited in.

Figure 8-25.
The man consults his watch.

To check it, he glances off screen...

at an object (the sundial)...

then looks back at his watch.

The sundial shot is called a "cutaway," because the editor literally cuts the main shot, inserts the different one, then cuts back again.

Reinforcing with Audio

Sounds added in postproduction both enhance individual shots and tie together whole sequences.

Sound effects

Effects from a sound library can be laid in to synchronize with on-screen subjects or suggest things, such as gulls, that are not seen.

Background tracks

Tracks of continuous noise, such as city traffic, can tie shots together and smooth out differences in their separate production sound tracks, **Figure 8-26**. These are called ***background tracks***.

Narration

Narration relates different shots to one another by, in effect, explaining why they belong together.

Techniques for Different Programs

The lighthouse example was suggested by a sequence in a feature movie—an obvious work of fiction. Conscientious documentary makers, on the other hand, might insist that their programs are not manipulations of reality, but are simply condensed versions of it. As we have seen, however, the moment they start selecting the material to show and the order in which to show it, they begin building a video world that is both less than, and different, from the real world.

The Lord of the Rings and the Star Wars movies are good examples of completely artificial "realities."

Video and "Truth"

Most viewers of "realistic" programs think that the video world is the same as the real one. This is partly because they do not know or care how video worlds are constructed, and partly because the resulting "reality" is so convincing. The ability of video to present a convincing "truth" is a powerful tool, and like any tool, it can be misused. To prevent this, responsible program producers need to know how far they can properly go in changing reality.

Types of Programs

Is altering reality unethical? That depends on the type of program. Fiction programs, news and sports, documentaries, educational programs,

Figure 8-26.
Because background noises in different camera angles do not match, they are often replaced by background tracks that run under all the edited shots.

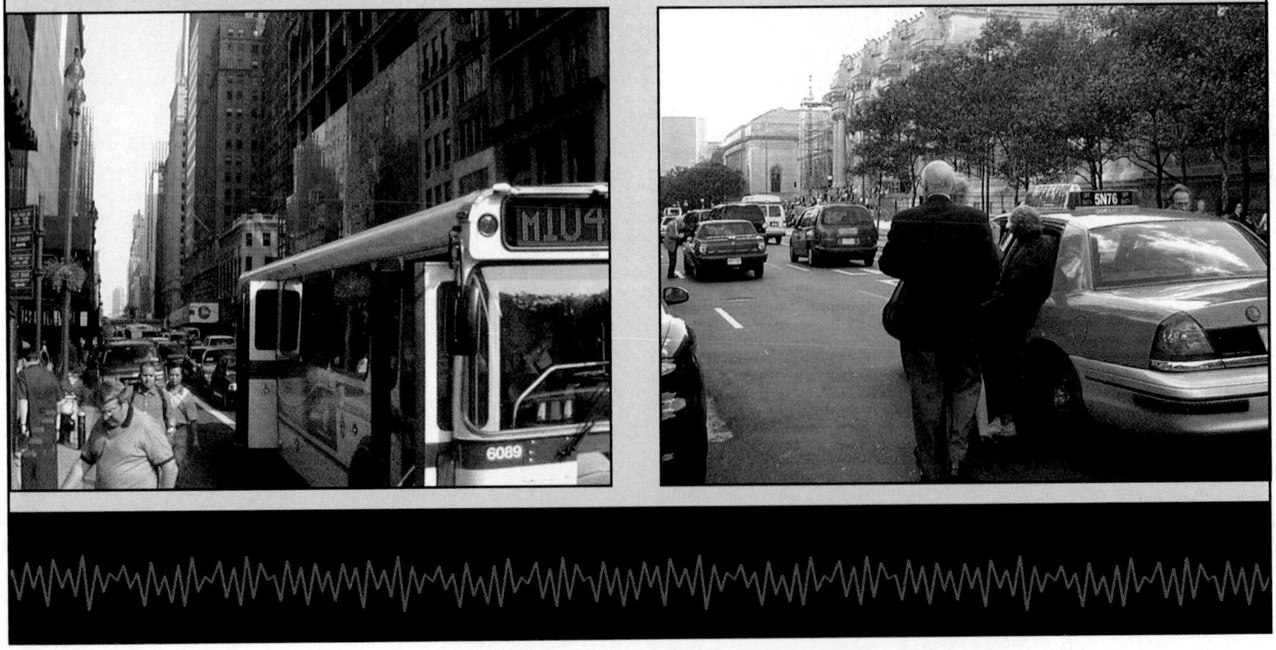

and commercials all demand different standards of truthfulness.

Story programs

Feature movies, TV series programs, and music videos are all made purely for entertainment. Viewers do not expect them to deliver facts or tell the "truth," but to present stories that are obviously invented. In programs like these, literally anything goes because viewers completely understand that they are not watching reality. In these programs, you can ethically exploit all the tricks available in creating video worlds.

Sports

Sports programs and reports are, perhaps, the closest to reality, especially when they are presented ***"live"***–that is, in real time. Even so, most productions have many cameras capturing images at once, and each camera is framing some subjects and framing off others, **Figure 8-27**. In the control booth, the director, at any given moment, is using one of those shots and excluding all the others. So even the most straightforward sports program alters reality by making decisions about what and when to shoot and what to select from the footage. The conscientious producer tries to present viewers with the most important action as seen from the most informative angles.

However, sports footage can be easily manipulated by selecting shots that fit the producer's agenda and supporting them with editorially weighted narration. For example, a nation's performance at the Olympic games can be "improved" by presenting all the wins and omitting the losses, or at least by excluding their worst moments.

News

Breaking (live) news is very much like sports, but many news reports concern events to which cameras were invited: interviews, **Figure 8-28**, press conferences, and so-called ***"media opportunities"*** set up specifically to be videotaped. Here, the people and organizations staging the events are also working to shape the "reality" presented.

Figure 8-28.
Many news features are interviews.

This practice is so common that it has a name: "spin," and changing reality to reverse damage is "spin control."

Whether staged or spontaneous, news reports are selected and edited by TV news departments. Television news tends to favor the more sensational stories.

A newsroom phrase for this practice is, "If it bleeds, it leads," meaning that the most violent or otherwise sensational events get the most prominent coverage.

Figure 8-27.
Sports program directors have several cameras shooting at once. (Corel)

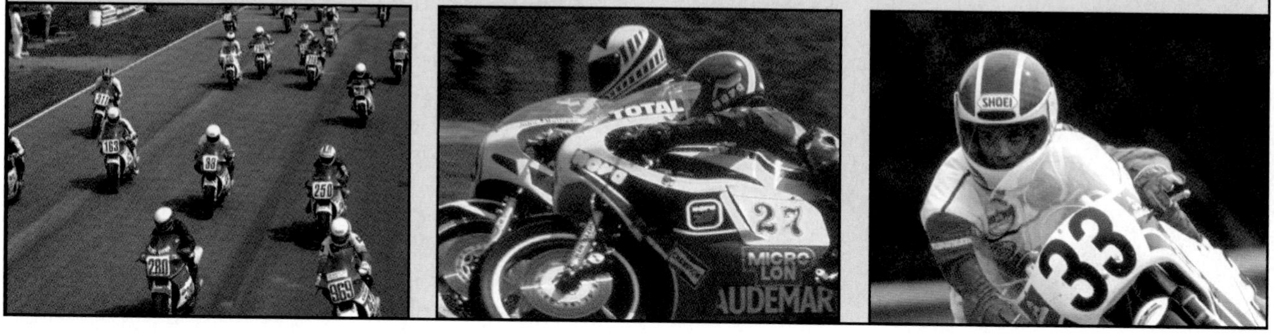

In addition, TV channels, and even whole media corporations, have been accused of distorting reality in support of their social and political agendas.

Documentary programs

Documentary programs aim to *document* reality, **Figure 8-29**. In their most rigorous form (often called ***"cinema verité"***) they present the essence of their subjects, painstakingly culled from dozens or even hundreds of hours of unstaged footage, often without the editorial comment of narration or the emotional support of music. But, as noted previously, the selection process must follow a plan and a plan represents a point of view. As a result, even the strictest documentary presents a video reality rather than genuine reality.

Figure 8-30.
Logging and anti logging documentaries could use many of the same shots, but for opposite purposes. (Sue Stinson)

Figure 8-29.
Animals and history are popular documentary subjects. (Corel)

Many programs could be called ***"editorial documentaries,"*** because they attempt to persuade viewers as well as inform them. For example, one program might urge the value of forest conservation while a rival program "proves" the importance and value of logging, with both videos using much of the same raw footage. See **Figure 8-30**.

Another term for documentaries with hidden agendas is *propaganda*.

The first great editorial documentary was Leni Riefenstahl's *Triumph of the Will*, made in Germany in the 1930s. Unfortunately, its editorial purpose was to celebrate Hitler and Nazism.

Whether or not propaganda is ethical is a complex (and important) debate that has never been satisfactorily resolved. See **Figure 8-31**.

Figure 8-31.
During World War II, American newsreels emphasized Allied successes. (Corel)

Commercials

Advertising videos, or ***commercials,*** use every tool in the video maker's kit to sell their products. Typically, they announce that they are *not* realistic through lavish use of computer graphics, extreme situations, and humor, **Figure 8-32**.

Commercials do not distort the obvious facts about their products because their manufacturers would be open to lawsuits or penalties from regulatory agencies. Nonetheless, they characteristically use video tricks such as wide-angle lenses to make cars look longer and sleeker, with roomier interiors. Colors may be intensified in postproduction and lengthy processes (such as mowing a lawn or painting a house) condensed by editing until they seem to take no time at all.

Figure 8-32
Commercials make extensive use of computer graphics. (Adobe)

Thirty- or 60-minute-long commercials disguised as programs are called "infomercials."

Political commercials, however, often distort facts—or at least misinterpret some of them and omit others that may be inconvenient for the candidate. Even when the facts are reliable, political commercials use every video trick available to sell their products: the candidates, **Figure 8-33**.

A commercial can distort reality by hinting that you will be more stylish in a certain brand of clothing or more important in a certain make of car, but these distortions do not always rely on the tricks of the video world.

Figure 8-33.
Political commercials work to "sell" the candidate to voters.

"Infotainment"

Cable channels are filled with "factual" videos that present information by frankly and openly shooting to a script and editing the footage to entertain as well as inform. These hybrid programs, often referred to as ***infotainment***, combine the characteristics of documentaries with the techniques of fiction movies. At their best, they do so in order to present information more clearly and vividly; but their makers never forget that these programs are designed first of all to entertain.

Since they are imitating (or at least evoking) reality rather than distilling it from actual footage, these hybrid programs use many video techniques. They shoot action that has been staged for the camera. They re-enact action

Figure 8-34.
History programs rely heavily on period graphics, creating movement by panning and zooming the camera.

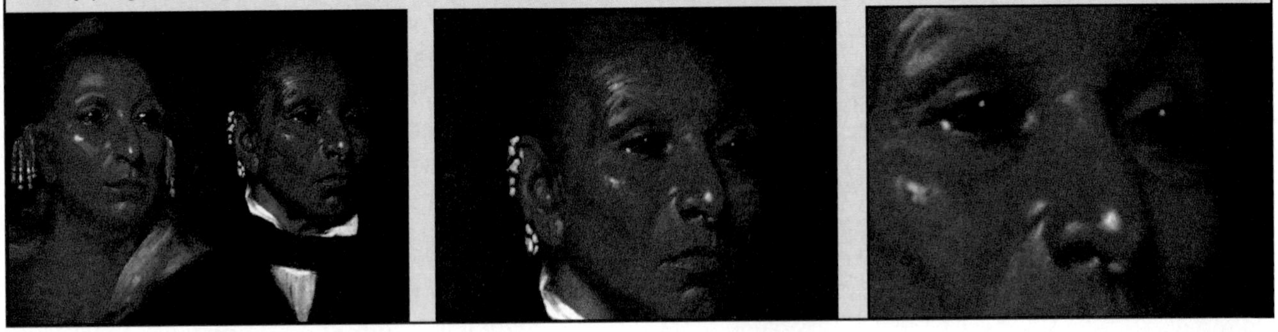

that happened in the past. They combine stock shots with actual footage. They use still photos and graphics where live action is not available, **Figure 8-34**. Most importantly, they edit together elements that occurred at different times and places.

These techniques are covered more thoroughly in Chapter 9, Program Development.

The Ethical Producer

The question remains: how close to the real world should an honest producer try to make the video world? There is no simple answer to this complex question, but some guidelines can be suggested.

- In a documentary, the edited material should represent the actual subject as closely as possible.
- In an infotainment program, the essentials of the subject should be accurate and truthful, though the material has been manipulated for presentation.
- Commercials should be honest where facts are concerned, though it is generally accepted that they may exaggerate the good qualities of the product being sold.
- Sports and news should be as accurate and unbiased as possible. However, it is traditional to present news features with a "slant," or "hook" to increase viewer interest.
- Fiction programs make no claim to show reality at all, so they can employ every technique in the video toolbox.

Creating a Plan

All of these program types begin with a blueprint for production—sometimes a treatment or storyboard, but usually a fully developed video script. The process of program development is the subject of the next chapter.

Chapter Review

Answer the following questions on a separate piece of paper. Do not write in this book.

1. Video space can be defined by two basic concepts. Which of these choices is *not* one of those concepts?
   A. The power of the frame.
   B. Closely defined height and breadth dimensions.
   C. The illusion of size and distance.
2. *True or False?* Objects that are smaller on the screen appear to be farther away from the camera.
3. ______ is the most powerful tool for shaping the video world.
4. The existence of something that is outside the frame can be implied by ______.
5. *True or False?* Fiction programs, documentaries, and commercials are all held to the same standard of truthfulness.

Technical Terms

Background track: An audio track of the characteristic sounds of an environment, such as ocean surf, city traffic, or restaurant noises.

Cinema verité: A style of documentary that presents its material with as little intervention by the program makers as possible.

Commercial: A very short program intended to sell a product, a person, or an idea.

Compositing: Combining foreground subjects with different backgrounds.

Composition: The purposeful arrangement of visual elements in a frame.

Documentary: A type of nonfiction program purporting to communicate information about a real-world topic.

Editorial documentary: A documentary that attempts to win viewers over to its position or point of view.

Frame: The border around the image. Also, to frame something is to include it in the image by placing it inside the frame.

Frame off: To frame off is to exclude something from an image by placing it outside the frame.

Infomercial: A program-length commercial masquerading as a regular program.

Infotainment: A form of documentary whose primary purpose is to entertain viewers.

Live: Being presented as it is transmitted from the recording video cameras. "Live on tape" means a recorded but unedited presentation of a live program.

Media opportunities: Events created specifically for the purpose of being covered by news organizations.

Narration: Spoken commentary on the sound track.

Parallel time streams: Two or more lines of action presented together, either by alternating pieces of the various streams, or by splitting the screen for simultaneous presentation.

Perspective: The simulation of depth in two-dimensional visual media.

Scale: The perception of an object's size by comparison to another object.

Sequence: A segment of a program, usually a few minutes long, consisting of related, organized material.

Shot: An uninterrupted recording by a video camera.

Spin: The management of information for somebody's benefit.

Stock shot: A shot purchased from a library of prerecorded footage for use in a program; collectively called "stock footage."

Take: A single attempt to record a shot.

Telephoto: A lens setting that magnifies subjects and reduces apparent depth.

Trimming: Removing unwanted material from the beginning and/or end of a shot.

Voice-over: Spoken commentary on the sound track from someone who is not in the image.

Wide-angle: A lens setting that reduces subjects in size and exaggerates apparent depth.

What sounds would you include in the background track for this shot?

9 Program Development

Objectives

After studying this chapter, you will be able to:

- Discuss a video program in terms of subject, objectives, audience, delivery system, scale, scope, length, and concept.
- Select appropriate formats for developing the program.
- Write a treatment for a short video program.
- Develop a storyboard illustrating a program sequence.

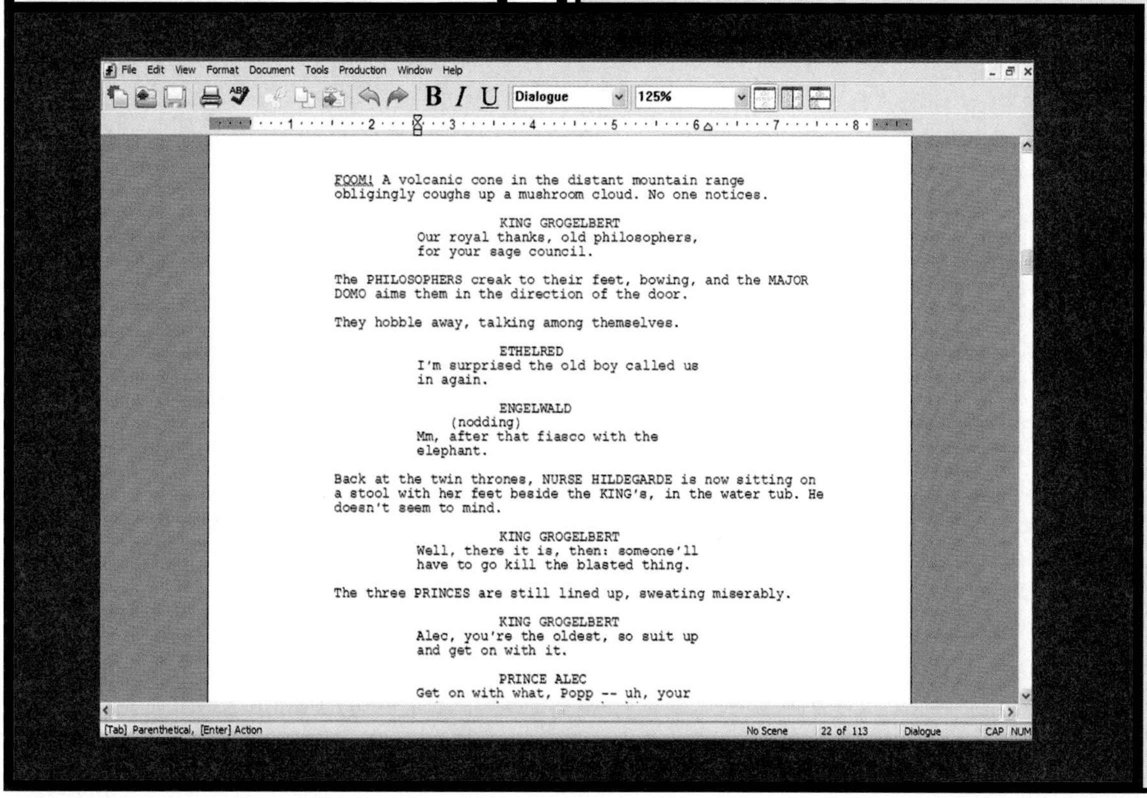

About Program Development

In Hollywood, a movie being prepared for production is said to be "in development." Why not just say that it is being "scripted" or simply "written"? The reason is that no matter how important a script may be, it is only an end product in a much larger creative process: the process of program development.

This chapter covers the fundamentals of this vital process. We will start with a common sense approach to defining the video program you wish to produce. Then we will look at various ways to create a design for your program, a design intended to guide you and your colleagues through production and postproduction.

If you propose to shoot a short simple video, you may be tempted to skip the development process. If, for example, you are making a music video about a garage rock band, you may feel that all you have to do is record the band playing a song and then edit the results. This approach is almost guaranteed to produce a less-than-satisfactory program. Even as simple a program as that garage band video can benefit from the process of development.

Defining the Program

You begin the development process by defining the program. The first step is to describe exactly the video that you want to make. This may sound obvious, but far too many programs suffer because they were poorly defined to begin with. By developing a detailed blueprint of your program, you provide the information you need to plan, shoot, and edit your video.

To illustrate the steps in defining a program we will use a promotional video for a fictional product. Imagine that you have been hired by Acme Power Tools, Inc., to produce a short program that explains their new cordless electric drill, the Sidewinder. Acme will distribute your video to hardware and building supply stores.

At first, the approach appears obvious: simply show what the Acme Sidewinder drill can do. But there is much more than that to defining this program. To draw a detailed blueprint, we need to specify the program's *subject*, *objectives*, *audience*, *delivery system*, *length*, and *budget.*

Subject

The first step is to identify the subject matter, and the easiest way to do this is by assigning your program a temporary working title that announces its topic. *Our Camping Trip, Ed and Darlene Get Married, Warehouse Operations, The Lions Club in Our Community*—each of these titles summarizes the content of a proposed video.

For your finished program, you may replace the literal working title with a more imaginative alternative.

Later, you will further limit and refine your subject; but even at this first step, your title may suggest that your topic is too unfocused. For instance, the title Warehouse Operations may tip you off that your topic is too broad for a single training program. To reduce it to manageable size, you might select one important part of warehouse operations and plan a program titled *How to Drive a Forklift.*

In the case of Acme Power Tools, Inc., your imaginary client, the subject/title is obvious: *The Sidewinder Cordless Drill,* **Figure 9-1**.

Figure 9-1.
The Sidewinder cordless drill.

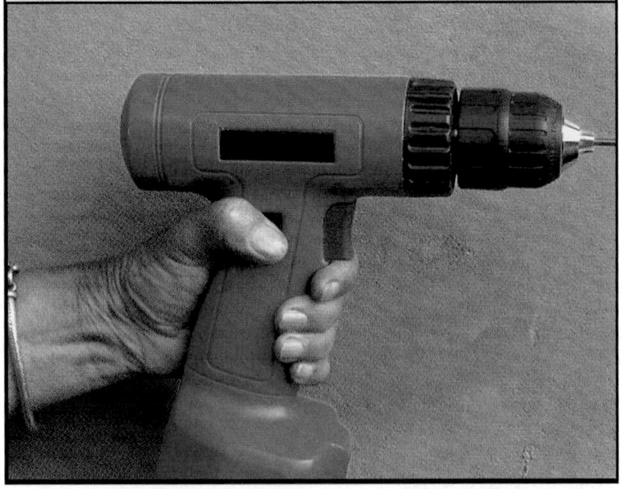

Objectives

The next step is to identify the client's objectives: what does Acme want this program to achieve? The obvious answer is to make customers buy Sidewinder drills. But that is only a start. Some potential buyers may not even think that cordless drills are very useful. Others may understand that such drills are worth considering, but they are leaning toward drills made by Acme's competitors. These facts suggest that two obvious objectives for your program are to persuade customers that:

- a cordless drill is a tool worth buying, and
- the Acme Sidewinder is the best cordless drill to buy.

Notice that these objectives are clear, specific, and simple; and they are phrased in terms of their effect on viewers. An objective like, "to demonstrate cordless drills" merely seeks to present information. By contrast, "to persuade customers that a cordless drill is worth buying," is an improvement because the objective in this form intends to actively move the viewer.

Even a personal video benefits from at least one clear objective. For instance, the objective for a child's birthday party video might be to communicate the love and care that went into creating the event.

Finally, these objectives are few in number and they fit together logically. To see why limiting objectives is important, imagine that Acme Power Tools initially wants two additional objectives: to increase customer interest in do-it-yourself projects in general, and to develop positive feelings toward the company. But you realize that general "do-it-yourselfing" would make the subject too broad to cover properly, and that the Acme public image is not directly relevant to product usefulness. You convince your client to limit the program's objectives to those that specifically sell the Sidewinder drill.

Audience

Now, to whom do you want to sell the drill? Different audiences will watch your video with different interests, different prejudices, different amounts and types of information. To communicate effectively, you must identify your target audience and address your video directly to it. To see how this works, imagine that you are preparing two programs: *How to Choose a Retirement Community*, an informational video for older people, and *Power on the Water*, a commercial aimed at speedboat buyers. See **Figure 9-2**. The senior audience will probably appreciate a straightforward, low-pressure presentation of their retirement options. The potential boat buyers will respond to dynamic, exciting shots of action on the water, backed by pulsing music and the howl of big engines.

Imagine what would happen if you delivered a loud and frantic video to the mature viewers and a steady, systematic presentation to the boat racers. Both groups would reject your message. To succeed, your video must be carefully tailored for its target audience.

To return to the Sidewinder cordless drill, pretend that in discussing the program with the Acme marketing department, you discover at least four different groups of potential viewers:

Figure 9-2.
Retirement community vs. power boating. (Corel)

- *Buyers for hardware and building supply chains.* These people will decide whether to offer the Sidewinder for sale, and if so, how to promote it against drills made by Acme's competitors.
- *Potential customers.* These people will decide whether they want to buy a cordless drill, and if so, whether the Acme Sidewinder is their best choice.
- *Female customers.* Unlike some males, these potential buyers are not dependably attracted to tools merely because they are intriguing adult toys. Instead, they want to know what the Sidewinder drill will do, how easily it will do it, and why doing this will be useful to them, **Figure 9-3**.

Figure 9-3.
Female customers want to know how well the product works.

- *Canadian customers.* Some of these viewers speak French rather than English.

Once you have identified your potential audience, you can make some informed decisions. First, you know that Acme wants to sell its drills in Canada, so you will need to plan for a second sound track in French. Secondly, store buyers have interests that are quite different from those of customers, so you eliminate them as primary targets. Next, you realize that it would be impractical to provide a separate program for female viewers, but you decide to address their interests and concerns as you aim your program at potential customers of both genders.

By comparing Acme's objectives to groups of possible viewers, you have decided that your primary audience consists of potential drill buyers of both genders in all of North America. As you continue developing your program, you will try to speak directly to this target audience.

Delivery System

The next question is, how and where will this audience see your video? The answer will determine many things about your program. Acme has asked for what is called a "point-of-sale" video, a program to be shown in stores to customers who are ready to decide right then and there whether to buy a Sidewinder drill. That situation requires a particular method of displaying your program, a particular ***delivery system***. It probably involves a video monitor or flat screen placed on a high stand or in a special Acme display kiosk in an aisle in a retail store, with customers strolling past it as they shop, **Figure 9-4**.

Figure 9-4.
A point-of-sale display.

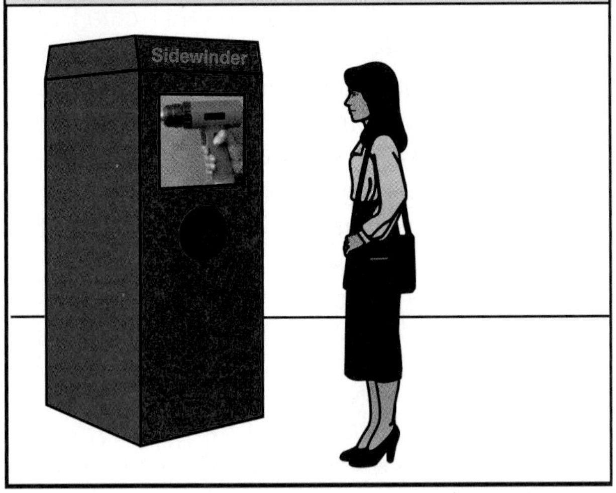

This delivery system imposes certain requirements on your program:

- It must be vivid enough to make passing shoppers stop and watch it.
- It must be visually simple, so that people can see it clearly on a small screen that may be several feet away.
- It must not depend too heavily on its sound track, since the audio may be degraded or even lost in the noise of the store.
- It must be nonlinear in its organization, so that customers need not watch it from the beginning in order to understand its essential message.

- Above all, it must be short—brief enough so that customers (who did not come to the store to watch TV) will remain for its whole message.

In contrast to a point-of-sale program, a training video designed to teach people how to use the Sidewinder might be watched by an employee in a quiet room or by a do-it-yourselfer in the comfort of a home den. With those very different delivery systems, these restrictions would not apply. Whatever form it may take, your program's delivery system (the situation in which it will be shown) should have a strong influence on its approach and style.

Program Length

The next step is to determine your video's length. Program running time may be governed by one or more of several factors:

- *Standard time units.* In broadcast and cable TV, for example, programs and commercials alike are always produced in standardized lengths, **Figure 9-5**.

Figure 9-5.
TV program segments are standard lengths, to accommodate commercials.

- *Resources.* As a rule, the longer a video runs, the more it costs to produce; so your program's length may be affected by the size of your budget.
- *Audience tolerance.* The length of the audience's attention span depends on the type of program you are making. Viewers might enjoy a movie for two hours or more. An infomercial about a line of products might hold them for 30 minutes at most. Training videos are hard to sustain for longer than 10 or 15 minutes, and commercials are too intense to hold up much longer than 60 seconds.
- *Subject matter.* Within the program types that are not standardized in length, the actual duration of a video may depend on the extensiveness of its content.

Superficially, subject matter might seem to be the first determiner of length, rather than the last. It would appear logical to allow a video to run as long as it takes to cover the chosen content with appropriate thoroughness. This is often how publishers set the lengths of nonfiction books such as this one. A book may have, say, 300 pages, or 600 or 1,200, depending on the extent of its subject. But a book is a random-access document. You can dip into it anywhere you please, read it at your own pace, pick it up and put it down when you feel like it. The length of a book is not critically important because the reader controls how much of it to absorb at any one time.

The video viewer, by contrast, cannot control the speed at which information is delivered, and most programs are designed to be watched continuously, at a single sitting. For these reasons, the lengths of video programs are influenced less by their subjects than by the other factors discussed previously.

Of course, if your intended delivery system is a computer-based interactive video or Web site, you will want to design it for a medium in which viewers do control access to the content.

Considering how these factors influence your Acme Sidewinder video, you realize that busy shoppers are not going to stand in front of a TV screen for even five minutes. On the other hand, you cannot demonstrate the major benefits of the new drill in a 30-second commercial. With this in mind, you set a program length of about three minutes.

Budget

Finally, you need to consider budget: the amount of money available to produce your program. Budget constraints will determine both what you can do and how you can do it.

What you can do

For the Sidewinder drill program, the original concept was to use fast motion techniques to show several women building an office building, using Sidewinder drills. However, you soon realize that a lengthy location shoot in a real construction site would be far too expensive, **Figure 9-6**.

Instead, you develop a less expensive concept, involving three performers in a kitchen.

Figure 9-6.
Women constructing an office building would be too expensive to produce. (Corel)

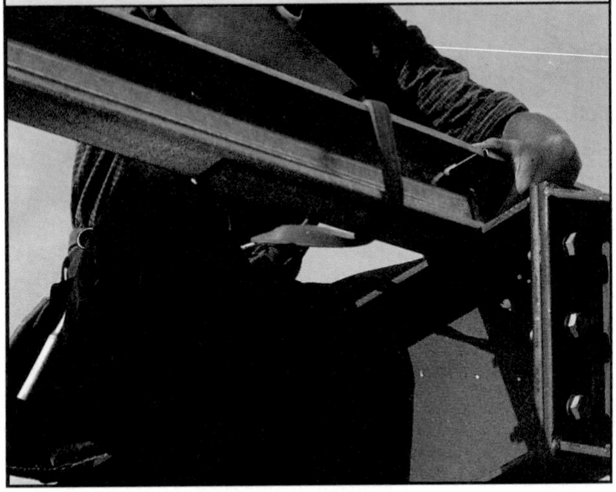

How you can do it

With any concept, there are often alternative ways to produce the program. For example, your first version of the kitchen concept might be to shoot the Sidewinder program in the elaborate kitchen of an expensive home. But when you add up the costs of renting the location, lighting the kitchen, and transporting cast, crew, and equipment, you realize that it will probably cost less to design a kitchen set and build it on a sound stage instead. However, when you cost out this plan, you discover that a newly designed set would still be too expensive. Instead, you can rent a stock kitchen set inexpensively from the studio where you will shoot. And so, by repeatedly revising your concept and estimating costs, you arrive at a plan that fits your budget.

Even the most expensive Hollywood productions must usually adjust the concept to fit the budget.

Selecting a Concept

When you have specified the objectives, the audience, the delivery system, and the length of your program, you have developed a nearly

Figure 9-7.
The subhead of this news story reveals the concept: David Slays Goliath.

McKinley Beats Fillmore High

Underdogs Rally to Defeat League Champs

Sparked by quarterback Charley Folsome, the McKinley Presidents rallied in the third quarter to power home two touchdowns that opened a lead Fillmore was never able to close. When the ground attack pursued in the first half failed against the bigger Fillmore line, Folsome took to the air, completing three passes to teammates Bruce Petit, Mickey Fleschner, and Norm Simas.

Boosters Club Donates Scoreboard

At a ceremony during the pep rally before Saturday's game, McKinley Boosters Club President Amanda Wang turned on the all digital

complete profile of your intended video. With these determining factors firmly in mind, you are *almost* ready to start developing your program. But you still need to come up with a program concept. A *concept* is an organizing principle, an idea that gives shape and meaning to your video. It determines what you include and how you treat it. You could say that the concept guides your approach to your subject or your perspective on it or simply your "angle," **Figure 9-7**. However you describe it, the concept of your video program enables you to make a coherent statement that your audience can understand and respond to. These program organizers are simple to demonstrate by examples, so the sidebar *Sample Program Concepts* suggests plausible concepts for various types of personal and professional videos.

In the case of the Sidewinder cordless drill, a possible concept is hidden in the already-selected objectives and audience. Acme Power Tools, Inc., wants to communicate the message that their drill is the best choice and they want to appeal to female buyers as well as male. Considering these two desires together, you develop a concept: *Get your own Sidewinder so your husband can have his back.*

With the concept in place, the whole program suggests itself: a group of brief scenes in which a husband keeps asking for his Sidewinder and his wife keeps promising to give it

Sample Program Concepts

To provide a clear idea of what program concepts are and how they can guide the video maker's approach to developing programs, here are six examples: three personal programs and three professional ones.

Personal Programs

Program type: Vacation video
Subject/working title: "Lake Omigosh Vacation."
Concept: Triumphing over rain.
Summary: Video focuses on comic results of trying to camp out during two solid weeks of bad weather.

Program type: Holiday
Subject/working title: "Easter Egg Hunt."
Concept: The great Easter egg deception.
Summary: Before letting the kids find the eggs, the parents repeatedly hide them in spots where the children have already looked.

Program type: Family oral history
Subject/working title: "Grandfather Reminisces."
Concept: Family connections survive despite long times and great distances.
Summary: Questions and Grandfather's answers focus on family continuity across three continents and two hundred years.

Grandfather reminisces.

Professional programs

Program type: Training video
Subject/working title: "Using Your Multiline Business Phone."
Concept: Conquering fear of buttons.
Summary: Humorous acknowledgment that business phones can be complex and frustrating evolves into the idea that a little study clears up the confusion.

Program type: Community service promotional
Subject/working title: "The McKinley Boosters."
Concept: Communities depend on volunteers.
Summary: The Boosters Club preserves and beautifies the town of McKinley.

Program type: Wedding video
Subject/working title: "The Anders/Goldstone Wedding."
Concept: As the glass flies apart, two lives come together.
Summary: Starting with a slow-motion shot of a goblet shattering under a cloth, the video moves from this symbol to documenting a traditional Jewish wedding.

back as soon as she's finished with it. In each scene, she uses the drill in a different type of project, demonstrating its simplicity and versatility as she does so.

Preparing a Treatment

If you are planning a short, personal video, you can probably start production as soon as you have settled on the content, length, and above all, concept. Most professional videos, by contrast, require that your informal design be transcribed onto paper as a script, a storyboard, or a treatment, **Figure 9-8**.

Figure 9-8.
A treatment is an outline in narrative form.

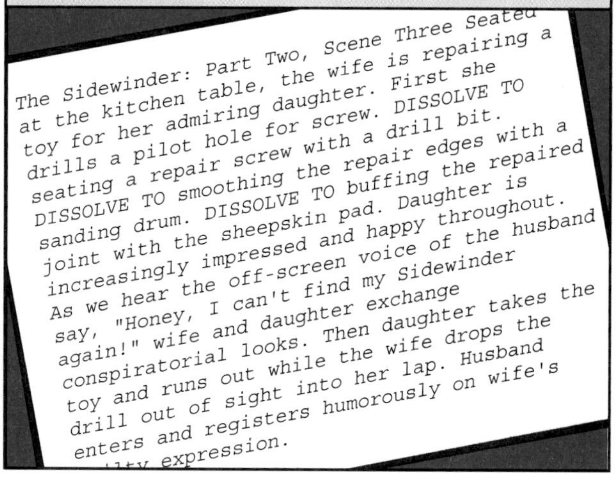

The Sidewinder: Part Two, Scene Three Seated at the kitchen table, the wife is repairing a toy for her admiring daughter. First she drills a pilot hole for screw. DISSOLVE TO seating a repair screw with a drill bit. DISSOLVE TO smoothing the repair edges with a sanding drum. DISSOLVE TO buffing the repaired joint with the sheepskin pad. Daughter is increasingly impressed and happy throughout. As we hear the off-screen voice of the husband say, "Honey, I can't find my Sidewinder again!" wife and daughter exchange conspiratorial looks. Then daughter takes the toy and runs out while the wife drops the drill out of sight into her lap. Husband enters and registers humorously on wife's expression.

Uses for Program Treatments

The simplest transcription is a ***treatment:*** a few paragraphs that explain the program's concept, its subject, its order of content presentation, and its style. A treatment serves two purposes: it allows you to see and evaluate the organization of your video, and it communicates your plan to others. By committing your ideas to paper, you can determine whether they are appropriate to your program and whether they flow smoothly and logically. By examining your treatment, you can spot problems and opportunities that might otherwise be overlooked.

A treatment is also useful in communicating your vision to other people, especially the colleagues who will help you produce your program and the clients who will pay for it. Without a treatment's overview of the program, your crew can only make one blind shot after another, without knowing how they should fit together and what they should achieve. As for the clients, few if any will underwrite your production without a clear idea of the program you propose to deliver.

Most clients demand a full script rather than just a summary treatment.

Levels of Treatment

Video program treatments have no fixed style or length. They can be a one-sentence statement of concept and content or a multi-paragraph synopsis. They can be an outline so detailed that it identifies every separate content component. Whatever their level of detail, program treatments attempt to convey the effect of finished videos. Here are samples of program treatments developed to three different levels of detail. Each is for the Sidewinder drill program.

A skeletal treatment

This treatment covers all three parts of the Sidewinder video, in the briefest possible form:

The Sidewinder Drill

Part One: A succession of quick scenes shows the many jobs performed by the drill. Part Two: Several vignettes in which a husband is frustrated because his wife is always using his Sidewinder drill. Part Three: After he presents her with her own Sidewinder, the two of them collaborate happily on a construction project.

A summary treatment

This excerpt from a summary-level treatment covers just one-third (Part Two) of the skeletal treatment in greater detail:

The Sidewinder, Part Two

Scene One: Husband asks where his Sidewinder is as we see wife using it to repair kitchen cabinet hinge. Scene Two: As wife assembles picnic table bench, husband appears and again asks where his Sidewinder is. Scene Three: Wife is repairing child's toy at kitchen table when she hears husband asking where drill is. As he appears in kitchen doorway, she hides drill in lap.

A detailed treatment

This excerpt from a detailed treatment covers only Part Two, Scene Three, as summarized in the previous version.

The Sidewinder: Part Two, Scene Three

Seated at the kitchen table, the wife is repairing a toy for her admiring daughter. First she drills a pilot hole for screw. DISSOLVE TO seating a repair screw with a drill bit. DISSOLVE TO smoothing the repair edges with a sanding drum. DISSOLVE TO buffing the repaired joint with the sheepskin pad. Daughter is increasingly impressed and happy throughout. As we hear the off-screen voice of the husband say, "Honey, I can't find my Sidewinder again!," wife and daughter exchange conspiratorial looks. Then daughter takes the toy and runs out while the wife drops the drill out of sight into her lap. Husband enters and registers humorously on wife's guilty expression.

The amount of detail in your own treatments will depend on how minutely you need to previsualize your program and how completely you want to communicate it to clients and colleagues.

Creating a Storyboard

The old saying claims that one picture is worth a thousand words, and this is often true in developing video programs. In graphic-based program design, a succession of pictures

A Storyboard Sequence

Here is a storyboard sequence as an advertising agency might create it for the Sidewinder Drill program.

1. SFX: drill.

2. SFX: drill continues.

3. SFX: drilling stops. **DISSOLVE TO:**

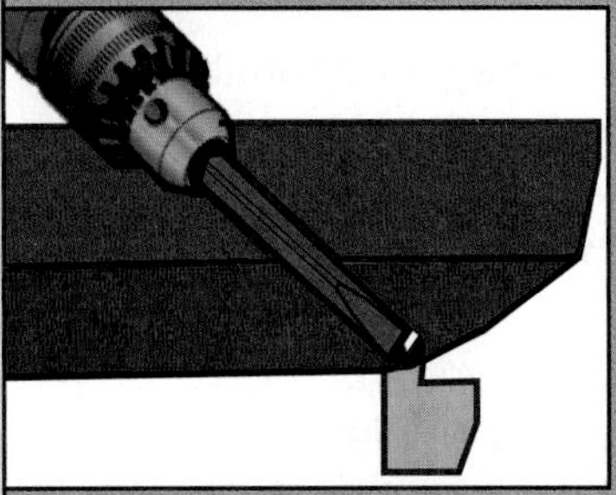

4. SFX: Slower sound of power screwdriver. **DISSOLVE TO:**

5. SFX: Abrasive sound of sanding. **DISSOLVE TO:**

6. SFX: Sanding continues. **DISSOLVE TO:**

7. SFX: Softer whine of sheepskin pad. Then, **HUSBAND (OS):** "Honey, I can't find my Sidewinder again!"

8. SFX: Drill noise stops.

9.

10. SFX: daughter's chair scraping back.

11. SFX: girl's footsteps.

12. HUSBAND (OS but closer): "Seems like every time I need that drill..."

13. HUSBAND: "It disappears."

14.

15. HUSBAND (meaningfully) "Doesn't it?"

resembling a comic book sketches all the important moments in the program. This script in picture form is called a ***storyboard***.

Storyboards got their name from the bulletin boards on which the drawings of scenes for animated cartoons often are pinned for inspection and editing.

Storyboard Uses

Storyboards have two main uses: to help others visualize the look of the eventual program and to preplan complex sequences shot-by-shot.

Visualization

Storyboards are particularly valuable for communicating content to clients and crew, because they present concrete images instead of the abstract words that describe them. Some people have less talent than others for thinking graphically. The problem is that they are often unaware of their inability to visualize, and so they indicate understanding of written descriptions when they really can't imagine them. The result can be serious miscommunication. A storyboard presents the images in previsualized form, along with captions containing dialogue, sound effects, and descriptions of the action.

Shot planning

Where complex visual sequences are involved, storyboarding can help you as well as others. By planning all camera shots in advance, you can see how well they will edit together and how clearly they will communicate their content. Many directors make extensive use of storyboards in their productions, especially in laying out highly complex action sequences like fights and chases.

Writing a Script

Another way to lay out a detailed production design is by writing a full ***script***. A script describes every sequence in your program, including both video and audio components. A script is especially useful if the program contains dialogue to be memorized and spoken by actors and/or voiceover text to be read by an off-screen narrator. Scripts are also a common alternative to storyboards for presenting programs to clients. They are especially valuable for planning and budgeting production, since they include every element of the program in a compact narrative form.

The Scripting Process

Writing a script can follow any system—or no system at all. In commercial and industrial production, however, the process of creating a script often breaks down into five stages:

1. Producer and client agree upon the program's content and concept.
2. A detailed content outline is written. This is often critiqued by the client and then revised by the writer.
3. A first draft script is written. This is the initial attempt to lay out a complete production script. Usually the client reviews this draft and orders revisions.
4. A revised draft of the script incorporates the client's changes.
5. The revised script draft is reviewed by the client and further changes are ordered. If all is going well, these changes do not require a complete third draft, but only a refinement of the second one.

Which Method Is Best?

Professional video makers employ all the forms of program development that we have covered, sometimes mixing and matching them as needed. For example, a program may be documented completely in script form, which is supplemented by storyboards of action sequences or other activities that demand precise visual preplanning. When you make fairly short, simple programs it is usually enough to write down your concept, develop a narrative treatment that covers the major components of your video, and perhaps storyboard critical sequences for camera angles and continuity. As your productions grow in scope and complexity, you will probably move up to fully scripted programs, with or without more extensive storyboarding. Whatever form or forms you choose, the result is the blueprint from which you construct your video.

A Nonfiction Script Format for Word Processing

The traditional nonfiction script layout consists of two vertical columns with the visuals on one side of the page and the dialogue, narration, and other audio on the other side. The problem is that this side-by-side layout can be inconvenient to use in a word processing program. For this reason, nonfiction scripts like the following excerpt are now often formatted over-and-under instead of side-by-side. (Some scriptwriting software, such as *Final Draft*, can automatically format two columns.)

25.
VIDEO Insert: Sanding the repair.
AUDIO SFX: Sanding sound.

26.
VIDEO Admiring daughter, as before.
AUDIO SFX: Sanding continues.

DISSOLVE TO:

27.
VIDEO Insert: Buffing the repair.
AUDIO **Husband (OS):** Honey, I can't find my Sidewinder again!
SFX (under) Buffing sound.

Scripting Programs

As you can see from any week's TV listings, video programs come in many different varieties, including fictional series and movies, news and sports features, documentaries, training programs, and commercials, to name just some of the more common types.

Some of these program types are discussed in Chapter 8, Video Communication.

The scripts for these program types have characteristic structures, each evolved through many years of trial and error. To thoroughly discuss each script structure would require a separate book for each one. We can, however, offer writing hints for a few of the more common types. These include story videos, training programs, documentaries, and news features.

Story Videos

Fiction videos are dramas (or comic dramas), and they typically employ dramatic structures that have been used for many centuries.

The Athenian Greeks had developed the art of dramatic structure by about 400 BC.

In general, dramatic story structure calls for an opening in which the story is introduced and launched, a middle section in which the story unfolds and conflict intensifies, and a concluding section in which the conflict reaches a critical point and is finally resolved. Though this "three-act" organization is not the only dramatic structure, Hollywood has found that it works equally well for short movies and long ones. So when you write a story script, you might try using this structure.

Story programs made for commercial TV are generally arranged so that an exciting moment precedes every commercial break. Considered as a whole, however, their action tends to divide into the same three general phases as other fiction programs.

Act one

Taking about one-fourth of the total story time, Act one introduces the characters, the setting, and the situation or event that starts the dramatic action. Here is a simple example.

Bob and Bill are good friends, except that Bob is involved with Nancy and Bill would like to be. A big dance is announced. Bill decides to win Nancy away from Bob and take her to the dance.

The dramatic action begins when Bill makes that decision, **Figure 9-9**.

Figure 9-9.
Bob and Bill are called the protagonist and antagonist in the conflict.

Act two

Consuming about half the program time, Act two shows the escalating conflict between opposing forces in the story.

Bill works to get Nancy to notice him. Bob discovers he has a rival and tries to get a commitment from Nancy. Taking advantage of the situation, Nancy starts playing Bob against Bill. Each one offers competing inducements like a dinner before the dance, a stretch limo, etc. Step by step, the conflict grows more intense, all the way up to the day of the dance. Not wanting to hurt either Bob or Bill , Nancy says yes to both of them–but how can she attend the dance with both escorts?

Act three

The last one-fourth of the story leads to the climax of the conflict and then shows how it is resolved.

Making an excuse to meet Bob at the dance, Nancy arrives with Bill, then uses one improvisation after another to shuttle back and forth between Bill and Bob. The crisis occurs when they discover the trick and confront each other. Going outside, Bob and Bill wrestle each other to a draw, ruining their clothes and inflicting considerable damage. Finally, with the honor of each one upheld, Bob and Bill go off as friends again, leaving Nancy alone at the dance.

In summary, then:

- Act one (1/4) introduces the characters and the situation that starts the conflict.
- Act two (1/2) intensifies the conflict with every event, building up tension.
- Act three (1/4) brings the conflict to a climax and then resolves the action into a satisfying conclusion, **Figure 9-10**.

Figure 9-10.
In graphic form, a dramatic action might look something like this.

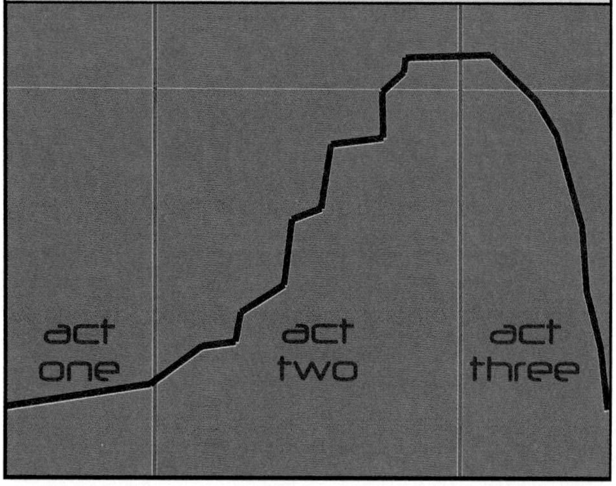

Notice that dramatic stories are almost always about *conflict*. Without a clash between opposite sides, the drama is missing from the story.

If the Bob and Bill video were to run 20 minutes, then act one would take about 5 minutes, act two would need around 10, and act three would unfold in the final 5. Though this popular three-act structure may seem to reduce stories to formulas, it is a perfectly sound way to organize a fiction video script; and since it has worked well for 2,500 years, it has certainly proven its effectiveness.

The three-act structure is not the *only* way to organize a story, so do not worry if your script does not follow it perfectly.

Training Programs

Programs intended to teach certain sets of knowledge and skills are easy to organize because the subject matter tends to do the job for you, **Figure 9-11**.

Some programs use chronological order. If your topic is, for example, how to assemble a model airplane, then the program will cover each construction step in the correct sequence. Other programs divide up the content by type. In a program on

Figure 9-11.
Training videos are used to teach job skills.

Figure 9-12.
NARRATOR: In this program, you'll learn about making fountain drinks, making sandwiches, and using the cash register.

how to clerk in a soda fountain, topic types might include making drinks like sodas and milk shakes, assembling sandwiches, and ringing up sales. Each would have its own section of the program.

The problem with training videos is that they cover the subject at the speed set by the director—a speed that will be too fast for some learners and two slow for others. With printed materials, learners can move at their own most comfortable pace, slowing down for difficult passages, flipping back to review materials already covered, and skimming things they already know.

Videos, however, move at a certain fixed pace, and (despite DVD chapters and fast-forward and reverse controls) learners are forced to move at that pace. Also, it is difficult to keep the content of a training video organized in the learner's mind. At various moments, viewers worry: *What did we already cover? Where does this part fit in the whole? What is left to learn?*

To address these difficulties, producers have developed three techniques so effective that they can be found in almost all educational programs. These techniques are *The three Ts*, *Signposts*, and *Lists*.

The three Ts

The easy way to remember the basics of training program structure is by recalling the so-called "three Ts:" **T**ell them what you are going to tell them; **T**ell them; **T**ell them what you have told them.

- *Tell them what you are going to tell them.* Begin the instructional body of the program by listing the topics and/or having them read aloud by the narrator, **Figure 9-12**. Because some viewers learn better aurally and others visually, it is generally best to use both titles and narration.

  In Figure 9-11, notice that the narrator's language closely follows the words on the title. If the narration used different language, such as, "creating drinks and sandwiches and processing customer payments," viewers would have to reconcile the differing words on the screen and on the track –distracting their attention from the program content. Even when you read them here on the page, it takes a moment to realize that the two different versions say the same thing:

  - making fountain drinks, making sandwiches, and using the cash register.
  - *creating drinks and sandwiches and processing customer payments.*

- *Tell them.* When you have listed the topics of the program, start presenting the live-action coverage of the first topic and continue through the training content, **Figure 9-13**.
- *Tell them what you have told them.* When you have finished presenting the training material, summarize what viewers have learned (often by repeating the opening organizer slide), **Figure 9-14**.

The three-T system organizes the material for viewers and continually reminds them of where they are in the presentation.

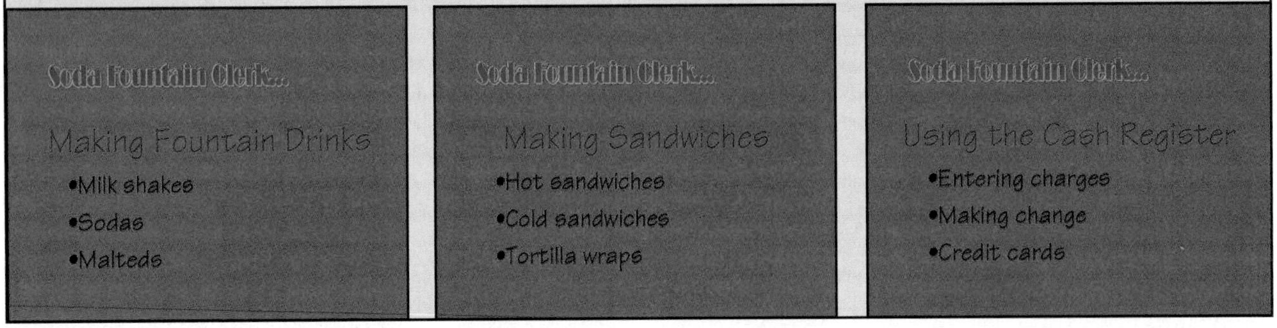

Figure 9-13.
Note that the headlines of the topic titles match the language of the "What You'll Learn" organizer title.

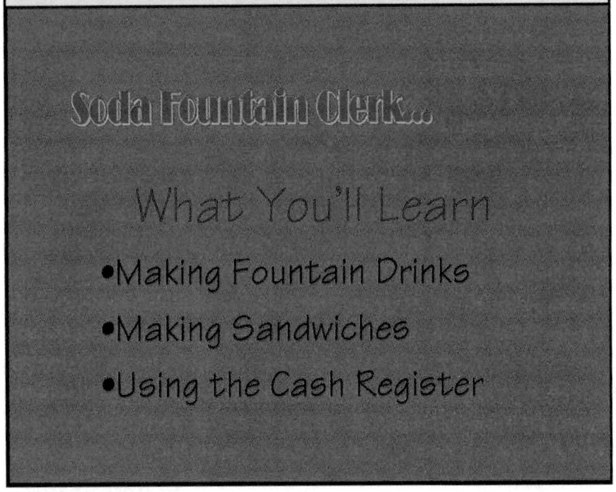

Figure 9-14.
NARRATOR: And so, we have covered making fountain drinks, making sandwiches, and using the cash register.

Signposts

Signposts are additional organizers that orient viewers by reminding them of where they have been and where they are going, **Figure 9-15**.

Figure 9-15.
Signposts within the program point in both directions.

Typically, they are created by repeating the introductory title with the new topic highlighted, **Figure 9-16**.

Lists

These title cards are examples of lists. Introductory lists present the major topics. Topic lists break out the key points within the topics. Buildups mark the progress through the topic content. Again, the narrator also delivers the key information, using the same language as the title. (See the sidebar, *Using Lists,* for more details about lists and buildups.)

Figure 9-16.
NARRATOR: Now that we've covered fountain drinks, let's look at making sandwiches.

Documentaries and News Features

A ***documentary*** is a program that selects, organizes, and presents factual material on a particular subject. The makers of so-called ***"cinema verité"*** documentaries claim to show events exactly as they unfold before the camcorder, without any script at all.

Using Lists

Training video lists are simply content outlines broken down and presented in separate parts, to help viewers organize program content.

Introductory lists set out the major sections of the program content.

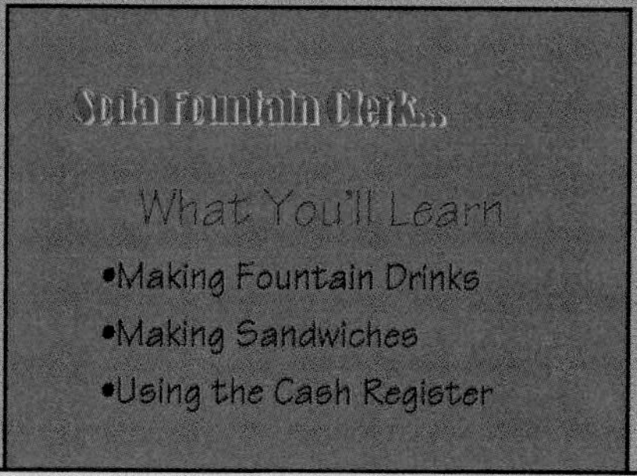

Topic lists include the key points in single topics.

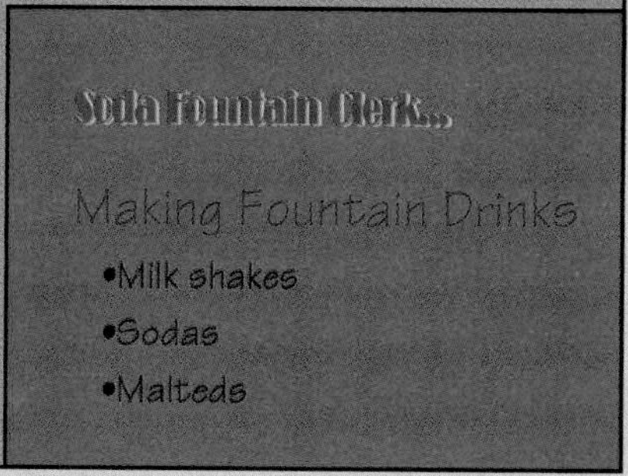

Buildups are titles that add a content line as each new subtopic is introduced. That way, the new subtopic, highlighted by a contrasting color, is seen in the context of the whole list.

The video starts with the program organizer list.

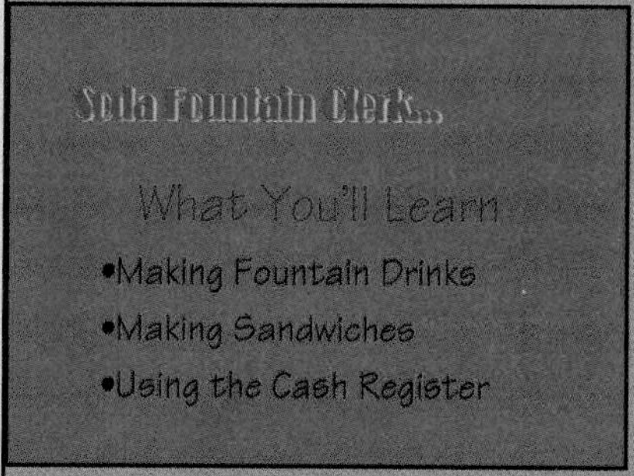

The program organizer list is repeated with just the first topic. It is highlighted in green.

The first topic organizer list appears.

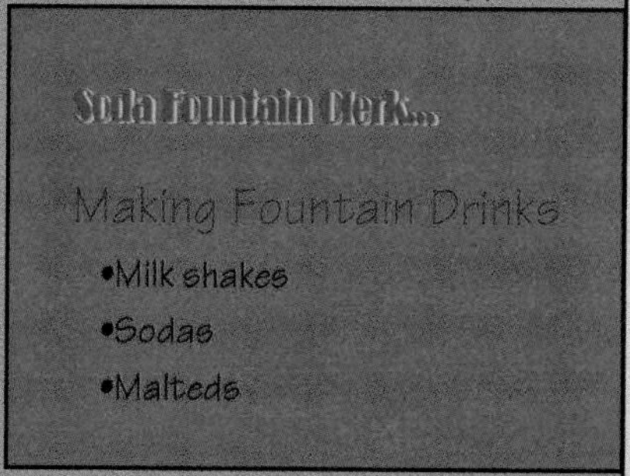

Subtopics on sodas and malteds disappear and the subtopic on milk-shakes is highlighted.

At the transition to the next subtopic, the second subtopic title is added and highlighted, while the first subtopic reverts to black.

The final subtopic appears, highlighted, and the previous ones are black.

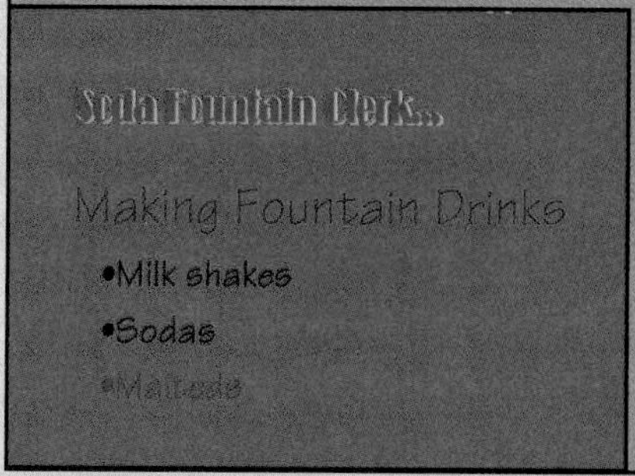

The program organizer list reappears with the second major topic added and highlighted.

The second topic organizer appears. This sequence is repeated throughout the program.

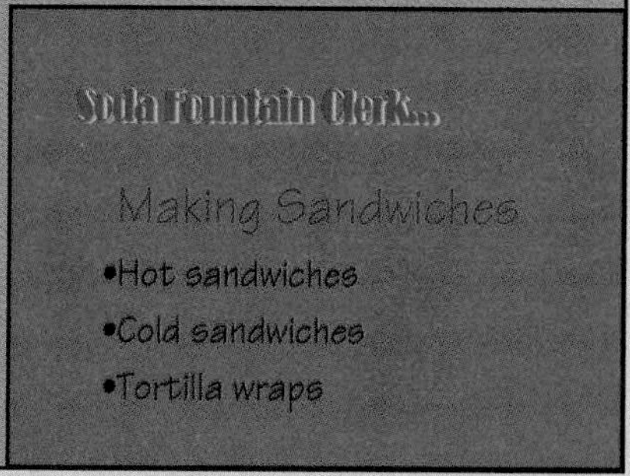

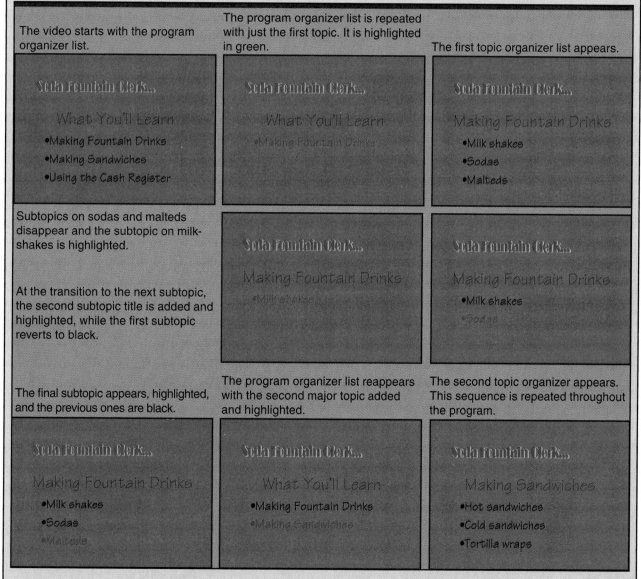

These titles are easily created in presentation software such as PowerPoint™, and then imported into the video during postproduction.

While it is true that some education and training videos do not lend themselves to this approach, it works well for most programs.

As explained in Chapter 8, even the most nonjudgmental verité documentary makers actually "script" their material informally by choosing what and what not to shoot, and then selecting, trimming, and sequencing the footage that appears in the finished program.

On the other hand, typical documentaries of the kind seen on cable channels and public television are usually fully written before shooting begins. These informational documentaries on history, **Figure 9-17**, biography, science and medicine, wildlife, travel, public affairs, and many other subjects are as tightly scripted as any story video or training program.

Between these extremes lie *news features*, both short "standup" pieces on daily news programs and longer segments on weekly news feature shows like *60 Minutes*.

In a *standup* feature, a reporter narrates the piece while appearing on camera, at the scene.

The producers of these features shape the raw footage and provide commentary that is often scripted after the fact. In some cases, extra material is written and videotaped to supplement the documentary footage. Interviews are conducted using questions written beforehand to enhance the subject matter. This complex process might be termed "interactive scripting."

Since *verité* documentaries are not formally scripted, we will focus on informational documentaries and news features.

Documentaries

As you prepare to write a script, remember that most documentaries are designed to be entertaining.

Because these documentaries are considered informational entertainment, they are often labeled "infotainment."

This means that you can employ many of the techniques developed for story programs, including the three-act structure discussed earlier. Some subjects lend themselves to conflict and suspense *(Will the expedition make it across the hostile desert?*). Many subjects unfold over time, so you can organize them as if they were stories (*How a frog egg becomes a tadpole and then a frog*).

Be aware, however, that fiction techniques are dishonest when they falsify the facts of documentary subjects. For example, if that supposedly hostile desert is really only one day's journey across, and the season is cool, it is misleading to present the desert as a dangerous antagonist.

Also, it is easy to falsify behavior. Suppose, for example, you show a seal suddenly raising her head in the air, **Figure 9-18**.

If you place that shot after a clip of a baby seal barking, you are probably honest. The two shots were not recorded in sequence, the pup may not be barking at anything important, and the baby may not even belong to this mother. Nonetheless, baby seals do bark and mother seals do notice their cries.

Figure 9-17.
American history is a popular documentary subject. (Corel)

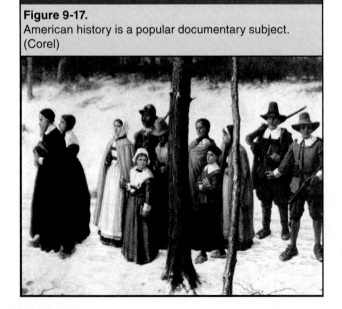

Figure 9-18.
The seal raises her head and looks screen left. (Corel)

Figure 9-19.
The program switches back and forth among four groups of animals. (Corel)

But suppose instead, that you do not have a shot of the seal pup, so you write voiceover narration: *Sensing that she has been away too long, the mother seal suddenly remembers her pup*. That is less than honest, because you do not really know what the mother was thinking, or even if she is capable of such thoughts.

In organizing a documentary subject, it helps to orient viewers at the beginning, so that they understand what the program is about and why they are looking at it. The three-T approach often works well, if it is handled informally. Instead of displaying title cards with topic build-ups, use voiceover narration for introductions and transitions from one topic to the next.

Cross cutting can be particularly effective, especially when you are following two or more sets of subjects. Suppose, for example, your topic is a year with the wildlife on a certain stretch of seashore. The main script organizer is the seasons, but within each one, you could switch back and forth among the seals, birds, otters, and even the dolphins offshore, **Figure 9-19**.

News features

Like some documentaries, news and sports features are often scripted both before, during, and after shooting. (This is especially true of the longer features made for "news magazine" shows.) To see how this works, we will follow the progress of a news feature about the Leibniz County Fair.

Fictional Leibniz County is the home of McKinley, a town to be introduced in Chapter 10.

Before shooting: Prior to sending out a news crew, the producer develops a ***concept*** for the feature (for more on concepts, see the *Sample Program Concepts* sidebar earlier in this chapter). The concept: *Ranching is the heart of Leibniz County economy*. The concept tells the reporter and crew to focus their shooting on the livestock events at the fair, **Figure 9-20**.

Figure 9-20.
The livestock barns at the Leibniz County Fair.

Figure 9-21.
The reporter tapes scripted narration.

Figure 9-23.
The edited feature is now entirely about the Jersey heifer contest. (Sue Stinson)

During shooting: At the fair, the news crew discovers that the dairy cow competition is both photogenic and easy to shoot, so the reporter writes (or at least outlines) and records standup narration to explain that dairy farming is a good example of Leibniz County ranching, **Figure 9-21**.

After shooting: Back in the editing suite, the producer determines that the best footage was shot during the judging of Jersey heifers, and furthermore, the competition generated suspense and human interest. So voiceover narration is written to focus attention on this, **Figure 9-22**.

If time and circumstances permit, the reporter is sent back to the fair to record on-camera narration and, if possible, interview the mother/daughter team that won the competition.

In this typical example, a feature script is roughed in before shooting, focused more narrowly during production according to the shooting opportunities offered, and finalized during postproduction to fit the concept as refined during editing. The result *looks* as if the production team set out originally to follow one competitor in the heifer judging, when (in fact) they did not know precisely what their feature was about until the postproduction phase, **Figure 9-23**.

In so-called *"breaking news"* such as crimes, accidents, and weather disasters, the scripting process is even more informal. In these cases, the production team gathers whatever footage it can, a reporter delivers an on-the-spot standup narration to explain it, and the producer writes

Figure 9-22.
The feature focuses on the Jersey heifer contest. (Sue Stinson)

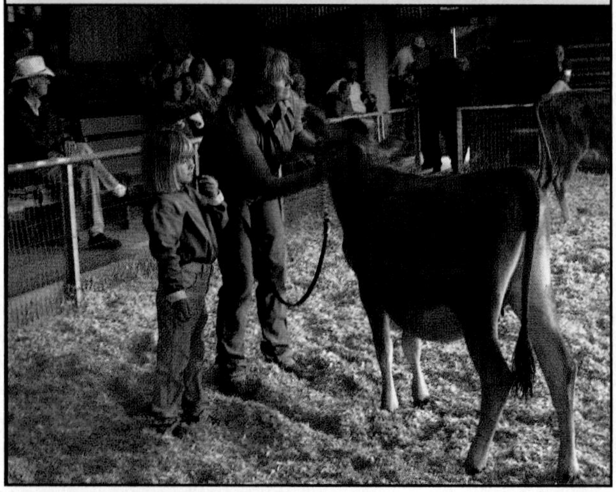

Figure 9-24.
A professional reporter can improvise from notes made at the scene of the breaking news.

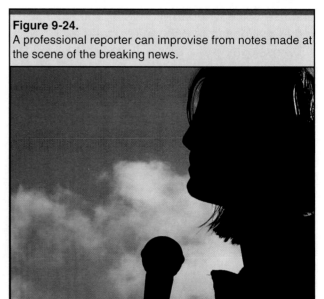

follow-up copy to be read by newscasters on the air, **Figure 9-24**.

Nonfiction Program Elements

The standup report delivered on-camera is just one of several components commonly used in documentaries and news features. These video, audio, and graphics resources have by now become standardized.

Video Resources

Since video is essentially a visual medium, the most important component should be what appears on the screen.

Production footage

Live-action footage of the subject typically provides the most compelling visuals.

Library footage

Where live action material is unobtainable (as in historical documentaries), archival footage can often be found in stock film rental libraries. Documentaries about 20th century wars are typically assembled from ***library footage***, **Figure 9-25**.

Figure 9-25.
Library footage from World War II. (Corel)

Reconstructed footage

Often it is possible to re-stage and videotape actions that happened in the past. When using ***reconstructed footage***, the ethical producer uses one of several methods to stylize the footage, to indicate that it is not real. (For more on shooting re-enacted events, see the sidebar.)

Audio Resources

Audio is especially important in documentary and news programs, sometimes more important than the accompanying video (see the sidebar, *Creating Wallpaper*).

Production sound

Nothing adds more realism to visuals than live, synchronized sound. The problem is that

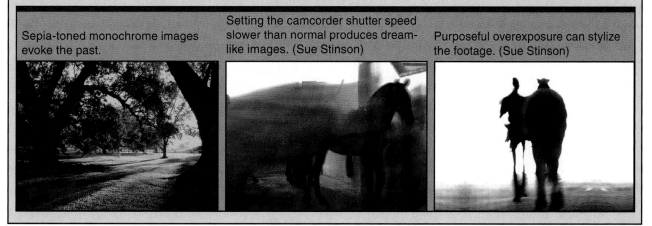

Shooting re-enacted events

Documentaries often dramatize past events by re-staging them. To make it clear that the results are not actual footage, editors typically distinguish them by altering them in postproduction. Here are a few of the more popular methods of processing staged footage.

Sepia-toned monochrome images evoke the past.

Setting the camcorder shutter speed slower than normal produces dream-like images. (Sue Stinson)

Purposeful overexposure can stylize the footage. (Sue Stinson)

Figure 9-26.
A background track of wind and wave sounds can tie together a variety of shots. (Corel).

Figure 9-27.
Most interviews now show only the subject. (Corel)

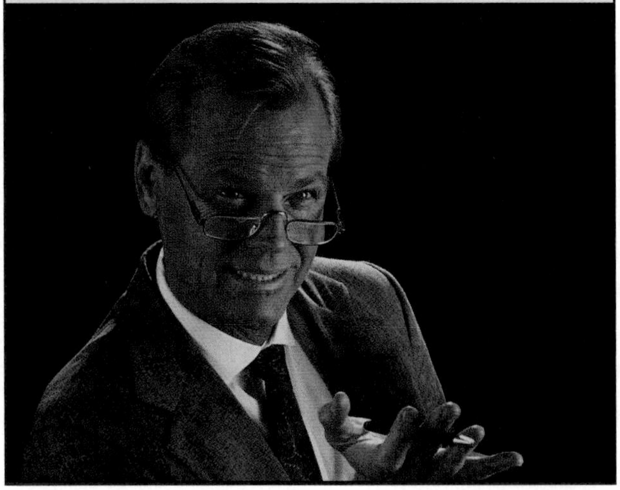

documentary shooting does not always allow the time or the opportunity for obtaining optimal sound quality. For this reason, it is especially important to record a track of general background sounds at each location, to use in smoothing out shot-to-shot differences in the production sound, **Figure 9-26**.

Sound effects

Sound effects add atmosphere and realism. In rostrum camera footage (see sidebar, *Storytelling with a Rostrum Camera*), sound effects can give the illusion that the events suggested by the still images are actually happening.

Interviews

Interviews are the backbone of many documentaries. Today, most interviews use an "invisible" interviewer. The reporter is never shown and the questions are edited out, **Figure 9-27**. Interviewees are coached to give answers that do not sound like responses to questions. For example, the reporter would not ask, "When did you start working with lions?" That might elicit an answer like "Ten years ago," which would make no sense when the question was removed in editing. Instead, the question is phrased, "Tell us how and when you started working with lions (and be sure to include *working with lions* in your answer). Now the interviewee might respond, "I started working with lions ten years ago and I've studied them ever since." That answer will be clear, even when the question is removed.

In many cases, the interview starts with the subject on camera, then switches to video footage of the topic, while the interview responses continue on the sound track as voice-over narration.

Sit-Down or on the Fly?

In both straight documentaries and docudramas, conventional interviews are often called "sit-downs," for obvious reasons.

To impart a more dynamic feeling to the essentially static interview format, directors often shoot them as "on-the-fly" interviews, in which the subject talks while doing something else.

In the most common type of on-the-fly interview, the subject is driving while talking. This setting is preferred for two reasons:

- Outside shots of the moving vehicle provide quick and easy cutaways,
- Though the vehicle is moving, the subject and videographer (often including lighting and sound equipment) are conveniently locked together because both are sitting in the front seats.

Lighting, miking, and camcorder operations are more difficult when the subject is moving around in a location.

On-camera presenter

In news features, especially, the reporter (***presenter***) often begins with on-screen narration, to introduce the visuals, then reverts to voiceover narration.

Sometimes this narration is a combination of field recording and studio material added later.

Narration

Whether the source is an interview subject, a reporter, or a professional voice-over performer, narration is the most important audio component, often carrying the majority of information. (See an example of this under the "Narration" heading in Chapter 7.)

Graphics Resources

In addition to such video resources as production footage, library footage, and reconstructed footage, programs may use such graphic resources as rostrum camera footage, maps and diagrams, and computer graphics.

Rostrum camera footage

A rostrum camera is a rig that allows you to videotape photographs and other two-dimensional objects, imparting an appearance of movement by zooming in and out and panning from one picture detail to another. Specialized and expensive, rostrum setups are being replaced by software that enables you to digitize images, pan and zoom around them with your computer, and then import the finished sequences into your program, **Figure 9-28**.

Creating "wallpaper"

Documentaries often need to convey abstract information—usually by voiceover narration. Since abstract concepts have no physical existence, they are impossible to videotape. But because the screen cannot be blank, the scriptwriter must invent visuals that appear to be plausibly connected to the narration—even if they do not show what is actually being talked about on the sound track. This type of seemingly (but not actually) related visual is often called ***wallpaper***.

To demonstrate wallpaper technique, here is part of a documentary on the Pacific island nation of Vanuatu.

NARRATOR: Lacking both natural resources and heavy manufacturing,

the people of Vanuatu have become experts at repairing and conserving

the infrastructure of their transportation system,

which—in a nation distributed among dozens of islands—

means ships and boats.

Here, workers restore a small inter-island freighter.

Obviously, the boat-painting sequence receives more attention than it deserves; but its job is to use up screen time while the narration discusses the abstract concept of transportation infrastructure.

Figure 9-28.
From the wide shot of the painting, the camera pans and zooms to frame the locomotive. (Corel)

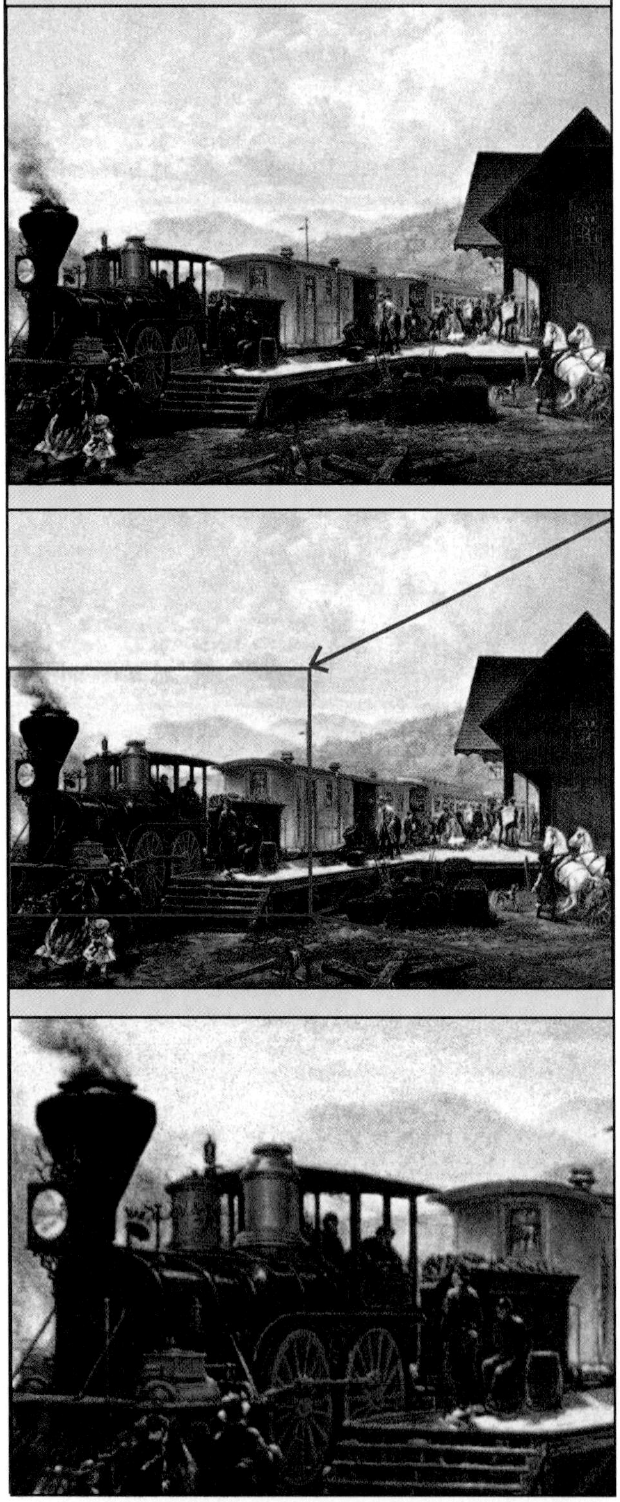

Storytelling with a rostrum camera

Ingenious rostrum camera techniques can illustrate complex stories with a minimum of graphic resources. Here, an engraving of an early 19th century sea battle...

A 19th century engraving scanned into a computer. (Corel)

...is used with voice-over narration and supporting sound effects to create an entire sequence.

Each red frame surrounds one of the images that follow.

Maps and diagrams

These common graphics are useful for orienting viewers and visualizing abstract ideas. As with photographs and artwork, these are now being digitized (or created in graphics software applications) and then set in motion by "virtual rostrum" software, **Figure 9-29**.

NARRATOR: The sea battle lasted seven hours and involved over 20 ships.

The cannon fire...
SFX: Cannon shot.

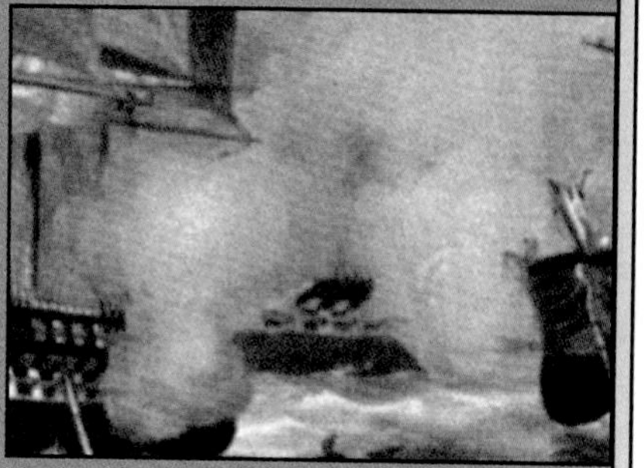

...from opposing frigates...
SFX: Cannon shot.

...was deafening.
SFX: Cannon shot.

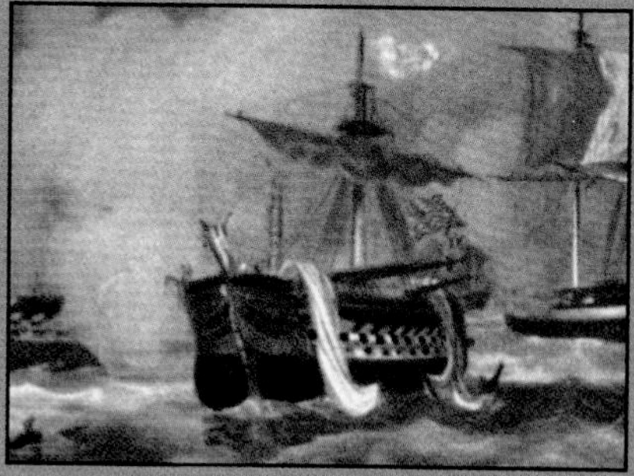

Dismasted early in the melee, the U.S.S. Indomitable wallowed helplessly...

...while some of her crew escaped in lifeboats. Other sailors, however...

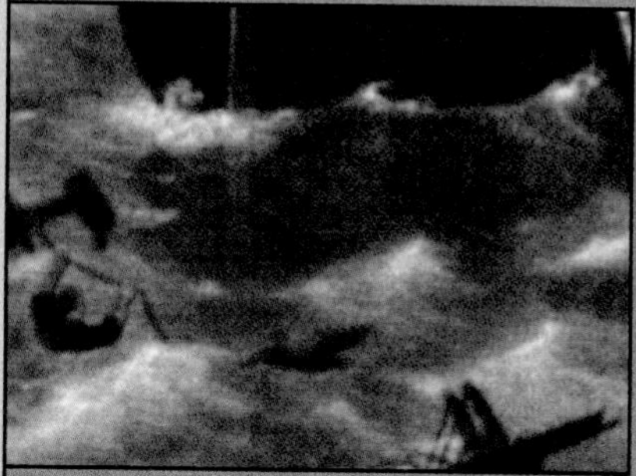

...were not so lucky.

By dusk, the fleet had taken a terrible pounding...

...but the Americans refused to strike their colors.

Notice that the last two images have been tinted blue in postproduction to suggest night. A sequence like this is scripted completely in advance, including both narration, images, and "camera" movement.

Computer graphics

Because documentary programs are often necessarily slower-moving and short on dramatic live action footage, lively computer graphics, titles, and transitions are often added to enhance the visual energy. In general, nonfiction programs of all types allow more varied and obvious effects than fiction programs, where excessive computer graphics distract viewer attention from the story.

From Development to Planning

With any type of video program, developing treatments, storyboards, and scripts is only the first half of preproduction. Once the program has been written, the next phase is the thorough planning needed to ensure a successful shoot.

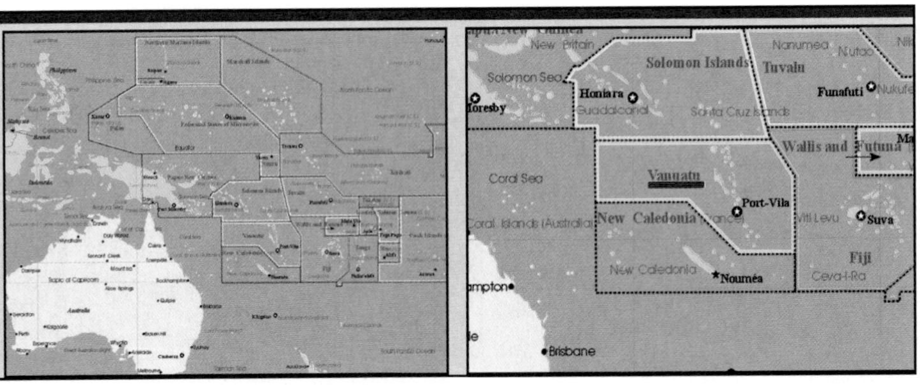

Figure 9-29. A broad area of the Pacific...

...zooms in to locate the nation of Vanuatu. (Corel)

Chapter Review

Answer the following questions on a separate piece of paper. Do not write in this book.

1. *True or False?* Program development is only needed for larger productions.
2. The document that describes a video program's concept, subject, order of content, and style is called a ______.
3. ______ and shot planning are the main uses for a storyboard.
4. In a three-act program structure, Act Two consumes about one-______ of the total running time.
5. The "three Ts" used in structuring training programs are "tell them what you are going to tell them," "tell them," and ______.

Technical Terms

Breaking news: News that is covered as it is happening (or soon afterward).

Buildup: A sequence of title cards, each one adding a new line of information.

Cinema verité: A type of documentary attempting extreme fidelity to actual events, with a minimum of intervention by its makers (often referred to as simply "*verité*").

Concept: The organizing principle behind an effective program. Often called an angle, perspective, or slant.

Conflict: The struggle between opposing sides that creates dramatic action.

Delivery system: The method (such as web site, TV monitor, or kiosk) by which a program will be presented, and the situation (alone at a desk, in a training room, in a crowded store) in which it will be watched.

Documentary: A program dealing with factual material.

Dramatic structure: The organization of a story to build interest and excitement.

Infotainment: Documentaries intended to entertain as much as to inform.

Library footage: Film or video collected, organized, and maintained to be rented for use in documentary programs (also called *stock footage*).

Presenter: An on-camera narrator who speaks directly to the viewer.

Reconstructed footage: Reenactments of past events so that they can be videotaped.

Script: A full written documentation of a program, including scenes, dialogue, narration, stage directions, and effects, that is formatted like a play.

Standup: A report presented on camera, usually by a reporter.

Storyboard: Program documentation in graphic panels, like a comic book, with or without dialogue, narration, stage directions, and effects.

Treatment: A written summary of a program, formatted as narrative prose, that may be as short as one paragraph or as long as a scene-by-scene description.

Wallpaper: Footage intended to take up screen time while the narration presents material that cannot be shown.

10 Production Planning

Objectives

After studying this chapter, you will be able to:

- Determine production requirements for a small-scale video production.
- Assemble people, equipment, and other resources for the production.
- Develop a production schedule.
- Calculate costs and budget the production.

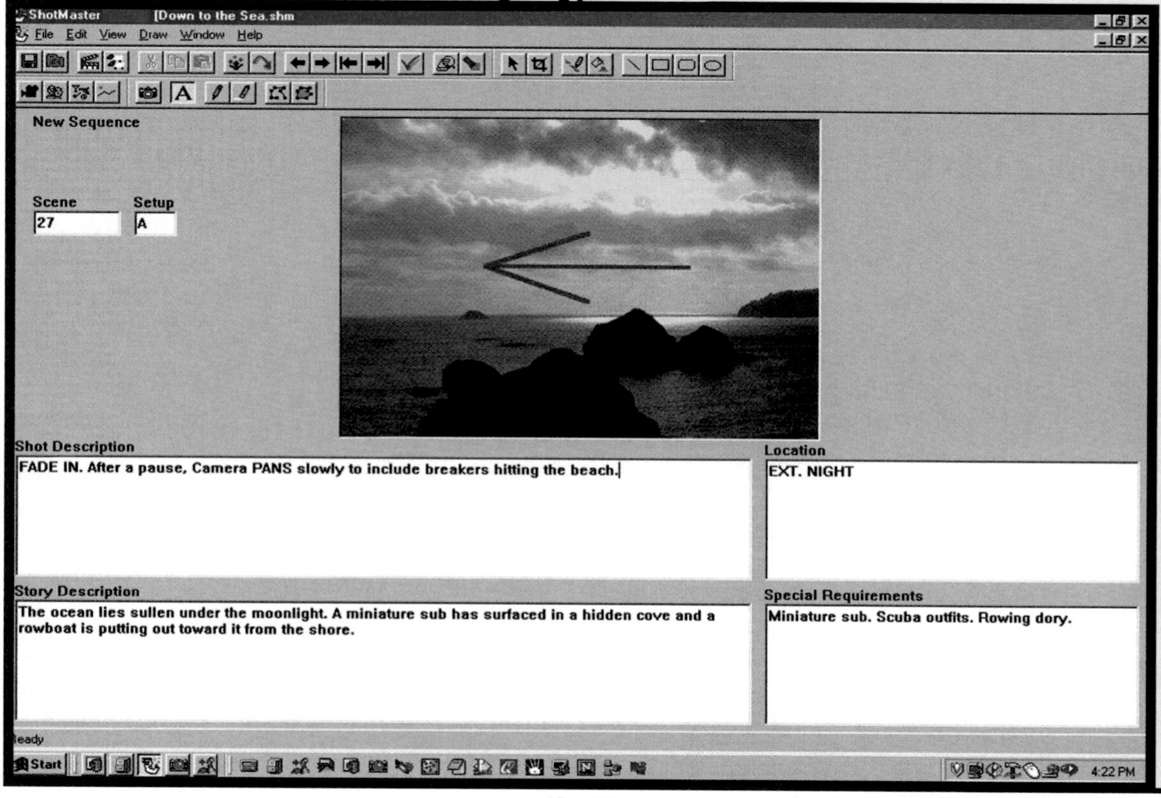

ShotMaster [Down to the Sea.shm]
File Edit View Draw Window Help
New Sequence
Scene
27
Setup
A
Shot Description
FADE IN. After a pause, Camera PANS slowly to include breakers hitting the beach.
Location
EXT. NIGHT
Story Description
The ocean lies sullen under the moonlight. A miniature sub has surfaced in a hidden cove and a rowboat is putting out toward it from the shore.
Special Requirements
Miniature sub. Scuba outfits. Rowing dory.
Start
4:22 PM

(Shot Master storyboard software, The Badham Company.)

About Production Planning

When the content of your proposed video is well developed, you are ready for *production planning*—the process that will organize and manage the complexities of shooting. Planning is critically important for two reasons: time and money. For example, you may have permission to shoot in a key location for only eight hours (time). Or, you may be renting equipment or paying crew and talent by the day (money). In professional video production the phrase, "time is money" is literally true because nearly everything (and everyone) is paid for by the week, day, or hour.

Thorough production planning can make the difference between a smooth, successful shoot and a disaster. To show you how to plan effectively, this chapter will take you through the process of preparing to produce a simple video.

The Big Moment

The importance of production planning becomes obvious if you think of an actual video shoot as similar to a theatrical performance. With a stage play, many people spend weeks conducting rehearsals, building sets, sewing costumes, and publicizing the show. All this lengthy preparation is for the two or three hours of the play's actual performance. During that critical period there should be no major mistakes, because all that preproduction effort has paid off in a show that works.

Though even a simple video may take longer than two or three hours to shoot, its production phase is much like a theatrical performance: a short, intense period when all the planning and preparation are rewarded. That planning involves:

- *People:* the cast and crew of the production.
- *Places:* the locations and/or sets where shooting will take place.
- *Things:* the equipment used to shoot the program, the props and costumes in it, and miscellaneous items like talent and location releases, **Figure 10-1**.
- *Plans:* the scheduling and other organizational processes that ensure a smooth shoot.
- *Budgets:* the predicted production costs.

The rest of this chapter examines each of these planning topics, in turn.

Figure 10-1.
Finding, renting, and scheduling this vintage car for a sunset shoot in the desert required extensive planning. (Corel)

People: Crew and Cast

In most professional video productions, the personnel involved demand the largest organizational effort and consume the majority of the budget. Some of these people appear in the production. Other equally important personnel work behind the camera.

The people who appear in video programs, whether actors, narrators, interviewees, or others, are often collectively called the talent.

If you make personal or small-scale professional programs, finding and recruiting people can be difficult because most amateurs lack the technical skills that video production demands and most producers lack the money to pay professional-level salaries. Here are some ideas for solving these problems when gathering people for low-budget videos.

Finding Production People

Even the simplest video project may be too much work to handle alone, so the first step is to find some assistants. The sidebar *Small Video Crews* describes three *crews*. The crew duties listed should give you an idea of how to assign responsibilities to your crew members.

Where do crew members come from? The first resource is always friends and family, especially for jobs that do not require too much technical re likely candidates for your crew.

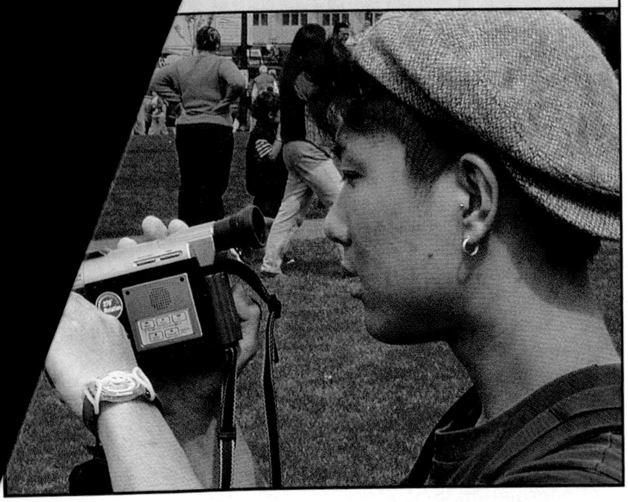

expertise. For example, a camera operator needs to know some techniques for obtaining good quality images, **Figure 10-2**, but a production manager can draw upon skills learned in almost any managerial position.

When you do need technical expertise, the first place to look is in community theater or local clubs that feature music groups. Both activities attract talented amateurs to design and operate lighting schemes and mix multitrack audio. If your project is for an organization—whether a business, a church, a school, or a community group—look for assistants within that organization. After all, group members are likely to be interested in the success of your project because their organization stands to benefit from it.

Figure 10-3.
Local theater groups are good sources of talent. (Corel)

In the case of the McKinley Boosters (introduced in the sidebar, *Case Study: The McKinley Boosters Club*), you might find a member who is interested in audio recording to become your sound person. A local amateur theater technician might want to learn video lighting. The business and professional members should offer any number of executives willing to manage the production (particularly if given a title like Producer). Other members might be called on to help provide transportation, hold reflectors during outdoor shooting, and perform other nontechnical duties. Still other club members, of course, will provide most of your on-camera talent: the cast, **Figure 10-3**.

Casting the Production

In personal and entry-level professional productions, you may not have much choice in casting the parts in your production. Your strategy is to make the most of the choices that you do have. To cast your show successfully:

- Do not always cast the obvious person. For roles (unlike "Father" or "Mother") that are not gender-specific, cast males and females even-handedly.
- In casting a real-world character, do not limit your choice to the actual person. For example, the current president of the McKinley Boosters Club is a nice man but not a very dynamic speaker. The former president, however, is a woman with a fine platform presence and an ability to read speeches convincingly. For these reasons, you cast her as the president and spokesperson in your video. (After all, by the time the video is completed and released, the current president will be out of office too, so it makes no difference whom you cast in the role.)
- Find out in advance whether people have other obligations that might conflict with your shooting schedule. No-show actors can bring your entire production to a halt.

In this book, "actor" refers to both genders.

Conducting Effective Auditio

Though you may have no choice
a personal family program or a prof
like a wedding, many video projec

Small Video Crews

Video production crews can be as large as 20 members or more, e small-scale projects, however, are crewed by just a few people — or ous reasons, members of small crews often do several jobs at on

2-Person Crew	
Title	**Responsibilities**
Producer/director/ videographer	**Producing:** Organizing and scheduling production, ca **Directing:** Determining camera setups and shaping the p **Videography:** Operating camera. **Lighting:** Designing and executing lighting with reflectors and/
Assistant	**Preproduction management:** Tracking cast members. **Sound recording:** Miking talent. Wielding the microphone boom. C with production audio mixer. **Camera/lighting assistant:** Helping transport and set up video equipn adjusting reflectors and lights as directed by producer. **Continuity:** Keeping track of shots and setups. Ensuring consistency of shot c action from shot to shot. Recording shot information for editor. **Transportation:** Taking people and equipment to and from locations, as necessary.

4-Person Crew	
Title	**Responsibilities**
Producer/director	**Producing:** Organizing and scheduling production. **Directing:** Determining camera setups and shaping the performances of cast.
Videographer	**Videography:** Operating camera. **Lighting:** Designing and executing lighting with reflectors and/or lighting units.
Video Crew	**Sound recording:** Miking talent. Wielding the microphone boom. Controlling sound levels with production audio mixer. **Camera/lighting assistant:** Setting and adjusting reflectors and lights as directed by producer.
Production Manager	**Continuity:** Keeping track of shots and setups. Ensuring consistency of shot contents and action from shot to shot. Recording shot information for editor. **Management:** Organization and tracking of shooting schedule and all contributing components. **Makeup/costumes/production design:** Designing and applying cast makeup. Selecting, fitting, and managing cast costumes. Determining overall visual effect of production. **Transportation:** Taking people and equipment to and from locations, as necessary.

10-Person Crew	
Title	**Responsibilities**
Producer	**Producing:** Organizing and scheduling production.
Director	**Directing:** Determining camera setups and shaping the performances of cast.
Videographer	**Videography:** Operating camera. **Lighting:** Designing and executing lighting with reflectors and/or lighting units.
Gaffer	**Lighting assistance:** Setting and adjusting reflectors and lights as directed by producer.
Asst. Camera/Grip	**Camera assistance:** Help with camera setups, pull focus, perform scene carpentry.
Sound Recordist	**Sound recording:** Miking talent. Wielding the microphone boom. Controlling sound levels with production audio mixer.
Script Supervisor	**Continuity:** Keeping track of shots and setups. Ensuring consistency of shot contents and action from shot to shot. Recording shot information for editor.
Production Manager	**Management:** Organization and tracking of shooting schedule and all contributing components. **Transportation:** Taking people and equipment to and from locations, as necessary.
Makeup/Costume	**Makeup:** Designing and applying cast makeup. **Costume:** Selecting, fitting, and managing cast costumes.
Production Designer	**Design:** Determining overall visual effect of production. **Props:** Dressing location and/or set shooting environments with properties.

Case Study: The McKinley Boosters Club

Throughout this chapter, we illustrate the discussion with a fictional organization called the McKinley Boosters Club.

In the small city of McKinley, the Boosters Club is an organization of business and professional men and women who meet for lunch weekly to discuss civic beautification and other urban improvements. The Boosters conduct fund-raisers for park benches, playground equipment, and landscaping projects. They work to save historic buildings and return them to productive use. On weekends, they spread out in squads to collect litter, clean up graffiti, and paint houses in low-income areas.

Now, they want a ten-minute video program to explain their activities, in order to recruit more members, raise money, and generally increase civic awareness of what they do. Imagine that, as a proud member of the McKinley Boosters Club, you have volunteered to produce this show for them.

President McKinley welcomes visitors to the Plaza.

to find appropriate people to fill on-screen roles. To match these roles to the right performers, you may wish to hold auditions at which people try out for parts in the program. You can simplify the tryout process by using these suggestions:

- Ensure privacy. Never ask people to read for a role in front of other applicants for the same part. Instead, have applicants wait in one room while you hold tryouts in another.
- Videotape all acting candidates to see how they look on screen. People often appear different on video, sometimes surprisingly so. Also, by recording tryouts you can later refresh your memory of each applicant (and they do tend to blur in your mind very quickly). Finally, you can edit the footage of various candidates to see how they look together. Suppose, for instance, that you have chosen your Romeo, but three different actors might make good Juliets. By intercutting audition shots of Romeo with footage of each different Juliet, you can see which combination works best.
- Tell actors not to look at the camcorder—and then see whether they can follow this direction. Some amateurs will spoil take after take by unconsciously looking into the lens.
- Give auditioners simple things to do, such as entering a room and sitting down or pouring and serving a glass of water. This will show you how naturally they can behave on camera.

Handling Talent

Auditions are psychologically punishing to all actors because they involve so much rejection, **Figure 10-4**. Only one applicant can be chosen for each role, so if ten people try out for a part, you will have to reject *90 percent* of them. In doing so, you appear to be rejecting them, not as performers but as people.

Think of it this way: if you turn down an applicant for a lathe operator position you are judging her metalworking skills. If you do not select a violin player for your orchestra you are evaluating his skills as an instrumentalist. But an actor's instrument is not a lathe or a violin; it is

Figure 10-4.
Auditioning can be stressful for performers.

his appearance, his voice, his personality—in short, his entire self. By rejecting the actor you may appear to reject the whole person. For this reason, it is important to support the people who audition for you.

First, give them adequate time. Even if you can see at once that applicants are unsuitable, give them time to show what they can do. It makes them feel that they have been taken seriously—and it often reveals talent that you may want for a later production.

Tell auditioners very simply what you want them to do. An audition is not the place for elaborate explanations of character. For example, if the role is that of an unsuitable job applicant, simply tell the actor something like this: "You want this job so badly that you can't help talking too much." A talented applicant will know what to do with this direction.

Finally, show that you are paying close attention to their efforts. Watch them as you listen to them read. Use their names frequently in conversation. Find something in every performance that you can praise.

By treating auditioners decently, you also enhance your own reputation. Whether a production is cast from a village's amateur theater crowd or the ranks of Hollywood professionals, the performing community is always small and the word gets around. By proving yourself sensitive to actors' feelings, you will make them wish to work for you, both now and in future productions.

Evaluating Audition Results

When auditions are over, you will need to select your cast. Some video makers systematically list and judge the pros and cons of each applicant, **Figure 10-5**; others rely completely on instinct for placing the best person in each role.

No matter how you cast your program, be sure to base your decisions on the audition videotapes. As noted above, the impressions people make on tape are often quite different from the way they appear in real life. People who are dynamic and forceful in person may seem dull on screen, while others who are not impressive in life are remarkably effective on tape. It is said that "the camera loves" these people; you can evaluate that affection only by studying the audition tapes.

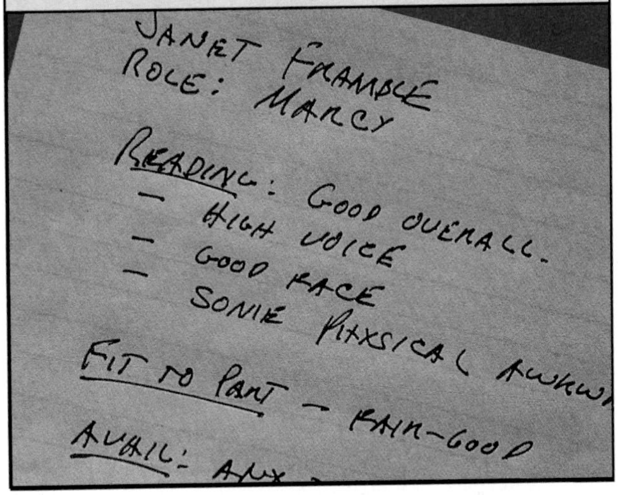

Figure 10-5.
Keep careful casting notes.

Places: Scouting Locations

In scouting a shooting location, you contact the local people whose help you will need, determine the suitability of the place for shooting, and look for potential problems that must be solved in order to conduct a successful shoot there.

Scouting People

First and foremost, you need to establish productive relationships with key people at any location. If they are on your side, these people can give you substantial help, but if they are uncooperative they can make life very difficult.

Cultivate the custodian or building engineer, who knows more about the facility than anyone else, including the locations of the electrical power box, spare furniture and other potential props and set decorations, **Figure 10-6**. For example, the McKinley Boosters Club can avoid high rental costs at their regular meeting room in the McKinley Arms Hotel by borrowing a lectern with the hotel logo on it and using it to stage the president's on-camera remarks at another location, outside the hotel. Who knows how to get hold of that lectern? The hotel's maintenance supervisor.

If your location is a public place, you may also need to coordinate with one or two other key people. See the executive in charge for permission to use the facility. For example, the Boosters Club might need to ask the hotel manager for permission to shoot establishing shots at a luncheon meeting. Often, management will

Managing Volunteer Helpers

Unreliable help is one of the biggest problems in producing personal or entry-level professional videos. If you have made videos with volunteers, you have probably experienced behavior like this:

- People arrive late and then announce that they must leave before the day's shooting is over.
- Crew members do not fulfill promises to do preproduction chores, such as finding props and costumes.
- Actors appear for later shooting sessions in clothes that do not match the outfits they wore in previous sessions.
- Crew members work slowly, or refuse to do jobs that do not appeal to them.
- Actors fail to memorize their lines, or, sometimes, even to read the script.
- Cast and crew members alike develop last-minute things to do instead, and fail to show up for shooting sessions.

Studio shoot, day one.

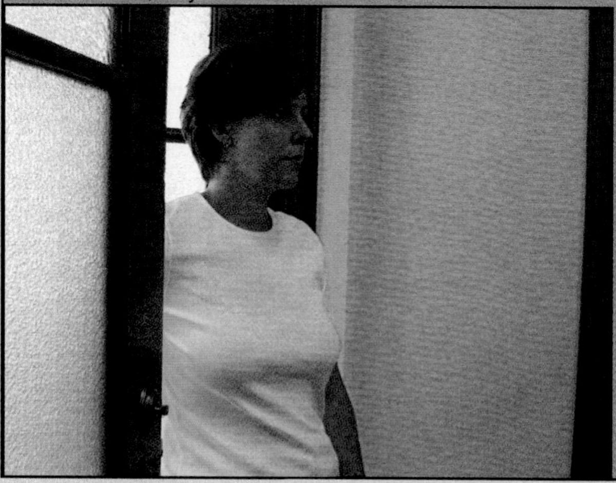

Day two (wrong shirt).

Why do volunteers often let the production down like this? Because they are not video professionals and do not pretend to be. Professionals have three powerful motivations for responsible behavior: they are being paid for their work, they will not be rehired if they develop a reputation for irresponsibility, and they enjoy their crafts and take personal pride in practicing them well. Amateurs, however, are motivated only by friendship with you, personal ego, and sometimes an interest in making a successful program.

Obtaining Loyalty

You may not be able to pay your cast and crew, but you can give them rewards that will help motivate them to work reliably and hard. These rewards are importance, attention, praise, credit, and ownership.

- *Importance.* Stress to cast members, in particular, that they are so essential to the production that when they are not present the production literally stops. So if they do not show up, many other volunteers like themselves are forced to stand idle. (Be careful in using this idea with certain actors, who may secretly enjoy this power to stop and start the whole production.)
- *Attention.* No matter how busy you may be, pay honest attention to the ideas and suggestions made by cast and crew. When people feel that their contributions are ignored, they lose their willingness to help in all areas.
- *Praise.* Notice and acknowledge good work. Praising actors is not too difficult, but your volunteer crew also needs recognition. A simple compliment like, "Great lighting setup — very realistic," may be all the pay that your lighting assistant will get.
- *Credit.* Before production starts, emphasize that cast and crew will receive on-screen credit for their contributions. People love to see their names in the credits.
- *Ownership.* Above all, create the idea that this is a team effort and the production belongs to everybody. In Shakespeare's Henry V, the king motivates his troops before a battle by calling them, "We few, we happy few, we band of brothers." That psychology works just as well today.

Figure 10-6.
Maintenance people know a building's systems.

Figure 10-7.
A secure place for recharging batteries is essential.

supply a person to accompany you throughout your shoot at the location, or at least to be on call when needed. For instance, the McKinley Arms Hotel might deputize the Assistant Manager for Banquets and Meetings to accompany the production crew in order to help out.

Consulting key people is important on personal projects as well. For instance, before moving furniture to get better camera angles on a birthday party, you might want to clear the idea with the party's host.

Scouting Facilities

You scout the facilities to ensure that you will have everything you need for production. First, check out the space where you will be shooting. Is there enough room? Can you create in it the on-screen environment you need (hotel room, office, dungeon, desert island)? Next, ensure that you will have access to a production area for storing equipment, costumes and props; for recharging camcorder and light batteries; for changing costumes, **Figure 10-7**. Is this area secure? Is there a phone that you are allowed to use?

Also, check the electrical service, starting with the power supply. Will the electricity fed to the shooting area power your camcorder, audio equipment, and above all, lighting? If not, can you obtain power by running a cable to the main service box? Verify that enough wall outlets are available for your use.

Tapping an electrical distribution box is strictly a job for a professional lighting person, unless a crew member happens to be a qualified electrician.

Finally, scout parking for cast and crew. Even a small production can involve a dozen cars that may have to be expensively garaged, or else parked far from the location.

Scouting Video Problems

Perhaps the most important scouting task is checking the location for camera problems, especially lighting difficulties. In many cases, the shooting area is too dark for good-quality video. By scouting, you can decide whether to bring in movie lights, increase the power of the available lighting, or even recommend moving to a better-lit area.

Another problem is excessive contrast: too much difference between the lightest and darkest parts of the scene. Camcorders cannot record high-quality video in environments where the contrast is extreme. In those conditions, the lightest areas "burn out" to glaring white or else the darkest areas "block up" to a solid black.

Take, for example, the video scene in which the mayor of McKinley praises the civic activities of the Boosters Club. This sequence is to be shot with the mayor sitting at his desk in his office. The problem is that the desk is backed by a wall of windows that are very bright. In this contrasty environment, the mayor at his desk will record as a black silhouette, **Figure 10-8A**.

By scouting for camera problems, you realize that you can place your camera where it will exclude the bright windows, as you can see from the office plan in **Figure 10-8B**. Simply direct the mayor to open the sequence facing forward and then swing his chair around to face the camera.

Figure 10-8.
A—In front of a bright window, the mayor is a silhouette. B—Repositioning the camera avoids the bright windows. C—Moving the desk allows extra camera angles.

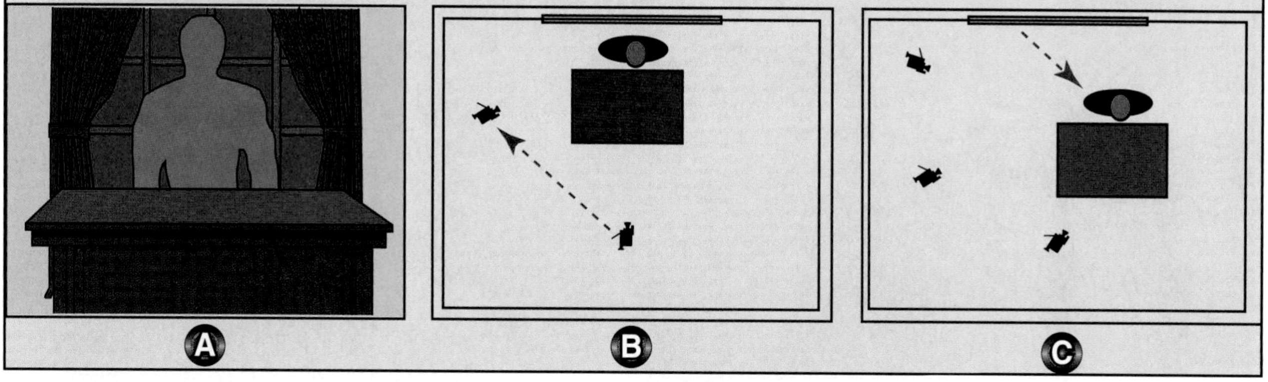

Another potential problem is mismatched light colors. In the mayor's office, light from a big desk lamp has a distinctly orange cast, while the window light is quite bluish. Noting this problem, you can plan to provide blue filtration to convert the lamp output to the color of daylight.

The same scouting trip will also reveal that you have a big problem in getting enough camera angles. Shooting at right angles to the window will give you one setup, but you will be unable to move to the side without running into the wall or else including the bright windows. After inspecting the location, you can solve the problem by planning to "cheat" the mayor's desk forward enough to permit a side angle on his face, **Figure 10-8C**. (The shift will not be noticeable to the viewers.)

Finally, many video problems arise because the camera must be placed too far from the subject. This is often true, for example, in wedding videos, where cameras and their operators are prohibited in certain parts of places of worship. By discovering that you will be far from the action, you can plan to bring a long lens or lens extender, plus a wireless microphone for the wedding couple, **Figure 10-9**.

Figure 10-9.
Wedding setups often require long lenses and wireless microphones.

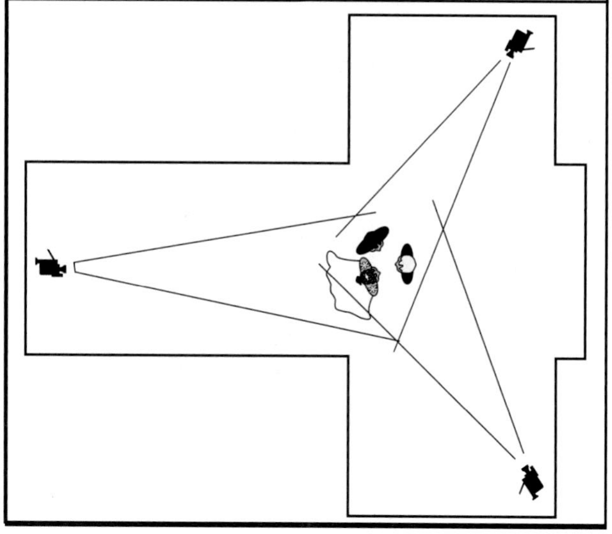

Scouting Audio Problems

A distant sound source is one of several problems in recording quality audio. Although so-called zoom microphones are available, they have nowhere near a lens' ability to zoom in for more detail and less background. If you catch the problem, you might provide wireless mikes, as suggested earlier. Or you may be able to feed sound from the facility's own public address system.

A second major audio problem is background noise, and here the McKinley Booster project is in serious trouble. Most of the shooting locations are outdoors at public buildings or in the town plaza. Traffic noise is almost impossible to eliminate, since it reflects from all sides. You can control the problem by staging important scenes with the worst traffic behind the camera and microphone. Alternatively, you could establish the scene in the noisy location and then tape all the close shots of talking in a matching but quieter background, **Figure 10-10**.

Wireless microphones can sometimes add as many sound problems as they solve, especially if they pick up interference. Auto ignitions and other electrical sources can cause static, and your microphones may share frequencies with other electronic devices. High-quality wireless microphones have fewer interference

Figure 10-10.
Moving the subject away from noisy backgrounds will often improve the audio quality. (Sue Stinson)

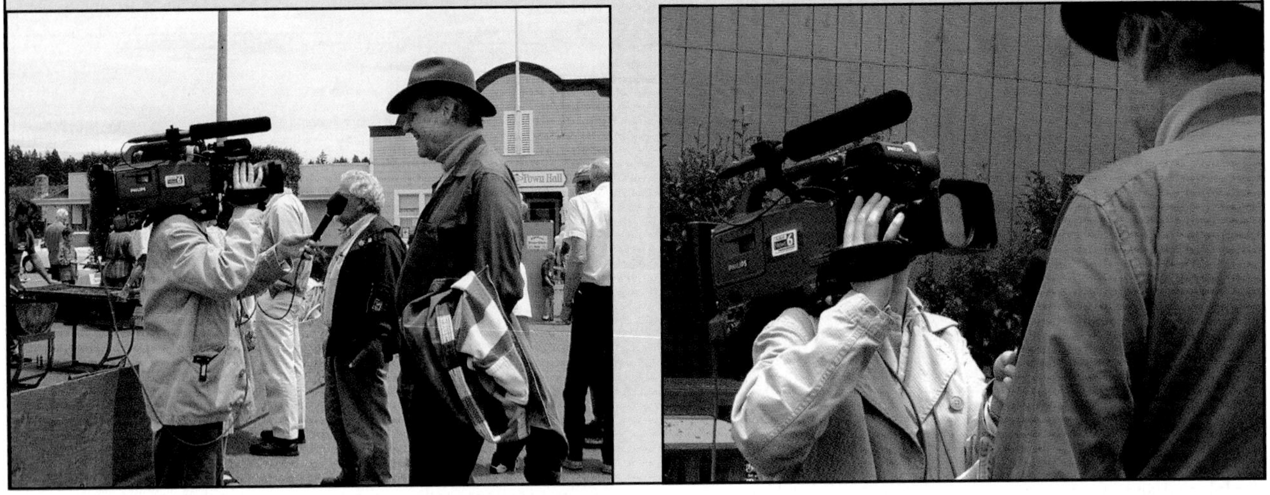

problems, but they can cost several hundred dollars apiece. So if you plan to use wireless mikes, take them on your scouting trips and test them at each location.

Things: Equipment and Supplies

After people and places, the most complex components of any production are *things*: cameras and lights and sound equipment, furniture, props, and set decorations. Without planning and careful management, looking after things can cause major problems.

Production Equipment

Where simple personal or entry-level professional productions are concerned, preproduction work with video equipment means ensuring that everything is present and in good working order. To do this, make a list of every piece of equipment, even if you have only a very simple video outfit. You would be surprised at how easy it is for small accessories and other items to be overlooked. Easily forgotten items include:

- Batteries and cables if you use external microphones.
- Camera quick-release components, **Figure 10-11**. One-half may still be attached to the tripod, but is the other half on the camcorder?
- Spare lamps for movie lights (since these special bulbs cannot be bought at just any supermarket).

Next, make sure that all your equipment is functioning properly by repairing problems as soon as you notice them. If the tripod lock is defective or if the zoom lens control is stiff, get it fixed before you forget it. It is all too common to arrive at a new shoot with equipment that has not been repaired.

Pay particular attention to power for the camcorder and for lights, if you use any. Check to see that all camera batteries are fully charged and that the charger is included in your gear for

Figure 10-11.
Quick release plates can become separated from their tripods.

use on location. Have plenty of industrial grade extension cords for movie lights.

Equipment planning also covers special items that you rent or buy for this shoot. Such items might include teleconverter accessories for shooting athletic events or special filters for creating optical effects.

Finally, ensure that you have an ample supply of videotapes ready for shooting. As explained in the sidebar, *Preparing Videotapes for Use*, tapes should be preblacked. Since this task must be completed in real time (a one-hour tape takes a full hour to black) it has to be done before actual production begins.

Sets, Properties, Costumes, and Makeup

Simple productions rarely involve actual set building, but you will often want to *dress* locations to make them look better on screen. To anticipate set-dressing needs, collect an assortment of potted plants, wall decorations, small furniture pieces, and other items to utilize if needed. It is also a good idea to have a roll of masking tape or double-faced tape.

Properties (usually shortened to "props") are all the items that are called for in the script: the glasses and silverware, the spectacles, the tire iron, the chocolate chip cookies—anything and everything that will appear in your program. To make sure you have every prop, make a list of them and check them off twice: once before you leave for location and again before you shoot. Nothing is more frustrating than being unable to shoot for an hour while someone runs all over town, frantically searching for a forgotten prop.

In the McKinley Boosters Club program, a Certificate of Merit is a key prop, because the president will be seen presenting it to a homeowner for

Preparing Videotapes for Use

To simplify editing later, you should prepare each new videotape for recording by a process called blacking the tape. This means recording the whole length of the tape, without any video or audio program signal. Blacking a tape does two important things. First, it lays down a continuous reference signal for locating and identifying shots.

In all digital formats, blacking the tape records time code, giving every individual frame a unique address. For instance, 00:22:12:25 identifies the single frame that is located zero hours, 22 minutes, 12 seconds, and 25 frames from the first frame of code placed on the tape.

Some digital camcorders reset time code to 00:00:00:00 following an interval of blank tape with no code. Preblacking the tape prevents this. Others continue the time count when recording resumes. If your camcorder has this feature, it is not necessary to black the tape to establish time code.

Time code is also laid down again with every shot recorded. The reason for prerecording it throughout the length of the tape is to provide readable code for any blank spots between recordings.

Second, blacking a tape lets you catch tape or camcorder problems in advance. By playing portions of the blacked tape, you can see (by the patterns on the screen) whether the tape is defective or the camcorder is failing to operate perfectly. The easiest way to black a tape is by placing it in a camcorder and pressing RECORD, with no video or audio signal coming in. To assure perfectly blank results from an amateur camcorder, black the tape with the lens cap on and a mini-plug converter inserted in the camera's external microphone jack. This will disable the built-in microphone.

If you have the required equipment, it is preferable to actually record a "reference black" color, rather than a blank signal. Most professional camcorders can generate their own reference black.

A miniplug is used to disable the camcorder's internal microphone.

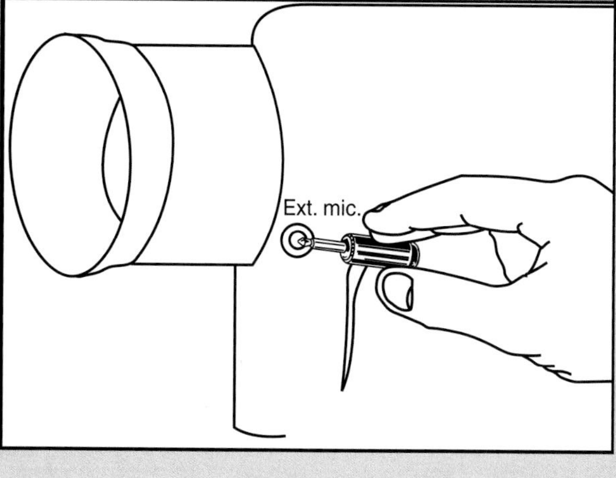

an exceptionally beautiful front yard. In this case, the actual certificate will not be ready in time for shooting, so you will need to fake one with a certificate form and a laser printer. From a distance, the camcorder will be unable to tell the difference.

In simple productions, costumes are usually provided by the actors, **Figure 10-12**, but you must include them in preplanning by using a costume inventory and check list. Before every scheduled shoot, call all cast members and remind them of the costume items for which they are responsible. (Perhaps the biggest source of delays in personal and small-scale professional production is actors who do not return in the same outfit they wore during the previous shooting session.)

Like costumes, makeup is usually the responsibility of individual actors. The sidebar *A Basic Makeup Kit* suggests a simple kit that you may want to assemble and have ready to use for touch-ups if needed.

Figure 10-12.
Performers often provide their own costumes.

Releases

Releases are legal documents granting you permission to include people, places, objects, and music in your program. Without releases, you can get into serious difficulties when your program is shown.

Hobbyist video makers rarely bother to obtain releases, and it is true that if you are making a program for *purely personal or family use*, you do not need to worry about permissions.

Professional video programs are different. If you are making your program for profit and/or public showing, you risk being prosecuted if you fail to get the necessary releases and rights. What does the phrase "*for profit and/or public showing*" mean? It includes one or more of the following uses for your program:

- You were paid for producing it.
- It will be offered for rental or sale.
- It will be transmitted publicly via broadcast, cable, satellite, the Internet, or any other method.
- It will be shown anywhere outside your home and those of family and close friends.

Note that a program qualifies if it is *either* for profit *or* public. A private corporate video is for profit because you were paid to make it. A showing outside your home is public even if no one receives compensation for it.

The McKinley Boosters Club video falls into the last category. Nobody will make a penny for working on this program and the club will not sell tapes or charge admission to showings. However, they will screen it all over town in order to develop support for the club's activities. For that reason, the Boosters Club must have all necessary releases on file before the program is shown.

Types of releases

Generally speaking, there are seven common types of releases and permissions: general, talent, minor, materials, location, stock photos and footage, and music.

U. S. copyright law is nationwide, but other forms of permission may be affected by state laws. Release forms sold in local camera stores will be appropriate for your location, but it is wise to have the language of each release form reviewed by an attorney.

General release: A general release is for people who will not be paid to appear in your video. This applies to subjects of documentaries and volunteer actors alike.

Talent release: A talent release is used for people who are paid to appear. (This is sometimes called a **model release** on commercially available forms.)

Minor release: A minor release is used when the subject is under 18 years of age. It includes provision for printed, signed permission granted by the minor's parent and/or legal guardian. (Often, a general release can be modified

A Basic Makeup Kit

Very little makeup is used in most personal and entry-level professional videos. Female cast members generally wear their normal street makeup. Men usually wear none at all. Nevertheless, you should have at least the following items to use for touching up your talent:

- Neutral powder for eliminating skin shine.
- Light- and medium-toned powders for minimizing skin imperfections.
- One or two powder brushes.
- Neutral blemish spotter in a variety of skin tones.
- Facial tissues.
- Mirror.
- Hairbrush, comb.
- Hair spray, to keep hair in place.
- Trim scissors.
- Small plastic tool or tackle box to hold everything.

If your programs include performers of different ethnicities, you will want to customize your powder and spotter colors to suit. In most larger cities, you can buy makeup at theatrical supply houses. In smaller cities and towns, try stores that sell dance supplies or party and Halloween costumes. You can obtain simple products at any drug or discount store makeup counter.

Some basic items for a makeup kit.

With your supplies assembled, here are some tips for using makeup in your programs:

- Use as little as possible.
- Fold tissues around collars to keep powder off costumes.
- Use makeup only to cover things up (shine, blemishes), not to alter a performer's appearance.
- Let women apply their own makeup and both women and men fix their own hair. They know themselves better than you do.
- Check the effect of makeup on an external color monitor before taping.

by adding a permission statement signed by a parent.)

Materials release: A materials release is for objects included in the program, often photographs or graphic materials. If you want to show a group portrait of the McKinley Boosters in 1926, you need to find out who owns it and execute a materials release for using it.

Location release: A location release allows you to shoot on private property.

Stock permission: You need permission to include still or moving images obtained from professional stock footage companies. Typically, permission for a specified use of the material is included when you buy the rights to the footage.

Music permission: In the same way, you must obtain permission to use copyrighted music, whether previously recorded or newly recorded for your program. It is not safe to assume that music is in the public domain. *Happy Birthday*, for example, is copyrighted.

See the sidebar *Music Permissions* for more details.

When to obtain releases

In many cases, people start out to make a program for purely personal use, so they do not obtain releases. When the video is finished, however, they decide that they want to show it

Music Permissions

It is vitally important to get permission to use copyrighted music in your video, particularly because over 25 "Music Rights Societies" worldwide monitor usage diligently and enforce copyrights aggressively. In the U.S., the two music rights clearing houses are ASCAP (The American Society of Composers, Authors, and Publishers) and BMI (Broadcast Music Incorporated). For useful links regarding music rights, look at the website of the National Music Publishers' Association (**www.nmpa.org**).

Release Forms

You may take either of two opposite approaches to release forms. Some people feel that unpaid amateurs, especially, may be intimidated or made suspicious by overly complicated language. A clear, simple release like the one below is preferred.

Performer release

My signature below attests that I, ________________________, have read this release and that I understand and agree to the following:

1. I permit recordings of my voice and likeness to be made in connection with a media project currently titled ______________________________ .
2. I authorize ______________________________, the proprietor(s) of this project, in perpetuity, to process, store, reproduce, distribute, and display recordings of my voice and likeness made in connection with this project, in whatever ways, forms, and media, and by whatever methods and technologies they may choose.
3. I waive, in perpetuity, the rights to any and all compensation for these recordings, other than such rights as may be set forth in other written agreements between myself and the proprietor(s) of this project.

____
Performer signature

____
Performer printed name

____
Date

____
Witness signature

____
Witness printed name

____
Date

publicly—perhaps by entering it in a contest or screening it for a local club. At this point, it may seem too late (or too much trouble) to secure releases. In this situation, some producers will enter or show the program anyway, thinking that participants will not care (or perhaps even know) that they are being publicly displayed.

This is not only illegal, but risky as well. It is far better to obtain all necessary releases and permissions before and during shooting, even if the program is not intended for public use and/or profit. In short, if in doubt, *get releases*.

The sidebar *Release Forms* includes sample releases.

When releases are not needed

If you are shooting on public property, you normally do not need permission to show buildings, people, etc., who happen to be in the background.

However, you may still need releases. Here are just a few situations that require them.

- You are shooting *into* private property (such as a residential backyard).
- Your program implies that someone is doing something unlawful or improper. For instance, if you show pedestrians in a night scene and voice-over narration says "many people out at this hour are engaged in illegal

Other people feel that a release should address every legal contingency. A sample of a comprehensive release appears below.

Release Agreement

Whereas, ______________________________ (the "Producer") is engaged in a project (the "Video"), and
Whereas, I, the undersigned, have agreed to appear in the Video, and
Whereas, I, understand that my voice, name, and image will be recorded by various mechanical and electrical means of all descriptions (such recordings, any piece thereof, the contents therein and all reproductions thereof, along with the utilization of my name, shall be collectively referred to herein as the "Released Subject Matter"),
Therefore, in exchange for $1.00, receipt of which is hereby acknowledged and whose sufficiency as consideration I affirm, I hereby freely and without restraint consent to and give unto the Producer and its agents or assigns or anyone authorized by the Producer, (collectively referred to herein as the "Releasees") the unrestrained right in perpetuity to own, utilize, or alter the Released Subject Matter, in any manner the Releasees may see fit and for any purpose whatsoever, all of the foregoing to be without limitation of any kind. Without limiting the generality of the foregoing, I hereby authorize the Releasees and grant unto them the unrestrained rights to utilize the Released Subject Matter in connection with the Video's advertising, publicity, public displays, and exhibitions. I hereby stipulate that the Released Subject Matter is the property of the Producer to do with as it will.
I hereby waive to the fullest extent that I may lawfully do so, any causes of action in law or equity I may have or may hereafter acquire against the Releasees or any of them for libel, slander, invasion of privacy, copyright or trademark violation, right of publicity, or false light arising out of or in connection with the utilization by the Releasees or another of the Released Subject Matter.
It is my intention that the above-mentioned consideration represents the sole compensation that I am entitled to receive in connection with any and all usages of the Released Subject Matter. I expressly stipulate that the Releasees may utilize the Released Subject Matter or not as they choose in their sole discretion without affecting the validity of this Release. This Release shall be governed by (*name of state*) law.
I hereby certify that I am over the age of eighteen, and that I have read, understood, and agreed to the foregoing.

Print Name	Signature Date
Address	City, State, Zip:
Phone No (______) ____________________	

Remember that these are only samples, and may not be appropriate for your area. *Do not use them without checking with an attorney.*

activities," you are indicating that the narration applies to the pedestrians.

- An exterior or interior location is put to fictitious use. For example, suppose you show a subject entering the street door of a restaurant, after which you cut to a restaurant set. You are saying, in effect, that your set is the interior of the actual restaurant, so you need to have permission to do so.

Examples could be multiplied, but the general principle is clear: *when in doubt, get releases.*

Plans: Production Logistics

Logistics is the overall task of organizing and supplying the shoot. In an elaborate production, this can be a formidable job; for simpler programs, the biggest tasks are creating a shooting schedule and providing support services.

Scheduling

Scheduling is the process of deciding which scenes to shoot in which order. The task seems

Figure 10-13.
Night shooting requires careful planning.

straightforward until the production manager runs into problems like these:

- For a scene between them, the Mayor cannot work on Wednesday and the Boosters Club President is unavailable on Thursday.
- The bulldozer for a dump cleanup sequence is available only on weekends, when it is not in use; but the city personnel who drive it do not work weekends.
- The statue illumination ceremony has to be shot at night, **Figure 10-13**.

Juggling several conflicting demands at once can make scheduling a frustrating business. Fortunately, there is a well-developed system for organizing a shoot. To sort out shooting priorities, the production manager lists certain characteristics for every scene in the script. (A scene is a single piece of action that happens in a single place at a single time.) Key characteristics of scenes include:

- *Places.* For obvious reasons, it is more efficient to shoot every scene that takes place at one location before moving on to the next location. Try to group scenes together by location, to minimize moving the production from one place to another, then back again.
- *People.* At each place, different performers may be needed at different times; so scenes in which the same actors perform together should be scheduled consecutively.
- *Things.* Properties and large items such as special automobiles may be available only briefly, so the scenes in which they appear should be grouped together.

Figure 10-14.
Downtown McKinley. The conditions before sunset called the "magic hour."

- *Conditions.* Daylight or darkness, sunshine or overcast, some natural conditions are called for in the script. You can sometimes shoot night scenes during the day and even vice versa, but the process is complex and expensive. So, for each location, schedule night or sunset scenes together, **Figure 10-14**.

A typical production breakdown will prioritize scenes as follows:

1\. All scenes at location X.
1.1 All location X scenes at special times, especially night.
1.1.1 All location X night scenes with performer A.

If key performers can work only at certain times, you may be forced to organize the shoot around their schedules, even if it means returning to the same locations.

Support Services

Even on small productions, support services can make a big difference in the morale and enthusiasm of cast and crew. Parking is especially important, and if the production is big enough to require supply trucks for lighting and other equipment, close and convenient parking is essential.

Food is equally important. It is good to have light snacks, coffee, and soft drinks available, especially for early morning or late evening shoots. In addition, meal breaks should be scheduled and adhered to. If public eating facilities are not easily available, meals should be provided.

Finally, attend to everyone's comfort. Since all production involves considerable waiting, especially for the performers, provide a comfortable lounging area (often called a "green room"). At interior locations, this may be just a room or other quiet area out of the way. Outdoors, it may be as simple as an area supplied with folding chairs and shaded by a tarpaulin. Restroom facilities are important too, and, if you are shooting at an outdoor location, may be required by local ordinance. If you are working, say, in the desert, you will need a water supply for washing as well as drinking, along with appropriate portable toilets.

Budgets: Production Costs

All video production costs money. Anticipating and controlling expenses is as vital to the success of a project as writing, directing, or videography. The formal process of identifying these production costs and allocating funds to pay them is *budgeting*. Typically, a project budget is created early in production planning and then refined and updated throughout the course of the production. Though budgets can sometimes be complex, the principles that govern them are simple:

- Identify every item that will cost money, including salaries, rentals, purchases, services, and rights.
- Estimate the amount of each item that will be needed.
- Determine the exact unit cost for each item.
- Multiply the item amount by the unit cost to determine the total cost for that item.
- Add the item totals together to determine the overall expected costs.

Figure 10-15.
Budget line items.

ITEM	AMOUNT	UNIT COST	TOTAL
actor	4 days	$300/day	$1,200
camera dolly	9 days	$120/day	$1,080
gaffer tape	2 rolls	$12/roll	$24

Figure 10-15 illustrates this process with three types of budget items: a salary, a rental, and a purchase. (Simpler projects will, of course, cost less.)

Standard Production Budgets

Even simple productions can have so many expenses that it is difficult to identify every one in advance. For this reason, most established production houses develop or buy preorganized budget breakdowns. These are usually computer spreadsheet or database programs developed for film and video budgeting, **Figure 10-16**.

Figure 10-16.
Summary page of a sample budget. (Movie Magic Budgeting, Creative Planet)

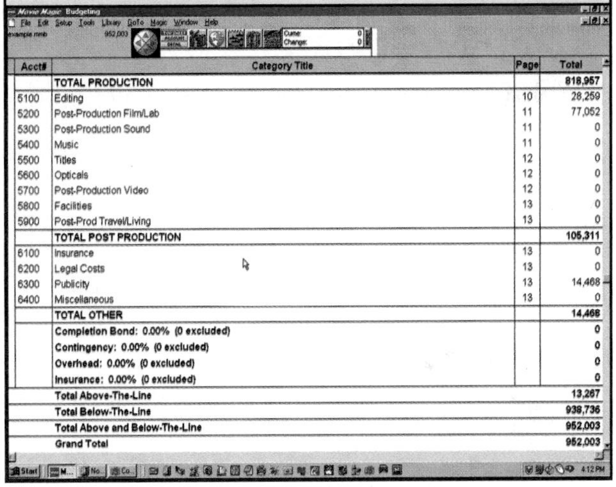

Acct#	Category Title	Page	Total
	TOTAL PRODUCTION		818,957
5100	Editing	10	28,259
5200	Post-Production Film/Lab	11	77,052
5300	Post-Production Sound	11	0
5400	Music	11	0
5500	Titles	12	0
5600	Opticals	12	0
5700	Post-Production Video	12	0
5800	Facilities	13	0
5900	Post-Prod Travel/Living	13	0
	TOTAL POST PRODUCTION		105,311
6100	Insurance	13	0
6200	Legal Costs	13	0
6300	Publicity	13	14,468
6400	Miscellaneous	13	0
	TOTAL OTHER		14,468
	Completion Bond: 0.00% (0 excluded)		0
	Contingency: 0.00% (0 excluded)		0
	Overhead: 0.00% (0 excluded)		0
	Insurance: 0.00% (0 excluded)		0
	Total Above-The-Line		13,267
	Total Below-The-Line		938,736
	Total Above and Below-The-Line		952,003
	Grand Total		952,003

To customize these standard forms, you delete line items that do not apply to your project, add any specialized items that may be unique to your production, and then fill in the amount and rate for every item. Typically, the software calculates totals, sales tax, and similar charges when applicable.

If you have some experience in developing spreadsheets or other databases, you may find it productive to build your own custom system for production planning, scheduling, and budgeting. In addition to spreadsheets and databases, some project management software packages include financial functions.

Contingency Funds

Since video production is always vulnerable to unexpected events and expenses, every budget should include a contingency fund: an amount of money set aside to cover unforeseen costs. A contingency amount is typically calculated as a percentage of the total production budget—anywhere from one percent to perhaps 15 percent. A five-percent contingency in a $10,000 budget would be $500, a relatively modest sum.

To determine how to set your contingency fund, evaluate the uncertainties (like weather) that might prolong production, and adjust your rate proportionally. For example, a training video on how to perform an assembly line procedure might be safe with a very small contingency budget item. But if you are shooting a nature program on the elusive mountain lion in its natural habitat, provide enough extra money for many additional shooting days.

Ready to Shoot

At this point you may think that planning even a simple video requires a great amount of work. To put preproduction in perspective, remember that many of the requirements covered in this chapter are unnecessary for short, simple videos—although all programs will benefit from a certain amount of systematic preparation. If you plan to make professional programs of even modest scope, you will find that careful production planning is essential to success.

Chapter Review

Answer the following questions on a separate piece of paper. Do not write in this book.

1. Actors, narrators, and other people who appear in video programs are collectively called the ______.
2. *True or False?* Local amateur theater companies are usually not a good source of talent for a video production.
3. Cast selection is most effectively done by viewing ______.
4. *True or False?* You must obtain releases for any program produced for public showing, even if there is no profit involved.
5. A production budget should always include a ______ fund.

Technical Terms

Blacking the tape: Prerecording an entire tape with a pure-black picture, no sound, and time code (or in analog formats, a control track).

Budgeting: Predicting the costs of every aspect of a production and allocating funds to cover it.

Crew: Production staff members who work behind the camera. In larger professional productions, the producer, director, and management staff are not considered "crew."

Dress: To add decorative items to a set or location. Such items are called "set dressing."

Talent: Every production member who performs for the camera.

11

Camera Systems

Objectives

After studying this chapter, you will be able to:

- Explain the functions of major camcorder controls.
- Describe the major camcorder support systems.
- Prepare camera equipment for shooting.
- Operate all camcorder systems.

About Camera Systems

The camcorder is, of course, the one indispensable tool for video makers. If you understand how it works and how to use it effectively, you are on your way to professional production. Camcorders include many different systems and controls. To make sense of all these features, it is helpful to separate the camcorder into its two fundamental halves.

The word "camcorder" is a combination of "camera" and "recorder," because the unit contains both components:

- The *camera section* converts incoming light into images and sound into audio, both coded in the form of electrical signals.
- The *recorder section* copies those signals onto magnetic tape or other storage media and then plays them back on demand. The recorder section works exactly like a VCR deck; indeed, you can connect a camcorder to a TV set and play back footage directly. Many camcorders have a two-position switch. In one position, the unit functions as a camera/recorder. In the other position, it becomes a VCR/editor.

Not only does a camcorder include two types of machines, it also records two completely different kinds of information:

- *Video (picture)* information is converted into the video signal by the imaging chip.
- *Audio (sound)* information is converted into the audio signal by a microphone, either built into or plugged into the camcorder. Each type of information has its own controls and requires its own procedures, so it will be helpful if you think of video and audio separately.

Audio recording, which is also performed by the camcorder, is covered in Chapter 16.

Camera Functions

The job of the camera half of a camcorder is to convert incoming light into coherent optical images (recognizable pictures) and then translate those optical images into an electrical video signal. Here is how the video signal is created:

- Light passes through the camcorder lens, which organizes it into a coherent image and projects it onto the surface of the chip, **Figure 11-1**.

The imaging chip is sometimes referred to simply as "the chip," despite the fact that it is only one of several solid-state chips in the camcorder.

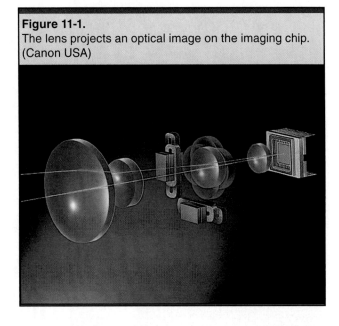

Figure 11-1.
The lens projects an optical image on the imaging chip. (Canon USA)

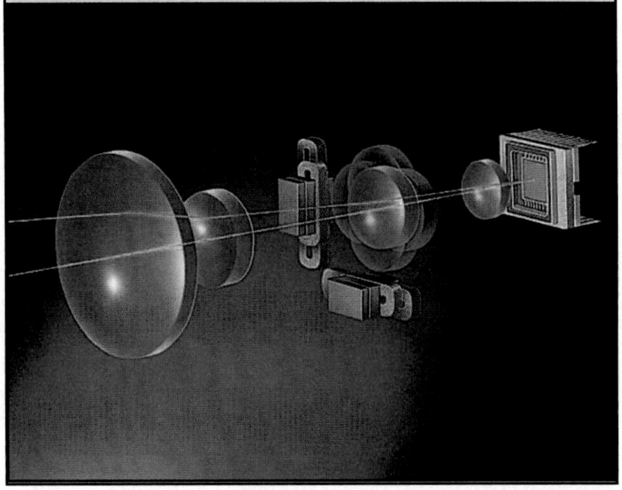

- The chip contains hundreds of thousands of microscopic sensors. Each sensor responds differently to the amount and color of light striking it, so each sensor creates a different electrical signal.
- The thousands of separate signals are processed by electronic circuits that combine them into the composite signal for one complete image called a frame.
- This process repeats 25 or 30 times each second, to create a video track coded in electronic form.

In the North American video display format (NTSC), each frame is electronically divided into two separate 1/60-second ***fields***, one containing all the even-numbered picture lines and the other all the odd ones. The two fields are combined to make one frame by a process called "interlacing."

Each frame of the video signal lasts only 1/30 second before it is replaced by the next frame. This means that the signal must be displayed as it is created or it will be lost forever.

One Chip or Three?

Most modern camcorders convert optical images into electronic signals by means of microchips called "CCD or CMOS" chips.

The imaging chip is a Charge-Coupled Device (CCD).

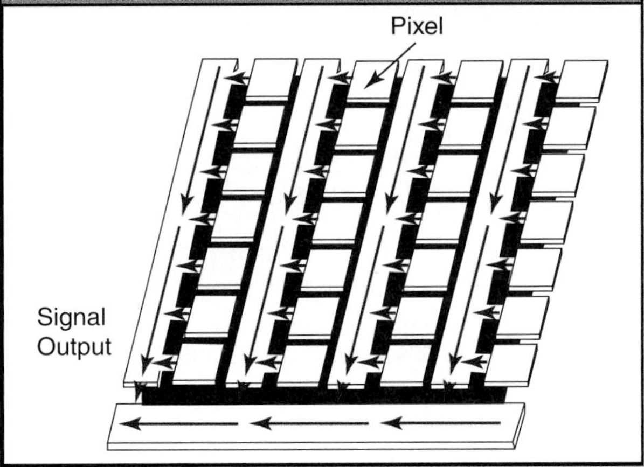

How a chip works

A chip is a grid of tiny solar cells—so small that 400,000 or more can fit on a rectangle about one-quarter inch wide, measured diagonally. Like all solar cells, each of these minute sensors converts light into electricity—the brighter the light, the stronger the current. Sixty times each second (in NTSC video), electronic circuits drain the energy generated in these cells and send the pattern of charges to the recording mechanism. It is this pattern of light and dark that makes the picture.

Recording colors

Since the strength of the current generated by cells depends on the brightness of the light, rather than its wavelength, these solar collectors cannot distinguish colors. To a cell, a certain amount of blue light "looks" much like the same amount of green or red light.

To record color information, each full-color picture unit is created by not one but three sensors working together. Each light collecting surface is covered with a different colored filter, red, green, or blue. As a result, each sensor creates a slightly different amount of current.

Though visible light contains a theoretically infinite number of different colors, they can all be created by combining different intensities of the primary hues: red, green, and blue. By analyzing the currents produced by the three sensors, the camcorder circuits deduce the color striking each sensor. When you consider that this analysis takes place 60 times per second on hundreds of thousands of cell triplets, you can appreciate the engineering miracle in every camcorder.

Every color in the spectrum can be created by combining the red, green, and blue primary colors.

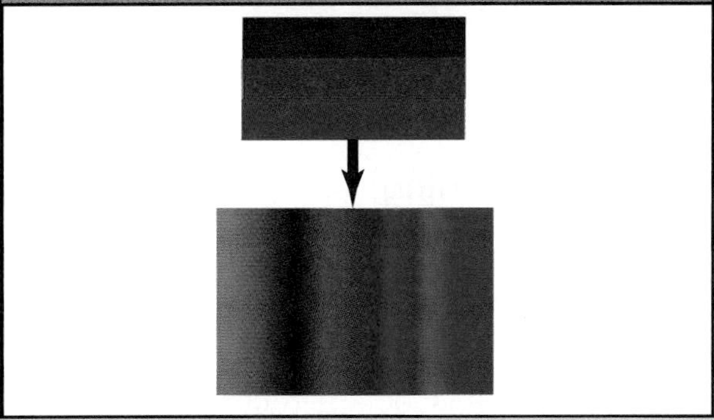

This three-cell design has drawbacks, too. Because it is so difficult to lay these near-microscopic color filters precisely over individual cells, they are not always perfectly positioned, and color inaccuracies can result. And because three cells are needed for every one color element, the chip has only one-third the resolution (sharpness) that would theoretically be possible if every cell were recording a separate part of the image. That is why the more sophisticated camcorders use not one chip but three.

Three-chip cameras

In a three-chip camcorder, the entire surface of each chip is covered with a single primary color filter: red, green, or blue. This approach eliminates filter registration problems and increases picture sharpness.

Three-chip camcorders have one problem: in order to direct the incoming light to three different surfaces, it must be divided into thirds by optical elements called beam splitters. Because some light is lost in this more complex processing, less gets through to the imaging chips. As a result, three-chip camcorders do not usually perform as well as their one-chip cousins do in very low light. In most professional applications, this is not a drawback because production companies typically light their scenes.

A three-chip imaging system. (Canon USA)

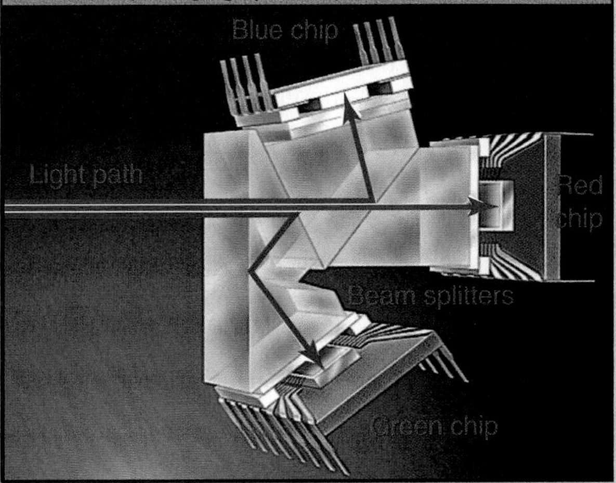

Recorder Functions

To preserve short-lived video images, engineers developed the videotape recorder to record them. Combined in one unit with the camera, the recorder forms the "corder" part of the camcorder package.

Interlaced scanning

Traditionally, video cameras produce "interlaced" frames, in which the odd-numbered lines of each frame are recorded as one field and the even-numbered lines as a second field.

Here is how a videotape recorder handles interlaced images:

- The electrically encoded information about each video field (one-half frame) is sent to a record head spinning on a drum in the recorder.
- The record head preserves the signal by selectively magnetizing metallic particles on the videotape that moves across its path.

VCR drums actually have at least two video record and two video playback heads. To create a two-field frame, each head records or plays a single field.

- When set to playback, the recorder passes the tape across the spinning drum again. This time, separate playback heads recreate field from the magnetically coded tape particles.
- Sent at a rate of 60 per second to a playback unit, these fields are interlaced line-by-line to create 30 complete images displayed on the surface of a picture tube or LCD display, **Figure 11-2**.

In short, the "corder" section of a camcorder is a very compact VCR, with all the usual tape controls. Because of their sophisticated recording features, camcorders are often used in editing, as either source or assembly decks.

Some digital camcorder systems record every line in the video picture, in order, instead of dividing odd and even lines into separate fields. This single-pass recording system is called "progressive scan."

Digital Camcorders

Analog camera systems have been largely replaced by cameras that record audio and video signals digitally. This change is vitally important to video postproduction, but it does not greatly alter the basic processes of shooting with the camcorder. In most ways, digital camcorders work much like analog models. The lens, focus, shutter, and aperture admit and process light into images, and the chip converts those light images to coded electrical signals. The recording system preserves the results on videotape.

Figure 11-2.
A standard TV image is created by "interlacing" the lines of Fields A and B. The lines are exaggerated to show the process more clearly.

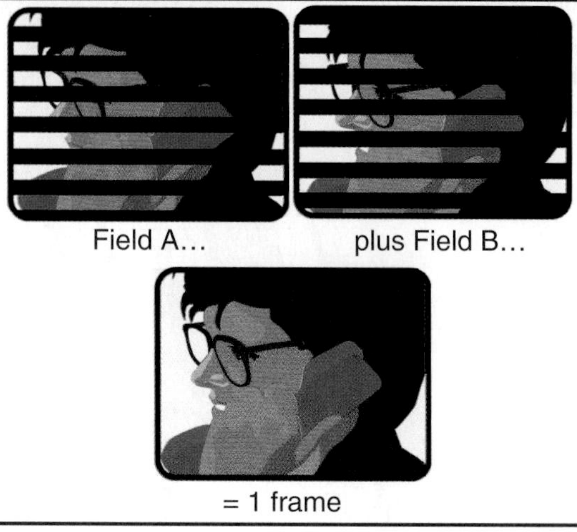

The one major difference is inside the camera, between the chip and the recording mechanism. There, an analog-to-digital *(A/D)* chip converts the electrical signal from the continuously varying voltage of an analog waveform to the characteristic on/off pulses of digitally coded current. The recorder section captures this signal as digital video.

In some ways, digital recording does affect the way a videographer uses the camcorder. For example, digital signal processing allows more extensive control over white balance and color. For this reason, it is sometimes possible to obtain acceptable image quality in working environments with mediocre lighting conditions, **Figure 11-3**.

Digital video also permits the recording of finer detail. The mini DV format that is popular in amateur and entry-level-professional camcorders captures images superior to similar analog formats. In fact, the resolution is so superior that mini DV digital camcorders are now used in many industrial and ENG (Electronic News Gathering) applications.

Figure 11-3.
Even severe color casts...

can often be corrected by a digital camcorder. (Corel)

Camera Support Systems

With the principal camcorder functions explained, we can survey the major equipment systems, starting from the ground up, with camcorder support. Camcorders are supported by *tripods*, *stabilizers*, or *dollies*.

Tripods

Tripods are by far the most common camera support. They have three legs instead of four because a three-point support will not wobble on uneven ground.

To see the problem with four legs, try leveling a card table outdoors.

To deploy a camera tripod, set up the feet, the legs, and the head, **Figure 11-4**, in that order. Most tripods have two sets of feet: soft, non-skid pads for smooth surfaces like floors and pointed pins for softer outdoor surfaces. It is easier to select one set or the other first, while the tripod is still collapsed. Unless you are setting up a low-angle shot with the tripod legs unextended, start by extending them fully. Different models use different methods for locking the leg sections.

Before you can adjust the tripod head, you need to level it on the legs. Professional tripods let you unlock the head, tilt it until a bubble level shows that the head is level, and then lock it, **Figure 11-5**.

Figure 11-4.
A tripod head designed for a lightweight video camcorder. (Manfrotto/Bogen)

With less expensive units, you must level the head by adjusting the legs. First, use the bubble level to locate the high side of the tripod; then point one leg directly at that side. Shorten the high-side leg until the bubble level indicates that the head is level.

Adjusting the Head

With the tripod set up and leveled, the next task is to set various controls so that the head operates smoothly when you move it and does not move by itself, **Figure 11-6**.

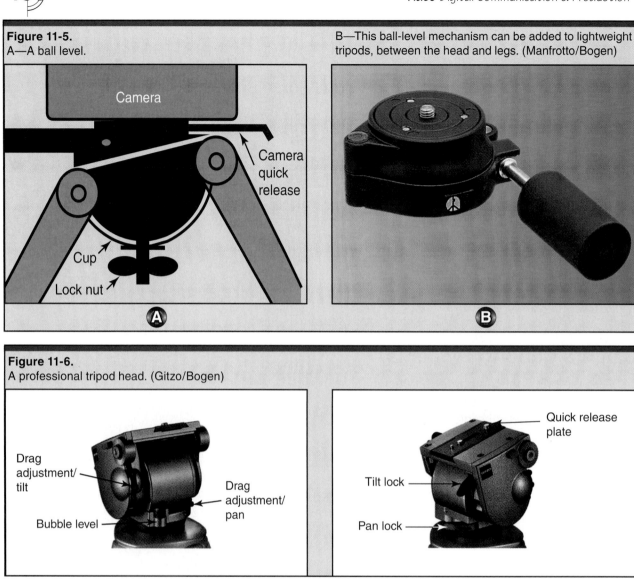

Figure 11-5.
A—A ball level.
B—This ball-level mechanism can be added to lightweight tripods, between the head and legs. (Manfrotto/Bogen)

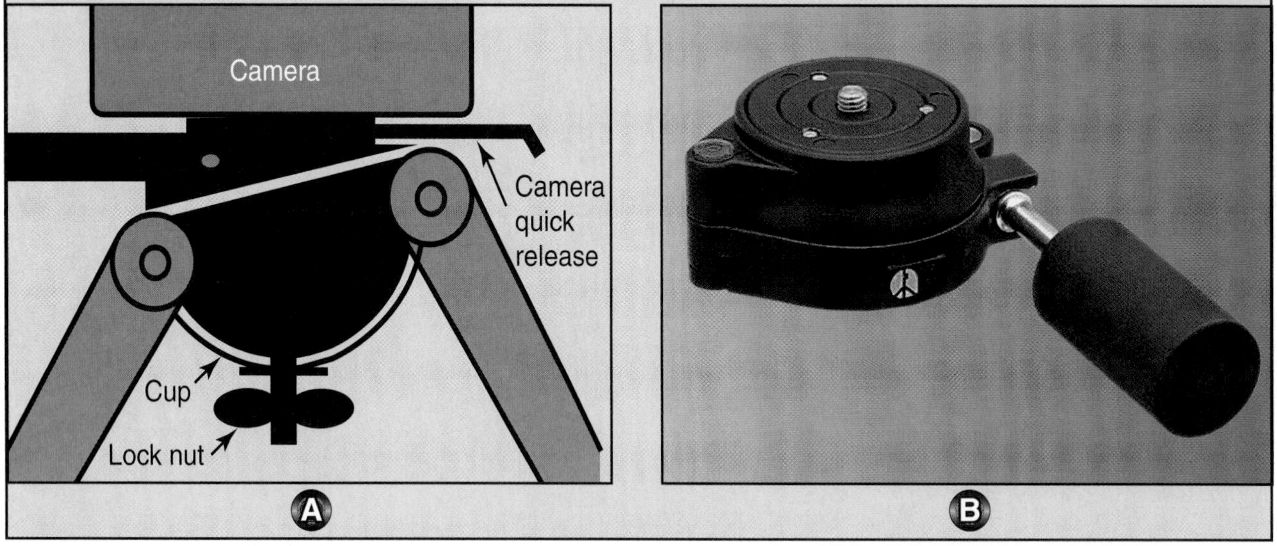

Figure 11-6.
A professional tripod head. (Gitzo/Bogen)

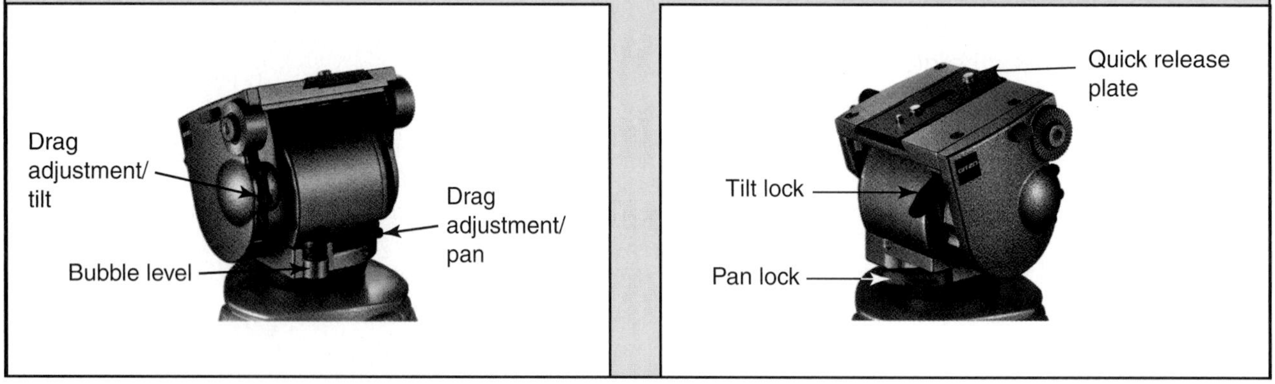

Quick release mechanism

If your camera and tripod connect via a quick release system, make sure that the camcorder and tripod head are tightly screwed to their docking plates. This helps prevent camera wobble.

Pan handle

Generally speaking, you want the pan handle installed on whichever side of the tripod head gives you easiest access to camcorder controls. Normally, the handle is installed on the right side because the viewfinder is on the left. However, if the camcorder's only record button is on the right-side hand grip, you may want to install the pan handle on the left.

Avoid using the camcorder's hand grip while working on a tripod. Turning the unit with the pan handle makes for smoother movement and less stress on the camcorder.

Camera balance

Some tripods allow you to balance the weight of the camcorder, either by sliding the unit forward or backward on the head, or by adjusting a built-in counterbalancing mechanism. Adjust the balance until your camera remains level even when the tilt lock and tilt drag controls are completely disengaged.

Tilt lock

A tilt lock prevents an unattended camcorder from tipping forward or backward. (Some tripod heads lack balancing mechanisms, and even those that have them may also provide a tilt lock, for extra security.) At the start of the shooting session, engage and disengage this lock to ensure that it is working properly.

Tilt and pan drag

You execute vertical and horizontal camera swings more smoothly if you are pushing against resistance (called ***drag***). Too little drag is ineffective, while too much makes for jerky movements. Adjust the horizontal drag control to your taste; then disengage the tilt lock and do the same for the vertical drag control. That way, you will not waste time on setup when the equipment is needed.

Stabilizers

If you will be using a *camera stabilizer,* you should preassemble, adjust, and test it at the beginning of the shooting session. Camera stabilizers are systems designed to smooth hand-held shots, **Figure 11-7**. The most popular types are available in sizes to fit every camera from consumer mini-camcorders to large professional models.

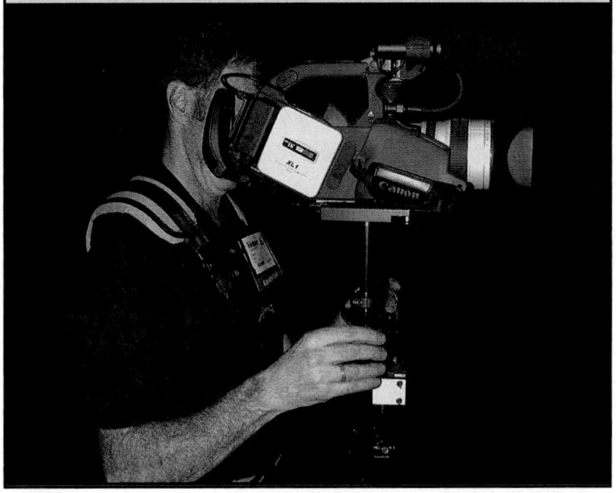

Figure 11-7.
A camera stabilizer in operation. (Hollywood Lite)

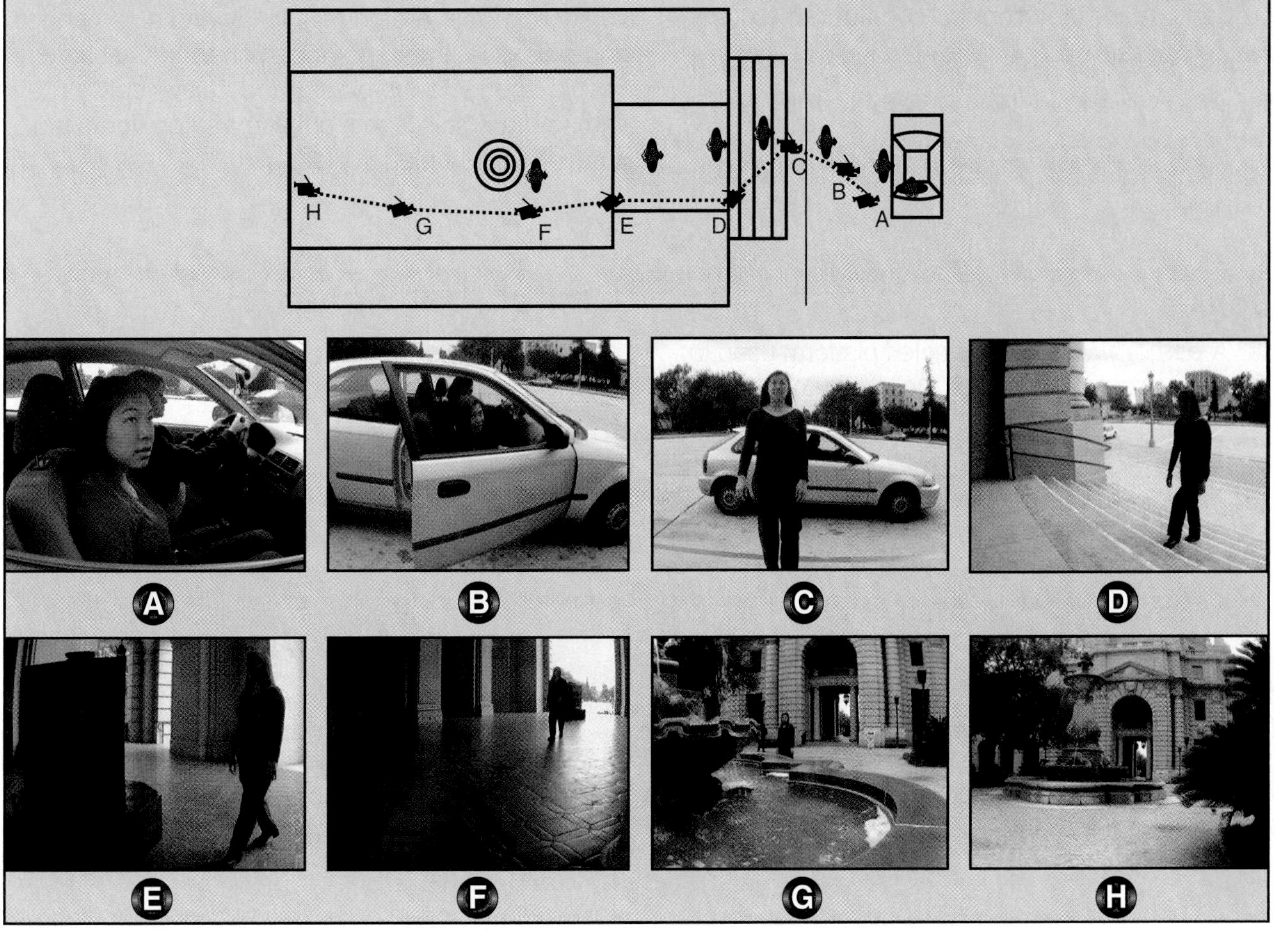

Figure 11-8.
The paths of the subject and the camera operator. A—The camera starts close to the car... B—pulls back as the subject emerges... C—climbs one step to the sidewalk... D—plus six steps to the archway... E—moves to one side... F—begins to draw ahead of the subject... G—who stops at the fountain... H—while the camera continues to the other end of the courtyard.

Unlike internal stabilizers, which steady the image being recorded, camera stabilizers are external hardware designed to steady the entire camcorder.

Though various camera stabilizers differ in their design and engineering, most of them consist of four parts:

- The head, which supports the camera.
- The counterweight system, which balances the camera.
- The pivot system, which allows the camera and counterweight to move freely.
- The view screen, which allows the operator to frame and follow action without touching the camera's internal viewfinder.

In some systems, the view screen and/or external battery is part of the counterbalance system.

Well-adjusted and competently used, stabilizers work amazingly well, delivering footage as steady as dolly shots, while allowing the camera operator to follow the action with a flexibility that dollies cannot match. See **Figure 11-8**.

On the other hand, stabilizers can be time-consuming to set up and to counterbalance, and they may need frequent fine-tuning to work properly. Also, they require considerable practice to operate.

In feature film production, camera stabilizer operators are specialists who are hired solely to provide stabilized shots. Often, they receive screen credit for their work.

Dollies

A ***camera dolly*** is a wheeled platform used to make smooth moving shots. Dollies are classified by two sets of criteria: design and wheel system. There are two major types of camera dolly designs. *Pedestal* dollies consist of a camera head mounted on a telescoping center column. The column, in turn, is supported by a wheeled base. Pedestal dollies can be smoothly moved up or down ("pedestaled") during a shot. *Cart* dollies are essentially rolling platforms fitted with tripods. Heavier, more expensive types allow booming via a crane arm that moves up and down, but simpler cart dollies generally lack this feature. Some ultra-light dollies consist of tripods supplied with wheels and mounted on rails, **Figure 11-9**.

Dollies may be fitted with two different types of wheels. Floor wheels are designed to roll directly on smooth surfaces like interior floors. Most systems allow the dolly to change the direction in mid-shot by swiveling as they roll. The most sophisticated systems have four-wheel steer, so that all four wheels can be turned at once and the dolly can "crab" sideways. When used with rails, the Fisher dolly (see Figure 11-9) maintains alignment using the small horizontal guide wheels visible in the illustration. Rail wheels, by contrast, work like train wheels, fitting over rails designed for the purpose. Floor wheel systems are quicker to set up (there are no rails to lay and align) and more flexible (the camera can change direction in mid-shot). On the other hand, rail wheel systems permit dolly shots outside and on floors too uneven for wheeled dollies.

Camcorder Setup

With camera support ready, the next task is to set up the camcorder itself, including the

Figure 11-9.
Dollies. A—Some track and dolly systems are extremely portable. (Microdolly Hollywood) B—A large studio dolly. (J.L. Fisher Corp.) C—A wheeled spreader can convert a lightweight tripod into a dolly. (Manfrotto/Bogen)

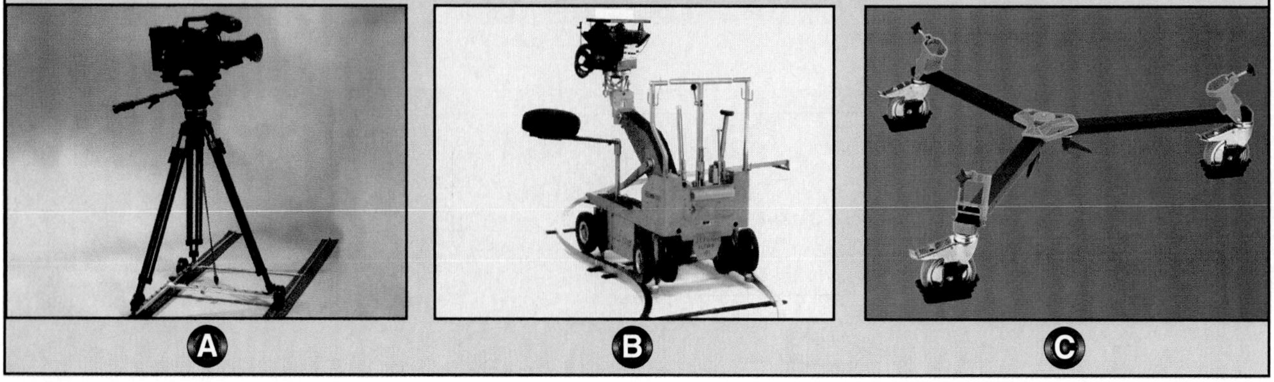

power supply, the external monitor (if present), and the videotapes.

To avoid repetition, the aperture and shutter systems are covered in the following chapter, though they too are camcorder components.

Power Supply

All major camcorder systems run on electrical power. Even the lens needs electricity for its zoom motor. Typically, a wall plug and cord supply 120-volt AC power to a charger/converter. The charger/converter changes the high-power AC to low-voltage DC power and sends it to the camcorder to run its systems and charge its battery.

Many countries use 240-volt AC power, rather than the 120-volt that is standard in North America. Many professional camera systems can be set for either voltage.

Though all power originates as 120-volt AC, you will operate your camcorder on battery power much of the time, except in studio-type situations where camera movement is restricted and wall power is conveniently close. Running on DC power, you may be using any of three different types of battery types. Brick batteries fit on

Figure 11-10.
Professional batteries mount externally. (Anton Bauer)

or in the camcorder itself, **Figure 11-10**. They are light and convenient, but their capacity is relatively small. Block batteries are carried in cases on shoulder straps. Their capacity is often greater but they are heavy and relatively clumsy. Belt batteries are as powerful as block batteries, but more convenient to wear because their weight is distributed around the videographer's waist.

Brick, block, or belt styles are all suitable for news, documentary, and event (wedding)

Camcorder Power Checklist

Before shooting, use this checklist to ensure that you will have reliable power.

AC power

1. Locate power outlets.
2. Supply power cord and extension as needed.
3. Hook up and test power supply to camcorder.

Battery power

1. Set up battery charger and check condition of all batteries.
2. Mount fully charged battery on camcorder and test.
3. Recharge second battery for later use.

To decide how many batteries you need for continuous shooting, determine the time required to fully recharge a completely discharged battery. Then calculate how long a battery lasts in a typical shooting situation, where you are not recording continuously. For example, if a battery typically lasts one hour, but recharging takes twice as long, you can shoot continuously if you have three batteries and two external chargers.

A two-charger system for batteries A, B, and C

Hour	1	2	3	4	5	6	7	8	9
In camera	A	B	C	A	B	C	A	B	C
In charger 1	B	A	A	C	C	B	B	A	A
In charger 2	C	C	B	B	A	A	C	C	B

videography, especially if you also use battery-powered lights. Whichever battery type you select, you should have an external recharger and enough spare batteries to continue shooting while recharging.

Viewfinder

The viewfinder shows you the world as your camcorder sees it, because all finders display essentially the same image that is recorded. Viewfinders come in two basic styles: internal and external.

An internal finder is basically a tiny monitor inside a housing on the camcorder. Internal viewfinders have several advantages. They are fitted with magnifying lenses to make critical focusing easier. These lenses can also be adjusted so that people with vision problems can see the finder clearly without glasses. Safe inside their housings, they are shielded from image-degrading light (like a bright light falling on a TV screen). Because they are actual TV tubes, they show color and contrast much as they will appear on playback.

External viewfinders, **Figure 11-11**, are small LCD (liquid crystal display) monitors attached to the camcorder body.

External finders also have advantages. Their larger image can be easier to see, except in very bright light. Because they can be seen from a wide range of angles, they permit greater flexibility in positioning the camcorder. By freeing the videographer from pressing the internal finder to one eye, they greatly reduce fatigue during lengthy shoots. Because they are obviously TV screens, the user tends to look *at* them rather than *through* them. This makes it easier to evaluate images as two-dimensional compositions, rather than as the real world viewed through a tiny window.

External viewfinders have some problems as well. Their color, contrast, and sharpness cannot equal those of a tube display, so they deliver a less accurate representation of the image being taped. Also, their picture is easily washed out in bright light.

Many external viewfinders can be fitted with hoods for outdoor use; and some newer screen designs can be viewed in bright light.

Whether internal or external, all viewfinders display information about what the camcorder is doing. Different models show different data, but the finder information in **Figure 11-12** is typical.

The information display is yet another reason for looking *at* the viewfinder rather than *through* it. Because it warns you that your battery is out of power or that you have neglected to insert a tape, the viewfinder display can often avert major problems.

Most camera viewfinders are adjustable. At the start of each shooting session, set up the:

- ***Displays.*** To minimize the text overlays that can obscure parts of the image, program your finder to show only the information you want.
- ***Brightness and contrast.*** Set these values to match the light in which you are shooting.

Figure 11-11.
An external LCD viewfinder.

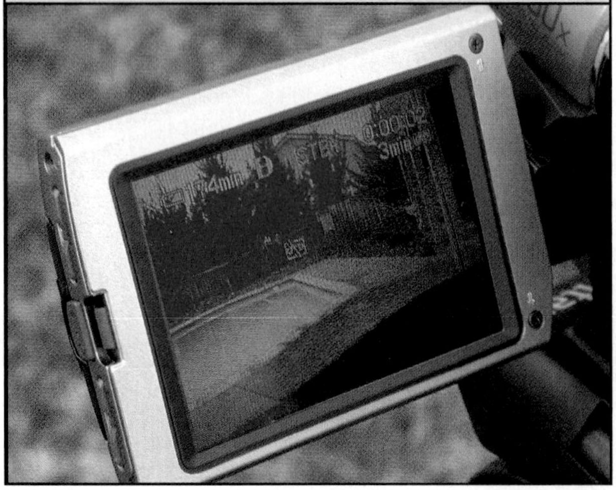

Figure 11-12.
Typical viewfinder information.

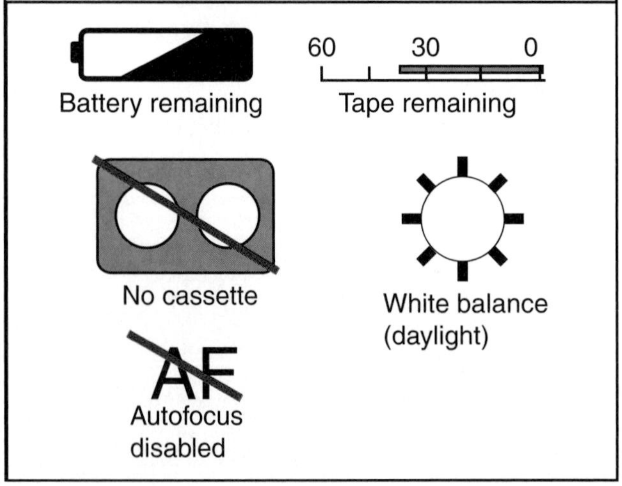

- ***Diopter correction.*** For internal finders, customize the focus to your eyesight by adjusting the diopter control until the finder screen is sharp.

In setting diopter correction, sharpen the text overlaid on the viewfinder rather than the image beneath it. Remember that you are not trying to focus the picture on the screen but the screen itself.

Monitor

If you are shooting at the advanced amateur level, an external color monitor is highly desirable; if your production is professional, it is indispensable. A standalone production monitor should have at least a nine-inch screen, and a 13-inch model is better, **Figure 11-13**. Camera-mounted LCD monitors are useful, but they are too small and low-resolution to use in evaluating video quality.

An external monitor performs two vital tasks. It lets you check exposure, lighting, color, and sharpness as shots are being taped; and it allows other people (like a director or script supervisor) to see exactly what is being recorded.

Separate monitors may be impractical for programs such as weddings and documentaries. Also, when making long or complex moving shots, you may need to disconnect the monitor cable from the camcorder.

To ensure accurate color display, the monitor should be set up at the start of each shooting session. All professional and many consumer camcorders can output standard color bars for use in adjusting the monitor. If your camcorder cannot generate bars, set it up as follows:

- Set the white balance on the camera, as discussed later in this chapter.
- Aim the camcorder at a standard color model card, obtainable at most larger camera stores.
- Adjust the monitor for most pleasing picture.

High-quality professional production monitors offer both AC and battery operation, and can be calibrated to the same standards as studio monitors. For less-elaborate productions, nine- or 13-inch consumer-grade monitors are available, **Figure 11-14**, at far lower cost, with AC/DC power and the composite and/or S-video inputs required for a quality picture.

Avoid monitors that have only RF connectors, since most camcorders cannot use RF cables.

Figure 11-13.
A reference monitor. (JVC Corporation)

Figure 11-14.
An AC/DC TV set can be used as a location monitor.

Videotape

Set up and test recording tapes as carefully as any other component of your videographer's kit. In particular, verify that the tape has been blacked, that it will record a signal properly, and that it is identified with a unique label. To double-check that a tape has been blacked, play back a few minutes of it from the start of the tape and verify that the screen is black and the time code readout is functioning.

Tapeless Video Recording

Eventually, all videotape will probably be replaced by one or more tapeless recording systems. Any review of the specifics of these systems will become obsolete almost immediately, due to rapid advances in computer storage capacity, speed, and miniaturization, along with advances in disc recording technology and new types of storage hardware. In general, however, the current trends are likely to continue.

Storage types

Three alternatives to tape recording are common:

- **Hard drives.** As hard drives grow ever smaller in size, larger in capacity, and faster in access time, they can be incorporated directly into camcorders (or in some cases, installed in computers, which are then fed real-time data directly from the camcorders).
- **Flash media.** Very small plug-in chips can store enormous amounts of data, then plugged into computers, for data transfer to a hard drive.
- **Discs.** Disc recording media (often smaller in size than DVD discs) can be burned directly in recorders built into the camera.

Advantages

Tapeless recording offers three powerful advantages over tape:

- **Random access.** Even digital tape is a linear system: to reach any point in a recording, you must roll through all the footage that precedes it on the tape. Tapeless recording permits instant access to any point in the data.
- **Archival storage.** Once the contents of hard or flash drives are burned onto disk, they form a nearly permanent record (direct disc recording, of course, omits this step). In addition, perfect quality backup copies can be quickly created for insurance. Even digital tape gradually loses its magnetic signal.
- **Instant editing.** The more powerful laptop computers become, the more they are used to edit high-quality video in the field. Tapeless recording eliminates the need to slowly transfer tape data to the computer before working on it.

Instant editing means that news and sports features can be ready for airing almost as soon as shooting finishes. On documentary and fiction program shoots, editors working in parallel with directors can examine footage and determine the need for re-takes or additional coverage.

Overall, the advantages of tapeless recording are unmistakable.

Preblacking recording tapes is discussed in Chapter 9. Procedures for preparing hard drive, disc, and flash memory recording media vary from one system to another.

Next, verify that the tape will record properly by taping any convenient subject and then playing it back. If the tape is defective, the horizontal lines and/or rolling picture will warn you of the problem before you record actual program footage. If you see any problems, mark the tape defective and discard at once.

Finally, make sure that the tape is properly labeled, as suggested in the sidebar, *Identifying Videotapes.* By systematically labeling every single tape, you can help prevent the disaster of destroying previously recorded material by mistakenly reusing a tape.

To further prevent taping over existing material, it is essential to start each session by "rolling down to raw stock." If it is a new tape, play from the head of the tape to about one minute in order to verify that it is blank. If it has footage on it, roll through the recorded material until you are certain that you are past the last shot and into blank tape.

In either case, begin your new recording session at that point on the tape.

White Balance

With the videotape prepared, the next task is to adjust the camcorder's various exposure systems to suit the shooting conditions. You begin this process by setting the white balance. ***White balance*** refers to the overall color cast of the light in which you are shooting. Outdoor light tends to be slightly bluish, **Figure 11-15**, while lightbulbs cast a warmer, redder light. To the human eye, a sheet of paper looks equally white in both kinds of light because the brain automatically compensates for differences in color tint. Unlike a brain, however, a camcorder can only record colors exactly as its chip and circuits

Identifying Videotapes

Many beginners are casual about labeling videocassettes, until they are taught a painful lesson by accidents like these: a used but unlabeled tape is reused, thereby erasing the original footage, or a tape is misplaced because its contents are not evident from its label. Here are some guidelines for the proper identification of miniature videocassettes.

Miniature face labels

Small format tapes offer very little writing space, so give the production a code instead of a title.

- Program work title: Prod. 013
- Tape type and number: Cam orig #03

Miniature spine labels

- Program work title: 013
- Tape type and number: Orig #03

Identifying the tape's purpose (camera original) may seem unnecessary when shooting; but later, when you edit the program, you will need to distinguish camera tapes from assembly tapes, work tapes, and other types.

Mini-DV cassette labels.

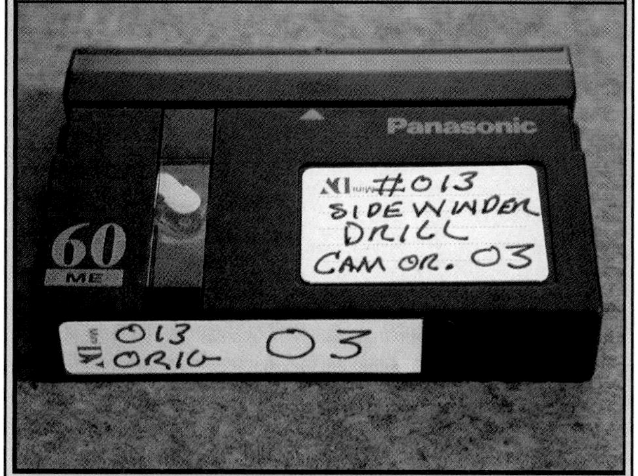

receive them. Without adjustable white balance, outdoor scenes would look too blue and/or indoor scenes too red. The videographer's job is to set the camera's white balance to match the color of the light at each new location.

Different levels of camcorders offer different degrees of white balance control. Simpler models set white balance automatically. Circuits in the camera analyze the incoming light, determine its overall color cast, and adjust color sensitivity to produce a tint-free white on playback.

Figure 11-15.
Outdoor light is naturally bluish.

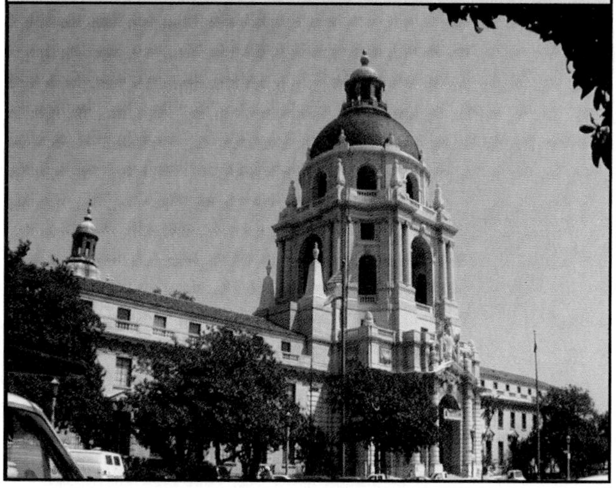

You do not have to think about white balance at all, but you must accept whatever setting the camcorder makes.

More flexible camcorders include a white balance switch for selecting outdoor, indoor, or fully automatic settings. Some cameras include a separate position for fluorescent light. With this type of white balance control, try using the automatic setting only when you are forced to shoot too fast to worry about subtleties of color balance. In many camcorders, the individual settings will usually deliver better-looking results.

Still more sophisticated camcorders include fully manual white balance. To use it, hold a white card so that its image fills the entire frame. See **Figure 11-16**.

Figure 11-16.
Setting white balance manually.

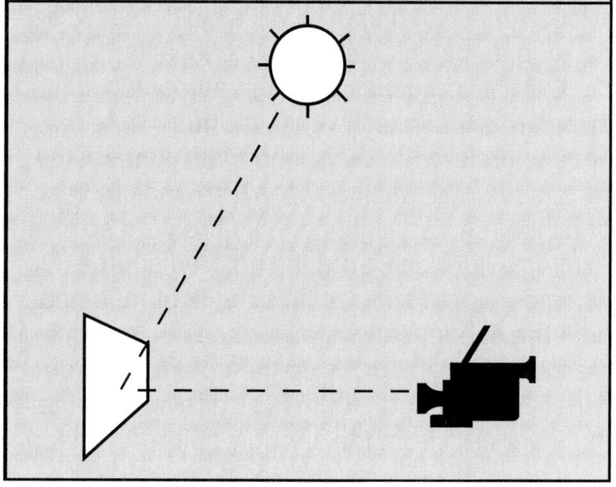

Color Temperature

The overall color cast of a light source is described as its ***color temperature***, and is expressed in degrees on the Kelvin temperature scale. The higher the color temperature, the bluer the light. Here are some typical light sources and their color temperatures ("K" is read as "degrees Kelvin").

Typical color temperatures.

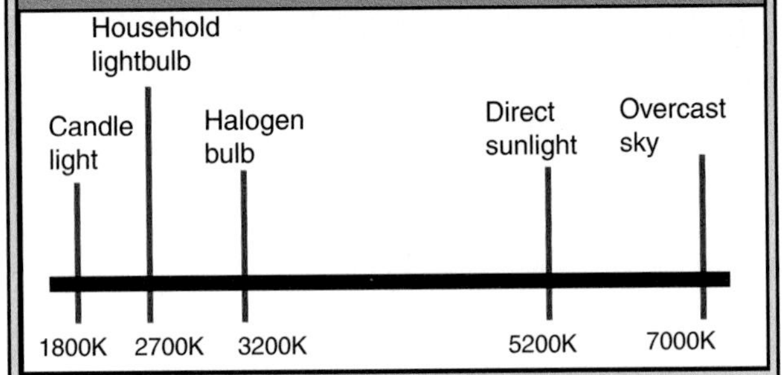

The daylight or outdoor white balance setting usually matches the 5000-5500K of sunlight and the indoor or incandescent setting matches the 3200K of halogen movie lights. (Almost all halogen bulbs, including work lights and home bulbs are 3200K.)

Many modern camcorders have separate white balance settings for fluorescent lights. Fluorescent light differs from both natural and incandescent (lightbulb) illumination. Try both fluorescent and outdoor settings when shooting in fluorescent light, observing the effect on a color monitor. With warm fluorescent tubes and/or yellowed light fixture diffusion panels, you may find that the indoor setting yields the most pleasing color. If you have full manual white balance, use it instead, to match the fluorescent lighting precisely.

When you activate the white balance circuitry, the camcorder reads the light reflected from the card and sets itself to record it as neutral white. A professional videographer will typically set white balance to match every shooting location.

Even if you usually use the automatic setting, there are situations in which manual white balance is preferable. When the main light source is an odd color, such as yellowish sodium vapor or bluish mercury lighting, hold the card so that it reflects light only from the colored source. Or, when a colored surface tints the light reflected onto the scene, set white balance while aligning the card so that it picks up the colored reflections. This method can reduce or even eliminate the color cast from the strongly colored surface.

The best white balance systems add fully-adjustable override controls. Once you have set the white balance, you can gradually make the overall tint warmer or cooler until you obtain exactly the color you prefer. When making adjustments, use a good external color monitor to check the effect. Whatever your system, it is essential to select (or at least check) the white balance setting each time you move to a new shooting location. Poor white balance can be corrected somewhat in postproduction, but the results are rarely as good as if the camera had been set correctly in the first place.

Lens Setup

The process of recording video images begins at the camcorder's lens, which gathers light rays reflected from subjects and focuses them on the imaging chip, **Figure 11-17**.

Most general purpose video camcorders are fitted with zoom lenses that allow you to change image sizes, either between shots or continuously while shooting.

Figure 11-17.
Cutaway of a zoom lens with optical stabilization. (Canon USA)

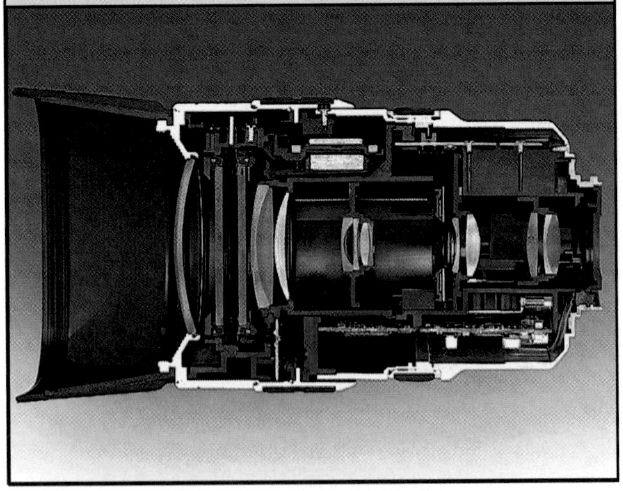

Lens Mount

On most cameras, the lens is permanently attached. A few more sophisticated models have lens mounts that permit you to replace one lens with another, **Figure 11-18**. However, changing lenses during a shooting session is uncommon, for two reasons. First, the alignment between a lens and the imaging chips behind it is so critical that it is often safer to mount and calibrate a lens in a studio or lab.

Figure 11-18.
A camcorder with interchangeable lenses. (Canon USA)

All camcorders with interchangeable lenses are three-chip designs. Certain models, notably from Canon, are designed to allow changing lenses in the field.

Second, lenses are usually changed to accommodate the demands of very different shooting situations. For example, you may exchange a general-purpose zoom lens (perhaps 5mm-60mm in some formats) for another zoom with a much longer range (such as 40mm-480mm) in order to shoot wildlife at great distances from the camera.

If your camcorder does have interchangeable lenses, you should always check to see that the lens is firmly attached to its mount, without any wiggle ("play").

Front Element

The next task is to set up the front end of your lens, including cleaning, installing a filter, and mounting a lens hood.

Cleaning

Visually inspect the lens to ensure that it is free of dust and smudges. Dust can be spritzed away with a blower bulb and brushed off with a lens brush. Smudges should be removed by cleaning, as described in the sidebar *Cleaning the Lens.* Any dirt on a lens will degrade its quality; in bright light, the depth of field (focus) of the lens may be so great that the front element (glass) of the lens itself is sharp, including all the dust and smudges on it. Also, a dirty lens increases light flare.

Filter

If your lens is not recessed behind a protective glass window (and if it is threaded to receive filters), you should mount a "transparent lens cap" called a UV, 1A, or skylight filter to keep the lens clean. In very bright lighting conditions, a ***neutral density filter*** can replace the clear model. Neutral density filters reduce the light intensity without affecting color. A ***polarizing filter***, **Figure 11-19**, can be installed to enhance the appearance of the sky. Never install more than one filter at a time, to avoid degrading image quality and showing the filter edge in the picture. A filter should be cleaned exactly like the lens itself.

Lens hood

If the lens is mounted outside the camera body and the front end is threaded, you may wish to mount a lens hood to help keep direct light off the lens. When selecting a lens hood, check it to verify that it does not appear in the picture when the lens is in the extreme wide angle zoom position. If it does, use the hood only for telephoto applications like sports and wildlife videography.

Figure 11-19.
A polarizing filter. Clouds in a blue sky... are enhanced by a polarizer. (Tiffen)

If you have mounted a filter, screw the lens hood onto the filter's front threads.

Focus and Aperture

For setup purposes, decide whether to use the autofocus system or focus the lens manually, **Figure 11-20**. Most camcorders default to autofocus, so you set manual focus by disabling the auto mechanism. In the same way, the camera defaults to autoexposure mode. If you wish to adjust the aperture manually, disable the autoexposure system. Make sure that you are in manual mode by looking for an exposure readout in the viewfinder. Some camcorders display actual f-stops (2.0, 2.8, 4.0, 5.6, etc.), while others display increments above or below the automatically determined aperture (+1, +2, +3; -1, -2, -3, etc.). Check your camcorder manual to learn whether each increment is one-half or one-third stop.

Figure 11-20.
Auto/manual focus controls. (Canon USA)

Zoom

Always check your zoom lens to verify that it is functioning correctly. If the zoom is motor-driven only, a malfunction means that the camera is essentially inoperative. If the zoom can be operated manually (usually by a ring and handle on the lens), it is less likely to malfunction.

Image stabilization

Image stabilization is a mechanism for automatically steadying shaky pictures. Stabilization systems use sophisticated circuitry to detect camera shake by comparing each frame with the one that preceded it. (Some systems analyze the actual camcorder motion rather than changes in the images.) If just some of the pixels are different from those in the preceding frame, the system concludes that the subject is moving and no compensation is required. In frames A and B in **Figure 11-21**, the pixels

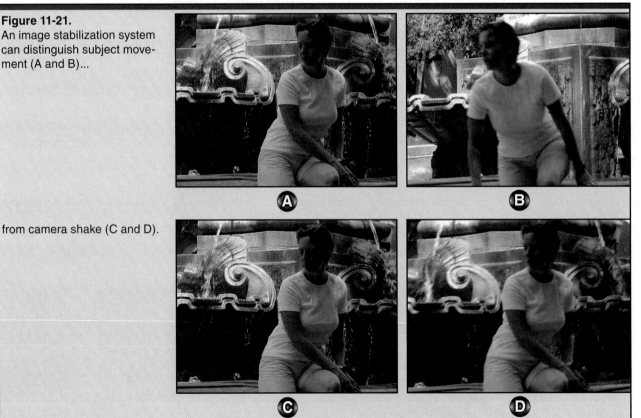

Figure 11-21.
An image stabilization system can distinguish subject movement (A and B)...

from camera shake (C and D).

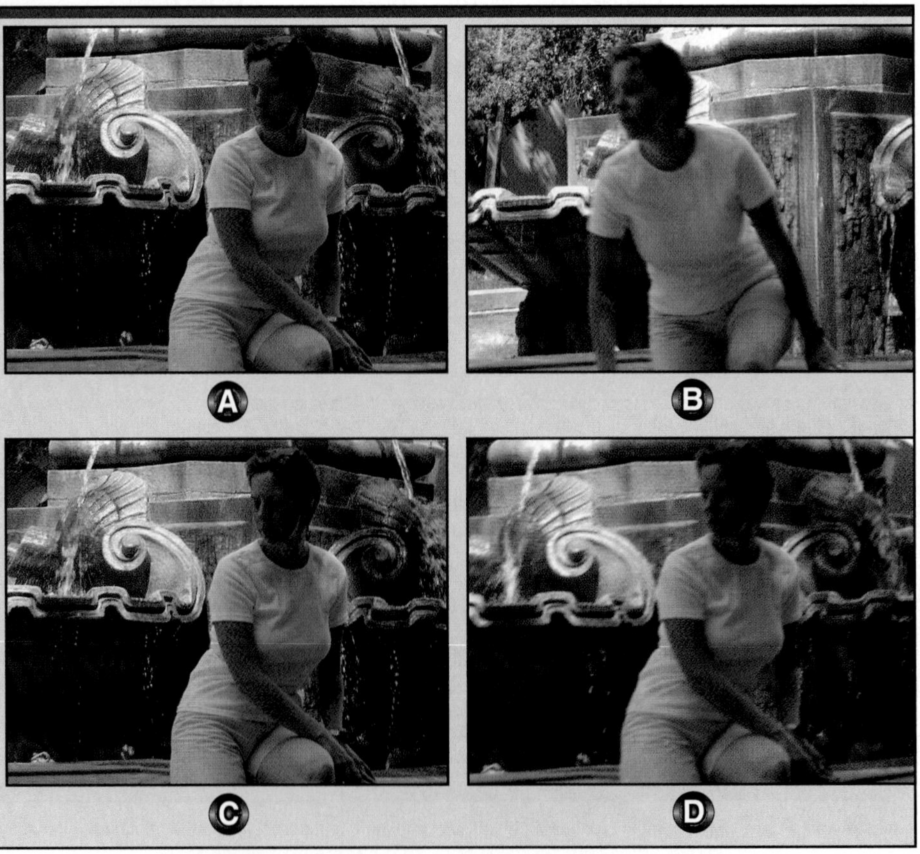

Figure 11-22.
An optical image-stabilizing system. (Canon USA)

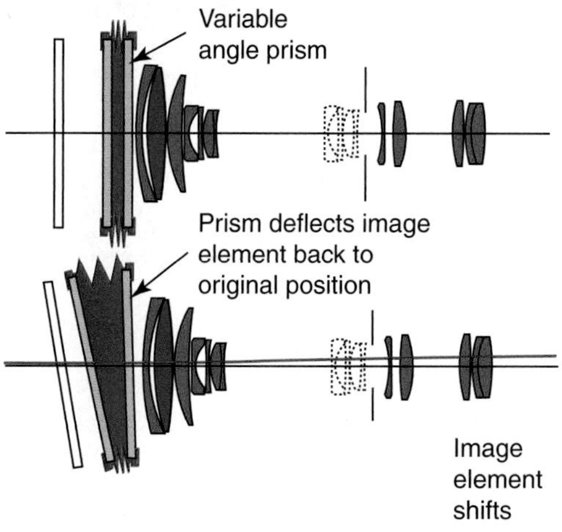

recording the subject have moved with her, but the background remains the same.

If all the pixels have shifted in the same direction, the system identifies camera shake and moves the pixels back to their original locations in the frame for recording. An image stabilizing system would recognize frame D of Figure 11-21 as *camera shake* because the entire image has shifted. (So-called "fuzzy logic" functions can recognize when all pixels are shifting because the camera is moving, rather than just shaking.)

Image stabilization should not be confused with camera stabilizers, which reduce camera shake rather than compensate for shaking's effect on the image.

There are two different systems of image stabilization, optical and electronic. Optical stabilization shifts the path of the light on its way through the camera lens, so that the same image strikes the chip. See **Figure 11-22**. Electronic stabilization records on only part of an oversize chip. As the image shakes, the active recording area shifts in the opposite direction to compensate.

Both systems work very well, and image stabilization is a highly desirable feature on any camcorder. If you have image stabilization, by all means use it for hand-held work.

Some manufacturers recommend disabling stabilization when the camcorder is mounted on a tripod. You should test your particular system and decide whether or not to do this.

All Systems Ready

With the major camcorder systems set up and checked out, you are ready to begin the actual process of videography. That is the subject of *Chapter 12, Camera Operation.*

Chapter Review

Answer the following questions on a separate piece of paper. Do not write in this book.

1. The imaging chip of a camcorder may be a CCD or a "______."
2. *True or False?* Digital video cameras use an A/Z converter to digitize the analog signal.
3. Tapeless video recording systems store information on a hard drive, ______, or disc.
4. An external color monitor should have at least a ______-inch screen.
5. *True or False?* Clean paper towels and distilled water should be used to clean the camera lens.

Technical Terms

Camera dolly: A rolling camera support. A cart dolly resembles a wagon; a pedestal dolly mounts the camera on a central post.

Color temperature: The overall color cast of nominally "white" light, expressed in degrees on the Kelvin scale. Sunlight color temperature (5200K) is cooler (more bluish) than halogen light color temperature (3200K).

Image stabilization: Compensation to minimize the effects of camera shake. Electronic stabilization shifts the image on the chip to counter movement; optical stabilization shifts parts of the lens instead.

Tripod: A three-legged camera support that permits leveling and turning the camera.

White balance: The camera setting selected to compensate for the color temperature of the light source that is illuminating the subject.

Cleaning the Lens

No matter how careful you are, you will have to clean your lens from time to time; so here are the supplies and techniques you will need. Cleaning supplies include lens fluid, lens cleaning tissues, and a blower bulb tipped with a soft brush. You can often find these items sold as a kit.

A combination brush and blower, with lens cleaning fluid.

In lens care, a good rule of thumb is, never do more than the minimum required to get the surface clean. To observe this rule, use the following methods in order, continuing to the next method only if the previous one fails to complete the job.

1. Making sure that your mouth is dry, blow short puffs of air across the lens to blow away loose dust.
2. Wipe the brush in concentric circles around the lens, starting from the center, while repeatedly squeezing the bulb to blow air through the brush.

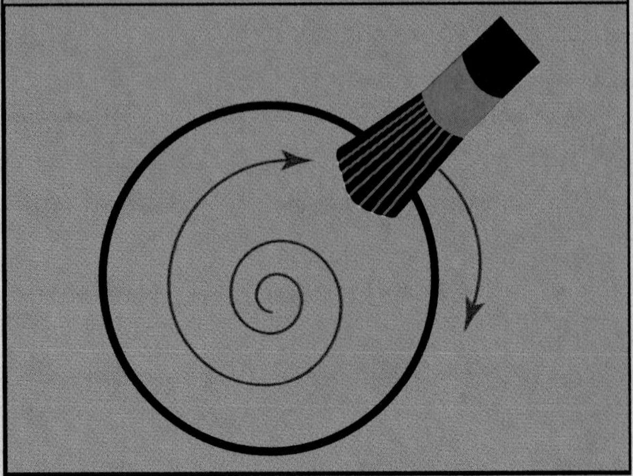
Brush the lens in an outward spiral pattern.

The lens is the most delicate component of a camcorder. (JVC Corporation)

3. Dispense a single drop of lens cleaning fluid onto the center of the lens; then crumple a sheet of lens tissue, grasp it between thumb and two fingers, and gently wipe the lens in a spiral motion, beginning in the center. When you reach the rim, crumple a second tissue and gently dry the lens, using the same spiral motion.

Never use facial tissues or paper towels instead of lens tissues, because they can leave lint on the lens or even scratch it. Also, avoid the cloths sold for cleaning eyeglasses if they have been chemically treated; however, microfiber lens cloths are effective and will not harm the lens coating.

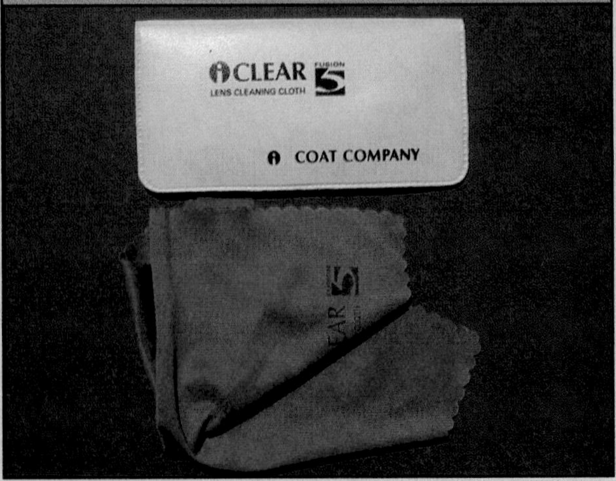

Used gently, microcloths intended for cleaning glasses are safe and effective on lenses.

12 Camera Operation

Objectives

After studying this chapter, you will be able to:

- Operate all camcorder systems.
- Establish and adjust camera settings as needed for each shot.
- Obtain pleasing pictorial compositions.

About Camera Operation

Just as gourmet foods are more than the recipes on which they are based, quality videography is greater than the sum of its recording techniques. The process is more art than science, and watching skilled videographers at work is like observing chefs in their kitchens. What makes them artists? What is the difference between a true videographer and a casual snapshooter? The answer has to do with the characteristics of the footage recorded: its quality, quantity, and usefulness.

Footage Quality

Professionally videotaped images display high technical and esthetic quality. That quality comes not only from technique but from attitude. Good videographers are fanatical about achieving the best images possible. This is their area of expertise and they love to utilize their skills and experience. They are never satisfied. Setup position, lighting, focus, and camera movement can always be improved. For a professional videographer there is no such thing as "good enough."

Footage Quantity

Amateurs typically videotape everything only once. They record a view or a piece of action and then move on to the next one. Videographers are just the opposite, often shooting far more material than will appear in the finished program. They know that when shots are less than adequate (or omitted accidentally), they often cannot be taped at some later date. A wise videographer ensures that everything is recorded, and recorded properly, the first time, **Figure 12-1**.

Footage Utility

Though good videographers are fanatical about quality and patient about shooting enough footage to achieve it, they never forget that the whole purpose of their art is to enhance the effectiveness of the overall program. Videographers are not working alone but reinforcing the intentions of the director, the writer, and in most professional work, the client. Working in teams does not mean that videographers cannot be creative, but they exercise their creativity in productive collaboration with their colleagues and in service to the needs of the production.

Figure 12-1.
A project can never have "too much" footage.

In short, professional quality videotaping results not only from skill with the hardware but from an attitude toward the work. Having said that, we can begin to look at the techniques of videography. We will cover working with lenses; setting focus; controlling aperture, shutter, filtration, and gain; moving the camera; and composing the shot.

Working with Lenses

Video recording starts at the camcorder lens, which turns light rays into visual images and determines the pictorial qualities that those images will display.

Lens Focal Length

The way in which a lens renders images is determined by its focal length. The *focal length* of a lens is the distance (expressed in millimeters) from the front element of the lens to the focal *point*, the point at which the light rays intersect on their way to the focal *plane*, where the chip is positioned, **Figure 12-2**.

Lenses are named according to focal length, for example, a "5mm lens" or a "75mm" lens. Originally, the focal length of every lens was fixed. To change focal lengths, a photographer had to dismount one lens and replace it with another. Today, however, almost all video optics are zoom lenses, whose focal lengths can be varied continuously between their shortest and longest positions. In the Mini-DV format, for instance, the focal length of a zoom lens might range from

Figure 12-2.
The principal dimensions used in describing lenses.

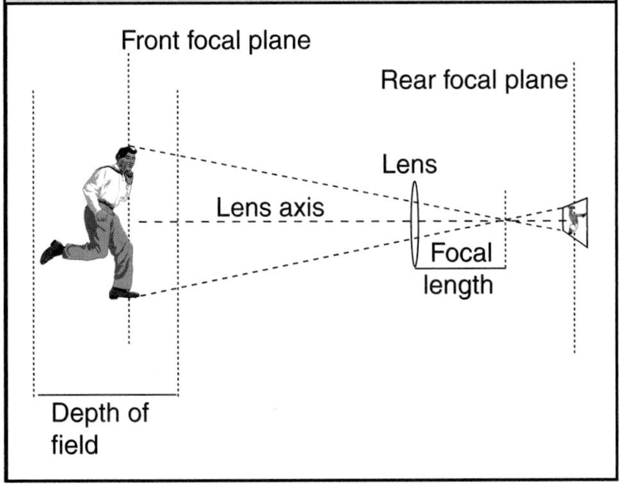

3mm to 48mm. Since the ratio of 48 to 3 is 16/1, this lens would be described as a "16-to-1 zoom."

As explained in the sidebar Wide Angle or Telephoto?, zoom ranges are dependent on the size of the imaging chip. Most of this discussion uses mini-DV focal lengths as examples.

Within a zoom range, the number of focal lengths is infinite. With a 3mm to 48mm zoom you could, theoretically, set the focal length at, say, 35mm or 35.2mm or 35.205mm (though you would need optical calibration equipment to do so). In actual production, however, professionals rarely refer to lenses by exact focal length. Instead, they divide them into three broad groups of focal lengths: normal, wide angle, and telephoto.

- Normal focal length lenses produce images that closely resemble human vision.
- Wide angle focal lengths include a wider field of view than normal lenses.
- Telephoto focal lengths show a narrower, more restricted field of view.

For proper contrast with "wide angle," long focal length lenses should probably be called "narrow angle" instead of telephoto.

Though wide angle, normal, and telephoto settings are actually just focal lengths within the zoom range of a single lens, they are usually discussed as if they were separate lenses. So when you read "telephoto lens" or "wide angle lens," remember that this typically refers to a setting on a zoom lens. As we discuss the optical qualities of wide, normal, and telephoto lenses, keep in mind that each category includes a whole range of individual focal lengths. In our mini-DV example, 3mm, 3.5mm, and 4mm are all different degrees of wide angle lens, and 14mm, 26mm, and 45mm are all telephoto focal lengths.

Finally, remember that the pictorial qualities of a particular focal length are dependent on the size of the camera chip on which the image is created. For example, the 8mm focal length that is considered "normal" on a 1/4″ chip would be an ultra-wide angle "fisheye" lens on a 35mm still camera. This is explained more fully in the sidebar *Wide Angle or Telephoto?*

Lens Characteristics

All lenses have three characteristics that affect the way the videographer uses them: angle of view, maximum aperture, and depth of field.

Angle of view

Every lens "sees" within a certain arc or angle of view, **Figure 12-3**:

- Wide angle lenses typically cover 70°-140°.
- Normal lenses cover 35°-70°.
- Telephoto lenses cover 18°-35°.

These distinctions are somewhat arbitrary — a "normal" lens that covers 69° is not significantly different from a "wide angle" lens that covers 71°. Because of their broader coverage, wide angle lenses are useful in confined shooting areas, where longer focal lengths would not cover the whole action area.

Figure 12-3.
Angles of view of typical lenses.

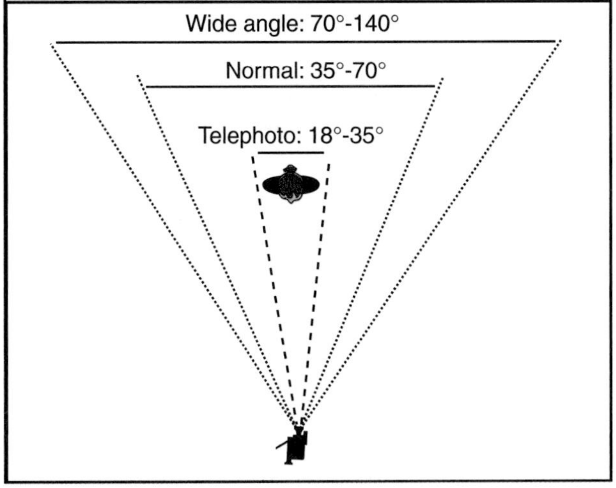

Wide Angle or Telephoto?

Whether a lens is wide angle, normal, or telephoto depends on the size of the imaging area for which it is intended. The image created by the lens must cover the whole area of the chip that is recording it. Since lens images are circular, while video frames are rectangular, the minimum image diameter must slightly exceed the diagonal of the frame, as you can see from the following illustration.

The diameter of the lens must be slightly greater than the diagonal of the frame.

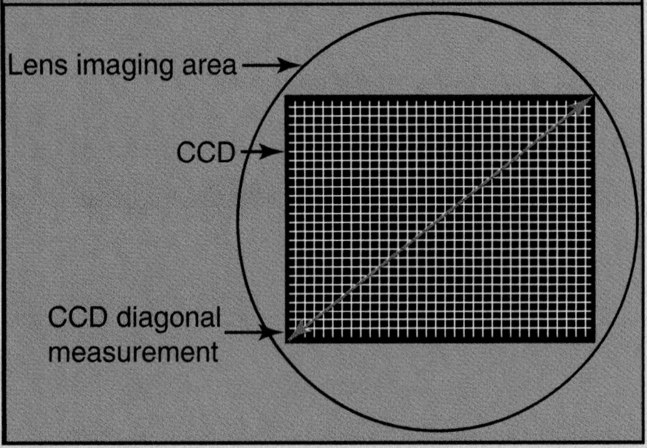

As a rule of thumb, a lens focal length slightly longer than the image diagonal produces a "normal" perspective. Shorter focal lengths yield wide angle images and longer ones create telephoto images. Since the visual perspective of any focal length depends on the area upon which the image is formed, a focal length that is wide angle on a frame of 35mm still film is telephoto on a 1/4″ chip. To illustrate this, here are typical focal lengths for three image formats:

- A nominally 1/4″ chip for a mini-DV camcorder.
- A 1/2″ chip for a full-size professional camcorder.
- A 24mm x 36mm frame of 35mm still film.

CCD chips for different visual formats.

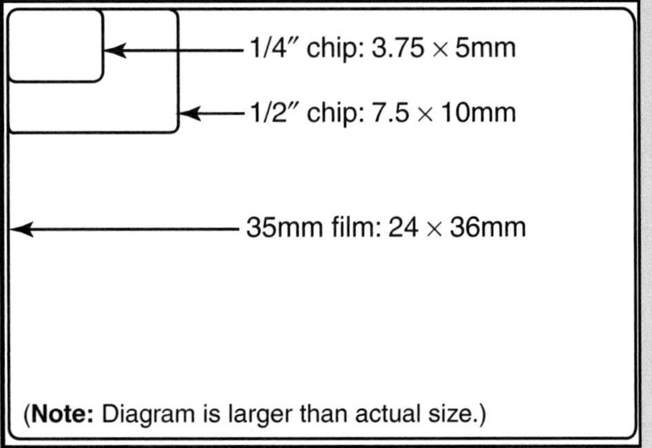

The still camera is included because many people are familiar with the lenses typically used on it.

By comparing 35mm focal lengths to those of different video chips, you can get a relative idea of how different focal lengths behave in different image formats. As you look at the chart, notice three things:

- The 35mm focal lengths vary somewhat from straight multiples (like 17.5mm, 35mm, 70mm, 140mm) because still camera lenses are more typically designed in slightly different focal lengths (18mm, 35mm, 70mm, 135mm).
- The comparisons are not exact because the 3-to-2 proportions of the 35mm frame are different from the 4-to-3 proportions of the video frame.
- Because of the mathematics of optical design, the difference (expressed in millimeters) between focal lengths is quite small at the wide angle end of the spectrum, but much larger at the telephoto end. For example, a 4mm wide angle lens produces an image twice as large as a 2mm, though the difference between them is only 2mm. At the telephoto end, a 48mm lens also has twice the magnification of a 24mm lens; but the increase in focal length required to double the image size is 24mm, rather than just 2mm.

1/4″ Chip Focal Length	1/2″ Chip Focal Length	35mm Focal Length	Lens Characteristics
2mm	4mm	18mm	extreme wide angle
3mm	6mm	28mm	wide angle
4mm	8mm	35mm	mild wide angle
6mm	12mm	45mm	short normal
8mm	16mm	55mm	normal
12mm	24mm	70mm	long normal
16mm	32mm	105mm	short telephoto
24mm	48mm	135mm	telephoto
32mm	64mm	200mm	long telephoto
48mm	96mm	300mm	very long telephoto

Maximum aperture

As a rule, normal and wide angle lenses have larger maximum apertures (lens openings) than telephoto designs. That is because the "speed" (light gathering ability) of a lens depends on both the lens diameter and the focal length. For a lens of any diameter, the shorter the focal length, the larger the maximum potential aperture.

In fixed focal length designs, wide angle lenses should theoretically have larger maximum apertures than normal lenses. Because of other constraints in designing wide angle lenses, however, normal focal lengths often have the largest maximum apertures.

Since a zoom lens has a fixed diameter but a variable focal length, its maximum aperture often will become smaller as the lens zooms from wide angle to telephoto. For example, a lens that can open to f/1.4 in the wide angle position may open only to f/2.8 at full telephoto. That is why you will often see identification markings on lenses like "f/1.4–2.8." This indicates f/1.4 is the maximum aperture at full wide angle, while f/2.8 is the maximum at full telephoto.

Some professional zoom lenses are designed to maintain the same maximum aperture throughout their zoom ranges.

Understanding how maximum apertures work in zoom lenses will help you use them effectively. In low light levels, for example, you may wish to restrict yourself to normal or wide angle lens settings, because their larger maximum apertures admit more light. They will yield better quality images in dim illumination.

Depth of field

Depth of field is the distance, in front of and behind the plane on which the lens is focused, in which objects in the image appear sharp, **Figure 12-4**. Like maximum aperture, depth of field grows smaller as the lens focal length grows longer. At any given aperture and distance from the subject, wide angle lenses have the deepest depth of field and telephoto focal lengths have the shallowest.

You need to be aware of these characteristics when following sports or other action in telephoto mode, and when setting focus for zooming. A scene that looks perfectly sharp in wide angle may turn soft when you zoom to telephoto. That is because the wide angle setting's greater depth of field includes the distant subjects, but the telephoto setting's shallower range does not.

The Pictorial Qualities of Lenses

The only true optical difference among lens focal lengths is angle of view: the arc within which a lens gathers light to form images. All other seeming differences are merely appearances. Nonetheless, those appearances are quite visible to viewers, and videographers use them to shape the images they create. We may

Figure 12-4.
A—Depth of field. B—In this telephoto shot, depth of field is limited to the area between the red lines.

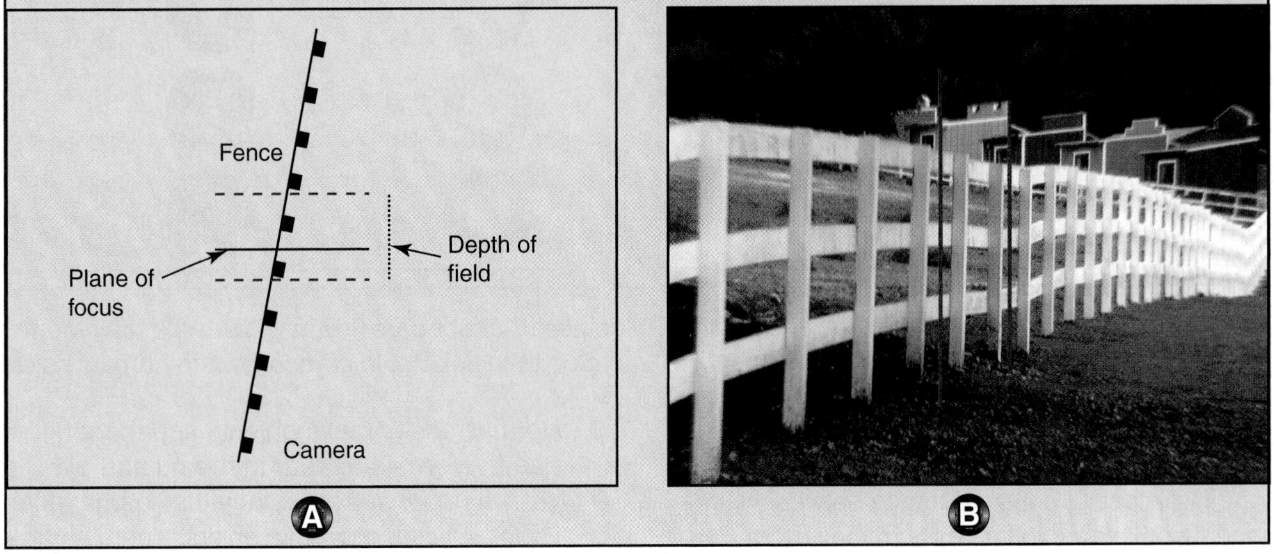

Apertures and f-stops

How much light gets through a lens depends on two factors:

- Its focal length (such as 10mm or 100mm or 200mm).
- Its maximum aperture: the full diameter of the lens opening.

The light-gathering ability of a lens is called its speed; a sensitive design is called a fast lens. The speed of a lens is determined by a simple formula:

$$\frac{\text{focal length}}{\text{maximum aperture}} = \text{lens speed}$$

According to this formula, if a lens has a focal length of 100mm and a diameter of 50mm, then its speed would be 100 divided by 50, or f/2.

$$\frac{100\text{mm}}{50\text{mm}} = f/2$$

(By convention, apertures are indicated with a lower-case “f” and referred to as f-stops.)

A 200mm lens with the same 50mm diameter would have a speed of f/4.

$$\frac{200\text{mm}}{50\text{mm}} = f/4$$

An iris diaphragm varies the effective aperture.

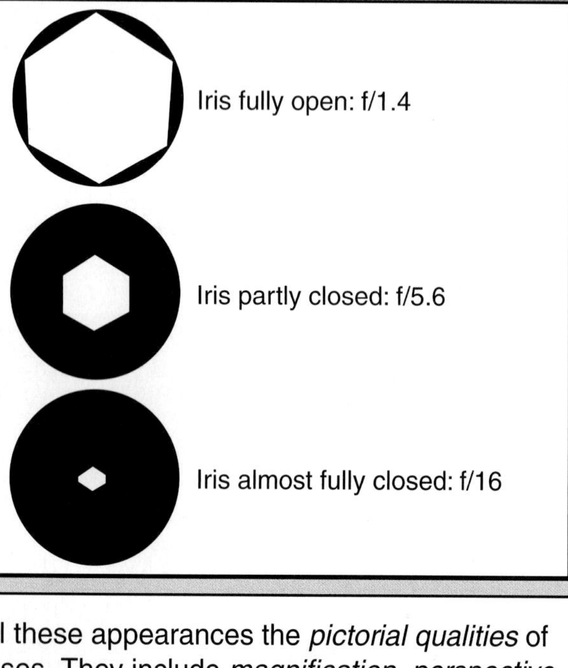

Aperture range

Every lens includes an internal iris diaphragm (“iris” for short) that can be adjusted to progressively reduce the diameter of the lens opening. For this reason, a lens has a whole range of apertures.

Each reduction in aperture changes the result of the equation, so:

$$\frac{100\text{mm}}{50\text{mm (iris wide open)}} = f/2$$

$$\frac{100\text{mm}}{12.5\text{mm (iris partially closed)}} = f/8$$

Apertures on most lenses have been standardized at f/1.4, f/2, f/2.8, f/4, f/5.6, f/8, f/11, f/16, and f/22. Though these numbers seem arbitrary, a moment’s study will reveal that they are all based on square roots.

Smaller equals higher

As you can see from the diagram, the smaller the f-stop number, the larger the aperture. Commonly, f/1.4 is the largest aperture and f/22 is the smallest (though certain specialized lenses can “stop down” to f/64). By another convention, the speed of a lens is expressed as its maximum aperture. Thus, an f/2 lens is said to be “two stops faster” than an f/4 lens because it includes the two extra stops (f/2.8 and f/2) below f/4.

Apertures grow smaller as their f-stop numbers grow bigger.

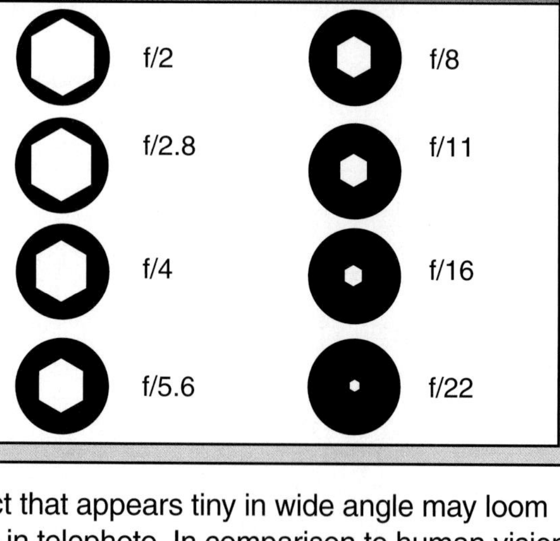

call these appearances the *pictorial qualities* of lenses. They include *magnification*, *perspective*, *movement*, and *distortion*.

Magnification

The most obvious difference between wide angle and telephoto lenses is magnification. An object that appears tiny in wide angle may loom huge in telephoto. In comparison to human vision:

- *Normal lenses* make objects seem about the size they would appear to the human eye.
- *Telephoto lenses* make objects seem bigger.
- *Wide angle lenses* make objects seem smaller.

Apparent magnification is an effect of angle of view, because every lens must fill the frame from edge to edge. Since the screen size does not change, filling it with a 5-degree view will make objects appear much larger than filling it with a 90-degree view. This concept is illustrated in **Figure 12-5**.

First-time camcorder users quickly discover that zooming from wide angle to telephoto enlarges the subject in the frame.

Perspective

Perspective is the illusion of depth in the picture, the apparent distance from the camcorder to the farthest part of the scene. Wide angle lenses exaggerate this virtual depth, making distances seem greater and faraway objects seem smaller. Telephoto lenses reduce apparent depth, seemingly bringing distant objects closer. **Figure 12-6** shows the same setup photographed with both wide angle and telephoto lenses. Notice that much of the scene is excluded from the telephoto shot because of the lens' narrow angle of view. Objects in the telephoto image are much larger than in the wide angle view, and the buildings appear squeezed together.

In Figure 12-6, the lens remains the same distance from the subject at both wide angle and telephoto settings. But if you move the camera you will achieve a quite different effect. **Figure 12-7** shows the setups used to shoot the two frames in Figure 12-8. In **Figure 12-8A**, the two subjects are photographed with a wide angle lens. In **Figure 12-8B**, the same subjects are photographed with a telephoto lens, but with the camera placed much farther away. As a result, the near subject is exactly the same size as in the previous wide angle setting, but the far subject appears much larger and closer.

Figure 12-5.
Wide angle and telephoto lens fields of view. Because the frame remains the same size, filling it with a telephoto view makes the subject larger.

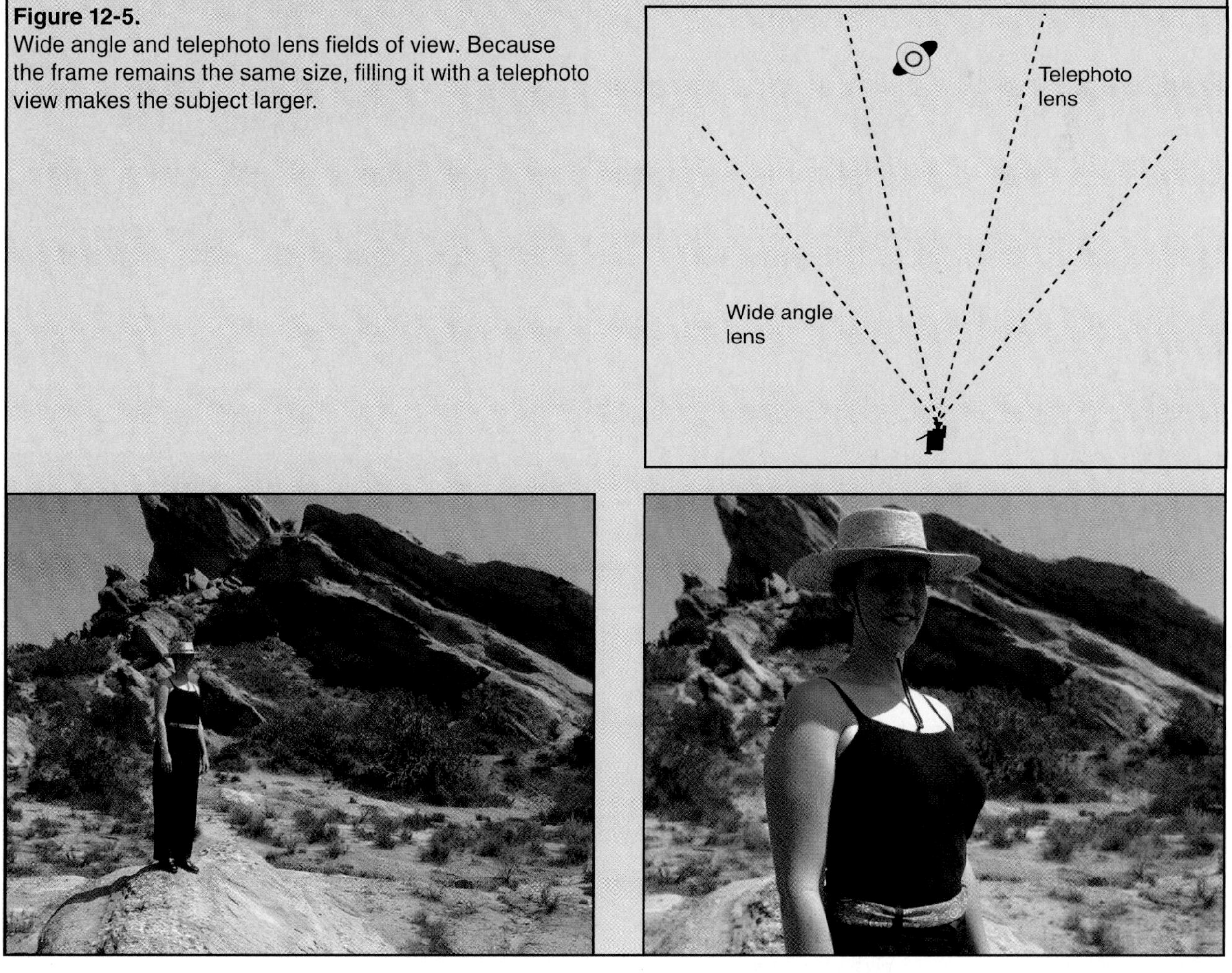

Figure 12-6.
Lenses change perspective. A—Wide angle lens. B—Telephoto lens.

Figure 12-7.
Setup used to photograph Figure 12-8.

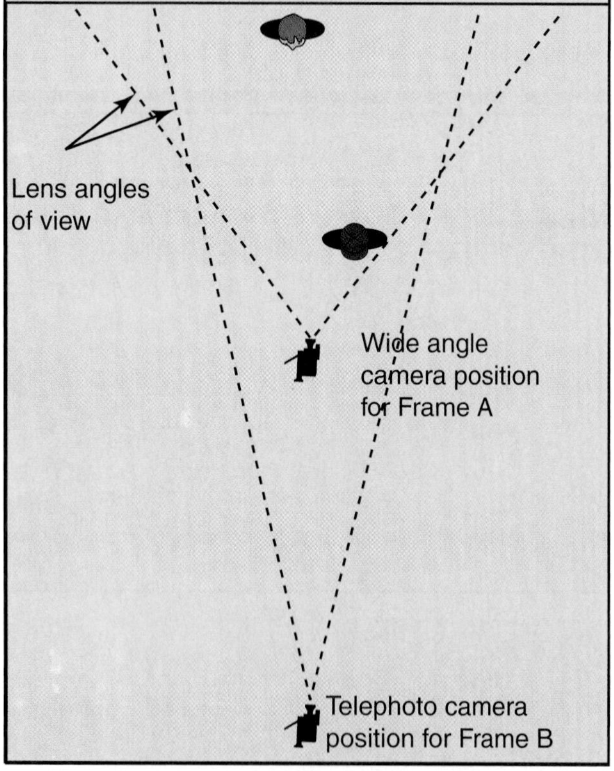

Figure 12-8.
A—Wide angle image. B—Telephoto image with camera moved to maintain the foreground subject size.

Movement

Because wide angle and telephoto lenses affect the apparent depth of an image, they also modify apparent movement toward or away from the lens. In wide angle shots, moving people and objects seem to rush quickly toward the camera or away, **Figure 12-9**. In telephoto shots however, fore-and-aft movement is minimized, and people do not change size dramatically as

Figure 12-9.
In a wide angle shot, both the truck and the sidewalk grow dramatically as they approach the camera.

they move forward or backward. For these reasons, wide angle lenses are useful for dynamic action scenes like fights and chases, while telephoto lenses convey a contrasting sense of distance and detachment.

Distortion

Since wide angle and telephoto lenses affect the way objects "shrink" and "grow," depending on their distance from the camera, they also can distort the shapes of those objects. Note how the wide angle lens in **Figure 12-10** makes the vehicle look longer and more powerful, while the telephoto shot seems to squeeze it together.

Be careful in using wide angle lenses on the human body. **Figure 12-11** shows what happens when wide angle perspective stretches arms and legs. For the same reason, try to avoid wide angle lenses in portraiture, since the extra depth can distort chins and noses unpleasantly.

Moderate telephoto settings are generally preferable for portraits of women. However, a mild wide angle lens can make a male portrait look more rugged.

Focal length and composition

To arrange images effectively with wide angle and telephoto lenses, first decide whether you want to create a composition in three dimensions or two. Wide angle lenses help you enhance the illusion of a third dimension in a composition by emphasizing apparent depth. Telephoto lenses let you organize compositions on the screen surface, rather than "behind" it. As you can see from **Figure 12-12**, both approaches can yield very effective compositions.

Finally, keep in mind that the lens focal lengths most like human vision are in the so-called normal range. When you want viewers to subconsciously think that they are looking directly at the action, rather than at a TV screen, normal focal lengths render images without the exaggerations and distortions that announce the presence of a lens.

Figure 12-10.
Lens distortion. A—The wide angle lens elongates the vehicle. B—The telephoto lens appears to squeeze it together.

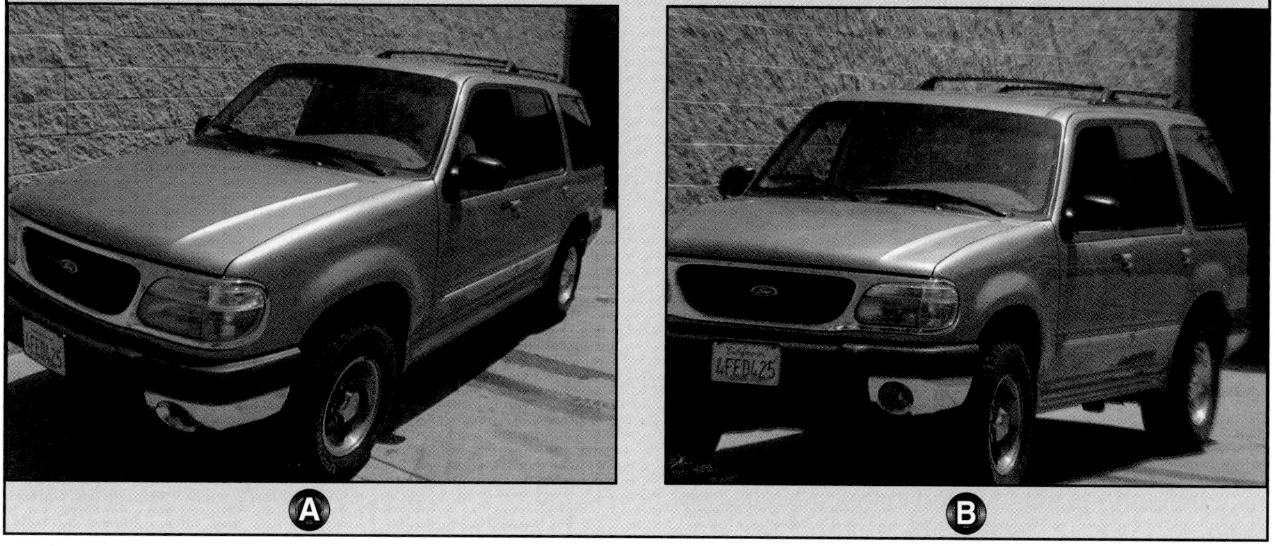

Figure 12-11.
Wide angle lenses can distort human proportions and make unpleasant-looking portraits.

Figure 12-12.
A—Choose a wide angle lens to enhance the illusion of depth. B—A telephoto lens creates the composition on the surface of the screen.

A

B

Operating the Zoom

As a rule, use the zoom sparingly, because the seconds required to move between the start and end compositions are sometimes a waste of screen time. When you zoom, try to set the shot up in such a way that the editor can remove the zoom by cutting directly from the start to the finish, **Figure 12-13**.

A zoom cannot be easily removed if it contains action. For example, if you shoot someone walking, removing the zoom will result in a jump cut, as you can see from **Figure 12-14**.

The solution is to follow the shot by reshooting the same action from a different angle, beginning before the zoom started in the previous shot, to keep the action continuous, **Figure 12-15**.

Zooming techniques

Skilled videographers often prefer to zoom manually, for better control. (If your motorized zoom lacks a manual mode, it probably offers variable speed operations.) Here is a technique for professional looking zooms: begin slowly, accelerate during the first half of the zoom, then decelerate during the second half. Finish the zoom slowly and come to a stop. Never correct a zoom after you have initially stopped by "bumping" it a bit closer or farther away. Either accept the end composition where you first halted or else retake the shot.

At one time, zooms were rarely used in feature films. Nowadays, though, the prejudice against them has lessened.

Optical vs. Pictorial Lens Qualities

Photographers know that wide angle and telephoto lenses render image magnification, perspective, movement, and distortion in markedly different ways. Optical engineers, however, insist that there is only one difference between wide angle and telephoto lenses: angle of view. All other seeming differences are merely illusions.

Which side is right? Both. On the one hand, the many photos in this chapter clearly show the pictorial differences between wide angle and telephoto images. On the other hand, if you enlarge the center of a wide angle image until it matches the angle of view of a telephoto, the perspective of the two shots will be almost identical.

In the photo panel, A is a wide angle scenic and B is a telephoto shot made from exactly the same spot. Photo C is the very center of the wide angle image, enlarged to match the telephoto picture (hence its relatively poor quality).

Studying the building doors and railings, notice that the scale and perspective in the blowup are essentially the same as those in the telephoto shot. From an optical standpoint, there is no difference between wide and long lenses except angle of view.

A—Wide angle shot. B—Telephoto shot. C—Center portion of the wide angle shot.

A B C

Figure 12-13.
With no action in the shot, it is easy to cut out the zoom from wide angle to telephoto.

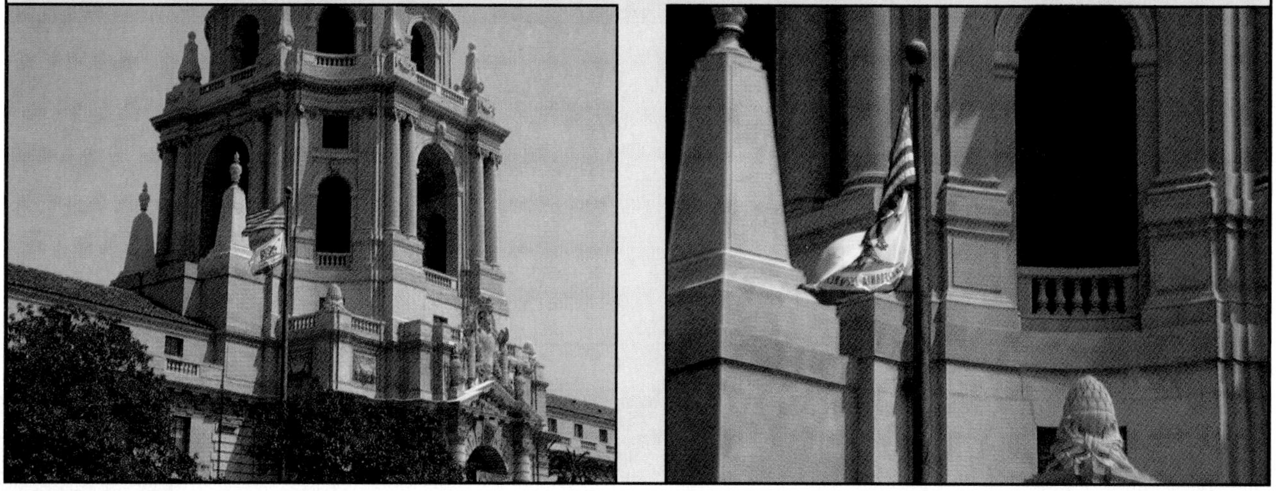

Digital Zoom

Digital zoom is an electronic feature that can magnify images beyond the ability of the lens to enlarge them. True zoom, called *optical zoom,* adjusts the lens to fill the entire imaging chip with a smaller and smaller angle of view. Because the entire chip is always used to form the image, the picture quality always remains the same. Digital zoom, by contrast, works by selecting a smaller and smaller percentage of the image on the chip and electronically enlarging it to fill the whole frame. Since the number

Figure 12-14.
Omitting the zoom from an action shot creates a jump cut.

Figure 12-15.
A—Before the zoom. B—Covering the action during the zoom. C—Shot A resumes after the zoom.

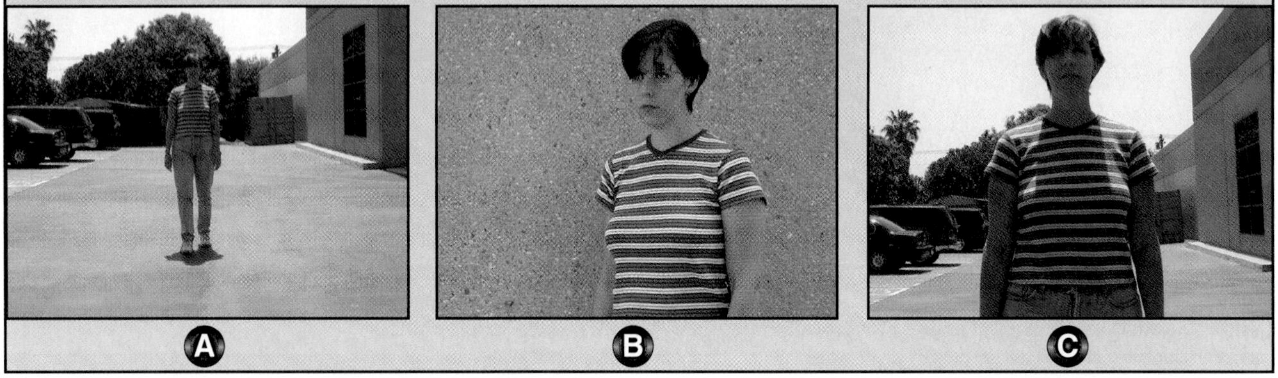

of pixels is fixed by the total number on the chip, using just part of the chip requires filling the frame with fewer pixels. The fewer the pixels, the worse the image quality, **Figure 12-16**.

Pixels (short for "picture elements") are the dots that make up the image.

Because image quality is degraded by digital zooming, this technique should be used only when subjects are too far away to be magnified adequately by optical means alone. Common applications for digital zoom include nature videography, shooting sports from the sidelines, and surveillance.

Figure 12-16.
Zooming in on a distant subject... optical zoom maintains full image quality... digital zoom degrades image quality.

Special Lens Filters

The front elements of professional lenses are machined with threads to receive screw-on filters. (The lenses of some compact cameras may not have filter threads.) Many filters used in film photography (such as starburst and color shift effects) are not employed as often in video because their effects can be supplied by software during the process of computerized editing.

As a rule, it is better to record the best possible unmodified images and then filter them digitally in postproduction. Any effect that modifies an original camera recording is permanent. Two types of filtration, however, can be achieved only in the camera: neutral density and polarization.

Neutral density filters

Gray-colored **neutral density filters** are used to cut down the incoming light, either to reduce the intensity of an exceptionally bright scene, or to force the aperture wider open in order to reduce depth of field.

Polarizing filters

A polarizing filter, or **polarizer**, is a sandwich of two rings. The back one threads onto the lens and the front one, holding the filter, rotates freely. By changing the orientation of the filter with respect to the lens, you can suppress small, bright glints of light (called **specular reflections**) from surfaces like metal, glass, water, and shiny paint. You can also reduce unwanted reflections on window glass, as shown in the accompanying illustration. Polarizers can also darken blue skies to make clouds stand out. A polarizer can do extra duty as a variable-strength neutral density filter. Generally, (though not always) the more you rotate the filter, the less light you admit to the lens.

A polarizing filter can reduce reflections, such as the glare off this store window.

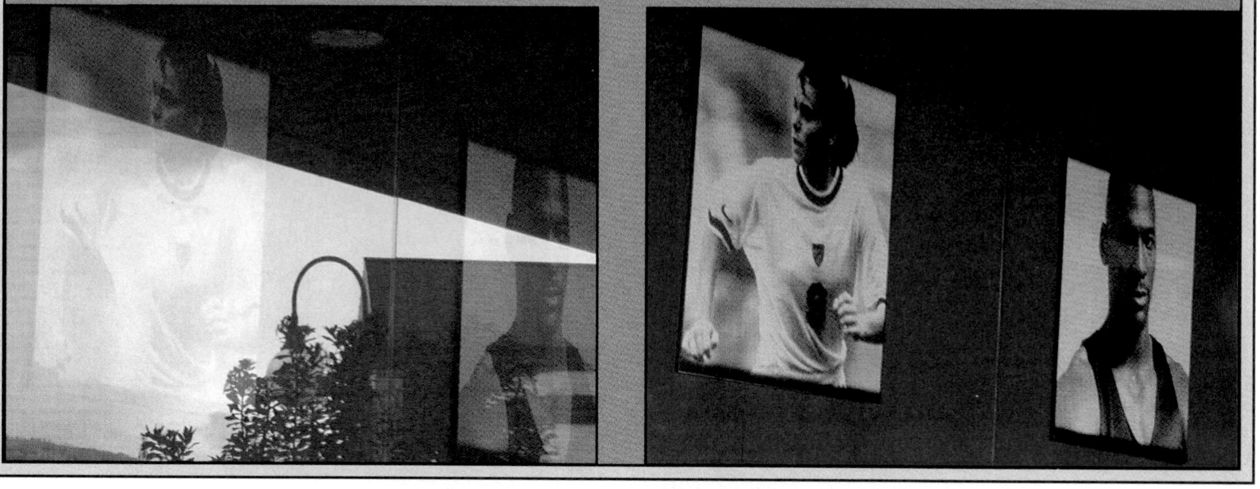

Setting Focus

Setting focus is the process of adjusting the lens so that subjects at a certain distance from the camera appear sharp on the focal plane. Since the imaging chip (CCD) is located on that plane, objects in focus there will be sharp in the video image. Of course, focus must be changeable because subjects may be videotaped at many different distances: a postcard six inches away from the lens, a group of people six feet away, a distant mountain six miles away.

Like your own eyes, optical systems cannot always keep everything from a few inches to the horizon in focus at the same time.

In very bright light and at wide angle lens settings, every object from the postcard to the mountain may be in reasonably sharp focus. On the other hand, when shooting in dim light with a telephoto lens, the depth of field (range of focus) may be as shallow as an inch or two. The reasons for this are explained in the sidebar *Depth of Field.* Almost all camcorders can compute and set focus automatically, and modern autofocus systems work remarkably well. But they are not perfect.

Autofocus

Autofocus is a camcorder system that adjusts the lens continuously and automatically

Depth of Field

Mathematically speaking, objects in an image are in perfect focus only if they lie on an imaginary plane surface that is precisely a certain distance from the lens. Objects that are behind or in front of that plane — even as little as a fraction of an inch — are not in theoretically perfect focus, as shown in the diagram. (Technically, this imaginary plane of perfect focus is called the "front focal plane" while the plane on which the chip lies is called the "rear focal plane.")

In reality, however, objects up to a certain distance behind and in front of the plane of "perfect" focus are also acceptably sharp, because they appear that way to viewers. This acceptably focused zone is called the depth of field. (Note that the depth of field in front of the plane of focus is shallower than the depth of field behind it, as illustrated. This is always the case.)

The depth of field in an image is determined by three factors working together: lens focal length, lens aperture, and the distance from the lens to the theoretical plane of focus.

Lens focal length

The longer the lens focal length, the shallower the depth of field. If you set up two camcorders with a 4mm lens and a 16mm lens, both at the same distance from the subject and using the same aperture (f-stop), the 4mm lens will yield a greater depth of field than the 16mm lens.

In theory, only a single mathematical plane is in perfect focus. In fact, the range of acceptable sharpness extends before and behind the plane.

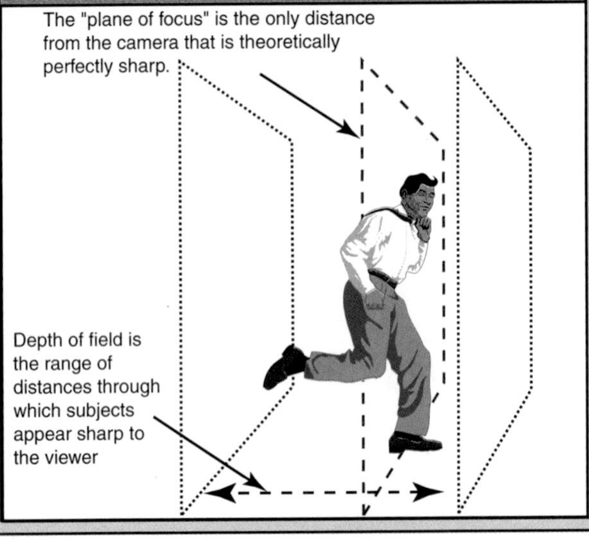

The "plane of focus" is the only distance from the camera that is theoretically perfectly sharp.
Depth of field is the range of distances through which subjects appear sharp to the viewer

The longer the lens focal length, the shallower the depth of field.

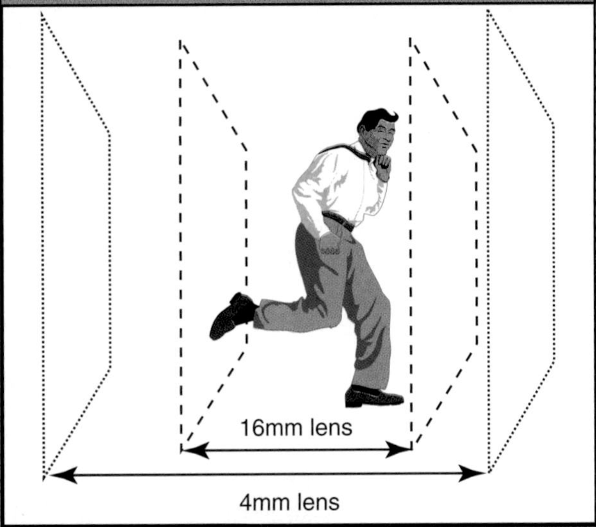

to keep objects in the image sharp. Though different camcorders employ different approaches to autofocus, most of them involve the same basic procedure. The system calculates the distance to objects inside a focusing zone within the image—usually the central portion—and a servo motor in the lens adjusts focus to the calculated distance. Most camcorders default to the autofocus setting, **Figure 12-17**.

Autofocus has several limitations that can make it unsuitable for professional shooting. If the subject is not within the focusing zone, it may be out of focus. If the camera and/or subject moves, the autofocus may not respond quickly enough to maintain uninterrupted sharpness. If unimportant objects move into the focusing zone, the system may focus on them, rather than on the center of

Figure 12-17.
Many camcorders use the center of the image for focusing. (Sue Stinson)

Lens aperture

The wider the lens aperture, the shallower the depth of field. If you set up those same two camcorders, both at the same distance from the subject and both using the same lens focal length, an f/16 aperture will produce the greater depth of field and an f/1.4 stop will produce the lesser.

Subject distance

The shorter the lens-to-subject distance, the shallower the depth of field. If your two camcorder lenses have both the same focal length and the same aperture, but one camera is set closer to the subject than the other, the more distant camera will realize the greater depth of field and the closer one the lesser.

Depth of field can be confusing to calculate because it is determined by focal length, aperture, and distance, all working at once. For example, some people think that if the depth of field produced by a telephoto setting is too shallow, they can fix the problem by switching to a wide angle setting with its inherently greater depth of field. The problem is that in order to match the view of the telephoto lens, they would have to move the wide angle lens closer to the subject. Since moving closer reduces depth of field, the eventual wide angle focus would be no deeper than the original telephoto focus.

The wider the aperture, the shallower the depth of field.

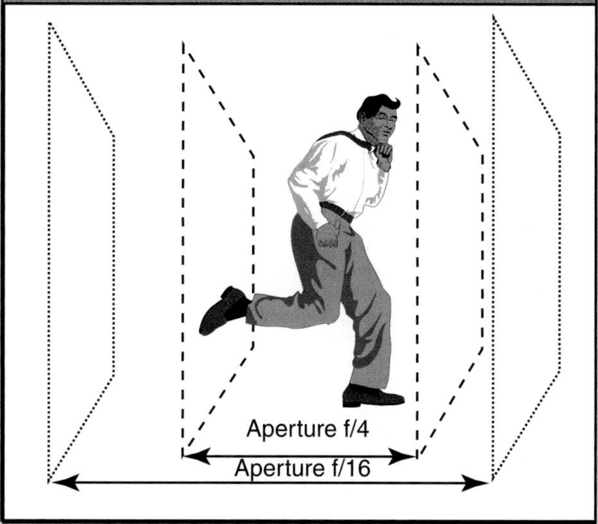

The shorter the distance, the shallower the depth of field.

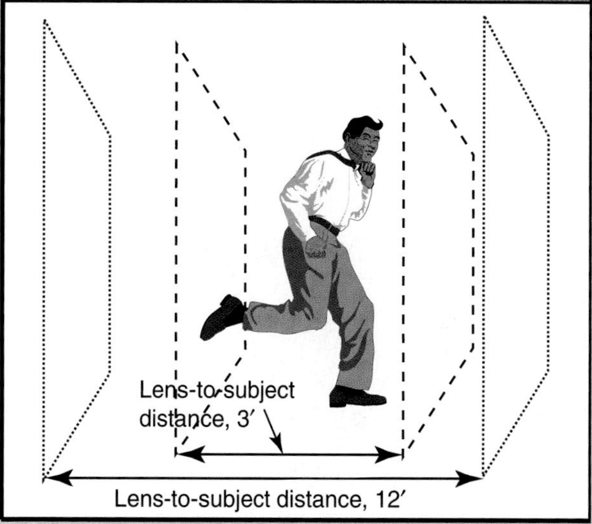

interest. For these reasons, better camcorders offer three alternate ways to maintain focus: autofocus lock, manual focus, and pulling focus.

Autofocus Lock

Autofocus lock uses the automatic mechanism to focus on a subject and then prevents that setting from changing, even if the subject moves. To use the autofocus lock system, focus on the subject, then engage the focus lock control on the camcorder.

When the shot is complete, remember to disable the focus-locking mechanism.

Locking the autofocus function is especially useful when the center of interest is outside the focusing area. Simply center the frame on the subject, focus the lens, lock the focus, and recompose the shot. The subject will stay sharp.

You can also use autofocus lock when unimportant foreground objects "steal" the focus from your subject matter. If you lock focus on the subject, the focusing system will ignore the intruding objects. In **Figure 12-18**, for example, the subject is taped as she moves along a sidewalk behind a series of hedges.

The autofocus system does not always work reliably during zooms, because the depth of field narrows as you zoom in. This means that a focus setting made at a wide angle focal length may lose sharpness as you zoom into a distant part of the scene. This is especially true of a zoom

Figure 12-18.
A—The subject is sharp as she starts to walk behind a hedge. B—As the foreground hedge fills the screen, the autofocus system shifts the focus to it, softening the subject. C—Locking the focus on the subject prevents this unwanted shift.

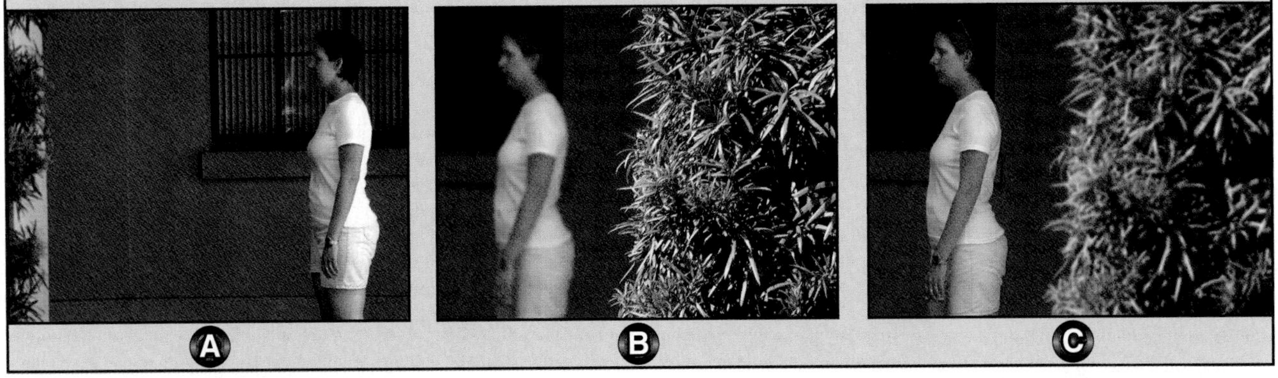

in, where the autofocus system has set focus in the wide angle start position, **Figure 12-19**. (In a zoom out, the start focus will be properly set and the system will usually adjust itself satisfactorily as the shot widens.)

For this reason, you should use the autofocus lock option to set up every zoom in, as shown in **Figure 12-20**. In most situations this procedure will keep your shot sharp throughout the range of the zoom.

Figure 12-19.
A scene focused at a wide angle setting... may not stay in focus at the end of a zoom-in.

Figure 12-20.
Zoom to full telephoto, allowing the autofocus system to focus. Lock the focus. Zoom out to your start position.

Manual Focus

In some situations, the autofocus lock method will not work because the autofocus mechanism fails to get a fix on the subject. This can happen in low light, when shooting through a window, or when the subject lacks features that the focusing system needs for reference. The procedure for focusing manually is similar to focusing for a zoom:

- Disable the autofocus system.
- Zoom to the full telephoto position to obtain the shallowest depth of field.
- Manually adjust the focus until the subject is sharp.
- Zoom back out to the desired composition.

Except in news and sports applications, most professional videographers use manual focus whenever practical.

In low light levels, a long zoom may require pulling focus, as described below.

Pulling Focus

Pulling focus is the technique of shifting manually from one focal setting to another during the shot.

This procedure is also called "follow focus," "racking focus," and other names.

Pulling focus is practical only when the camera has an external zoom lens fitted with a focus ring. It usually requires an assistant to revolve that ring to change focus. To prepare a shot for pulling focus:

- Wrap a narrow piece of white tape around the lens barrel, next to the focus ring.
- Focus the lens at the starting position and mark a reference line on the tape.
- Repeat the process for each additional focus position, **Figure 12-21**.

Focus pulling is often used to keep a moving subject sharp. To do this when the subject approaches the camera, set focus points at the far, middle, and near points of the movement. As the subject moves toward the camera, shift the focus continuously, using the reference marks on the tape.

Figure 12-21.
White tape marked with interim focus points.

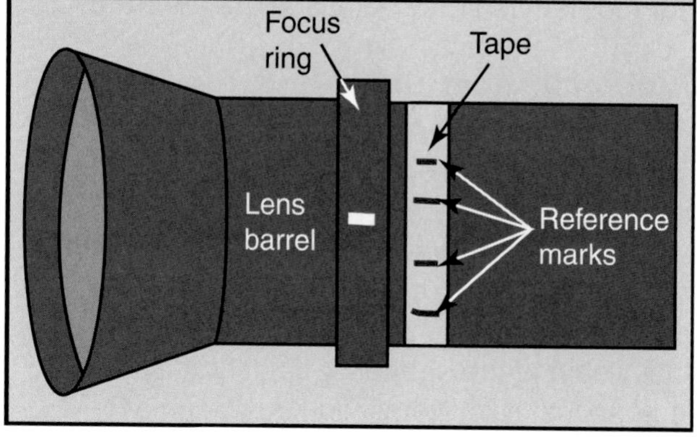

Controlling Exposure

Setting exposure means literally *exposing* the camcorder circuits to the exact amount of light required to form a quality image. If the imaging chip receives too much light, the picture will be too light (overexposed) or even "burnt out" to white. With too little light, the image will be too dark (underexposed) or "blocked up" to black. Sometimes the camcorder cannot handle the ultra-bright light in an environment like a beach or a ski slope. On the other hand, when taping at night or in a dark room, there may be too little light to form a good image. To solve the problems of too much or too little light, the camcorder relies on four exposure control systems: *aperture, shutter, gain,* and *filtration.*

Aperture

As explained in the section on lenses, the aperture is the hole through which light enters the camcorder. The bigger this hole, the more light gets in. The aperture is set by an *iris,* a circular opening inside the camera lens that can be expanded to nearly the full diameter of the lens or contracted until it is almost completely closed.

The aperture is usually polygonal rather than truly circular. That is why lens flares (from shooting into the sun or other bright lights) appear as multiple polygons.

Aperture Exposure Control

The autoexposure system built into every camcorder works by adjusting the aperture.

Automatic aperture

In auto mode, your camcorder sets exposure by analyzing the intensity of the incoming light and automatically changing the size of the lens opening, so that the same amount of light reaches the chip, regardless of the light level outside. Autoexposure systems adjust aperture only, leaving the shutter at its default speed. Autoexposure systems work so well that you can use them much of the time, as long as you avoid panning too quickly from, say, a very bright scene to a dark one.

When light levels are too different, try restaging the action to avoid the light change, or get a cutaway shot to replace the badly exposed transitional section when you edit the scene later.

Manual aperture

Many amateur and all professional camcorders also allow you to set the aperture manually. One way to do this is by locking the exposure:

- In auto mode, frame the scene that you wish to expose for.
- Lock the aperture.
- Make the shot.

Figure 12-22 shows a typical exposure problem that can be solved by this method. Notice that it is very similar to the focus problem discussed previously.

You can also set aperture in full manual mode to fine-tune overall exposure. This is especially useful when backlighting is a problem or when the center of the frame is much brighter than the rest of the image. Excessively bright areas are more irritating, visually, than overly dark ones. For this reason, it is often better to correct backlighting problems by lighting the foreground subject or moving it away from the bright background. However, when it is the subject that is too bright you can expose for it and allow the darker areas to "block up." To do this:

- Zoom in until the bright subject fills the frame.
- Lock the autoexposure.
- Zoom back out to the original composition.
- Make the shot.

Programmed aperture

Most camcorders offer programmed aperture settings to address various exposure problems. Though different models offer different programs, two of the more popular aperture-based options are backlight and spotlight.

Backlight. Backlight programs automatically open the aperture a certain amount to increase the exposure for dark subjects in front of bright backgrounds. Unfortunately, many backlight programs have only one setting. If it doesn't match the situation exactly, the results may not be satisfactory. If your camcorder has manual aperture control, it is better to compensate for backlighting by increasing exposure in small steps until you are satisfied with the result.

Spotlight. The spotlight program is the opposite of backlight. It is intended for theater performances and other situations in which the subject is brightly illuminated against a much darker background. The spotlight program reduces the contrast between the two elements, and improves the picture considerably.

Figure 12-22.
Original exposure. Autoexposure increases exposure for the column, overexposing the subject. Manual aperture retains original exposure.

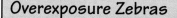

Overexposure Zebras

Many professional cameras warn the operator when part or all of the scene is too bright to record properly. Typically, this warning overlays the too-bright portions of the image in the viewfinder with a striped pattern. For this reason, it is called a zebra.

The most common form of zebra warning appears when part of the image is brighter than 100% of the camera's ability to record it. Another type is programmed to appear when the center of the image is too bright to record Caucasian flesh tones properly (about 70% of maximum brightness). Most professional camcorders offer both settings.

A striped overlay identifies the overexposed portion of the image.

Some cameras can be set to offer warnings in the viewfinder when skin tones are overexposed. (JVC)

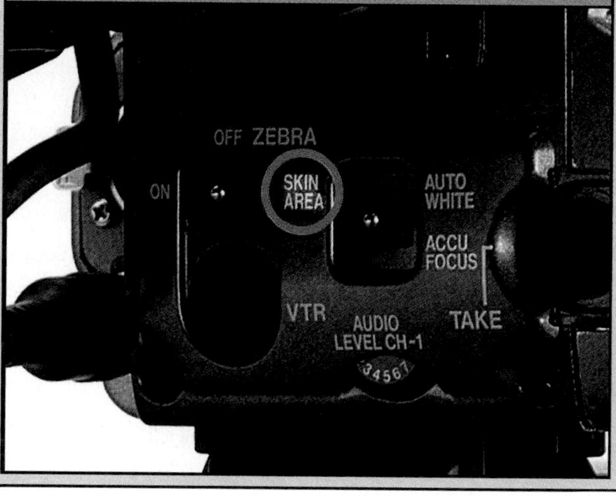

Shutter

While the aperture controls the size of the opening through which light is admitted to the camcorder, the *shutter* controls the amount of time allotted to forming each video frame (image). In a film camera, the shutter is an actual window that opens and shuts for each image. In video, however, the shutter is a set of circuits that control how long the incoming light is allowed to build up electrical charges on the chip before they are removed and sent on to be processed and recorded.

A few professional video cameras do use mechanical shutters, but not as film cameras employ them.

NTSC video is displayed at 30 frames per second, with every frame made up of two interlaced fields. This means that the effective normal shutter speed is 1/60 second. That is, when the shutter circuits are in "normal" mode, they allow the chip to build up electrical charges for each field for 1/60 second before collecting them and starting over.

Some interlaced systems combine the information from adjacent scan lines. This electronic processing can increase the effective exposure from 1/60 to 1/30 second. In terms of recording movement, however, the shutter speed is still 1/60.

Shutter speed usually remains locked at 1/60 second so that exposure control is handled entirely by the aperture. There are special situations, however, in which you may want to change shutter speed. For this purpose, many camcorders offer shutter control programs.

Programmed Shutter

Like exposure programs, shutter programs adjust one or more camcorder settings for you automatically. All you do is select the effect you want and the circuitry does the rest. Shutter programs have three purposes: forcing open the aperture, freezing rapid motion, and compensating for extreme light levels.

Portrait

The portrait program increases shutter speed moderately in order to force the aperture wider open. The wider aperture reduces depth of field so you can keep your portrait or still life subject sharp in the foreground while throwing a distracting background out of focus, **Figure 12-23**.

Figure 12-23.
The portrait program opens the aperture to throw backgrounds out of focus. (Sue Stinson)

As explained in the sidebar *Apertures and f-stops*, each time you double shutter speed, the aperture opens by one f-stop to compensate.

Sports

The sports program increases shutter speed more than the portrait program in order to sharpen fast action that might be blurred at the normal shutter speed. Sharpening the image improves detail but can make movement look jerky.

High light levels

The high-light-level program (often called "sand and snow" or some other proprietary term) increases shutter speed when outside conditions are so bright that the smallest aperture still admits too much light. Since each doubling of the shutter speed reduces incoming light by one full f-stop, a moderate shutter increase (say, to 1/125 second) is usually enough to compensate for the brightest scene. See **Figure 12-24**.

Most professional camcorders control excessive brightness by reducing the gain setting, enabling built-in neutral density filtration, or both.

Low light levels

Some shutters can be set slower than normal, typically at 1/15 of a second. This allows the CCD four times as long to form an image in very low light. The drawback is an obvious blurring of the action; and since the circuitry must record each frame twice for normal playback speed, the resulting movement is somewhat jerky. In general, a slow shutter should be used only when the light is otherwise too dim to record an image. A slow shutter can also create an eerie, dreamlike effect for music videos and other special-purpose applications.

Manual Shutter

Some camcorders allow you to change shutter speed manually. For instance, you can manually control depth of field in a portrait shot more precisely than with an automatic portrait program. Very high shutter speeds are often useful for scientific analysis and similar purposes.

As you manually adjust shutter speed, keep two important points in mind. First, increasing shutter speed always widens the aperture proportionately (in order to maintain the same exposure). This can seriously reduce depth of field, making it hard to hold focus at long telephoto settings. Second, shutter speeds faster than

Figure 12-24.
The high-light-level program reduces overexposure. (Sue Stinson)

about 1/125 second jerk and strobe too much to yield normal-looking movement. Because of these motion problems, exposure is adjusted less often by manual and programmed *shutter* settings than by *aperture* controls.

Gain

Gain is the electronic amplification of the video signal. In most camcorders this "video volume" can be turned up to compensate for low light levels. Unlike aperture and shutter, gain does not affect the amount of light used to form the image. Instead, it operates *after* the image has been coded as an electronic signal by increasing the strength of that signal. Gain control can operate in several different ways, depending on the sophistication of the camcorder:

- In simple units, the gain turns on automatically whenever light levels drop too low.
- Better models allow you to turn the gain on and off manually.
- Still more sophisticated cameras let you select different amounts of gain compensation, to suit lighting conditions precisely.

Professional video cameras are often designed to operate in normal light with a certain amount of gain enabled. That way, the videographer can compensate for excessively bright conditions by reducing the gain instead of using filtration.

Gain control is useful where the situation is otherwise too dark for recording; but it cannot improve the poor quality of the image. It can only brighten it, and the result is often washed out color, coarse detail, and video noise that appears on the screen as "snow," **Figure 12-25**.

Filtration

The final way to regulate exposure is by placing a neutral density filter over the lens. This is a gray filter that reduces the brightness of the light without changing its color. Popular neutral density filters are offered in shades labeled ND3, ND6, ND9, and ND12, which reduce incoming light by 1, 2, 3, and 4 f-stops, respectively. In unusually bright shooting conditions, neutral density filters enable the camera to capture a good quality image. Used at normal light levels, they can reduce depth of field by widening the aperture without increasing the shutter speed.

Filters for bright conditions

Very bright sunshine in reflective locations can overpower your camcorder's chip and circuitry. The result is flares of white, and smeary colors that "bloom" beyond their proper borders. In these conditions, an appropriate neutral density filter will reduce the incoming light to a level that the electronic circuitry can handle, **Figure 12-26**.

Filters for aperture control

Even when brightness is not excessive, you may want to reduce the incoming light in order to force open the aperture. You could do this by increasing shutter speed, as explained previously, but fast shutters often result in undesirable stroboscopic effects. To widen the aperture

Figure 12-25.
Gain control circuits can lighten the image, but the quality is often poor.

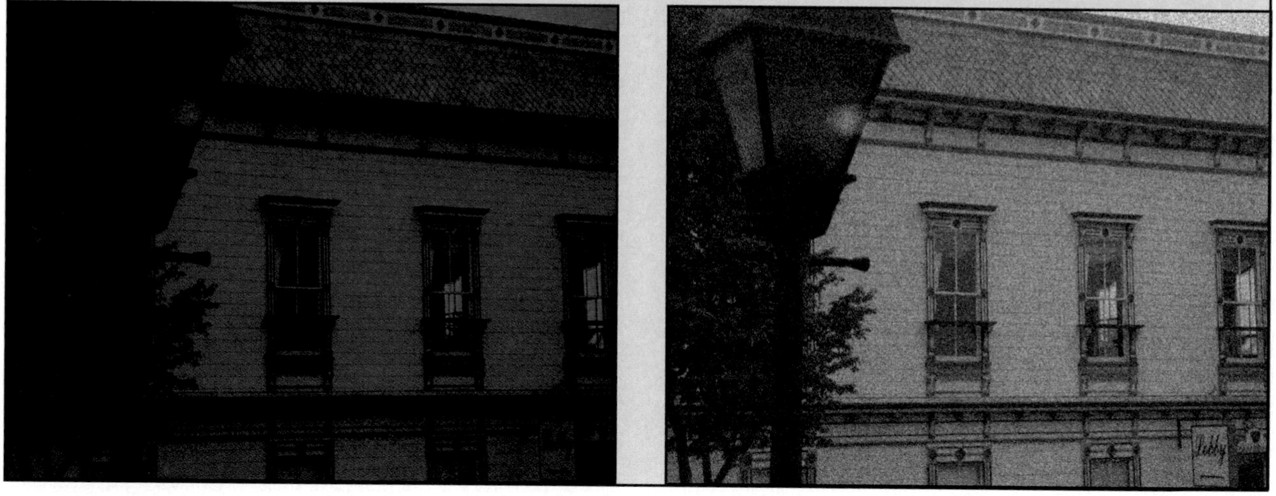

Figure 12-26.
Overexposure without filtration. With an ND3 (one-stop) neutral density filter. With an ND6 (two-stop) neutral density filter. (Sue Stinson)

without changing shutter speed, use a neutral density filter instead.

There are two reasons for working at a wider aperture: first, the reduced depth of field lets you separate your subject from the background by keeping it sharp, while the area behind it is soft. Next, almost all photographic lenses record their best images at apertures in the middle of their range. If, for example, the apertures of a lens range from f/2 (widest) to f/22 (narrowest), the lens will perform best between about f/5.6 and f/11. Since even average sunshine may require a very small aperture, a neutral density filter can improve picture quality and help control depth of field.

If, for instance, the autoexposure system sets the lens aperture to f/22, an ND9 neutral density filter over the lens will reduce the aperture three f-stops, to f/8, where the depth of field is extensive but not excessive and the optical quality of the lens is at its best.

Since the gain control is enabled only in low light levels and most filters reduce light levels, the two controls are rarely employed at the same time.

Operating the Camera

Managing the camcorder capably is so important that fine camera operators are highly regarded. No matter how good the acting, costumes, sets, or lighting, a shot can be no better than the competence with which it is recorded. Basically, a camera operator's job is to frame (compose) the shot well, then to move the camera as needed to maintain an effective composition, **Figure 12-27**.

Figure 12-27.
A professional videographer at work.

Framing the Shot

Framing a shot means creating a pleasing composition within the frame. The key to success is your ability to see the viewfinder not as a window but as a picture within the frame.

Visual composition is covered in Chapter 5.

Correcting

To compensate for movement within the image, the camera must move as well. Professional camera operators rarely lock the tripod's pan and tilt mechanisms. Even when the shot is supposed to be fixed, they are constantly moving the camera slightly, to maintain a good composition. In a skilled operator's hands, this camera movement is so smooth and unobtrusive that it is invisible to the audience. To achieve this fluid movement, you aim the lens much as a hunter aims at a constantly moving target.

Factors Affecting Exposure

Like focus, exposure is the result of several factors working together. In order of importance, these factors are aperture (f-stop), shutter speed, filtration, and gain. Filtration and gain operate only when you enable them, but aperture and shutter are always working.

A fixed amount of light

The key to understanding exposure is the idea that to make a good quality image, the camcorder must always receive exactly the same amount of light. Whether the original light is on a beach lit by the sun at noon or by a fire at midnight, the light that eventually reaches the imaging chip must be at the same brightness level, even if the noon sunlight is 250 times as bright as the firelight.

The most common problem is to reduce excessively bright light so that it does not overwhelm the sensitive chip. To do this, the camcorder uses first the aperture and then, if necessary, the shutter.

Aperture

At its widest aperture (such as f/1.4) the lens admits nearly all the light that strikes it. If the raw light is too bright for the chip, the aperture grows smaller and smaller. Each full stop reduction in the aperture size is called one f-stop (or just "stop"), and each smaller stop admits one-half as much light as the next wider one. The usual f-stops are 1.4, 2, 2.8, 4, 5.6, 8, 11, 16, and 22 (to control very bright light, some camcorder lenses will "stop down" to f/32). Most aperture control systems can change in partial stops, closing, say, from f/2 to f/2.4, which is half way to the next full stop, f/2.8. More sophisticated models divide stops into one-third stop, rather than one-half stop increments.

Shutter

In situations where the aperture alone cannot fully control exposure, the shutter is brought into play. Typical shutter speeds are 1/60 second, 1/125, 1/250, 1/500, 1/1000, 1/2000, 1/4000, 1/8000. Each time you double the shutter speed, you reduce the light by half, because you allow the chip half as much time to form an image before it is removed and processed.

Aperture plus shutter equals exposure

The amount of light used to form an image depends first on how much is allowed to reach the chip (aperture) and how long the chip is allowed to soak it up (shutter speed). A one-stop decrease in aperture reduces the light by half and so does each doubling of shutter speed.

To see how aperture and shutter work together, imagine a sort of equation:

Aperture + shutter = correct exposure

Pretend that "correct exposure" has a value of 10, so,

Aperture + shutter = 10

Next, give the aperture f/1.4 a value of 1, f/2 a value of 2, f/2.8 a value of 3, and so on, increasing the value by 1 for each full stop.

In the same way, give shutter speed 1/60 second a value of 1, 1/125 a value of 2, and so on, also increasing the value of each doubled speed by a value of 1. You can value both f-stops and shutter speeds in multiples of 1 because each change represents the same reduction in light. With these values, you can see that 1 + 9 = 10, 2 + 8 =10, 3 + 7 = 10, 4 + 6 = 10, 5 + 5 = 10. If you translate these numerical values back into apertures and shutter speeds, the result would look like this for bright sunlight:

Aperture		Shutter		Total light
1 (f/1.4)	+	9 (1/16000)	=	10
2 (f/2.0)	+	8 (1/8000)	=	10
3 (f/2.8)	+	7 (1/4000)	=	10
4 (f/4.0)	+	6 (1/2000)	=	10
5 (f/5.6)	+	5 1/1000)	=	10
6 (f/8.0)	+	4 (1/500)	=	10
7 (f/11.0)	+	3 (1/250)	=	10
8 (f/16.0)	+	2 (1/125)	=	10
9 (f/22.0)	+	1 (1/60)	=	10

In a darker environment like a living room at night, the available exposure combinations might resemble this:

Aperture		Shutter		Total light
1 (f/1.4)	+	9 (1/250)	=	10
2 (f/2.0)	+	8 (1/125)	=	10
3 (f/2.8)	+	7 (1/60)	=	10

In this darker location, f-stops smaller than f/2.8 and shutter speeds faster than 1/250 are not available because there is not enough light to allow them.

As you review these examples, keep in mind that "aperture plus shutter speed equals exposure" is not a true mathematical equation, but only a rough analogy.

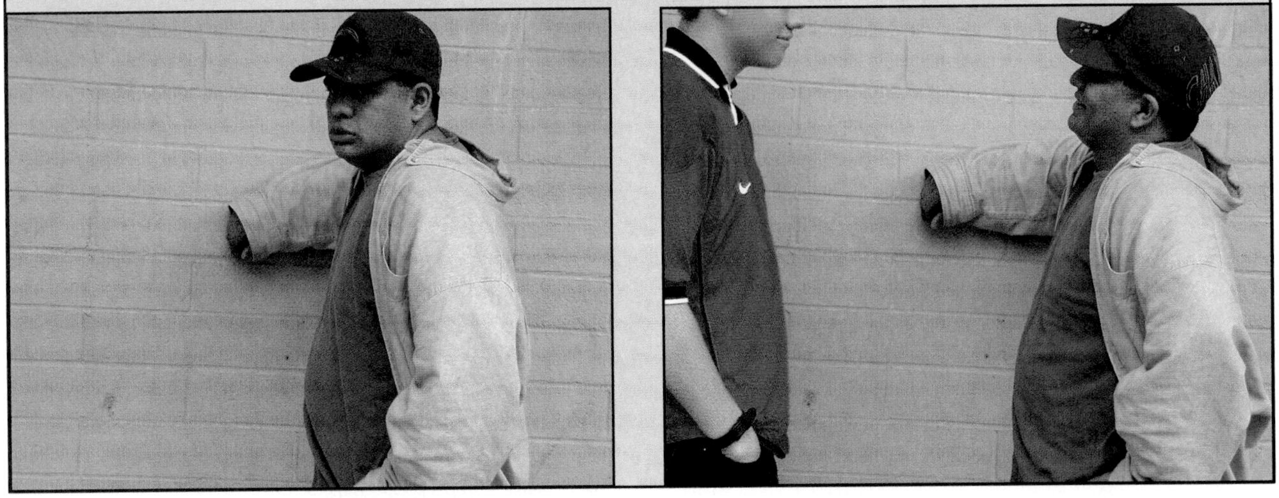

Figure 12-28.
Without camera compensation, the new subject spoils the composition.

But instead of following the central subject, you "aim" at the entire composition within the frame.

Compensation

Frequently, you correct the composition to compensate for added or subtracted elements. Suppose you have a composition in which a second subject joins a first subject. **Figure 12-28** shows the effect without camera compensation. As you can see, failure to reframe for the second subject ruins the composition. **Figure 12-29** shows correction for the second subject.

Look room

When framing people, allow more room on the side toward which they are facing. This avoids the claustrophobic feeling that they are boxed in by the frame. **Figure 12-30** compares two different compositions of a subject. Since people move as they speak, a common application for the floating frame is maintaining adequate look room.

Lead room

When subjects move so that the camera too must move to hold them in frame, look room is called lead room. The idea is exactly the same: allow extra room on the side of the frame toward which people or objects are moving, to prevent them from crowding the frame line, **Figure 12-31**.

Framing Moving Shots

Maintaining lead room will keep a moving subject properly framed, but it will not guarantee a good composition. To do that, you must

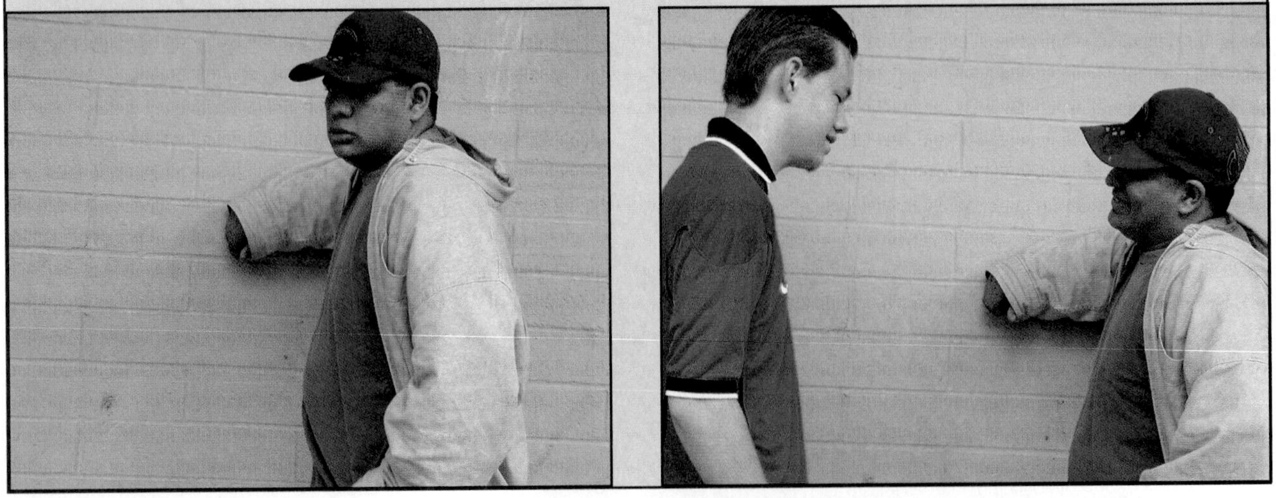

Figure 12-29.
Camera compensation corrects the shot.

Figure 12-30. Composition with inadequate look room. Composition with good look room.

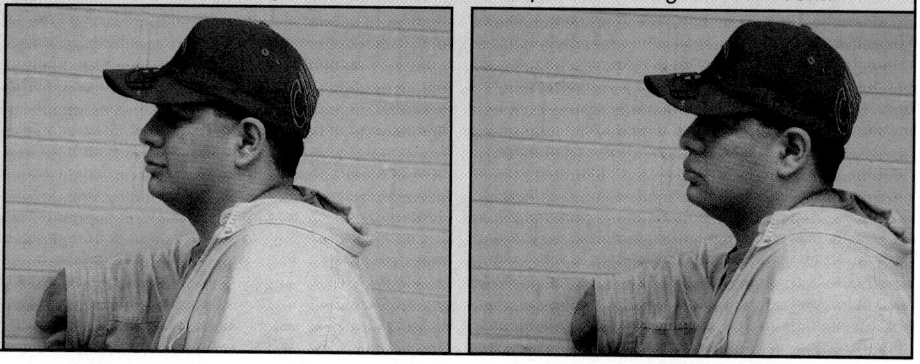

Figure 12-31. The subject is "crowding the frame." Centering the subject improves the composition. Ideally, the subject should be slightly "behind" the center line of the frame.

consider the background as well as the subject. To ensure good composition overall, set up an effective start composition that accommodates both subject and background. Preframe an equal or better ending composition with the subject blocked to complete the move at the designated spot.

In video as in theater, a subject's predesigned and rehearsed movements are called ***blocking****.*

Moving the Camera

Composing moving shots naturally requires moving the camcorder, and this is not always as simple as it sounds. The operator must keep the move smooth, while maintaining a good frame and operating camera features like the zoom control. Whether pivoting the camera on its tripod or rolling on a dolly, the key to a professional-looking move is variable speed. Start the move slowly and then accelerate to the desired maximum speed; then slow smoothly and gradually to a stop.

Panning and tilting

When rotating the camera horizontally (panning) or vertically (tilting), follow these suggestions:

- Set the drag (resistance) on the tripod head so that you are pushing against it hard enough to smooth the move, but not hard enough to cause jerks in the motion.
- Position your body at the midpoint of the movement. By twisting first toward the start of the movement and then toward the end, you can make smooth pans of 180 degrees. See **Figure 12-32**.
- Use an external monitor, either camera-mounted or stand-alone. By not looking through the built-in viewfinder, you can position yourself far enough away from the camera to avoid impeding its free movement.

Dollying

These suggestions apply equally to moving the camera on a dolly. The main problem with dolly shots is keeping perfect pace with the moving subject so that the lead room remains constant. It is very distracting to have a moving subject shifting forward and backward in relation to the frame. To make dollying effective, remember that even the fastest-moving subject is standing still in relation to the frame. To give the shot a feeling of dynamic movement, be sure

Figure 12-32.
Start facing the center of the pan and twist your upper body to frame start of the pan. You can pan up to 180°.

to include a distinctive background. On screen, it is the background that will fly by, not the subject.

Hand-holding

All the suggestions for smooth panning and dollying apply as well to hand-held moving shots. The challenge with these shots is to keep them smooth. Here are some tips for professional hand-holding:

- Use a wide angle lens setting to minimize the appearance of camera shake.
- Use an external view screen. This keeps your head from transmitting bumps to the camera, and it lets you use your peripheral vision to watch where you are going.
- Keep your arms away from your sides so that your bent elbows can act as shock absorbers.
- Walk with knees slightly flexed and try to glide along rather than striding.
- Carry the camcorder as if it were a very full and very hot cup of coffee.

If you combine these techniques with a good lens stabilization system, you can get hand-held shots that rival professional dolly work.

Roll Camera

At this point, you have worked your way through the basics of professional videography, and now you are ready to shoot. Just one final reminder: always roll at least five seconds of tape before signaling to begin the action; and keep on rolling at least five seconds after the action has ended or moved off-screen. The editor will thank you.

Chapter Review

Answer the following questions on a separate piece of paper. Do not write in this book.

1. A ______ focal length lens produces images close to the human field of vision.
   A. wide angle
   B. normal
   C. telephoto
2. Depth of field grows______ as the lens focal length grows longer.
3. *True or False?* The only difference between wide angle and telephoto lenses, from an optical standpoint, is angle of view.
4. ______ is the size of the opening through which light enters the camcorder.
5. To reduce the amount of light reaching the camcorder's CCD, you can use a ______.

Technical Terms

Angle of view: The breadth of a lens' field of coverage, expressed as an arc of a circle, such as "10°."

Aperture: The opening in the lens that admits light. The iris diaphragm can vary the aperture from fully open to almost or completely closed.

Autofocus: The camcorder system that automatically adjusts the lens to keep subjects sharp.

Blocking: The predetermined movements that a subject makes during a shot. Camera movement is also blocked.

Depth of field: The distance range, near-to-far, within which subjects appear sharp in the image.

Digital zooming: Increasing the subject size by filling the frame with only the central part of the image.

f-stop: A particular aperture. Most lenses are designed with preset f-stops of f/1.4, f/2, f/2.8, f/4, f/5.6, f/8, f/11, f/16, and f/22.

Focal length: Technically, one design parameter of a lens, expressed in millimeters (4mm, 40mm). Informally, the name of any particular lens, such as, "a 40mm lens."

Gain: The electronic amplification of the signal made from an image, in order to increase its brightness.

Iris diaphragm (iris): A mechanism inside a lens (usually a ring of overlapping blades) that varies the size of the lens opening (aperture).

Magnification: The apparent increase or decrease in subject size of an image, compared to the same subject as seen by the human eye. Telephoto lenses magnify subjects; wide angle lenses reduce them.

Optical zoom: Changing a lens' angle of view (wide angle to telephoto) continuously by moving internal parts of the lens.

Pulling focus (following focus, racking focus): Changing the lens focus during a shot to keep a moving subject sharp.

Setting focus: Adjusting the lens to make the subject appear clear and sharp.

Shutter: The electronic circuitry that determines how long each frame of picture will accumulate on the imaging chip before processing. The standard shutter speed is 1/60 per second for NTSC format video.

Specular reflections: Hard, bright reflections from surfaces such as water, glass, metal, and automobile paint that create points of light on the image. Often controllable by a polarizer.

Speed: The light-gathering ability of a lens, expressed as its maximum aperture. Thus, an f/1.4 lens is "two stops faster" (more light sensitive) than an f/2.8 lens.

Sockets on the back of a camcorder provide connecting points for audio cables, DC power, and earphones.

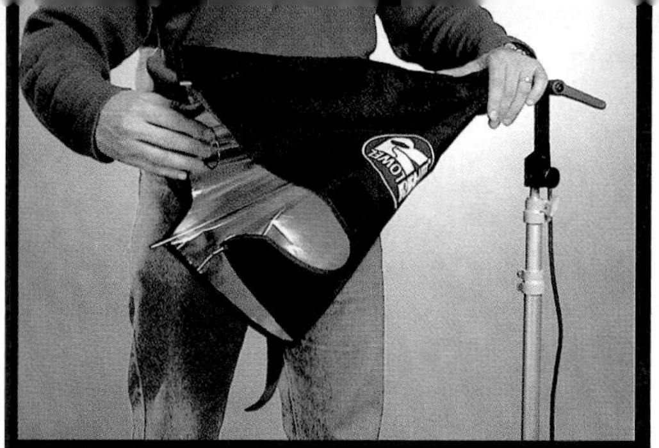

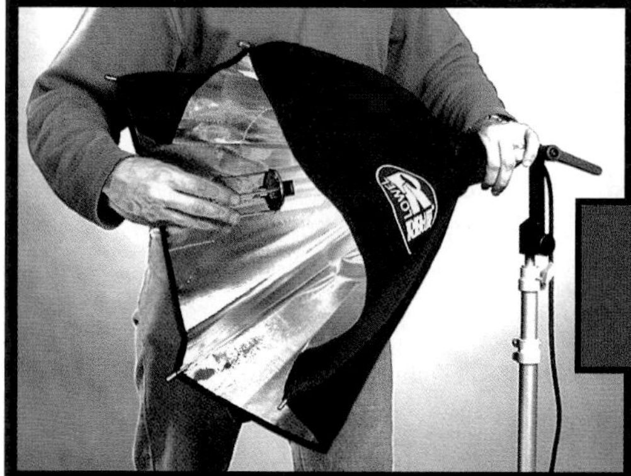

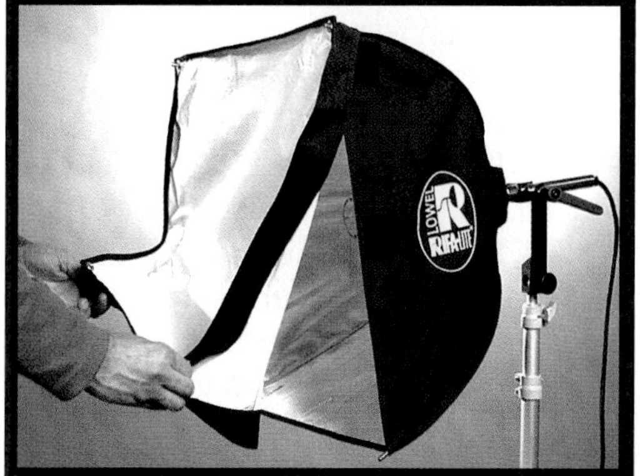

13

Lighting Tools

Objectives

After studying this chapter, you will be able to:

- Identify the principal components of video lighting.
- Demonstrate the characteristics of different lighting instruments.
- Explain the functions of different lights and accessories.

(Lowel Corporation)

About Lighting Tools

Considered from one perspective, video is about *light*. You may think you are taping beautiful backgrounds, exciting action, and expressive performers, but you are not. What you are actually recording is only the light that bounces off your subjects, streams through the lens, and strikes the imaging chip. Since that light is all that your camera sees, you can shape video images by controlling the light that illuminates them.

This discussion of lighting equipment is the first of three chapters on video lighting. The next chapter covers the principles of lighting design and the final chapter brings tools and designs together in a survey of typical lighting applications.

Because camcorders can capture a satisfactory image in almost any quantity and quality of illumination, it is possible to make videos without paying much attention to lighting. But if you ignore lighting, you neglect one of your most powerful tools. One difference between a casual shooter and a video artist is that the beginner merely *records* images while the videographer *creates* them—in part, by managing light.

Video lighting is the art of creating images with light. To do this, of course, you need to understand the special tools used to create lighting designs. These include tools for *controlling available light*, tools for *adding light*, *lamps*, and *accessories*.

Since it is difficult to discuss lighting equipment, Figure 13-1, without mentioning its uses, this chapter anticipates some of the content in the next two.

Tools for Available Light

In thinking about video lighting, it helps to distinguish the light you may provide from the illumination that is already present at the scene. In many circumstances, indoors as well as out, ***available light*** will provide much or all of your video lighting.

"Available light" is whatever illumination already exists at the location you are preparing to light for video.

Controlling available light means modifying and redirecting it to improve the quality of the image. In working with available light, you control its *quantity*, *direction*, and *color*.

Controlling Quantity

Controlling the quantity of available light means reducing it when it is too intense. (By definition, you cannot *increase* the amount of light that is naturally there.) In reducing light quantity, your principal tools are flags, screens, silks, and filters:

Flags are flat, opaque cards, usually constructed of metal, thin plywood, or fabric on a metal frame. Most are rectangular, ranging from squares to long, thin paddles. (Flags are also used extensively to control spill from video lights.)

Reflectors can serve as flags when set to block light rather than reflect it.

Foam board, used mainly for reflectors, also works well as a flag. Flags may be held by crew members or clamped onto stands. Flags, of course, completely block the path of directional light, **Figure 13-2**.

Figure 13-1.
Lighting equipment and accessories. A—Reflectors control available light. (Bogen Manfrotto) B—Spotlights add light. (Lowel Corporation) C—Accessories include reference monitors. (JVC Corporation)

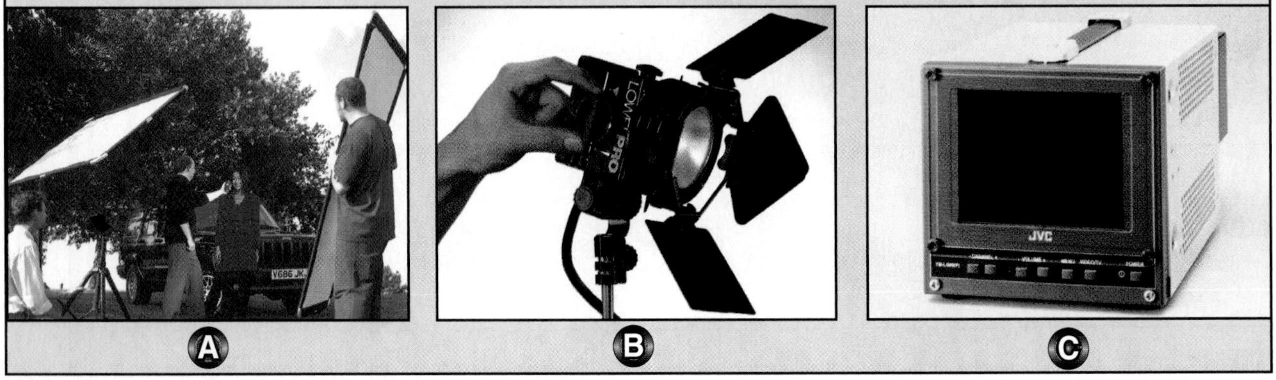

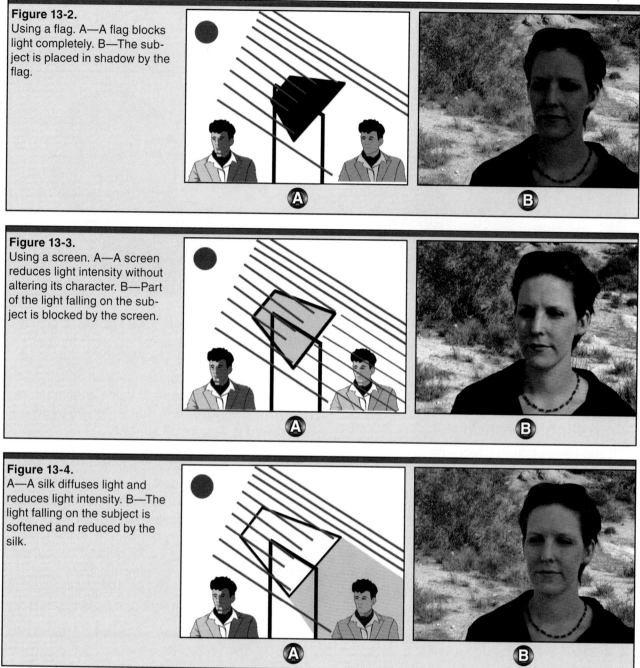

Figure 13-2.
Using a flag. A—A flag blocks light completely. B—The subject is placed in shadow by the flag.

Figure 13-3.
Using a screen. A—A screen reduces light intensity without altering its character. B—Part of the light falling on the subject is blocked by the screen.

Figure 13-4.
A—A silk diffuses light and reduces light intensity. B—The light falling on the subject is softened and reduced by the silk.

Screens are large expanses of plastic mesh stretched on frames. Small screens can be held by stands or by crew members, but most are mounted on support structures, **Figure 13-3**. Unlike flags, screens block only part of the light; and unlike silks, they do not change the light's directionality or character. The screening material can be obtained in different densities, depending on how much light reduction is needed.

Silks look and work like screens, except that they are sheets of thin white synthetic fabric.

Actual *silk* fabric is not used on "silks" anymore.

Like screens, silks reduce the intensity of light on the subject. But as you can see from **Figure 13-4**, silks also diffuse the light, changing it from a single, directional beam to an overall glow. Because silks diffuse the light as well as reduce it, they change the light source's quality as well as its quantity.

Neutral density filters are sheets of gray-tinted plastic large enough to cover entire windows. Daylight is typically much brighter than interior lighting. To prevent on-screen windows from "burning out" to blank white, neutral density filter plastic is applied over the windows, **Figure 13-5**.

Figure 13-5.
A neutral density filter reduces incoming light. (For comparison, the window's side panels are unfiltered.)

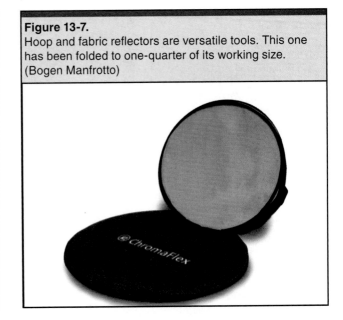

Neutral density filtration material is supplied in several strengths, reducing light intensity by one, two, three, or four f-stops, coded "ND3," "ND6," "ND9," and "ND12."

Controlling Direction

If you have enough direct light to cast shadows, you can multiply its effect by bouncing it back on your subjects.

Reflectors are large boards with bright surfaces that are positioned to receive light from directional sources and aim it back at subjects. Large studio-style reflectors are made of stiff materials and mounted on yoked stands, **Figure 13-6**. Because of their weight, they are stable in winds, but awkward to handle.

Figure 13-7.
Hoop and fabric reflectors are versatile tools. This one has been folded to one-quarter of its working size. (Bogen Manfrotto)

Hand-held reflectors have just the opposite advantages and disadvantages: light and easy to handle, they tend to wobble in breezes, making the light on the subject waver unsuitably. The lightest models are made of fabric stretched on wire hoops, **Figure 13-7**. These can be folded and stored in very small spaces.

Reflectors come in four main types:

Hard metallic: Highly directional, these silver reflectors throw a light beam a long distance and/or fill a small subject area. They can be painfully bright when directed at an actor's face, **Figure 13-8**.

Soft metallic: Less directional, these silver reflectors still throw light a long distance, but disperse it over a wider area, **Figure 13-9**.

White: Extremely diffuse, these reflectors provide a soft, even fill light that does not hurt

Figure 13-6.
A heavy-duty studio-style reflector. (Lowel Corporation)

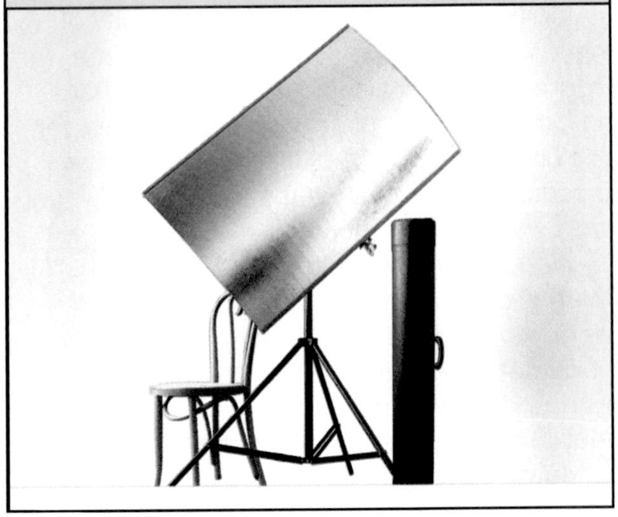

Figure 13-8.
Hard reflectors reflect light over long distances, but can be hard on subjects' eyes.

Figure 13-9.
Soft reflectors are a compromise between hard metallic and white.

Figure 13-10.
White reflectors deliver the softest light.

Figure 13-11.
Gold-tinted reflectors are useful for warming up cool light. (Bogen Manfrotto)

actors' eyes, **Figure 13-10**. The most common white reflector is made of foam board — a sheet of rigid plastic foam laminated between two sheets of white paper. Stiff, light, and disposable, foam board is an indispensable lighting tool.

Tinted: Metal, cloth, or foam board reflectors can be obtained in colors, notably gold tone for metallic surfaces and pale amber or pale blue for paper. By changing the color of the reflected light, these reflectors can warm up flesh tones, simulate moonlight, and deliver other specialized tints, **Figure 13-11**.

Controlling Color

Tinted reflectors offer only one way to change the color of available light. More commonly, the light is tinted by large plastic sheet filters placed over windows.

Filters placed directly on the camera lens change all of the light entering the camera. Sheet filters, however, affect only the light sources that they are covering.

Window filtering gels are used to change color temperature to match halogen movie lights. If your interior lights and camcorder are balanced for incandescent color temperature (3200K), the daylight streaming through visible windows will look too blue. Orange-tinted color correction filters positioned outside these windows will warm up the cool daylight to match the interior lights. These filters are often referred to by their code designation: "85."

Figure 13-12.
Color correction window filtration. The outside light is too bright to be shown on-screen.

Figure 13-13.
Combined in a single filter sheet, neutral density and color correction match the interior light.

The color temperature 3200K is read as "3,200 degrees Kelvin." The Kelvin scale was named for its inventor.

By itself, a color correction filter can seldom reduce the outdoor light *intensity* to the level of the indoor lighting. For this reason, color-correcting filters are used alone when the window light contributes to the illumination, but the actual windows will not appear on-camera, **Figure 13-12**.

If the windows *will* appear in the frame, both light intensity and color must be controlled, so color correction tinting is added to neutral density filtration, **Figure 13-13**. These combination filters are designated 85ND-3, -6, -9, or -12, depending on how much they reduce the light .

You can also change the color of available artificial light (lamps and fluorescents) already in use at a location. Sheet filters can be placed over fluorescent fixtures to match the warmer color of movie lights, **Figure 13-14**; or tube-shaped filters can be slipped over the fluorescent lamps themselves. Sometimes, the existing fluorescent tubes can be replaced with special lamps that supply matching color temperature.

As fluorescent video lights become more versatile, color-filtering of daylight and/or ambient fluorescent lighting is required less often. Instead, the camcorder white balance is set for fluorescent color temperature.

Tools for Adding Light

Tools for adding light are, of course, lighting ***instruments***. Sometimes, you will use them to supplement available light. In other situations, you will exclude ambient illumination and design with video lights exclusively.

Technically, a video light is called a "lighting instrument," Figure 13-15, the light source inside is the "lamp," and several units are, collectively, "lights."

Professional lighting instruments can seem confusing because there are so many to choose from. The lights of one manufacturer are designed differently from those of others; and each manufacturer offers several competing families of lighting instruments with different

Figure 13-14.
Blue filters, mounted in frames, change the color temperature of these halogen lights to daylight. (Lowel Corporation)

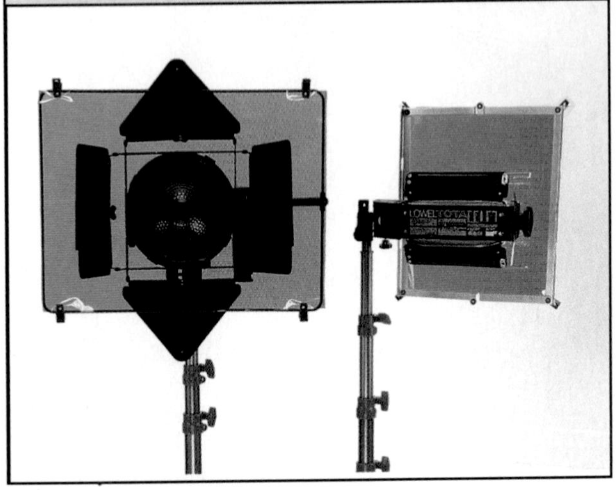

Figure 13-15.
The parts of a typical lighting instrument. (Lowel Corporation)

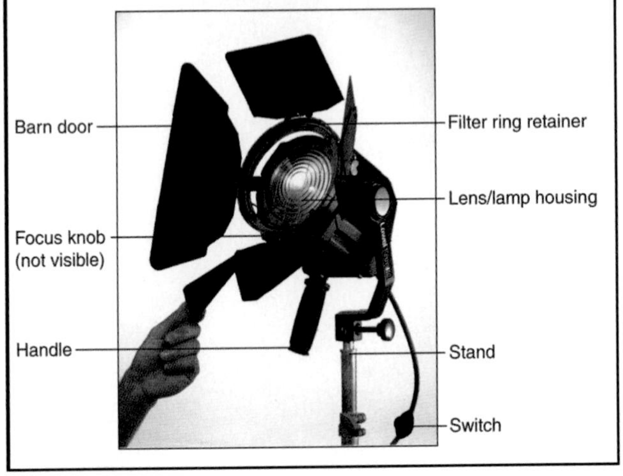

Figure 13-16.
Three different lighting kits offered by one manufacturer. (Lowel Corporation)

characteristics engineered for different applications, **Figure 13-16**.

In practice, many lighting designers eventually standardize on a particular make and family of lights, simply because those are they ones they happened to have mastered.

For all their individual differences in features and accessories, most lighting instruments fall into one of the following categories:

- spotlights.
- floodlights (broads and scoops).
- softlights (umbrellas, soft boxes, pans).
- on-camera lights.
- practicals.

We will look at each type, in turn.

Spotlights

Spotlights ("spots" for short) have small, intense light sources that produce a highly directional light pattern. This means that the light illuminates relatively small areas, casts distinct, hard-edged shadows, and is easily masked by barn doors or flags, **Figure 13-17**. Small source spotlights are often used as key (main) lights on

Figure 13-17.
Spotlights are available in different sizes. (Lowel Corporation)

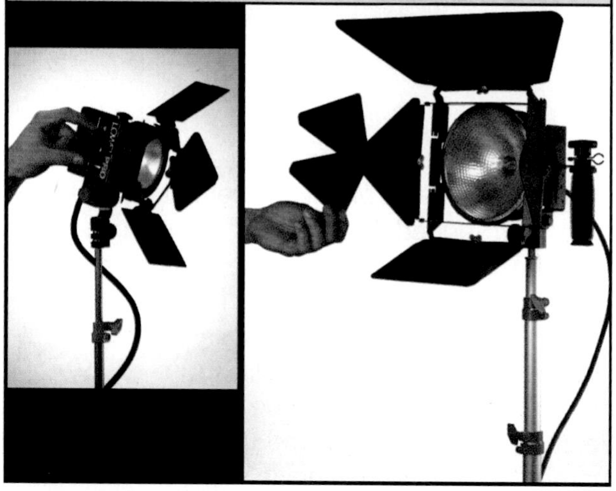

Figure 13-18.
A—A spotlight beam pattern in full spot focus. B—The same pattern in flood focus, with a larger diameter, lower intensity, and softer edge.

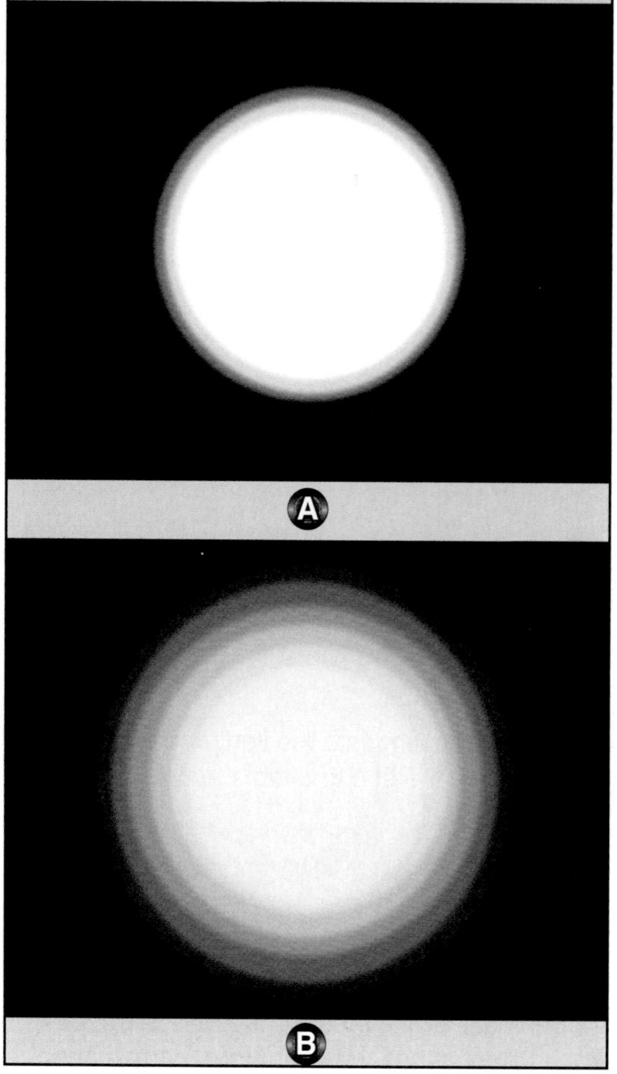

Fresnel Lenses

Fresnel lenses are named for their inventor, the French engineer Jean Augustin Fresnel ("Freh-**nell**")

In a simple lens, called a plano-convex (p-c) or flat-curved lens, light enters the flat surface and exits the curved surface, refracted to create a beam of parallel light rays.

A p-c lens (A) compared to a Fresnel lens (B) with the same size and optical qualities.

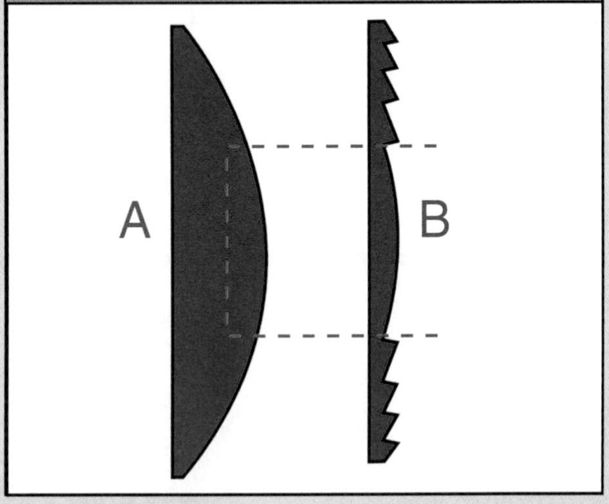

In 1822, a huge lens was needed for a new lighthouse called Cardovan Tower on the Gironde River, but a p-c lens massive enough for a lighthouse would be impossibly heavy because of all the glass between its flat and curved surfaces.

M. Fresnel realized that most of that interior glass was not needed because all the refraction was being done by the curved surface. So he designed a lens with that surface divided into concentric rings and pancaked to reduce the interior glass. Because the curvature of each ring is identical to the same area on a p-c lens, the light refraction is also identical (though the result is not quite as sharp). Lighthouses to this day employ M. Fresnel's innovation (the "F" is always capitalized to commemorate him), usually in multiple lenses.

The multi-faced Fresnel lens in the lighthouse at McKinleystone. (Sue Stinson)

Facing a similar problem, the theatrical lighting industry long ago adopted the concentric ring design and the Fresnel spotlight was born.

A Fresnel lens in a spotlight (behind a wire grid filter holder). (Lowel Corporation)

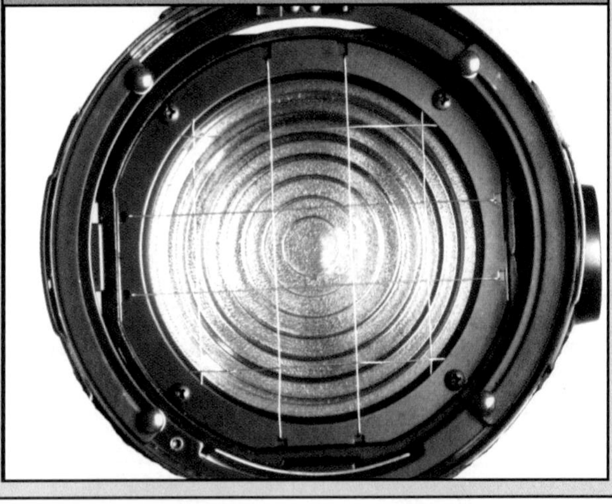

subjects, or to simulate the light from sources ranging from the sun to a table lamp.

The most effective spotlights are fitted with Fresnel lenses, for even better light control; but these units are often somewhat larger and heavier.

A spotlight can be focused by moving its lamp forward or backward in relationship to the reflector behind it and, if present, the lens in front. Focusing changes the beam from narrow and intense to wider and less intense. In the "spot" position the light is bright, concentrated, and relatively hard-edged. In the "flood" position, the light beam is larger and less intense, with more gradual falloff at the edges, **Figure 13-18**.

The advantage of small-source lights is that they are very easy to control, **Figure 13-19**. On the other hand, small-source light beams are relatively harsh and several used together throw unrealistic multiple shadows.

Figure 13-19.
Focusing a small-source light. (Lowel Corporation)

Floodlights

These larger lights are commonly used to fill in shadows (typically created by spotlights) and to light backgrounds. Compared with spotlights, floodlights have the opposite advantages and drawbacks. On the positive side, their light is somewhat softer and less directional; their shadows are fainter, and their beam pattern is broad and even. On the negative side, their intensity is less at any given part of the beam pattern and they are more difficult to control by masking. The most common floodlights are broads and scoops.

Broads are shallow rectangular pans that are small and portable, **Figure 13-20**. Because their light is moderately directional, they can be fitted with barn doors and frames to hold diffusion or filters.

Figure 13-20.
A "broad" floodlight. (Lowel Corporation)

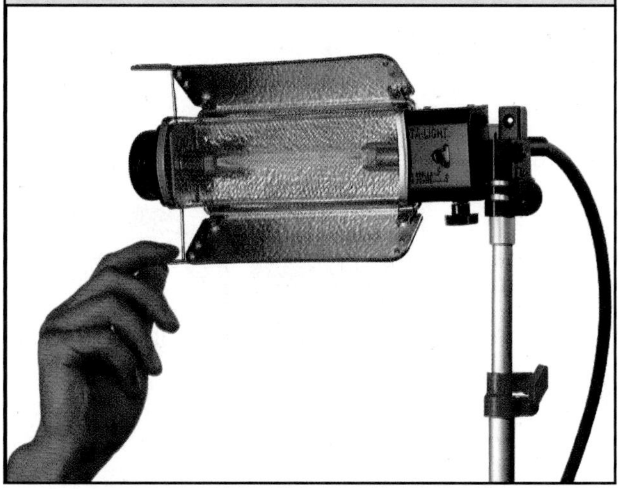

Scoops are large bowl-shaped reflectors, used mainly in TV studios.

Softlights

Softlights include umbrellas, softboxes, and pans. Softlights are very large light sources —ranging from one- to four-feet square—that deliver almost shadowless, directionless illumination. Softlights are easy to use because they do not throw distinct shadows. On the other hand, their light output is so soft-edged that it is difficult or nearly impossible to mask.

Umbrellas are really umbrella *frames* that can be fitted with different types of cloth, **Figure 13-21**. Some covers contain silver threads for greater reflectivity. Others are translucent white "silk."

Typically, a light is mounted at the base of the umbrella, which is pointed at the subject. For additional diffusion, the white umbrella can be turned around so that the spotlight is shining through it at the subject, **Figure 13-22**.

Softboxes are assemblies of fabric over wire frames, **Figure 13-23**. Some types enclose small lights. Others use proprietary lamps. Softboxes are about the size of umbrellas and throw similar light patterns.

Some softboxes are hybrids: fabric boxes supported by umbrella structures, **Figure 13-24**.

Pans are very large light sources fitted with fluorescent tubes, **Figure 13-25**. When covered with diffusion material, pans are—at least theoretically—the softest lights available, because of their exceptionally large size. Also, their fluorescent lamps use less power and emit less heat than halogens.

Camera lights

Miniature spotlights, **Figure 13-26**, are available to clip onto camcorders. Some models are fitted with a diffusion disc and barn doors. Other types of ***on-camera lights*** have multiple heads and lamps, so that you can vary both their light intensity and beam spread.

Practicals

So-called "***practicals***," **Figure 13-27**, are lights that will be seen by viewers, like table lamps or wall sconces. In set lighting, you will sometimes replace

Figure 13-21.
An umbrella with silver-thread cloth. (Lowel Corporation)

Figure 13-22.
A silk fabric umbrella can be reversed for even softer lighting. (Lowel Corporation)

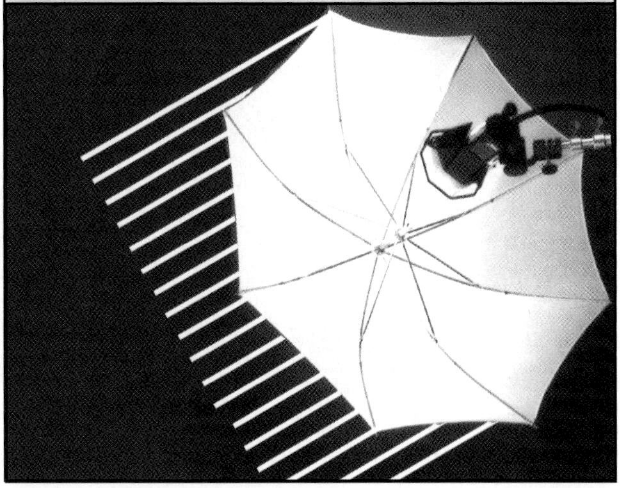

Figure 13-23.
A softbox. (Lowel Corporation)

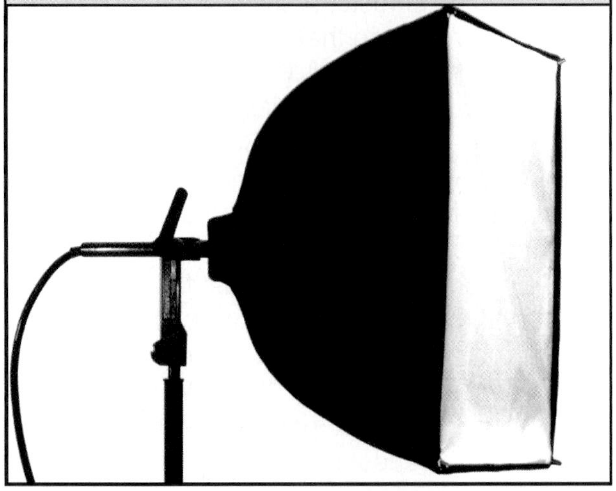

Figure 13-24.
An umbrella-framed softlight. (Photoflex)

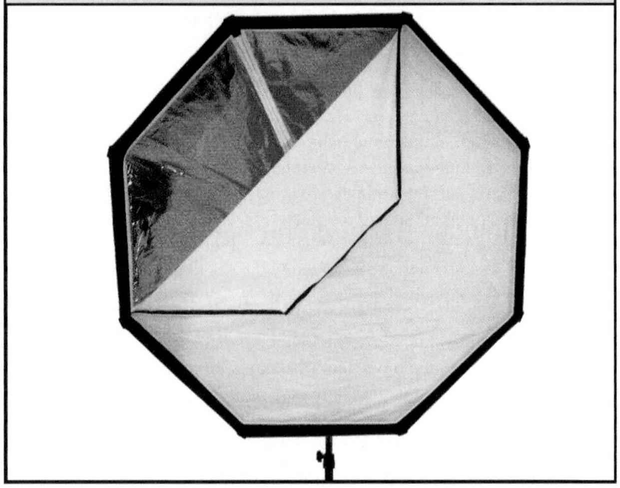

Figure 13-25.
A fluorescent pan, with its external ballast functioning as counterweight. (Lowel Corporation)

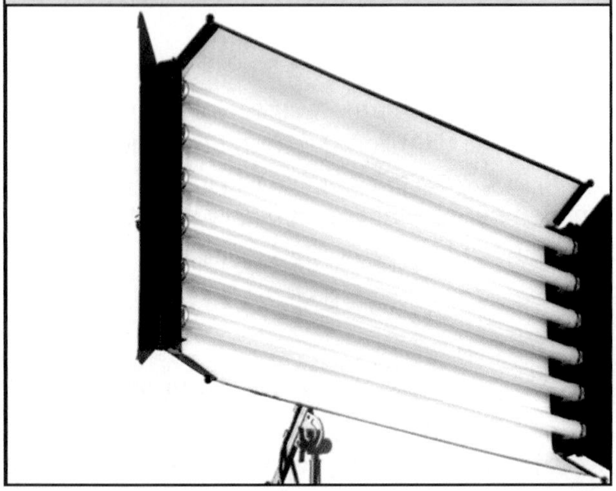

Figure 13-26.
A camera light with a barn door and diffusion disc that can be swung into the light path. (Lowel Corporation)

Figure 13-27.
The reading light is a practical.

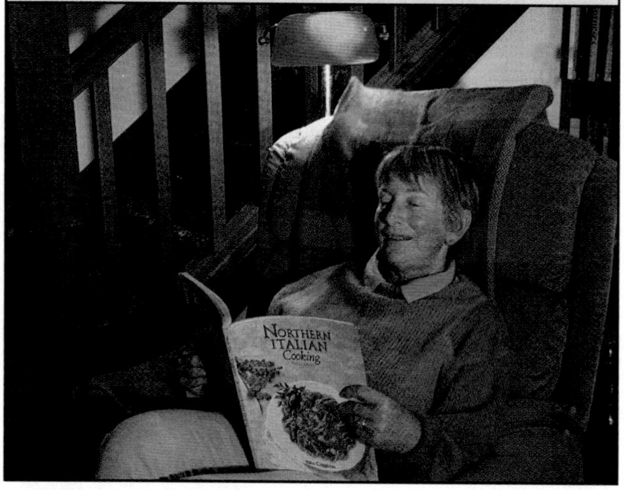

Figure 13-28.
A professional halogen lamp.

the lamp in a working practical to adjust its intensity or color temperature, or both. Practicals are easy to color-balance because 3200K halogen bulbs can be obtained with the medium screw-type bases used for household lighting.

The term practical comes from the theater, where on-stage lights that actors can actually turn on and off require special wiring that must be placed or removed with each scene change.

Lamps

The lightbulbs used in video lighting are called lamps, and there are four basic types: incandescent, halogen, fluorescent, and LED.

Incandescent lamps

Incandescent lamps contain metal filaments glowing in a near-vacuum. They are universally found in household lighting. Ordinary incandescents have a color temperature (relative whiteness) of 2700K-2800K, while the special models sold for photographic lighting burn at 3200K, the white balance of a camcorder's "indoor" or "incandescent" setting. Older lighting instruments take incandescent lamps, but they have been largely replaced by halogen units.

Halogen lamps

Halogen lamps contain metal filaments in a special mixture of gasses. All halogen lamps burn at 3200K. Unlike incandescent light, halogen illumination does not grow more yellow as the lamps age. Halogen lamps are also smaller and emit more light, watt-for-watt, than incandescents, **Figure 13-28**. They are generally less expensive than 3200K incandescents, though more expensive than household lamps. Halogen lamps have one potential safety problem: a tendency to shatter.

For this reason, always follow precautions in working with them. Never use a halogen instrument that does not place the lamp behind a safety glass. Never touch a lamp with your fingers. Your natural skin oil can etch the quartz envelope of the lamp. When the lamp heats up, it can shatter at the etched area.

Figure 13-29.
A household reflector-style halogen lamp.

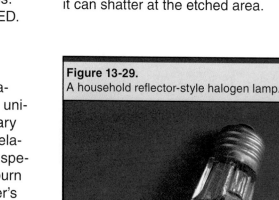

To change a halogen lamp, wait until it has cooled before removing it. Then wrap the replacement lamp in a facial tissue or use the foam sheet that is usually packed with it. Some types of halogen lamps are enclosed in a second outer envelope that can be touched. Others are mounted within integral reflectors, which can also be handled safely, **Figure 13-29**.

Fluorescent Lamps

Most camcorders have white balance settings optimized for standard ***fluorescent lamps***, and you can often obtain satisfactory video colors this way. But if you use fluorescents as movie lights, you will want to obtain specialized tubes that produce more pleasing color. When mixing fluorescent and halogen lights, use tubes rated at 3,000K. If your whole lighting setup is fluorescent, you may get better results with "sunshine" fluorescent lights, whose 5000K color temperature closely matches that of direct sunlight, **Figure 13-30**.

Figure 13-30.
When the available light is fluorescent, it is often better to supplement it with fluorescent video lights. (Lowel Corporation)

When choosing fluorescent tubes, note that some manufacturers provide a "Color Rendering Index" (CRI), a scale from 1–100 that rates the ability of a fluorescent tube to deliver natural-looking color. For example, inexpensive "shop light" tubes may have an index of 67. Better quality lamps have indexes in the 70's. The best sunlight tubes have a CRI rating over 90. For more consistent output, make sure that all your tubes are the same make and model and all are replaced at the same time.

Figure 13-31.
This compact scoop light uses three 26-watt fluorescent lamps. (Equipment Emporium, Inc.)

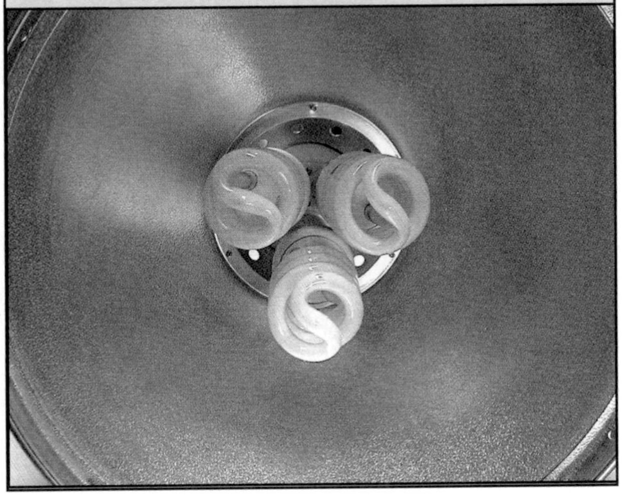

Recently, small spiral-design ("curly") fluorescent lamps intended for video lighting have become available. Fitted with medium screw bases like household bulbs, they can be used as practicals, or ganged in special lighting instruments, **Figure 13-31**.

LEDs

LEDs (Light-Emitting Diodes) are a recently developed form of lighting. The LED arrays are exceptionally light and compact, **Figure 13-32**, and they consume relatively little power. While they are now employed mainly for special purposes, their use is growing more widespread.

Figure 13-32.
An LED camcorder light. (LitePanels)

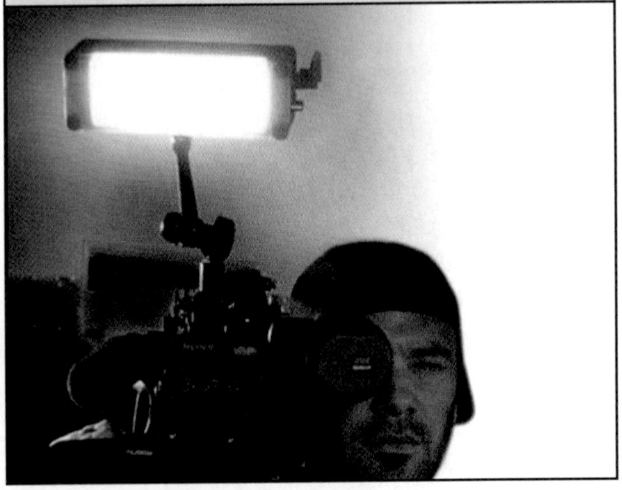

Lighting Accessories

Lighting accessories such as flags and filters have already been mentioned in passing. Following is a more extensive survey of the items that commonly complete a lighting director's kit.

Accessories for Lighting Instruments

The beams from small source lights (spotlights and narrow floods) can be modified by accessories that fit on the front of the instrument to mask, reduce, or diffuse the light.

Masking

The highly directional beams of spotlights can be masked to produce hard-edged cutoff patterns. Masking accessories include barn doors and flags.

Barn doors are pairs of metal flaps hinged to the sides and/or top and bottom edges of the lamp. Their front edges are moved into and out of the light path to control the beam edge, **Figure 13-33**.

Flags, as noted earlier, are opaque cards — usually mounted on stands—that block part of the light beam, **Figure 13-34**.

As a rule, the farther a masking device is placed from the light, the harder-edged is its shadow. Barn doors attached to the light throw softer shadows than flags mounted on century stands placed closer to the subjects.

Figure 13-34.
Small flags on movable arms can be positioned more flexibly than barn doors. (Lowel Corporation)

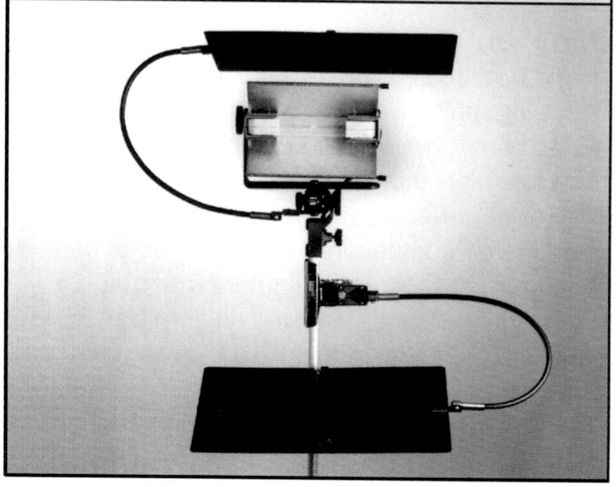

Light reduction

Spotlights can be fitted with circular frames covered with nets (actually made of screen wire) to cut down the light intensity, **Figure 13-35**. The most common nets are:

Singles: one-layer nets for moderate light reduction,

Doubles: two-layer versions for stronger reduction,

Halfs: frames with one-half netted and the other half clear. By rotating a half net, you can selectively reduce the illumination on just part of the area lit by the beam.

Figure 13-33.
All four barn doors on this spotlight can be shaped for better control. (Lowel Corporation)

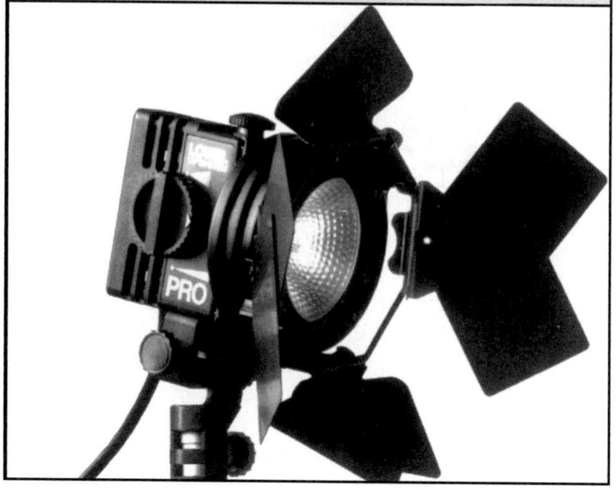

Figure 13-35.
This graduated screen uses 2, 1, and 0 layers of net. (Lowel Corporation)

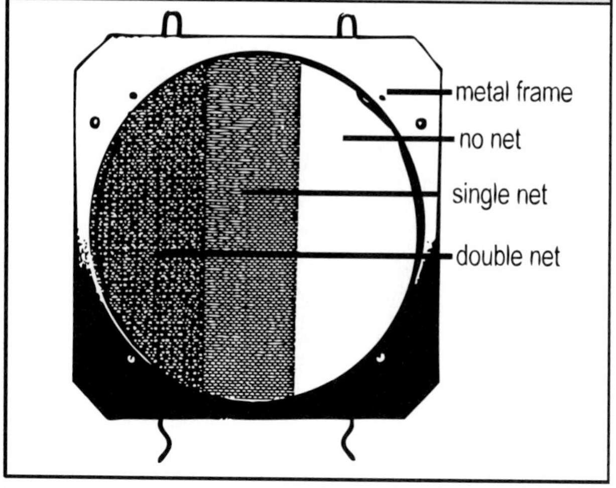

Half nets are also supplied in one- and two-layer versions, so a "half double" is a two-layered half-circle net.

Diffusion

All light beams can be softened by placing ***diffusion*** material in front of them. Like filters, circles of milk-white plastic can be installed in metal rings and placed in holders at the front of spotlights.

For a more diffuse effect, white "spun glass" sheeting can be secured to the edges of barn doors or hung in front of lights, **Figure 13-36**.

Umbrellas and softboxes seldom benefit from additional diffusion, but large pans are often covered with spun glass. White cloth sheeting is especially effective on fluorescent softlights, since the lights emit very little heat.

Figure 13-36.
Clamps or clothes pins can clip filters or diffusion to barn doors. (Lowel Corporation)

Figure 13-37.
A number 80 blue filter converts halogen color temperature to daylight. (Lowel Corporation)

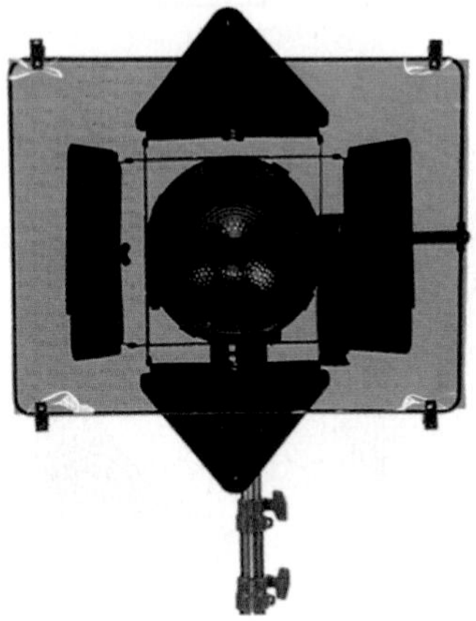

Color

Lighting instruments are sometimes gelled with colored plastic to simulate firelight, moonlight, or other effects, or simply to warm or cool the light on an area. Earlier, we noted the use of orange filtration to convert daylight to incandescent. Similarly, blue filters are often placed in front of halogen lights to convert their color temperature to daylight, **Figure 13-37**. (Usually, these blue filters are coded number 80.)

Other Lighting Accessories

In addition to accessories placed on or in front of the lighting instruments, a wide variety of other lighting tools are available. Here is a summary of the more common accessories.

Century stands

A ***century stand***—"C-stand" for short—has folding legs, a telescoping central column, and a universal clamp that will grip flat items or pipe arms, **Figure 13-38**. Each is supplied with one pipe arm, ending in a second universal clamp. Century stands will hold lighting instruments, flags, reflectors, diffusion, cloth drapes, or just about anything else. They are, perhaps, the most versatile accessory in the lighting director's kit, and large productions may carry dozens of them.

C-stands are widely obtainable from Internet vendors.

Figure 13-38.
A century stand will support any lightweight object, in any position.

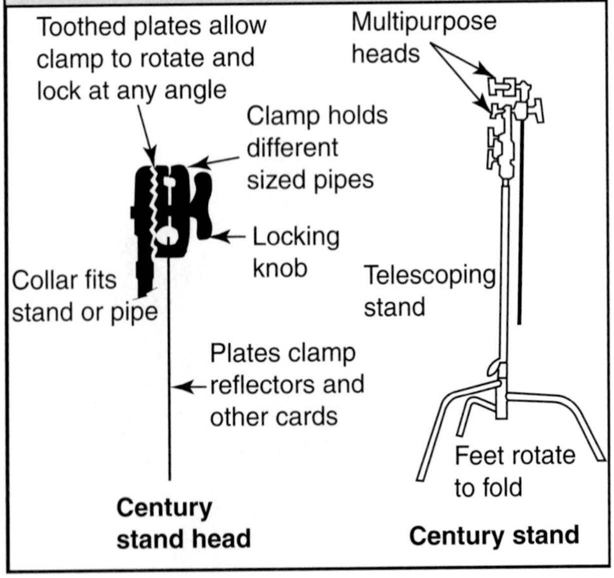

Figure 13-39.
For demonstration purposes, this spotlight carries a variety of special-purpose cookies. (Lowel Corporation)

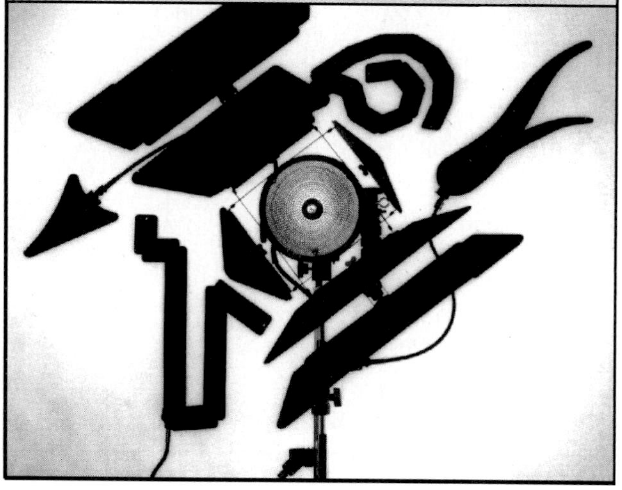

Cookies

Specially cut sheets, called "***cookies***," are placed in front of spotlights to create patterns on walls and other surfaces. Cookies can simulate leaves, bare tree branches, Venetian blinds, or any other patterned required, **Figure 13-39**. Cookies are covered in detail in Chapter 14.

Mounts

Light stands are the most common accessories for supporting lights. To illuminate larger areas, horizontal overhead lighting pipes spanning the center of a set may be supported by heavy stands and serve as mounts for hanging several instruments. Lightweight instruments can be attached to walls with gaffer tape, and small units can be clamped almost anywhere, **Figure 13-40**.

Electrical Accessories

Cables

For location lighting, extension cables are essential, **Figure 13-41**. Make sure that your cables are safe:

- Wire gauge should be no thinner than no. 14; no. 12 or even no. 10 is better. (The lower the number, the thicker the wire.)
- Extensions should not exceed 25 feet in length.
- Use single plugs rather than multiple outlet boxes, to avoid overloading circuits.
- Use high-visibility yellow or orange cables for safety.
- Tape loose cables to the floor, for additional safety. Tangled in a foot, a cable can topple a light on a stand, with disagreeable results.

Dimmers

A dimmer can be used to reduce the output of a light. Dimming a light warms up its color temperature, so dimming can be used to adjust color as well as output.

Gloves

Leather gloves are a must around hot lights. Find the heaviest industrial grade models you can, because incandescent lighting instruments can heat up to hundreds of degrees.

Gaffer tape

Gaffer tape looks just like ordinary duct tape but costs five times as much and works five times as well. It is sticky enough to tape lightweight instruments to walls and other surfaces. At the

Figure 13-40.
Mounting lights. A—Lights are mounted on stands or overhead pipes. B—Lightweight units can be taped to a wall. C—Clamps work best with small, lightweight units. (Lowel Corporation)

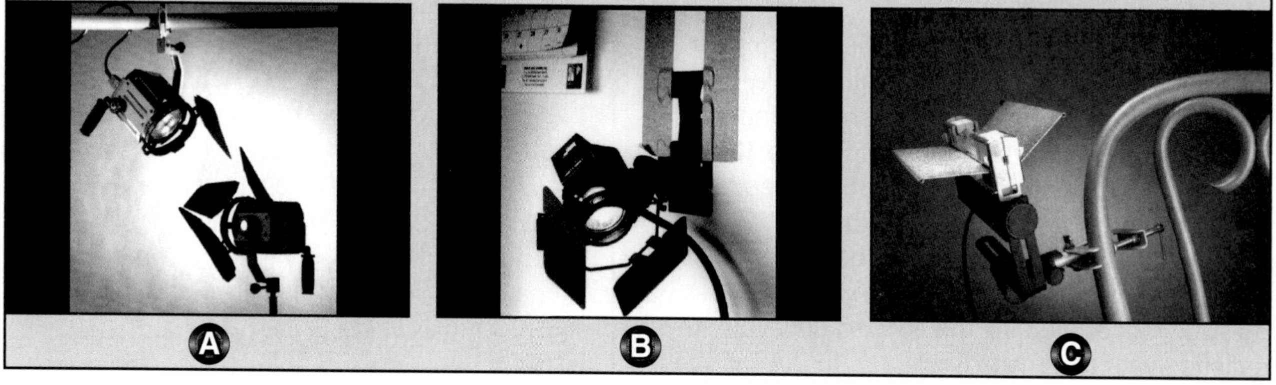

Figure 13-41.
An appropriate cable for location lighting.

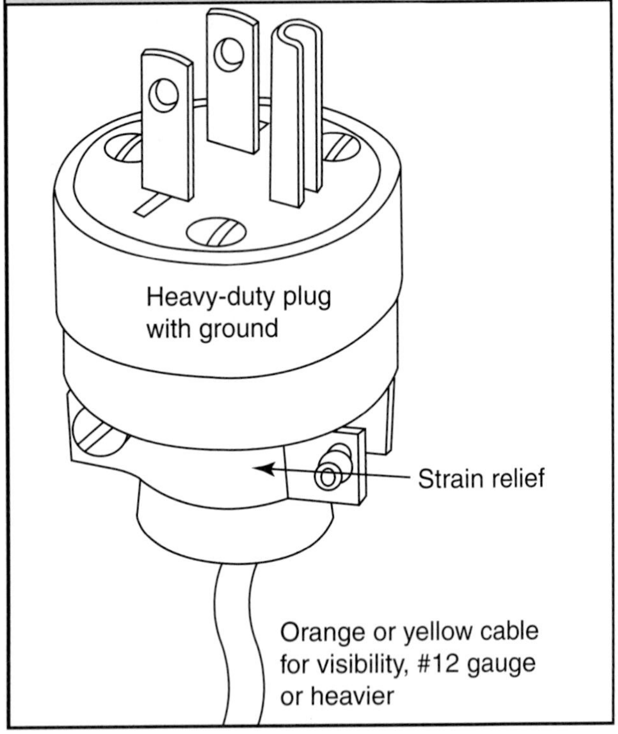

end of a shoot, it can be peeled from the wall (with patience) without lifting paint. Gaffer tape is available from video mail order companies.

Gaffer tape is a special high-performance version of duct tape. It was originally offered for lighting use in 1959 by Ross Lowel, the founder of the Lowel Corporation.

Lighting Evaluation Tools

In addition to an educated eye, two tools are useful for assessing a lighting design: a monitor and a light meter. Meters can be indispensable in certain situations, and a good monitor is essential for almost *all* types of shooting.

Monitors

No professional videographer likes to work without a reference monitor: a high-quality display that has been calibrated to show exactly what the camcorder is recording, **Figure 13-42**.

A good monitor delivers "what you see is what you get" information about the lighting. With it, you can light a set using basic common sense: simply check the effect on the screen as you work and adjust the lighting until you are satisfied with the image.

Figure 13-42.
A professional reference monitor. (JVC Corporation)

Light meters

Even with a reference monitor, a light meter is a useful measurement tool. You can use it to regulate ***contrast*** by checking the relative brightness of different image areas and adjusting light levels until the meter indicates the desired contrast ratio.

You can also use a light meter for checking overall brightness levels, in order to calculate the f-stop at which your lens is working. This helps you determine whether to use neutral density filters (to reduce light) or gain control (to amplify low-light signals).

Incident meters, like the studio model in **Figure 13-43**, are aimed at the lights, measuring illumination directly.

Figure 13-43.
This incident light meter is calibrated in "foot candle" units. (Mamyia/Sekonic)

Reflective meters are aimed at the subject, measuring the light that is reflected from the subject to the camera lens.

Color temperature meters precisely measure the relative blueness or redness of "white" light at any point in a scene. This information is useful if you are working with a mixture of light sources, so that you can decide which ones to adjust for color, and how much adjustment to make.

Some lighting directors like the precision afforded by light meters, while others seldom use them, except in specialized and/or very difficult lighting situations.

Figure 13-44.
A halogen work light.

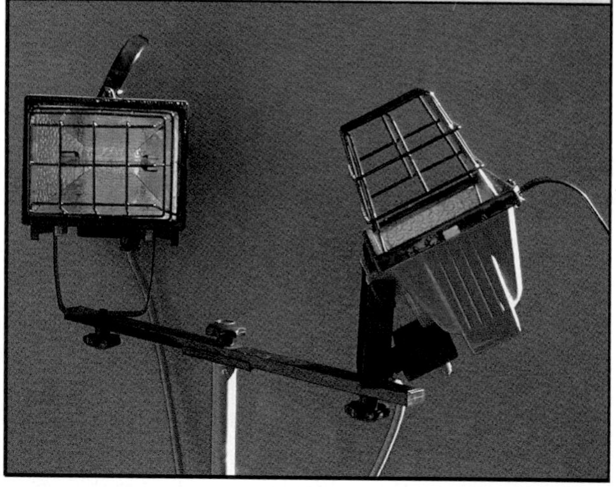

Inexpensive Lighting Equipment

Although professional lighting equipment is generally worth the money it costs, you can sometimes buy (or build) less expensive alternatives. This section covers the more common items in the "guerrilla lighting" kit.

Lighting Instruments

Video lighting benefits from several facts:

- All halogen lighting has a color temperature of 3200K, whether intended for video use or not.
- Better quality fluorescent tubes can simulate outdoor light with up to 90% accuracy.
- Both halogen and fluorescent lamps are now available with standard medium screw bases, for use in any household fixture.

In effect, this means that your choice of lights is limited only by your imagination. Here are a few suggestions.

Halogen work lights

Halogen work lights, **Figure 13-44**, are available at hardware and building supply stores at reasonable prices. Intended for light industrial use, these fixtures meet all code requirements for lamp safety, cord protection, and switches. Depending on your needs, you can select wattages from 150 to 1,000, single or twin heads, heads fitted with separately switchable lamps, and lights on telescoping floor stands, short legs, or clamps.

When shopping for work lights look for these desirable features:

- Long support columns (some units stand only about five feet high).
- Handles for moving the heads (to avoid touching the hot units).
- Heads that swivel up and down as well as back and forth.

For convenience, you may wish to install a switch on the power cord so that you can turn the lights on and off without reaching up to the heads.

Single work lights on spring clamps are useful for backlighting. Wrap the clamp jaws with duct tape (or dip them in liquid plastic) to avoid marring surfaces.

Reflector Scoops

Lightweight scoops fitted with clamps make useful instruments for screw-based halogen or fluorescent lamps. Though relatively flimsy, their inexpensive cost allows for easy replacement and their very light weight makes them ideal for clamping to door moldings and ceiling grids.

Household reflector lamps

Halogen lamps with built-in reflectors (designated "R" or "PAR" lamps) are available in a wide variety of diameters and wattages. Where beam edge control is not important, they can be used as crude but effective spotlights.

Light pans

If you are handy at construction, you can mount several individually switched fluorescent fixtures in shallow plywood trays to construct very effective pan lights that use less power and emit far less heat than halogens. Be sure to buy units with better quality ballast, since "shop light" fixtures may not adequately power the high-accuracy fluorescent tubes required for good color temperature.

Reflectors

Reflectors are easily constructed by taping aluminum foil over foam board sheets or light plywood rectangles (24 x 36-inch sheets is a good size). To create the hardest surface with the longest throw and brightest light, use spray adhesive to glue the foil, shiny side up, to the board. For a slightly softer surface, place the matte side of the foil up instead. For a still more diffuse light, ball up the foil lightly and then spread it flat again before gluing. This will break up the surface into thousands of tiny facets that will refract the light and diffuse the overall reflection, **Figure 13-45**.

For soft fill effects, any white cardboard will do; but white foam board is more rigid. For greater permanence, invest in one-inch board, rather than the thinner type. If you occasionally need to warm up the reflected light, choose a board with gold paper on one side.

The most common type of reflector combines a hard surface (shiny or matte aluminum) on one side with a soft surface (faceted aluminum or white) on the other side.

Folding auto windshield sunshades made of fabric-covered hoops are inexpensive. They are available in both silver and white.

Figure 13-45.
The surfaces of shiny, matte, and crinkled aluminum foil.

Flags

Foam board can be used for flags, and plywood reflectors can also be used to mask off unwanted light. However, the most versatile flag shape is a long, slim rectangle of thin plywood. (Plywood holds up better than foam board when clamped in a century stand.) Flags should be painted black, to minimize light bouncing off them, **Figure 13-46**.

Figure 13-46.
A typical flag.

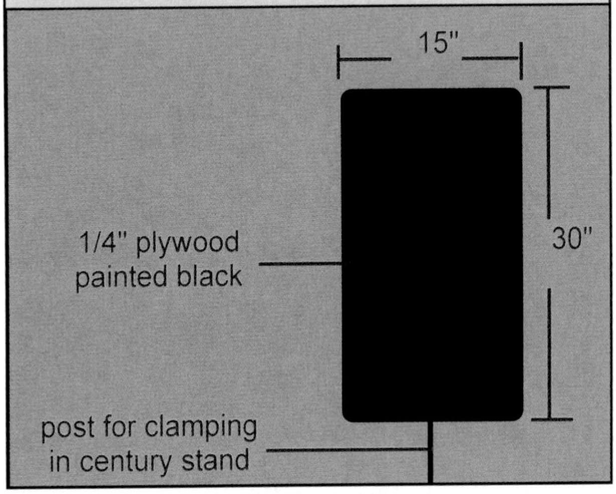

Screens and Silks

Screens reduce the intensity of light and silks diffuse it. Hand-held models are inexpensive enough to buy, and you can find them at most larger camera stores. Stretched on wire hoops, these models fold into small areas for storage and transport. For larger models, you

Figure 13-47.
A frame made of inexpensive plastic pipe.

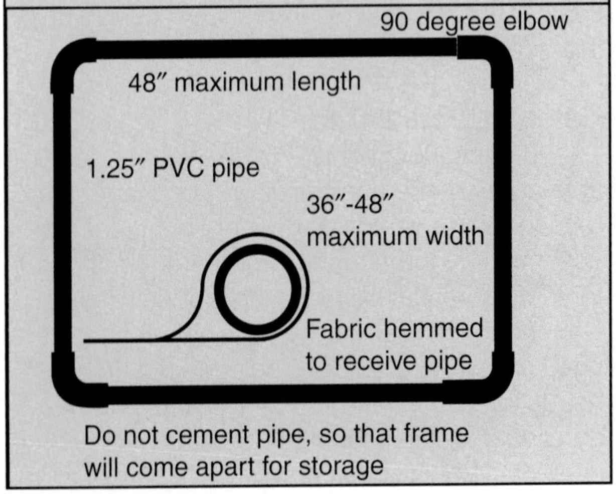

will need to construct a mounting frame. **Figure 13-47** diagrams a frame of plastic pipe that disassembles for storage.

Old bedsheets make excellent silks. For screens, plant nurseries and garden supply stores carry black plastic netting in large rolls. You can select the density of the screening, or double-layer lighter screens to block more light.

Weights

The great enemy of outdoor lighting control is wind, especially with screens and large reflectors. To hold down C-stands or screening frames, weights are essential accessories. Sandbags are cheap and effective weights, but they are heavy, dirty, and tend to leak. For this reason, some production companies use weights filled with water instead, **Figure 13-48**. By filling and emptying the bags at the location, they avoid transporting most of the weight.

Commercial water weights are available from production supply houses, but the collapsible plastic water bags sold for camping are inexpensive and work just as well. They are typically supplied with handles and hooks, to help hang them on light and reflector stands. If you need to improvise, one-gallon drinking water bottles are available at any supermarket at very low cost. They can be secured to stand bases with duct tape or bungee cords.

Figure 13-48.
Water bags and even plastic drinking water bottles make excellent weights.

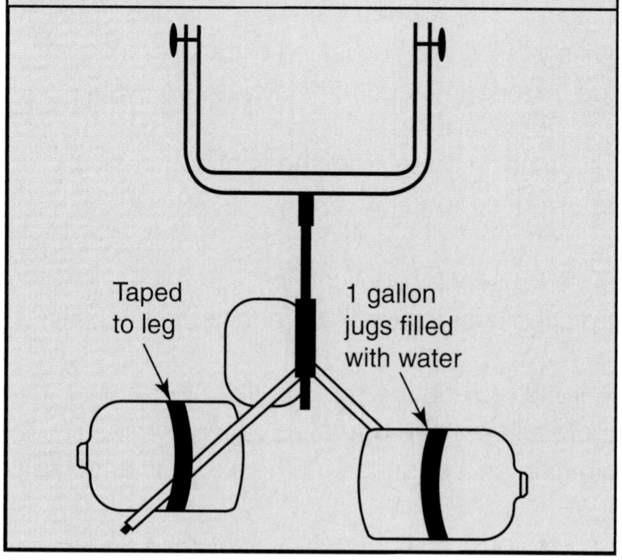

Reference Monitor

Professional reference monitors are expensive, but very good 13-inch color TV sets are widely available for much less, **Figure 13-49**. Many will operate on wall power, 12-volt auto plug, or batteries. Make sure the set accepts composite video input (identifiable as a yellow jack for an RCA plug).

Figure 13-49.
An AC/DC combination monitor/player is a useful location accessory.

Designing with Lighting Equipment

This chapter has covered most of the lighting equipment used for location and small studio shooting. The following chapter explains how this equipment is used to create lighting designs, and the final lighting chapter shows how to employ these tools in a variety of real-world lighting situations.

Chapter Review

Answer the following questions on a separate piece of paper. Do not write in this book.

1. In many situations, ______ light will provide most or all of your lighting.
2. ______ reflectors provide a soft, even fill light.
3. ______ cast distinct, hard-edged shadows.
4. *True or False?* Flags are mounted in pairs on a lighting instrument and adjusted to mask portions of the beam.
5. Low-cost lighting can be obtained by using ______ work lights purchased at a hardware store.

Technical Terms

Available light: The natural and/or artificial light that already exists at a location.

Barn doors: Metal flaps in sets of two or four, attached to the front of a spotlight to control the edges of the beam.

Broad: Small, portable floodlights in the form of shallow rectangular pans.

Century stand (C-stand): A telescoping floor stand fitted with a clamp and usually an adjustable arm, for supporting lights and accessories.

Color temperature meter: A light meter that measures the relative blueness or redness of nominally "white" light.

Contrast: The difference between the lightest and darkest parts of an image, expressed as a ratio (e.g. "four-to-one").

Cookie: A sheet cut into a specific pattern and placed in the beam of a light to throw distinctive shadows such as leaves or blinds.

Diffusion: White spun glass or plastic sheeting placed in the light path to soften and disperse it.

Filter: In lighting, a sheet of colored or gray-tinted plastic placed over lights or windows to modify their light.

Flag: A flat piece of opaque metal, wood, or foam board placed to mask off part of a light beam.

Fluorescent lamp: A lamp that emits light from the electrically charged gasses it contains.

Halogen lamp: A lamp with a filament and halogen gas enclosed in an envelope of transparent quartz.

Incandescent lamp: A lamp with a filament enclosed, in a near-vacuum, by a glass envelope. Also called a "bulb."

Incident meter: A light meter that measures illumination as it comes from the light sources.

Instrument: A unit of lighting hardware, such as a spotlight or floodlight.

Lamp: The actual bulb in a lighting instrument.

LED (Light Emitting Diode): An electronic light source for special applications.

Neutral density filter: In lighting, a gray sheet filter placed over windows to reduce the intensity of the light coming through them.

On-camera light: A small light mounted on the camera to provide foreground fill.

Practical: A lighting instrument that is included in shots and may be operated by the actors.

Reflective meter: A light meter that measures illumination as it bounces off the subjects and into the camera lens.

Reflector: A large silver, white, or colored surface used to bounce light onto a subject or scene.

Scoop: A type of floodlight used mainly in TV studios.

Screen: A mesh material that reduces light intensity without markedly changing its character.

Silk: A fabric material that reduces both light intensity and directionality, producing a soft, directionless illumination.

Softlight: A lamp or small light enclosed in a large fabric box, which greatly diffuses the light.

Spotlight: A small-source lighting instrument that produces a narrow, hard-edged light pattern.

14 Lighting Design

Objectives

After studying this chapter, you will be able to:

- Explain the three standards of quality video lighting.
- Recognize the major styles of lighting designs.
- Employ effective strategies in approaching lighting design problems.
- Follow proper procedures in working with lighting equipment.

(Photoflex)

The Development of Video Lighting

Historically, film and TV scenes have been lit quite differently. Since video is a hybrid of both older technologies, it takes something from each and adds techniques of its own.

Film lighting

Despite its many advantages, film has historically suffered from two major lighting problems: its relative insensitivity and its need to be processed and printed for viewing. The relatively slow speed (light sensitivity) of film meant that sets needed to be lit with many lighting instruments that consumed considerable power. Although today's color film is more sensitive than it was when film lighting techniques were developed, these techniques are still widely practiced, so that film lighting tends to be a relatively elaborate procedure.

Because film must be processed and printed before it can be viewed, it is impossible to see the precise effect of film lighting during shooting.

On most film productions today, the camera images are viewed on video monitors. Unlike reference monitors for video shooting, however, these displays cannot show the lighting exactly as it will look on film.

To overcome this inconvenience, professional cinematographers and lighting gaffers become experts at predicting how their lighting designs will look on film. To help in this process, they use specialized light meters to measure incident and reflected light levels at important points in the scene. Nonetheless, even with these sophisticated metering instruments and today's sensitive film stocks, film lighting is an art that can take years of practice to master.

The *gaffer* is the head of the lighting crew on a shoot. The term is derived from an old British word meaning "grandfather," or simply, "old man."

Television lighting

Studio-based television also presents lighting problems. For many years, the tube-type cameras employed could not handle ***contrast*** competently. This meant that the darkest parts of the image had to be no less than about one-third as bright as the lightest parts. Moreover, television programs are recorded more or less continuously by multiple cameras shooting from different directions and often covering different parts of the set. For this reason, every subject, place, action, and camera setup must be lit at the same time. Although technical progress has allowed television studio lighting to improve substantially, the demands of low-contrast lighting and full-set coverage can still make television lighting more flat and bland than film lighting.

TV studios use low-contrast lighting. (Corel)

Video lighting

Video is a true hybrid: a medium that combines the shooting techniques of film with the recording technology of television. Camcorders fitted with modern imaging chips are more sensitive to light than many standard film stocks and accept much wider contrast ratios than classic television cameras. Like TV lighting directors, videographers can see exactly what their images look like as they are setting them up and shooting them. For this reason, learning to light for video is easier than mastering film lighting. In effect, if it looks good on a quality reference monitor, it is good.

Perhaps the fairest assessment might be that today, all three visual media can support sophisticated lighting. While film and television continue using traditional methods with technical improvements, video lighting has evolved its own approach.

About Lighting Design

The previous chapter shows what lighting tools can do, and the next chapter puts them to work. Between them, this chapter explains the ideas behind video lighting design—the *basics*, *styles*, *principles*, and *procedures* that apply to all lighting assignments. With these design fundamentals reviewed and summarized, we will be ready to move on to the final lighting chapter and apply them to typical real-world lighting applications.

Lighting Standards

Whatever else it may add to a production, professional video lighting begins by delivering high-quality images. Without this technical quality, the subtler contributions of style and mood would be beside the point. To achieve high technical standards, images must be created with light of the proper *quantity*, *contrast*, and *color*, **Figure 14-1**.

Light Quantity

The camera's imaging chip (or chips) must receive exactly the right quantity of light in order to form a high-quality image. Too much light makes colors smear and "bloom" and forces the aperture to very small f-stops (openings). Too little light makes a murky picture with dull colors and little detail, and may push the electronic gain compensation up to a level that creates grainy, noisy images. The first task, then, is to decrease or increase the quantity of light to suit the needs of the camera.

The videography chapters show how to adjust the camera to suit the light. This chapter shows just the opposite: how to adjust the light to suit the camera.

Decreasing the quantity

Outdoors, you often need to control the light by reducing its quantity (intensity). To do this, you can either block or screen it, **Figure 14-2**.

To *block* light entirely, you place an opaque shield such as a lighting ***flag*** between the light

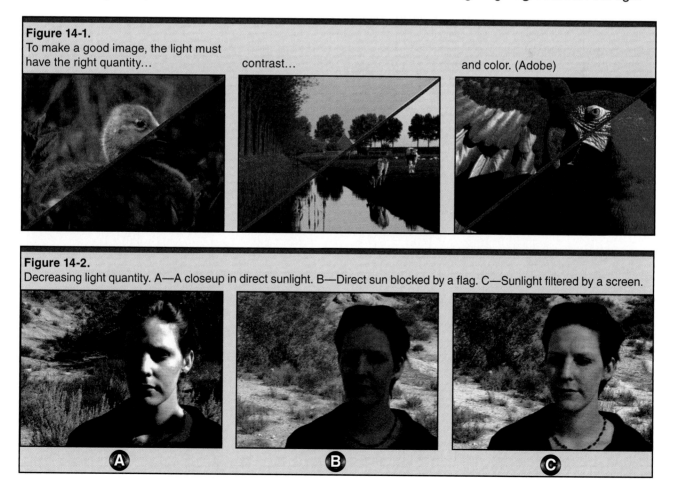

Figure 14-1.
To make a good image, the light must have the right quantity... contrast... and color. (Adobe)

Figure 14-2.
Decreasing light quantity. A—A closeup in direct sunlight. B—Direct sun blocked by a flag. C—Sunlight filtered by a screen.

A B C

source and the subject. This technique works when there is enough light from other sources to illuminate the blocked area of the subject.

To *screen* the light instead of blocking it, you place a large framed screen of mesh or cloth between the direct sunlight and the subject.

For consistency, some illustrations are repeated in these three lighting chapters.

Increasing the quantity

Often, the problem is not too much light but too little, so your job is to increase the quantity of illumination. You have two options: outdoors, you generally redirect the light, making double use of the illumination available. Indoors, you typically add lighting to the available illumination.

To *redirect* light that is already present, you use ***reflectors***. Reflectors are quick, simple solutions for outdoor lighting especially, and they can be used indoors as well. See **Figure 14-3**.

To *add* light to the available illumination means using one or more lighting instruments. Video lighting can be as simple as an on-camera fill light or as elaborate as a set lit with 20 instruments. In professional video work, almost all interior locations are lit for shooting, sometimes by supplementing the light already available at the location.

Light Contrast

After quantity, the biggest lighting problem is contrast. ***Contrast*** is the difference in brightness between the lightest and darkest areas in the image.

Lighting contrast is expressed as a ratio. A "four-to-one" ratio indicates that the lightest areas are four times as bright as the darkest ones.

Too little contrast makes subjects appear flat and bland. Too much contrast makes light areas appear stark white or dark areas solid black, or, in extreme cases, both at once, **Figure 14-4**. Regardless of the lighting mood or style, a well-lit image has an even range of brightness levels:

- True black in the darkest areas.
- Shadows with details visible in them.
- Several gradations of brightness in the middle range.
- Highlights that retain some detail.
- True white in the lightest areas.

Figure 14-3.
Using a reflector. A—Direct sunlight. B—Shadow filled with a reflector.

Figure 14-4.
Lighting contrast. A—This contrast is too great. B—This contrast is too small. C—This contrast is just right. (Sue Stinson)

Sidebar: Estimating Contrast

The human eye is not reliable in judging contrast for video because we can distinguish brightness levels up to a ratio as great as 100-to-1, while even the best video systems can achieve ratios only about ten percent as wide as that. For this reason, it is important to check contrast on a monitor. Conventional monitors are unreliable in bright light, and even the best small LCD screens cannot display contrast accurately.

If your monitors are not showing contrast very well, you can estimate the contrast of the image to be recorded by using an old photographer's trick:

1. Look at the scene to be taped.
2. Close one eye completely.
3. Squint the other eye until it is almost shut and its eyelashes form a sort of filter.
4. Look at the darkest parts of the scene.

If you cannot make out details in the shadow areas, the contrast is too great. Though this improvisation is no substitute for an accurate monitor, it can give you a rough approximation of the scene contrast as the camcorder will record it.

As described previously, you control contrast by adjusting the light quantity, either lightening the dark areas by filling them in with more light or darkening the bright areas by screening them. You can also reduce contrast by diffusing the light—scattering the straight-line rays of a light source so that they bounce around the subject and soften or eliminate the shadows. You can diffuse direct light by dispersing it through cloth or plastic sheeting; reflected light can be diffused by using white reflectors instead of metallic ones. Lighting instruments can be fitted with plastic, cloth, or spun glass ***diffusion*** material to scatter the light.

Light Color

To control the quality of light, you need to manage not only its brightness and contrast, but also its color. Commonly, this means setting and maintaining white balance. There are special situations, however, in which you will want to exercise highly selective control over light color.

White balance is a function of color temperature, which is discussed in Chapter 11.

For example, many locations require mixing light sources of different color temperatures, whether combining incandescent and fluorescent lights or managing daylight coming through a window, **Figure 14-5**.

In other situations, you may want to suggest the warmth of a sunset or fireplace. Or you may need to simulate cool moonlight to make daytime footage look like night. Whether creating special effects, balancing multiple color tints, or simply achieving good white balance, you are always dealing with the color of light.

Figure 14-5.
The window light coming from the right is bluer than the supplementary movie lights.

Recording a Standard Image

In managing the quantity, contrast, and color of light, your goal is to record a "standard" video image: an image in which both shadows and highlights retain full detail, and apparent "white" remains the same from shot to shot. To put it another way, you want to record an image that does not need correcting in editing.

Before digital postproduction, many effects were best created in the camera. For example, outdoor "night" scenes shot in daylight were underexposed, with the white balance purposely set to incandescent, in order to give all the footage a bluish, "moonlight" cast.

Today, however, you can control image characteristics in the edit bay, so it is best to start with a neutral, high-quality image. If you do not like an effect added in editing, you can easily return to the original footage and try again. If you created that

Figure 14-6.
Simulating a police car's flashing red light.

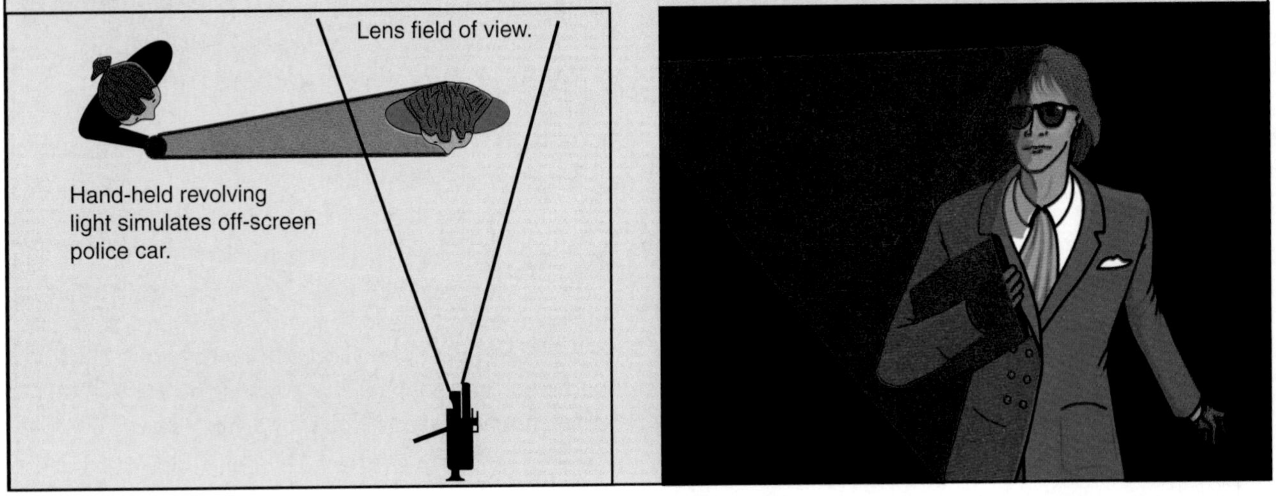

effect in the camera, however, you are stuck with whatever you recorded, like it or not. Moonlit nights, sunsets, and other overall color tints are easy to add during editing. Contrast and brightness can be adjusted to change day shots to night.

This rule is even more important with certain camera effects that cannot be undone by digital postproduction.

On the other hand, if you have colored light on just *one part* of your scene, it may well be easier to create the effect while shooting than to add it in postproduction. For example, the flashing red light of an off-screen "police car" colors only the left side of the subject in **Figure 14-6**.

Lighting Character

If you master the technical aspects of good lighting, your images may be well-illuminated, attractively colored, and pleasing to look at, but they may not contribute much to the program. To pull its weight artistically, lighting must also have character. Lighting character is the set of traits that individualizes the look of the scene by giving it a distinctive *style*.

Lighting Styles

Over the centuries, painters, stage designers, photographers, and cinematographers have evolved many approaches to lighting. Though there is no standard classification, we may distinguish several basic lighting styles and label them *naturalism*, *realism*, *pictorial realism*, *magic realism*, and *expressionism*.

Although these terms are not universally agreed upon, they are common in critical writings about art, theater, film, and photography.

Classifying lighting styles is not just a theoretical exercise. By choosing a basic approach to lighting and then customizing it to your personal taste, you can give your videography a consistent "look and feel," a distinctive quality that enhances its effectiveness and increases its professionalism.

Naturalism

Naturalism is the lighting style that most closely imitates real-world conditions.

Figure 14-7.
Naturalistic lighting. (Photoflex)

A naturalistic scene does not look "lit." Instead, a naturalistic image makes you feel that you are looking directly at the scene rather than at an image of it on a screen. Naturalistic lighting uses ***available light*** as much as possible and does not light anything to achieve an obvious effect, **Figure 14-7**. Naturalism is a useful style if you want your program to have the look of a straightforward documentary.

Realism

Realism is naturalism heightened and intensified somewhat to add dramatic effect. Where naturalism records actual lighting conditions as far as possible, realism *recreates* actual-looking conditions through lighting. The average viewer may not notice it as lighting, but it is visible to an informed eye, **Figure 14-8**. Realism is the most common style for TV programs, whether police dramas shot on location or comedies taped in a studio. If you make program videos, you will probably work most often in the realistic style.

Figure 14-8.
Although this room seems lit by natural daylight, the "hot spots" in the upper right reveal the lights above the frame border. (Sue Stinson)

Pictorial Realism

Pictorial realism is a "painterly" style, in which the lighting effects are frankly visible, although they still imitate real-world lighting such as lamps, streetlights, headlights, sunlight, and moonlight, **Figure 14-9**. Through most of Hollywood's history, until the 1960s, pictorial realism was the most popular lighting style. The simpler alternative of plain realism was largely confined to exterior shooting, where the lighting equipment and techniques needed for pictorial realism were often impractical.

Figure 14-9.
The dramatic modeling and strong rim light are characteristic of pictorial realism. (Corel)

At the height of the silent film period, so-called "Rembrandt lighting" was highly regarded; and during the era of classic studio sound films, an influential textbook on cinematography was titled Painting with Light.

Although pictorial realism sometimes looks artificial to modern viewers, it was accepted as realistic by the original audiences of classic Hollywood films. Today, pictorial realism is often used in commercials and music videos, where a purposely theatrical feel is required.

Magic Realism

Digital video image processing has led to a hybrid style that combines elements of realism, pictorial realism, and expressionism. Because its overall look is superficially realistic, but with a heightened, intensified quality, we may borrow a literary term and name it ***magic realism***.

Though not always manipulated by computer, images in magic realism typically look processed: colors may appear slightly more intense or more uniform than normal.

The hotel interiors in Stanley Kubrick's film, "The Shining," are lit in the magic realism style.

Different picture components may be sharp or blurred in ways that do not match the focusing characteristics of lenses. Perspective may be

Figure 14-10.
Magic realism. A—A cemetery processed in a magic realism style. (Sue Stinson) B—This dockside image gets its eerie quality from the combination of twilight available light and the harsh work light on the pier. (Sue Stinson)

A

B

skewed. Overall, the images seem to be combinations of photography and illustration, **Figure 14-10**.

Though magic realism is often created in postproduction, it can also be achieved through lighting, often by using extra rim lighting for an ethereal, dreamlike look.

Expressionism

Expressionism is the most frankly theatrical of the major lighting styles, and does not pretend to reflect reality at all. It is lighting for its own sake, in which any effect is acceptable as long as it *expresses* and intensifies the feeling of the scene. Often, though not always, expressionistic lighting is hyper-dramatic, even extreme.

Figure 14-11.
The Cabinet of Dr. Caligari.

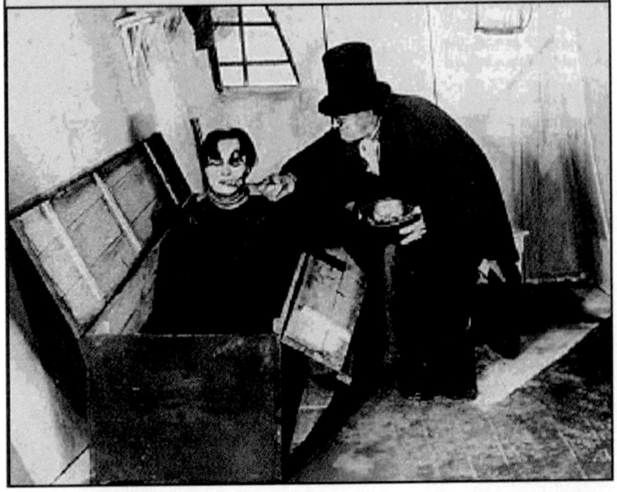

Expressionism began as a movement in painting and theater in Europe, and then was adopted by film makers, notably in Germany. German directors brought expressionistic lighting and camera work to Hollywood in the 1920s and 1930s. Universal Studios' horror films such as *Frankenstein* exploited expressionistic lighting to good effect. Today, expressionism, like pictorial realism and magic realism, is often used for commercials (especially humorous ones) and music videos.

The most influential expressionistic film is perhaps *The Cabinet of Dr. Caligari.* Ironically, because of contemporary limitations in film emulsions and lighting techniques, the expressionistic design of this film was created more with scene paint than with light. See Figure 14-11.

Lighting Strategies

Technical standards and graphic style are relevant to all video lighting, but the *strategies* you employ to achieve them will vary from one application to the next. As you approach each lighting situation, you need to decide which common lighting objectives are most important, which approach will do the job most effectively, and whether the program demands a "high-key" or "low-key" appearance. Once you have made these decisions, you are ready to start creating your design.

An Anti-style

In addition to the other styles, there is a visual look that might be called an "anti-style," in which the images constantly remind the audience that they are looking at a video or film. We have placed this anti-style by itself because it involves not only lighting but every other aspect of videography as well.

This anti-style exploits the characteristics and limitations of the video medium:

- Shots are purposely set up so that light sources will strike the lens and create intentional flares.
- Dark scenes are shot in low light levels with the gain circuits enabled for a murky, grainy look.
- Exteriors are shot in available light. If light levels are too high, colors smear. If the contrast is too great, faces turn into silhouettes.
- Light levels are not equalized so that the auto-exposure system is visibly adjusting exposure as the camera moves.

This anti-style is similar to naturalism, in that it strives to convince viewers that they are looking at unprocessed reality. The difference is that naturalism says, "here is reality" while anti-style insists, "here is a video *recording* of reality."

Some viewers think that this anti-style is an intrinsically honest approach to recording video images. Others find it pretentious and irritating. Either way, it can be powerfully effective in the right type of program.

Intentional sun flare. (Sue Stinson)

Ultra low-light shot. (Sue Stinson)

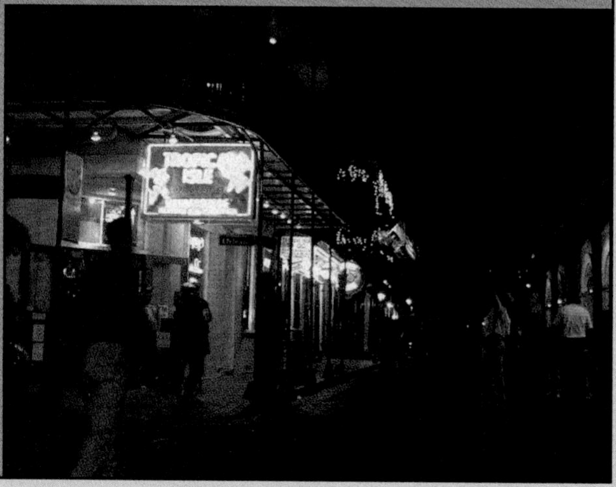

Identifying Basic Objectives

In building your designs, you usually want to achieve three objectives: motivate the lighting, light the subject (not the space), and enhance the feeling of depth in the image.

Motivate the lighting

The real world is full of light sources: the sun or the overcast sky, the lighting in a room, the special light of neon signs, computer monitors, headlights, candles, and numberless others. To make your lighting look "real," you generally try to imitate the light sources that would normally be present.

In practical terms, this means, for example, that you would not create scary cross lighting in an office washed by ceiling fluorescents.

So your first task is usually to decide what sources would naturally light the scene and then imitate them with your video lights.

Light the subject

Images are created by light reflected into the lens from subjects in the frame. For this reason, you do not have to light every cubic inch of your location, but only those areas containing subject matter. To light just the areas that will show on screen, follow this procedure:

- Study the performers' positions and movement ("blocking").
- Light the areas that they will use, paying special attention to places where they will stay, or at least pause. In other words, keep them adequately illuminated throughout their

Key, Fill, Rim, and Background

Most lighting designs are combinations of four basic functions—tasks so fundamental that they are part of every lighting style and present in most lighting applications. We cover these functions in detail in a later section; but because we also refer to them here, you need to know what they are.

Key light

The **key light** is the main illumination on the subject. Typically, it is the brightest light, and it mimics real-world light sources—usually the actual lights in the room or at the scene.

The key light provides the main illumination.

Fill light

The **fill light** is the secondary illumination on the subject. Usually placed opposite the key light, it fills in shadows to reduce contrast created by the key light.

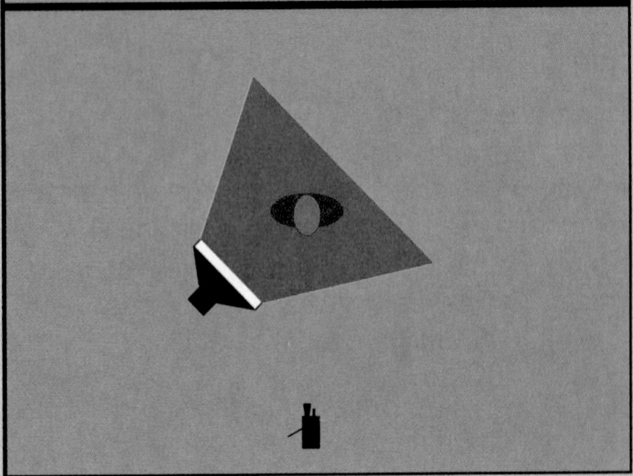

The fill light fills in shadows.

Rim light

The **rim light** is placed behind and well above the subject to separate the subject from the background by splashing a rim of light on head and shoulders.

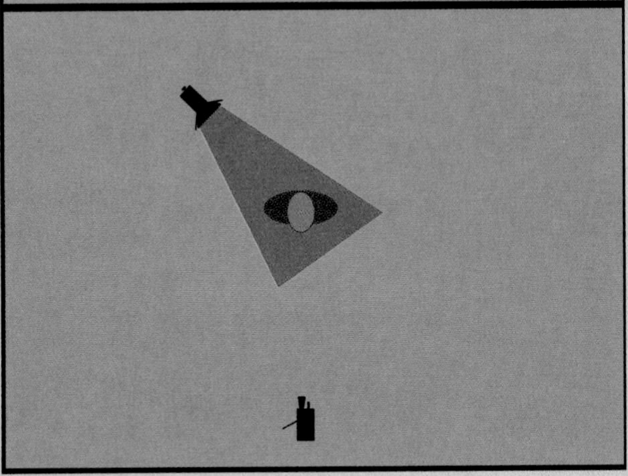

The rim light separates subject from background.

Rim light is also commonly called "back light." This term is not used here, to avoid confusion with *background* light.

Background light

The *background light* (or lights) raises the light level of the walls or other backing, adds visual interest, and enhances apparent depth by highlighting background features.

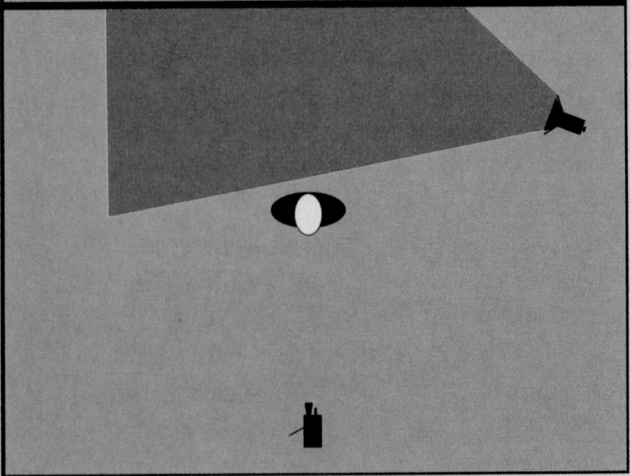

The background light illuminates the background.

Figure 14-12.
A lighting plan for a moving subject.

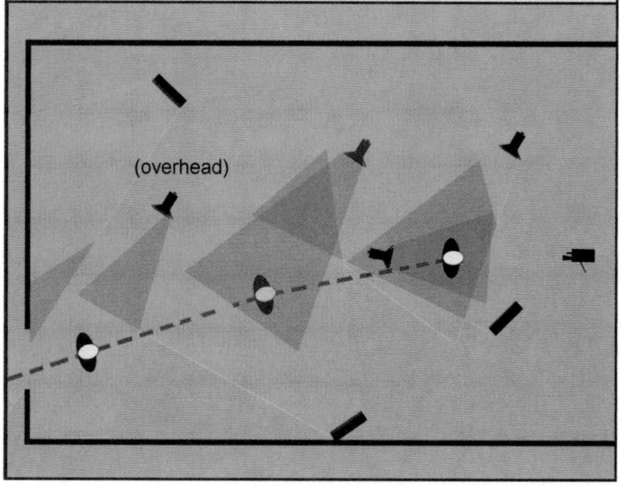

movement, and carefully lit at spots where important action happens.

- Light parts of the background (including furniture and any floor or ceiling areas) that will also appear in the image.

In **Figure 14-12**, the subject is lit by key, fill, and rim lights at the rear entrance, the center, and the forward stopping point. Two fill lights provide general area lighting.

The rear key light also helps wash the background. It is very common for lights to do double duty.

In this example, the left key light has been mounted overhead because its light stand would be in the wide shot. If there is no place to conceal lights, you may want to provide basic lighting for the entire area, adding or adjusting instruments as needed later to optimize the lighting for setups that focus on specific places within it.

This example is covered in greater detail in the next chapter.

Light to enhance depth

Where videographers use composition to enhance the illusion of depth, lighting directors use light accents. To suggest the third dimension, you will often create highlights deep in the real-world space that you are lighting. These accents guide viewers' eyes past foreground subjects and "into" the scene.

Though you usually want to achieve all three of these objectives, there are special cases in which you do not. For example:

- In the expressionistic lighting style, ***motivated lighting*** is not required or necessarily desirable.
- In some wide shots, so much of the environment is on-screen that your lighting must cover all of it.
- To create compositions on the picture plane, you may purposely avoid background light accents that call attention to depth.

Videographers contribute to these purposely flat images by using telephoto lens settings and minimizing the usual visual cues of perspective.

Selecting a Lighting Approach

With your objectives for lighting a scene in mind, you are ready to choose a specific approach to the task. In any lighting environment (except, of course, a studio) you have three different options in lighting a scene. You can work with the light available, you can start with the available light and supplement it with video lighting, or you can neutralize the available light and design exclusively with video lights. Your decision will depend on the lighting existing at the scene, the style you have adopted, the equipment at your disposal, and the amount of electrical power available.

Working with available light

Except for news and documentaries, most available light shooting takes place outdoors. If the day is overcast, the main concern is color

Figure 14-13.
The dark, textured wall makes a good background. Note the white reflector for additional fill. (Sue Stinson)

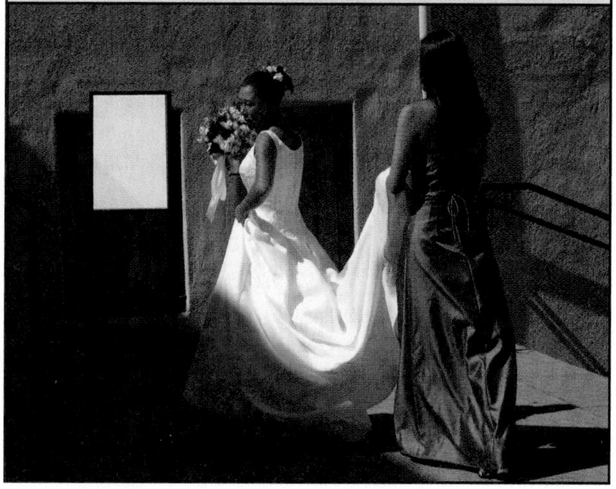

Figure 14-14.
A large silver reflector on a frame. (Photoflex)

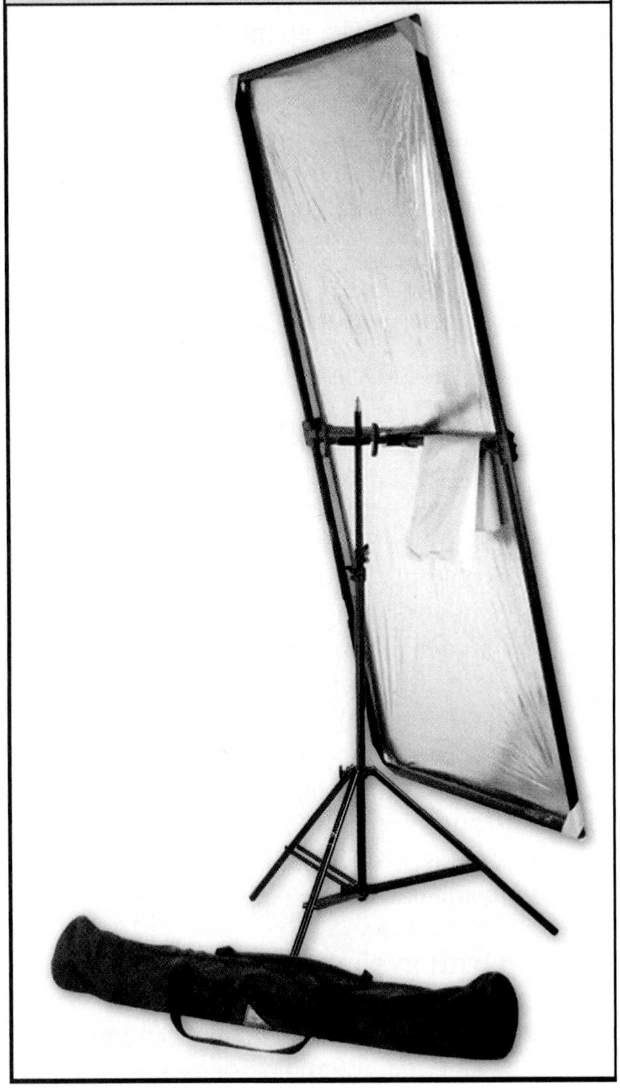

Figure 14-15.
A medium-size silk. (Photoflex)

temperature, which is addressed by adjusting camera white balance. In direct sunlight, however, each camera setup offers you three tactics that you can use alone or in combination.

Select the light. When the action permits, the easiest method is to find backgrounds and subject positions that naturally yield high-quality images, **Figure 14-13**.

This may mean placing subjects in existing shade or else in sunlight with a darker background behind them to prevent severe backlighting. If the ground is sand, light soil, or concrete, the light reflected from it may provide natural subject fill light. Sometimes, the ambient bounce light is so strong that you can use the sun as a rim light and the bounce light as a soft key and fill.

Be careful on grass, which can reflect its green color onto your subjects.

Reflect the light. The workhorses of outdoor lighting are reflectors, which can be used for key, fill, or backlighting—or in any combination. Reflectors are versatile and easy to use. On the down side, the light from hard reflectors can look harsh, and soft reflectors must be used close to subjects, which make wide shots difficult to light, **Figure 14-14**.

Filter the light. Screens and silks are able to control larger areas, but big ones require complicated frames and large crews to manage them. Units on small frames (like the one suggested in the previous chapter) can only screen one or perhaps two subjects at once. Like soft reflectors, screens and silks must be used close to subjects, so they cannot be employed in wide shots, **Figure 14-15**.

In outdoor shooting, you will typically find yourself using two or all three of these tactics at once.

Adding to available light

The second possible approach is to start with the available light, then supplement it with video lights. Outdoors, ***camera lights*** can be used to fill facial shadows or simply to enhance underlit subjects. You can call the viewer's attention to your center of interest by making it just a bit brighter than objects around it, **Figure 14-16**.

In Figure 14-16, notice how the extra front light slightly emphasizes the subject, without seeming to throw an obvious spotlight on him. (You can, however, see the subject's shadow on the wall.)

The existing-plus-video lighting approach comes into its own indoors, where electrical

Figure 14-16.
Adding fill light. A—An on-camera light. (Lowel Corporation) B—With available light alone, the lighting is murky. C—An on-camera light adds front fill.

power is available. Since all camcorders can record a good-quality image in well-lit interiors, the easiest approach is often to start with available light and enhance it with video lighting to complete your design.

Typically, indoor available light comes from windows, ceiling fixtures or grids, and lamps. In planning your tactics, you need to assign roles to each of the light sources that are available. Ceiling light is usually used for fill and backgrounds, but window light can be either key, fill, or background (depending on its intensity, quality, and position). Once you have given a light source its task to perform, you complete the design with video lighting.

Complementing an available key light. In **Figure 14-17**, the woman is lit primarily ("keyed") by soft light from an off-screen sliding glass door. (General ambient light provides the fill.) To enhance visual interest, while preserving the natural-looking design, an added ***practical*** (on-screen) light provides rim light.

Complementing an available fill light. In **Figure 14-18**, by contrast, the subject is lit primarily by soft fill from overhead fluorescent

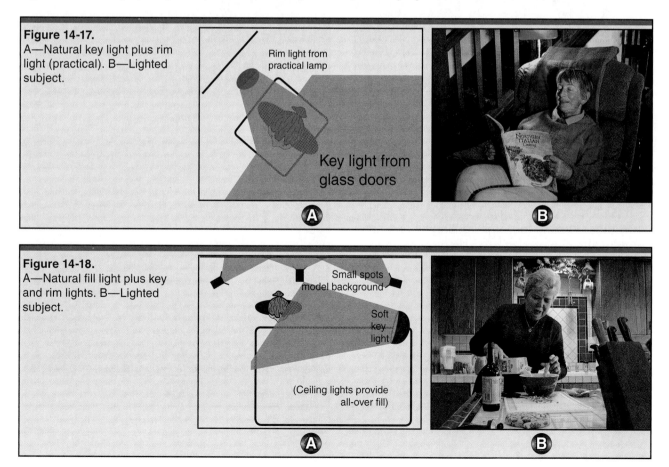

Figure 14-17.
A—Natural key light plus rim light (practical). B—Lighted subject.

Figure 14-18.
A—Natural fill light plus key and rim lights. B—Lighted subject.

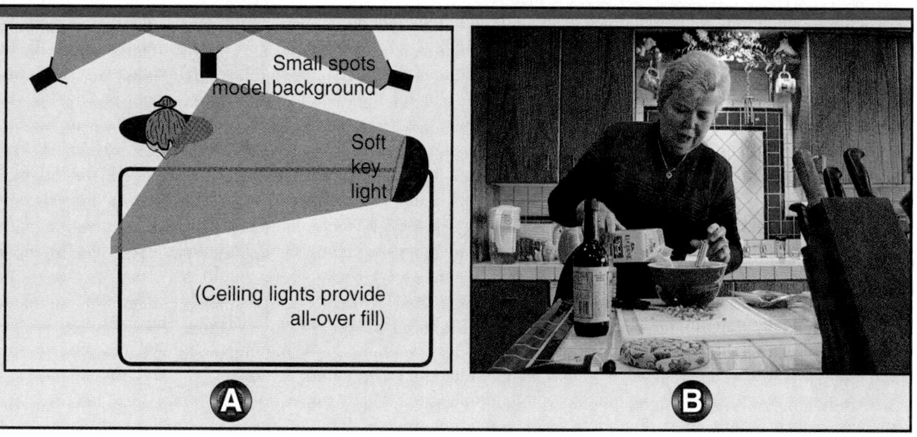

lights. In this setup, a soft key light has been added, placed to the side and at eye-level; and small overhead lights provide modeling for the background—note the hot spot on the flowers.

Using video lights only

Sometimes the existing light is difficult to manage, or simply wrong for your needs. The simplest solution in such cases is to neutralize the available light entirely and create your lighting design strictly with video lights.

You neutralize available light by masking it (as with curtains) or simply turning it off.

This strategy offers the tightest control over the lighting but demands the most resources. Before going this route, make sure that you have enough equipment to execute your design, enough power to drive it, and enough time and personnel to rig it and later dismantle and pack it. See **Figure 14-19**.

Figure 14-19.
Lighting this large interior required six softlights. (Photoflex)

It is difficult to generalize, but possibly safe to say that a typical four-unit lighting kit can handle up to two subjects at once, in a room up to perhaps 15 feet square.

Choosing a Lighting "Key"

When you have settled upon your objectives and your overall approach, your final strategic option is to pick a "key" for the lighting's overall look, **Figure 14-20**. A ***high-key*** image is composed mainly of lighter values, punctuated with

Figure 14-20.
A—High-key lighting. B—Low-key lighting. C—"Medium key" lighting. (Corel)

occasional dark areas. A ***low-key*** image is the opposite: mainly dark, but usually with a lighter center of interest. A ***"medium-key"*** image (to coin a new name for the purpose) has highlights and shadows more or less evenly distributed.

High-key and low-key are not themselves styles. It is perfectly possible to use either approach in naturalism, realism, pictorial realism, magic realism, and expressionism. However, it is easier to create high- or low-key designs with spotlights, because they allow greater control over light intensity, direction, and masking. Most training, promotional, and other nonfiction programs lean toward the more cheerful look of high-key lighting, without formally adopting it as a design approach.

Lighting Procedures

Moving from the general toward the specific, through lighting *standards*, *styles*, and *strategies*, we come now to lighting *procedures*: the methods you use to deploy your lighting equipment. Since each lighting tool is handled essentially the same way in most applications, it is easier to discuss equipment procedures together, rather than repeat information for every separate lighting situation.

Using Small-source Lights

Spotlights and the small ***floodlights*** called broads are the backbones of many lighting kits. These ***small-source*** lights are popular because of their portability, versatility, and small size. Both types grow very hot in use, so be sure to have leather work gloves handy.

Spotlights

Spotlights are used when you need precise control over the light path. To shape the beam, it helps to follow a regular sequence.

- Raise and aim the light. Always extend the upper sections of the light stand first, to keep as much weight as possible down low.
- Adjust the beam intensity by moving the light and stand toward or away from the subject (as long as they remain out of frame).
- At the desired position, fine-tune light intensity by operating the lamp focus control. Flood the light for lower intensity; use the spot focus for higher.
- If edge masking will not be an issue, soften the beam by installing a milky plastic diffusion filter in front of the lamp, or clipping spun glass diffusion to the ***barn doors***.

The hard-edged spot focus is easier to mask than the flood position. If edge control is wanted:

- Try setting the focus to full spot and adjusting beam intensity with screens.
- Mask the beam edges by adjusting the barn doors. If, as often happens, they will not manage all four sides, supplement the barn doors with clip-on or freestanding flags. (Move the flags toward the light for softer edges or away for harder ones.)
- Check the subject for overly bright areas (often a problem when the light is at an oblique angle to a surface like a wall). Use half-screens to reduce hot spots, **Figure 14-21**.

High-key/Low-key Examples

In classic Hollywood films, high-key lighting was used for comedies, romantic movies, and action pictures. Low-key lighting was employed for crime and mystery films, intense dramas, and horror. The so-called "film noir" style of the 1930s and 1940s executed low-key lighting in the pictorial realism style, with great effectiveness.

High- and low-key approaches are easier to study in black and white films. To select two examples, the comedy *Some Like It Hot* is generally high-key, and the drama *Dr. Strangelove* is low-key. (These classic films, among the best in Hollywood history, are rewarding to view just as movies.)

In video, high- and low-key lighting has been less common, because until recent years, TV did not offer enough range of contrast to permit it. (Prints of films intended for broadcast were often made with purposely reduced contrast, to accommodate the demands of the medium.)

With today's improved video technology, look for low-key lighting in TV dramas and high-key designs in situation comedies. Studio talk shows and news programs are generally "medium-key," while commercials and music videos often carry high-key/low-key differences to extremes not seen since classic Hollywood.

Figure 14-21.
Some spotlights use proprietary diffusion screens. (Lowel Corporation)

If a somewhat warmer color temperature is acceptable, you can also reduce intensity by putting the light on a dimmer.

Though this sequence works well in general, you will often have to go back and forth among focus, barn doors, and screens and diffusion, to get exactly the right effect.

Broads

Small and rectangular, ***broad*** floodlights emit light that is directional enough to cast shadows, **Figure 14-22**.

The procedure for using them is simple:

- Set up the stand and unit.
- Adjust the barn doors (but do not expect precise edge control).

Figure 14-22.
A lightweight broad for location use. (Lowel Corporation)

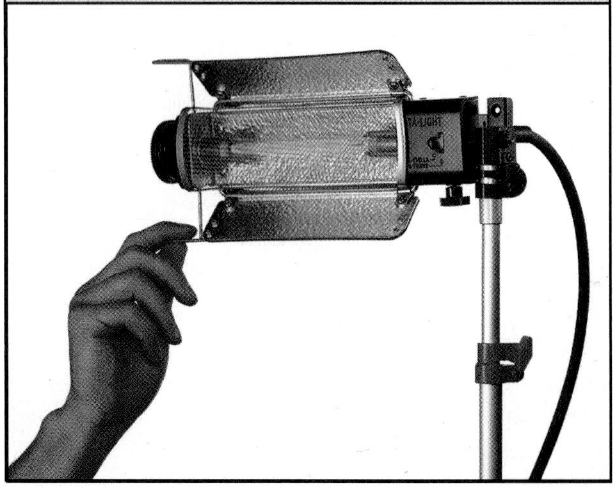

- Add framed or clipped-on diffusion if desired.
- Adjust intensity by moving the unit (there is no focus control).

Broads are compromise units: not as soft as large source lights, but very small and handy. Also, they are useful for background lighting, where true ***large-source*** lights lack a long enough throw.

A light's "throw" is either a) its maximum effective light-to-subject distance, or b) that distance in any particular setup.

Using Large-source Lights

Umbrellas, softboxes, and pans are easy to use but hard to control. With all three types of units,

- Set them up and aim them at the subject.
- Move them forward or back to adjust light intensity.
- Observe their effect on the background and adjust accordingly.

It is difficult to keep large-source light from spilling onto nearby backgrounds. This can be a plus if you are working with very few units and need them to cover background as well as subject. Otherwise, try moving your subjects forward from the background and resetting the lights to reduce spill.

Umbrellas

With umbrellas, you can adjust the light quality by changing the cloth cover: silver threaded for a harder effect, plain white for soft, translucent light if you wish to aim the umbrella at the subject

Figure 14-23.
Some umbrellas are fitted to specially designed lights. (Photoflex)

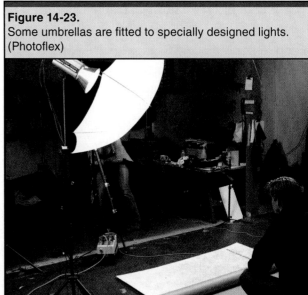

and shoot the light through it. Umbrellas can be used with small spotlights, but their extra control is wasted in this application. Small broads work well because they do not form a large shadow in the center of the light path, **Figure 14-23**.

Softboxes

The more sophisticated softboxes have internal baffles and/or layers of fabric, for fine-tuning the light quality. Some models also have shallow black grids that can be applied to their fronts to limit side spill. Some softboxes accept general purpose lights, but others use proprietary lamps and housings, **Figure 14-24**.

Figure 14-24.
This softbox has internal baffles for smoother light distribution. (Photoflex)

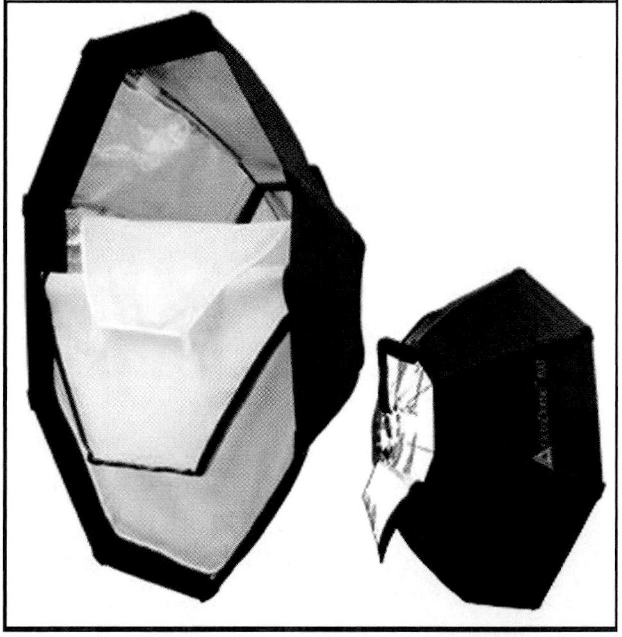

Figure 14-25.
Some light pans are designed to include their carrying cases. (Lowel Corporation)

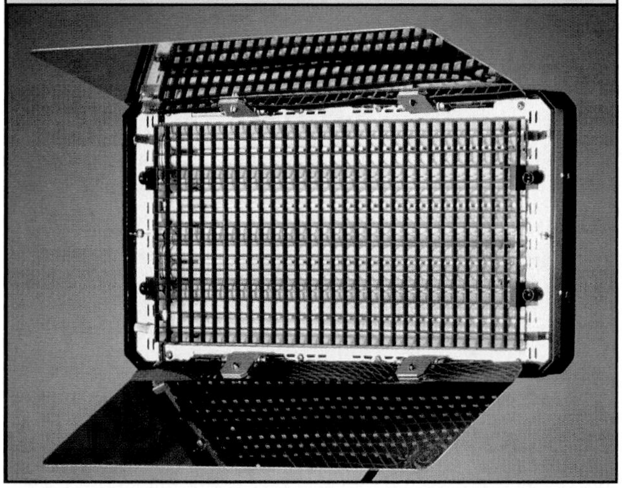

Pans

Fluorescent ***pans*** are desirable where electrical power is limited and/or where heat buildup is an issue. They also work well with available fluorescent light or daylight. Otherwise, their bulk and clumsiness fit them best for studio use, **Figure 14-25**.

In fact, all large-source lights are bulky and hard to work with in tight locations. Where a soft light is wanted, it is sometimes better to use diffusion with small, maneuverable spots and broads.

Using Cookies

A ***cookie*** (a short form of the obscure word "cukaloris") is usually a sheet of opaque material with cutouts, mounted in front of a light to throw

Figure 14-26.
Creating a background with a cookie. A—This blank wall lacks character. B—A leaf cookie adds an attractive pattern.

A

B

patterns on the background. Some cookies throw recognizable shadows, while others simply vary the light on background walls. With an artist's knife and a sheet of foam board, you can create any pattern you need, **Figure 14-26**.

Here are a few of the more popular cookie patterns.

Bare branch

A common wall pattern suggests the outline of a leafless tree, **Figure 14-27**.

Figure 14-27.
A bare branch cookie.

Leaves

A leaf pattern cookie, (as shown in Figure 14-26) can throw the shadow of leaves on subjects (which makes it useful outdoors as well as inside). Keep the leaf cutouts small, for best results.

If you have permission to cut an actual tree branch, you can clamp it horizontally with a century stand to make a very effective cookie, whose leaves can actually move in the "wind" created by gently shaking the branch.

Window blinds

Venetian blind patterns are easy to make. They are typically cut on the diagonal to throw slanting patterns on backgrounds, **Figure 14-28**.

Some lighting directors carry an actual small blind, which they suspend from a century stand in front of a spotlight and adjust to obtain the desired effect.

Figure 14-28.
A blind cookie.

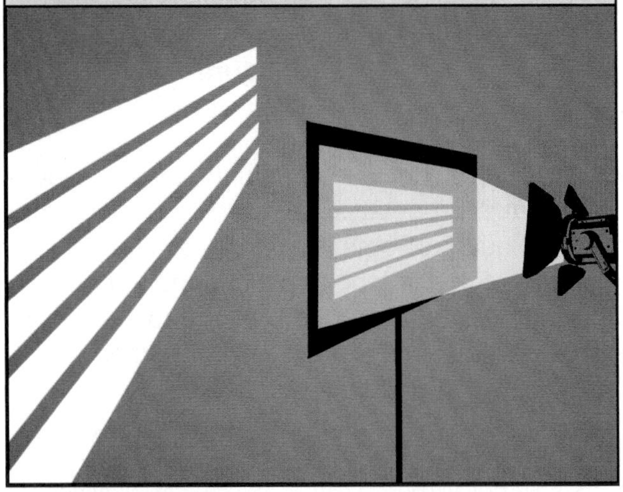

Studio cookies

Cookies can be used in TV studios to add interest to the plain curtains often used as backings. Simple variegated patterns break up the surface, and cutouts can suggest other backgrounds, **Figure 14-29**.

Figure 14-29.
A city skyline cookie.

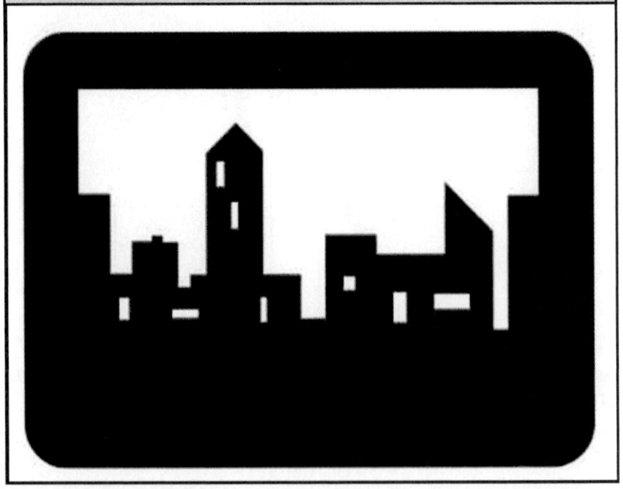

Pattern projections in TV studios are sometimes handled by "gobos," miniature cookies placed inside theatrical lighting instruments called ellipsoidal spotlights.

Special purpose cookies

You can cut a cookie to throw any shadow you need for a particular one-time effect. A gallows is a simple (if melodramatic) example, **Figure 14-30**.

Figure 14-30.
A gallows cookie.

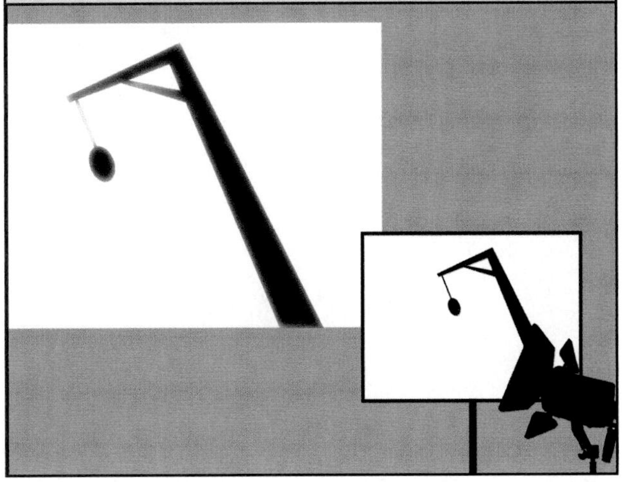

Cookies with actual images are most believable in the styles of pictorial realism, magic realism, and expressionism. In naturalistic lighting, they tend to look excessively theatrical. Abstract pattern cookies, however, work in any style to add subtle variety to backgrounds.

Cookie Procedures

To use a cookie, mount the sheet on the arm clamp of a century stand and focus a spotlight to throw a sharp, distinct shadow through it. (Floodlights do not yield good results and softboxes do not work.) Control the character of the shadow by the cookie's distance from the light. The closer the cookie is to the light, the bigger the shadow and the softer its edges. For smaller, sharp-edged patterns, place the cookie as far in front of the light source as practical. This setup sequence usually works well:

- Set up both the cookie and the light behind it.
- Watching the background, change the distance between light and cookie to obtain the desired edge sharpness.
- Moving the cookie and the light together (to preserve the distance set between them) adjust the size of the shadow by changing the throw (distance) to the background.

Wide opaque borders around cookie patterns allow the sheets to mask the beam edges when placed farther from the light.

Using Reflectors

Reflectors are so versatile that they are used both indoors and out. They are usually managed in the same way as lights, regardless of the particular lighting situation.

Outdoors, reflectors on stands should be weighted down, because they are essentially sails just waiting to blow away. Use rigid reflectors, if practical, because flexible cloth models ripple in the breeze and throw moving reflections on subjects.

Focusing reflectors

You can adjust the intensity of any reflector by moving it toward or away from the subject. (Be very careful to keep hard reflectors well back from subjects to avoid hurting their eyes with the hard light beam.) With some units, you can also reduce the light level by aiming the beam slightly away from the subject, so that only the edge of the beam is effective. Other units can be focused by flexing the panel, **Figure 14-31**.

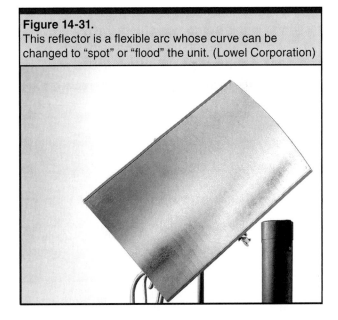

Figure 14-31.
This reflector is a flexible arc whose curve can be changed to "spot" or "flood" the unit. (Lowel Corporation)

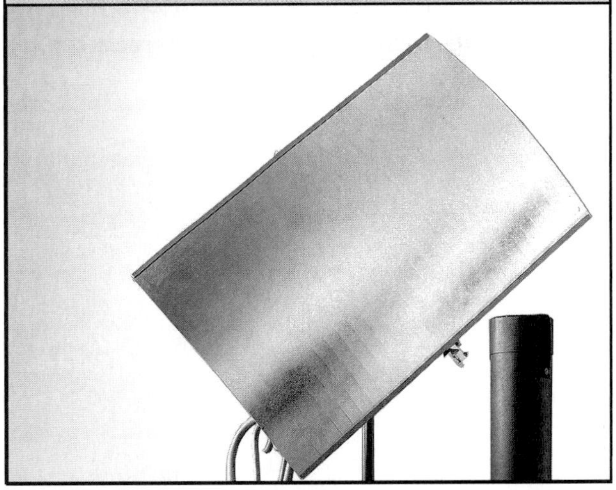

As you adjust fill levels by moving reflectors toward or away from their subjects, watch your monitor carefully to avoid overfilling.

Reflector lighting functions

You can use reflectors exactly as you would lights, for key, fill, rim, and background lighting.

Key light. Typically, you key with a reflector when you wish to use the sun as a rim (back) light.

Figure 14-32.
Here, the reflector is the key light and the sun is the rim light.

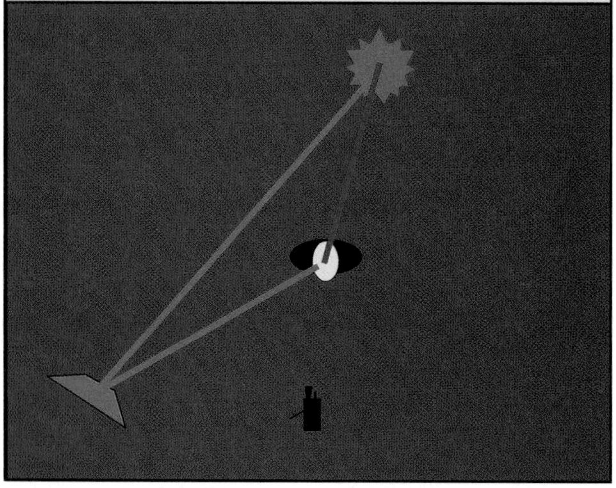

Figure 14-34.
If the unit can be placed high enough, the reflector can add rim light to the sun's key light.

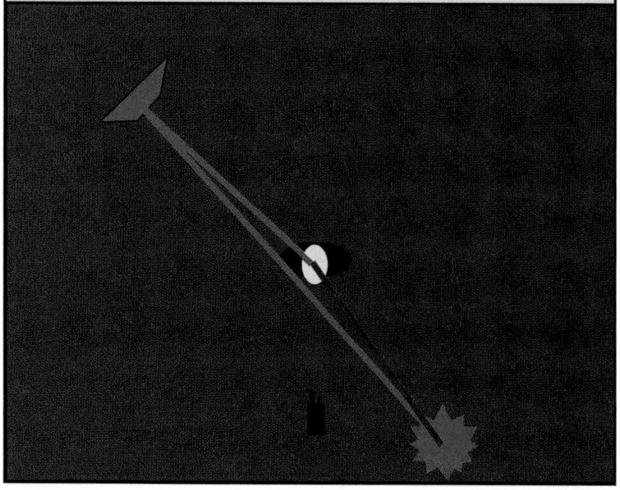

To do this, it is usually best to place the reflector on the side opposite the sun, **Figure 14-32**.

In these diagrams, direct sunlight is red; reflected light is yellow.

Fill light. Reflectors are used most often as fill lights. In this setup, the sun provides the key light and the reflector, on the opposite side of the subject, adds the fill, **Figure 14-33**.

Figure 14-33.
With the reflector as fill and the sun as key light, both are in front of the subject.

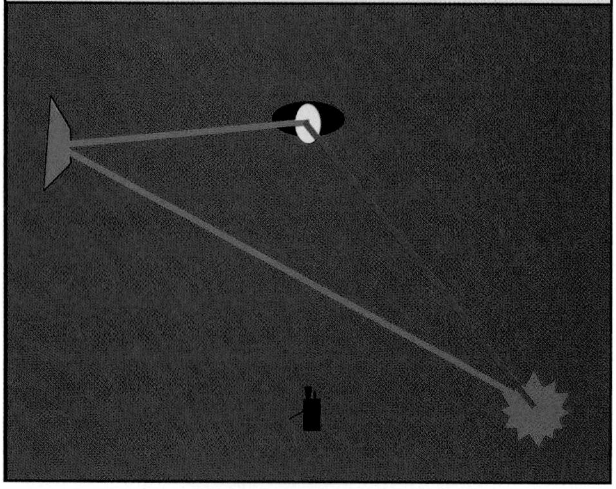

For closeups, subjects can sometimes hold their own reflectors, below the frame line.

Rim light. Rim light is difficult to add with a reflector because the light source should be much higher than the subject. If you can get a crew member up on a ladder that is off-camera, you can make it work. A hard aluminum reflector works best here, and there is no danger of hurting the subject's eyes, **Figure 14-34**.

Depending on the height of the camera angle, nearly horizontal rim lighting can sometimes be effective.

Background light. As explained earlier, the easiest way to prevent backlight exposure problems is to set the subject in sunshine, in front of a shaded background. If this makes the image seem dull and lacking in depth, use a reflector to rake the background with light, **Figure 14-35**.

Figure 14-35.
With the sun as key light, the reflector splashes light on the wall behind the subject.

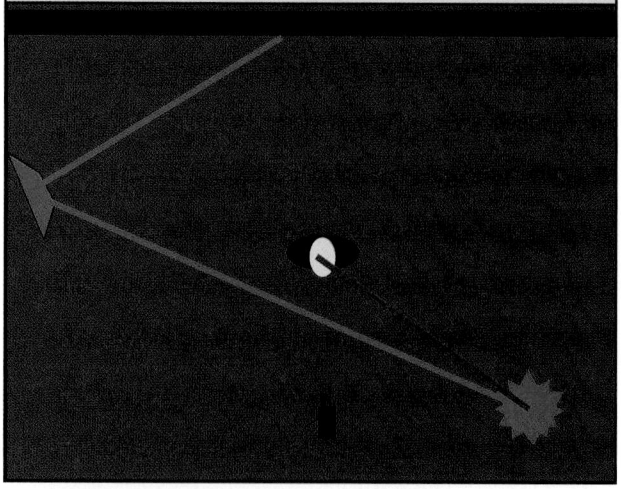

Reflectors indoors

Indoors, reflectors are used mainly for fill. They can be so effective that you can light a single subject with just one spotlight and a large reflector opposite for fill, **Figure 14-36**.

If you have two lights, you can use the second one as a rim light or a background light.

Figure 14-36.
The softlight (foreground, right) and the gold reflector provide all the subject lighting. (Photoflex)

Using Screens and Silks

Screens and silks have more limited uses because they must be placed close to subjects, making them unusable in wide shots.

You can solve this problem by ending wide shots with subjects still far enough away so that the lack of light filtering is not obvious, and the closeups seem to match the lighting of the wide shots.

Screens are useful for moderating directional sunlight to reduce excess contrast, but they have limitations. Large, framed models are clumsy and time-consuming to place. They are most useful in static situations like stand-up narration or interviews in which subjects remain in the same spot for extended periods, **Figure 14-37**.

Silks are less useful in this application because they soften and diffuse the light as well as reduce it. That will change the key light quality so that the closeup lighting will not match that of the longer shots. However, silks can be very effective indoors, when an extremely large light source is needed for absolutely shadowless light.

Figure 14-37.
The diffusion screens...

are out of frame in the video images. (Photoflex)

From Theory to Practice

Now that you have surveyed lighting tools and reviewed the basics of lighting design, you are ready to apply your knowledge to practical lighting applications. These lighting applications are the subject of the next chapter.

Chapter Review

Answer the following questions on a separate piece of paper. Do not write in this book.

1. Lighting _____ is the difference in brightness between the darkest and lightest parts of the image.
2. *True or False?* Lighting effects created in the camera, such as "moonlight" can be undone in editing, if necessary.
3. Naturalism, pictorial realism, and expressionism are examples of lighting _____.
4. _____-key images are mainly dark, but usually have a lighter center of interest.
5. A _____ softens and diffuses light, as well as reducing its brightness.

Technical Terms

For easier reference, some definitions are repeated from Chapter 13.

Available light: The natural and/or artificial light that already exists at a location.

Background light: A light splashed on a wall or other backing to lighten it and add visual interest.

Barn doors: Metal flaps, in sets of two or four, attached to the fronts of spotlights to control the edges of the beam.

Broad: A small, rectangular, open light used mainly for fill and background lighting.

Camera light: A small light mounted on the camera to provide foreground fill.

Contrast: The difference between the lightest and darkest parts of an image, expressed as a ratio ("four-to-one").

Cookie: A sheet cut into a specific pattern and placed in the beam of a light to throw distinctive shadows, such as leaves or blinds.

Diffusion: White spun glass or plastic sheeting placed in the light path to soften and disperse it.

Expressionism: A lighting style that adds a heightened emotional effect, without regard for lighting motivation.

Fill light: The light that lightens shadows created by the main (key) light.

Flag: A flat piece of opaque metal, wood, or foam board placed to mask off part of a light beam.

Floodlight: A large-source instrument that typically lacks a lens; used for lighting wide areas.

Gaffer: The chief lighting technician on a shoot.

High key: Lighting in which much of the image is light, with darker accents.

Key light: The principal light on a subject.

Large-source: Describing a lighting instrument, such as a scoop or other floodlight, with a big front area from which light is emitted. Light from large-source instruments is relatively soft and diffuse.

Low-key: Lighting in which much of the image is dark, with lighter accents.

Magic realism: A lighting style that creates a dreamy or unearthly effect, often enhanced digitally in postproduction.

Medium key: Lighting in which the image neither light nor dark tones dominate the image (a term used only in this book).

Motivated lighting: Lighting that imitates real-world light sources at the location.

Naturalism: a lighting style that imitates real-world lighting so closely that it is invisible to most viewers.

Pan: In lighting, a large, flat instrument fitted with fluorescent or other long tube lamps.

Pictorial realism: A lighting style in which lighting, though motivated, is exaggerated for a somewhat theatrical effect.

Practical: An instrument that is included in shots and may be operated by the actors.

Realism: A lighting style that looks like real-world lighting, though it is slightly enhanced for pictorial effect.

Reflector: A large silver, white, or colored surface used to bounce light onto a subject or scene.

Rim light: A light placed high and behind a subject to create a rim of light on head and shoulders, to help separate subject and background.

Screen: A mesh material that reduces light intensity without markedly changing its character.

Silk: A fabric material that both reduces light intensity and directionality, producing a soft, directionless illumination.

Small-source: Describing a lighting instrument such as a spotlight, with a small front area from which light is emitted. Light from small-source instruments is generally hard-edged and tightly focused.

Spotlight: A small-source lighting instrument that produces a narrow, hard-edged light pattern.

Umbrella: A silver- or fabric-covered umbrella frame used to reflect light onto subjects from its concave side, or to filter it through its convex side.

15 Lighting Applications

Objectives

After studying this chapter, you will be able to:

- Light subjects in both “classic” and “natural” modes.
- Light typical small and large interiors.
- Solve common lighting problems.
- Light interior and exterior night scenes.
- Light frequently encountered assignments.

(Bogen Manfrotto)

About Lighting Applications

This chapter takes the lighting tools and design principles covered in the two preceding chapters and puts them to work in real-world situations. We will see how to light subjects, locations, and night scenes; how to solve common lighting problems; and how to approach several types of frequently encountered lighting assignments.

There is some necessary content overlap among the three lighting chapters.

Lighting Subjects

Most of your time, of course, will be spent lighting people; and from the simplest production to the most elaborate, there are only two basic approaches to this task: classic studio lighting and soft "natural" lighting.

Classic Studio Lighting

Classic studio lighting uses three lights on the subject and usually one or more on the background, **Figure 15-1**. This is often called ***three-point lighting*** despite the frequent use of additional instruments.

Figure 15-1.
A classic lighting setup.

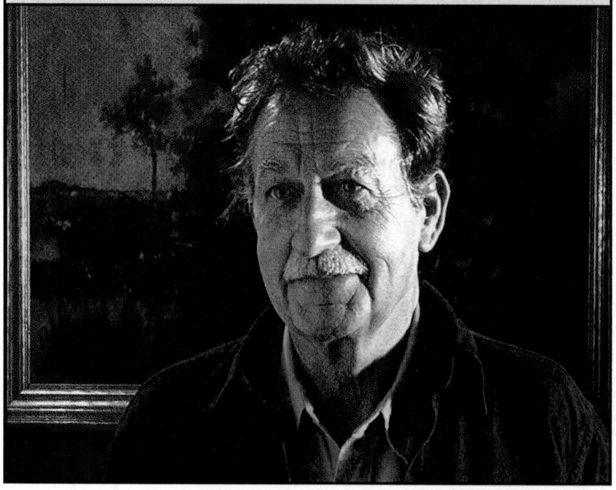

Key light

The ***key light*** provides the main illumination, typically mimicking an actual light source like a lamp or ceiling fixture. It is often placed at about 4:30 and 15°–30° higher than the subject's face, **Figure 15-2**.

Lighting on a "clock"

For convenience, the horizontal placement of lights is often described in terms of a clock face:

- The subject is at the center, facing the 6 o'clock position.
- The camcorder is at 6 o'clock, facing the center.
- The lights are at various "hours" around the clock face.
- The background, if shown, is at the top.

A four-light setup diagramed on a clock face, with the back light at 2:30, the key light at 5:00, the camera at 6:00, the fill light at 8:00, and the rim light at 11:00.

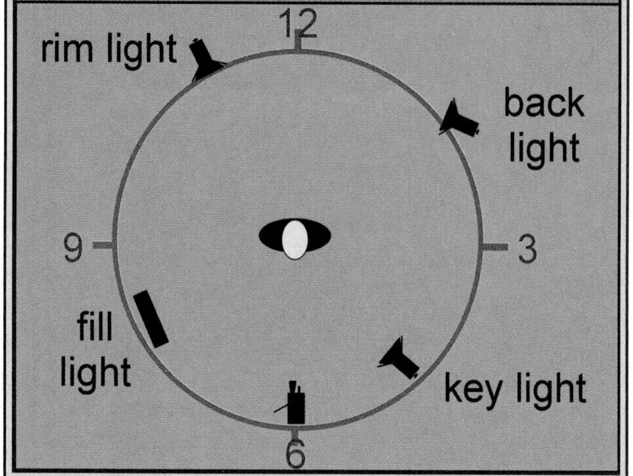

Although this diagram shows the key on the right and the fill on the left, their positions are just as often reversed.

The key light is typically a ***spotlight***, so the hard-edged beam is often softened with a sheet of spun glass clipped to the barn doors. Even so, it throws distinct shadows on the subject's cheek, upper lip, and neck.

Fill light

The ***fill light*** literally fills in the shadows created by the key light, **Figure 15-3**. Placed opposite the key light, the fill is often farther to the side and not as high as the key. That is to help reduce the cheek, lip, and neck shadows.

How completely the fill light moderates these shadows depends on the setting and mood of the scene. In a cheerful interior, the shadows might be slight; in an atmospheric night scene,

Figure 15-2.
Key light. A—Horizontal key light placement. B—Vertical key light placement. C—The effect of the key light.

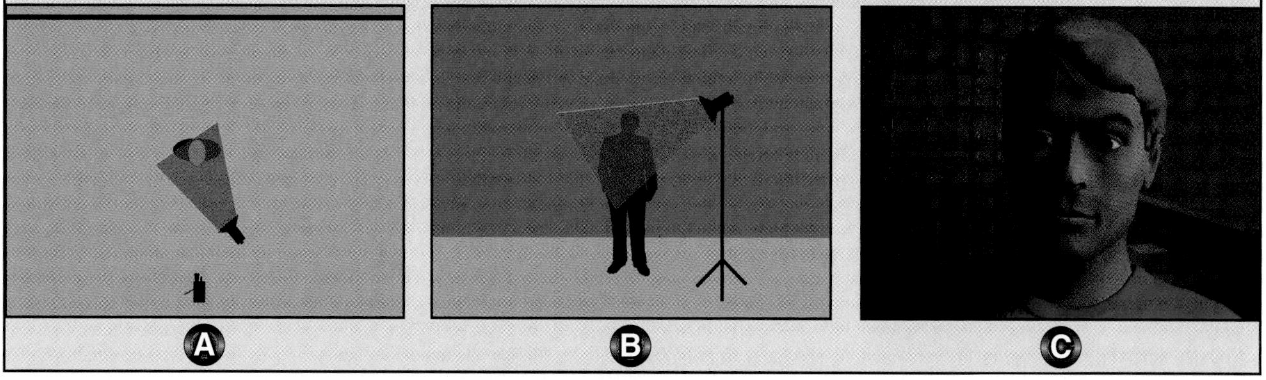

Figure 15-3.
Fill light. A—Horizontal fill light placement. B—Vertical fill light placement. C—The effect of the fill light.

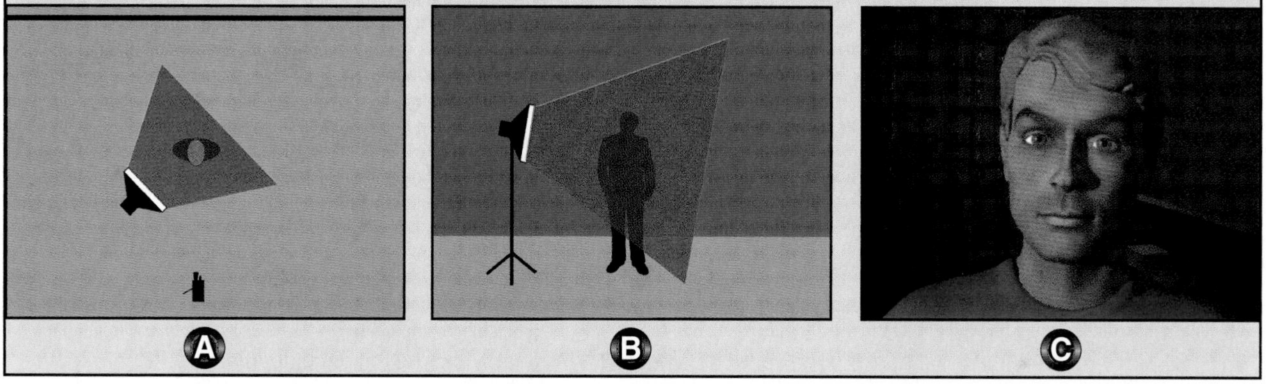

Figure 15-4.
Rim light. A—Horizontal rim light placement. B—Vertical rim light placement. C—The effect of the rim light.

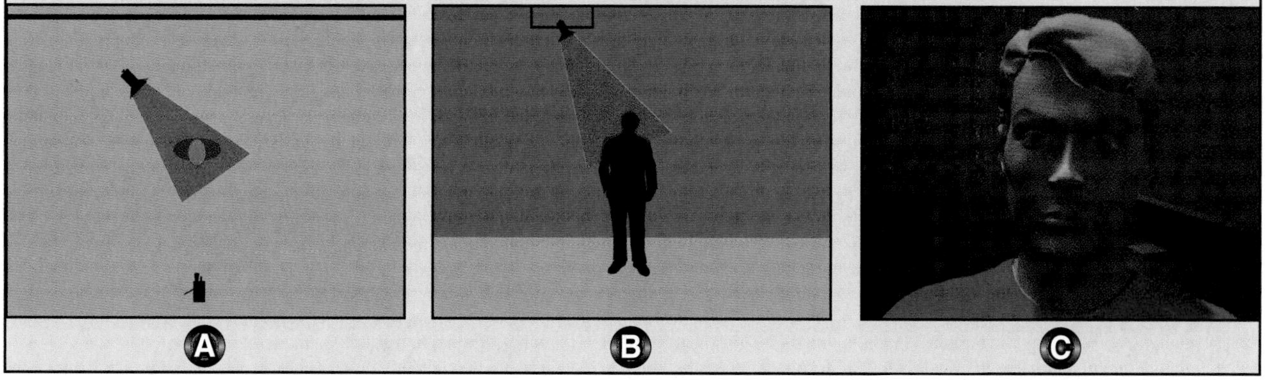

they might be so deep as to obscure details within them. In any case, the fill light should not be bright enough to make the subject lose the "modeling" that creates the illusion of depth.

Rim (back) light

The ***rim light*** is typically behind the subject and placed quite high, **Figure 15-4**. If however, its light stand appears in the shot, you can move the rim light aside until it clears.

Rim lights are frequently mounted overhead on clamps or stands with lateral arms. The brightness of the rim light depends mainly on the lighting style: pronounced for pictorial realism and moderate for realism. For naturalism, the rim light is just barely bright enough to visually separate subject from background. In some instances, it is omitted entirely.

The brightness of key and fill lights is adjusted by moving the lights toward or away from the subject. Rim light, however, may be controlled by a dimmer, since the warming effect of dimming a light is usually acceptable in this application.

Background light

Like the key light, the ***background light*** is usually "motivated,"—that is, it mimics light that would naturally fall on the walls or other background, like a wall lamp, a window light, or spill from a room light, **Figure 15-5**. When working with just a few lights, you can usually achieve background lighting by directing spill from the key and/or fill lights.

Background light intensity should be adjusted so that subject and background seem lit by the same environment, but the subject is slightly brighter. Two or more background lights may be needed to do the job.

Background lights often produce less intense effects because the lighting instruments must be placed well away from the background to keep them out of the frame.

With the four lights in place, we can build a complete lighting setup, **Figure 15-6**.

Though developed for classic pictorial realism, this basic scheme can be used with any of the four major lighting styles, **Figure 15-7**.

The basic lighting setup demonstrated here uses four lights and covers only a space about the size of a single action area. In large shooting locations, the lighting can involve many more instruments, but they tend to be deployed in multiples of these basic layouts.

An "action area" is a spot within a location that is fully lit because important activity takes place there.

"Natural" lighting

Because three-point lighting can look somewhat theatrical, many situations call for a more natural, "unlit" appearance. The key to this approach is soft light. Spots and broads can be used if heavily diffused, but large sources such as umbrellas or softboxes are often easier to work with.

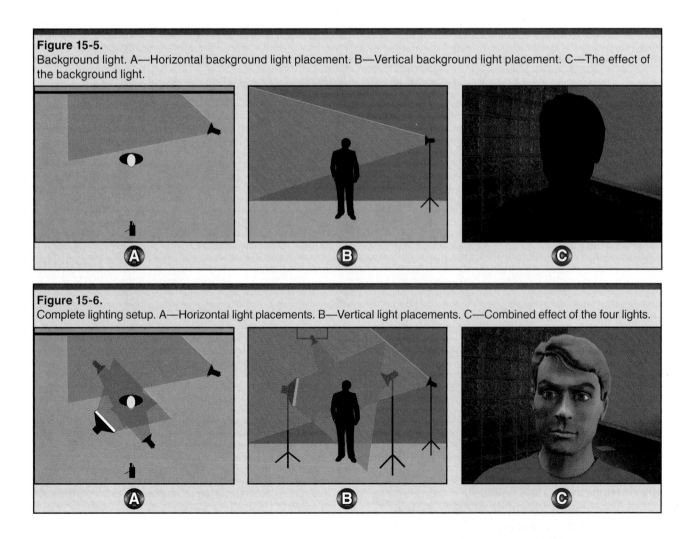

Figure 15-5.
Background light. A—Horizontal background light placement. B—Vertical background light placement. C—The effect of the background light.

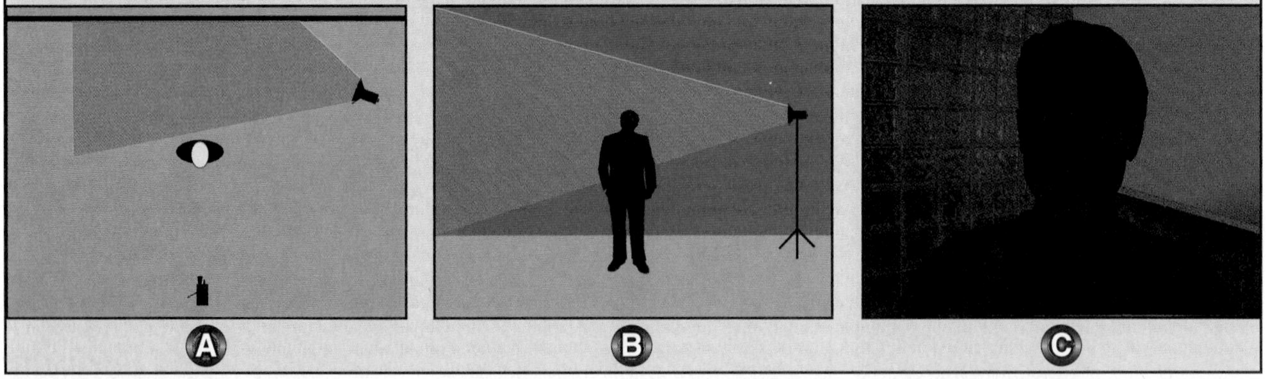

Figure 15-6.
Complete lighting setup. A—Horizontal light placements. B—Vertical light placements. C—Combined effect of the four lights.

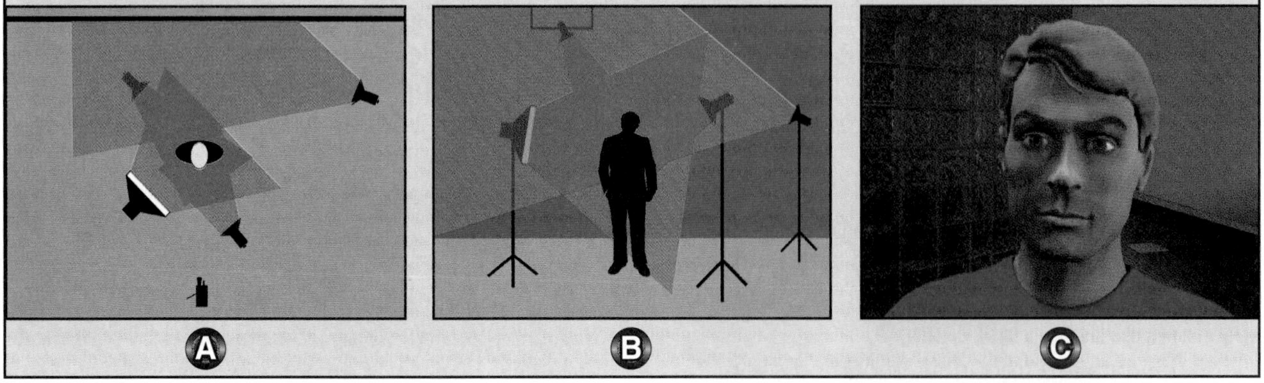

Figure 15-7.
A—Naturalism. B—Realism. C—Pictorial realism. D—Expressionism.

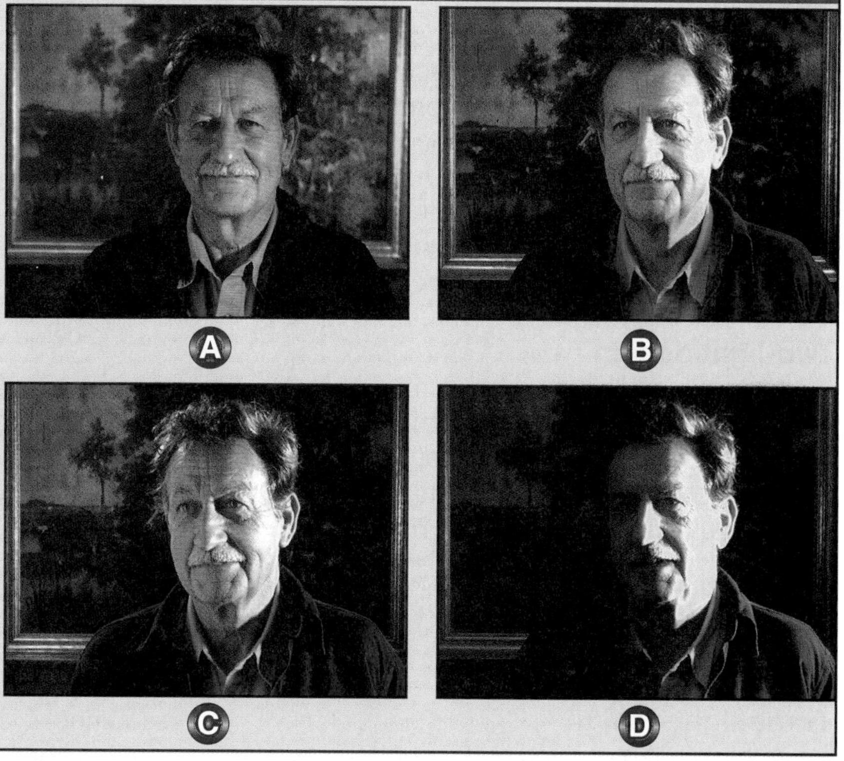

"Rugged" vs. "Glamorous" Lighting

How you position and diffuse your lighting instruments often depends on whether you wish to emphasize facial modeling for a so-called "rugged" look or whether you prefer to de-emphasize it for "glamour."

Rugged lighting exaggerates the planes and angles of the face and emphasizes skin texture. To do this:

- Keep the key light high, for more pronounced shadows.
- Reduce or omit key light ***diffusion***. The harder the beam, the more it emphasizes skin and other textures.
- Avoid over-filling, to retain enough shadows for pronounced facial sculpting.

"Rugged" lighting.

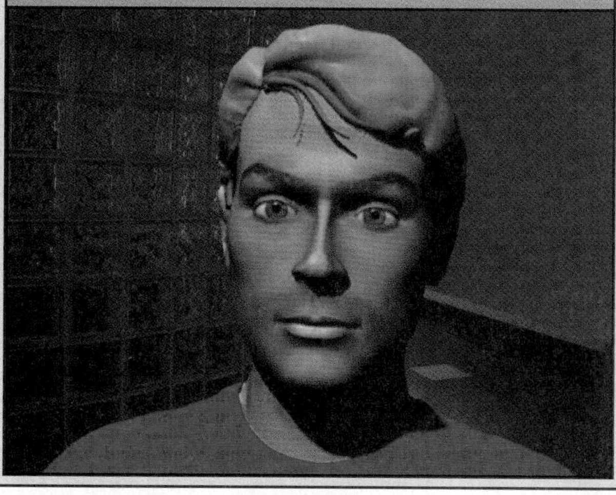

Glamorous lighting uses exactly the opposite approach:

- Place the key light for moderate shadows.
- Use considerable diffusion (or a ***softlight***) to minimize skin texture.
- Add fill light until the shadows are relatively faint (but avoid over-filling the neck area, to downplay aging skin).
- Use a generous rim light to accent hair.

"Glamorous" lighting.

The videographer contributes to the effect by using wide-angle lens settings for rugged lighting and telephoto for glamour.

One-light design

For close shots, a single soft source can deliver satisfactory lighting, especially when paired with a ***reflector*** for additional fill light. The subject should be close enough to the background so that the light spill can model it. Keep the unit low to simulate window light, and place it at about 8:30 on the clock, so that it "wraps around" the subject's face, **Figure 15-8**.

Two-light design

A second softlight provides a more versatile fill source. With this design, it often helps to place key and fill lights at the heights you would use for three-point lighting. With two lights, you can move the fill around as far as 3 o'clock, **Figure 15-9**.

Alternately, you may wish to continue to use the reflector for fill, and bring up the background with the second light.

Three-light design

A third light gives you better control over both fill and background, **Figure 15-10**.

Studying the light plan in Figure 15-10, note that:

- The subject's distance from the background, permits the two to be lit separately.
- The fill light is about three times as far as the key light from the subject.
- The background light is far enough to the side so that the hot spot created by the near edge of its beam and the overlapping key light spill (indicated in red) is outside the frame.

The natural style is very popular for lighting interviews because the lighting can match the location and because the backgrounds are frequently close behind the subjects. Also, soft lighting is fast and easy to work with.

Lighting Backgrounds

If you have the room and the lighting resources, it is often best to move subjects away from backgrounds so that you can light them separately. Spots or floods are effective for

Figure 15-8.
One-light design. A—Horizontal light placement. B—Vertical light placement. C—The effect of the design.

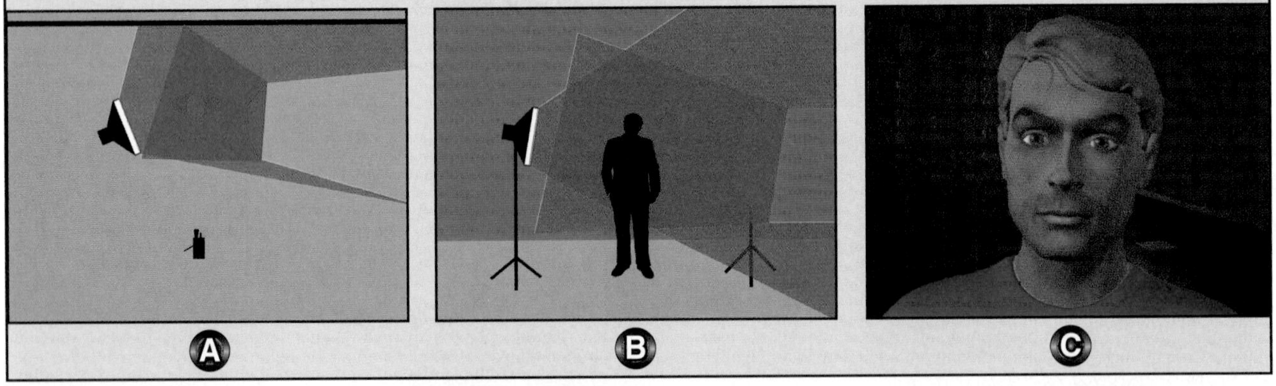

Figure 15-9.
Two-light design. A—Horizontal light placement. B—Vertical light placement. C—The effect of the design.

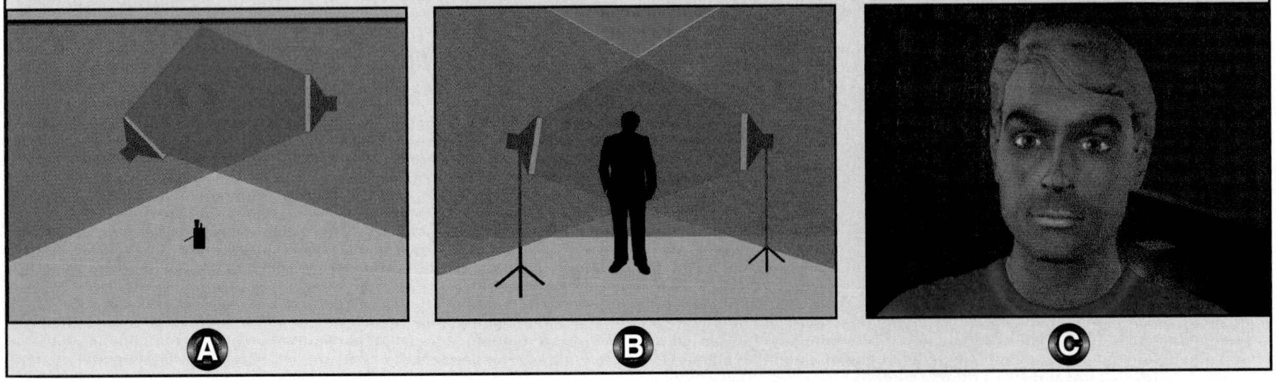

Figure 15-10.
Three-light design. A—Horizontal light placement. B—Vertical light placement. C—The effect of the design.

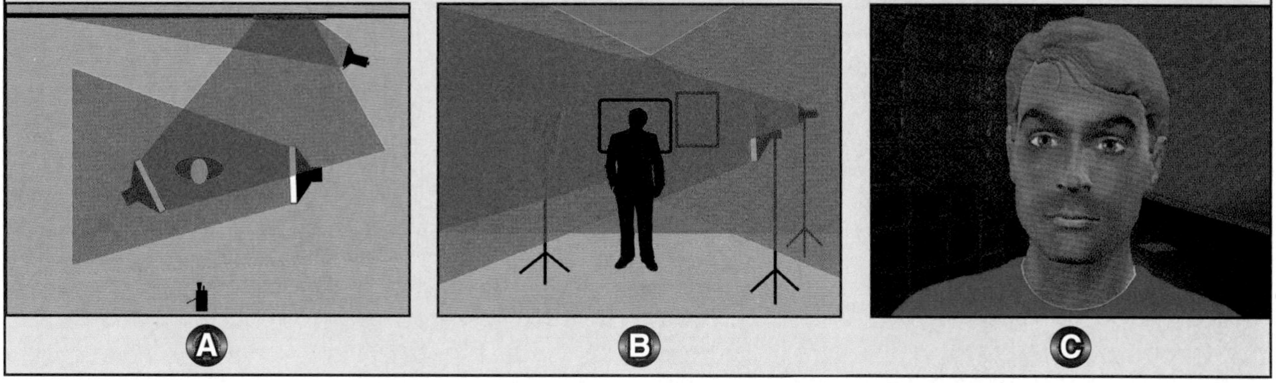

background lighting because their longer throw allows them to be placed far enough to the sides to remain out of the frame.

As noted elsewhere, a light's throw is the distance between the instrument and the subject or background that it is lighting.

Lighting for exposure

When the background is too dark, the image loses apparent depth, so you may want to wash some light on it to make it more visible. Be careful not to place too many highlights, or to make the background too bright, to avoid distracting attention from the subject(s) in the foreground.

Lighting for texture

You can often make dimensional surfaces (such as plaster or fabric) more interesting by bringing out their texture. For good cross lighting, place spots or broads as close to the backing as possible and rake the light across the surface.

Lighting for depth

Sometimes, you can enhance depth by highlighting surfaces (like furniture) in front of the background as well. Generally speaking, lights hung above the frame work effectively.

Adjusting intensity

When lights are placed close to a background, the beam is much "hotter" near the light. To even out the light pattern, use a half or double half screen positioned in the spotlight's filter holder so as to reduce light output on the side near the wall.

Outdoors, background lighting is usually created using hard-surface aluminum reflectors, Figure 15-11.

Figure 15-11.
A hard aluminum reflector. (Bogen Manfrotto)

Lighting Locations

Few video productions are shot on sound stages (except those made for cable or broadcast), so your interiors are likely to be locations. Every location presents challenges, and meeting them offers the satisfaction that comes from successful problem solving.

Small Interiors

Small interiors are usually difficult to work in, for several reasons, **Figure 15-12**.

Cramped quarters

Lights require room, not only for the instruments and stands, but for the throws of the lights (remember: you reduce light intensity by moving the unit away from the subject). Spots and broads are favored for cramped interiors because of their small size.

Figure 15-12.
This interior has problems with space, window light, and irregular ceilings.

Figure 15-13.
Although it delivers the light output of 375 watts of halogen light, this three-lamp fluorescent is only 78 watts. (Equipment Emporium, Inc.)

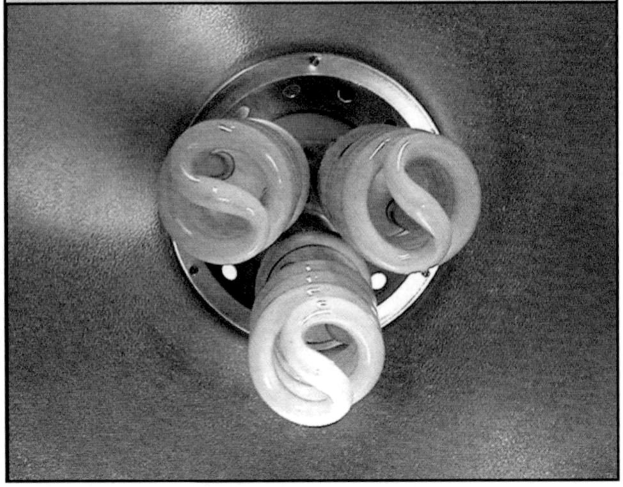

Spots are especially useful in tight quarters because they can be focused to vary the light output without moving the unit. Their intensity can be further reduced by using screens.

Power supplies

Small interiors are often located in homes or other private buildings where electrical circuits are typically only 15 amps and an entire room may be served by just one circuit. When working with inadequate power, high-efficiency ***fluorescent lamps*** draw less power per unit output than halogens. Consider, especially, the compact units with screw-base lamps, **Figure 15-13**.

Background spill

Small interiors make it difficult to keep subject light from spilling onto the background. Here again, using the more controllable spots and broads can help minimize the problem. On the other hand, the gentle spill from umbrellas or softboxes often makes a very agreeable background light.

Ceiling bounce

The low ceilings of many interiors can actually be a plus, because they make it easy to bounce fill light down onto subjects and background. Too much ceiling bounce, however, puts shadows under subjects' eyes and looks like institutional grid lighting.

Hiding lights

Small interiors often make it difficult to keep the lights out of the frame.

Study your monitor very carefully for cables, which have a way of creeping into the shot.

Light stands work well in front and to the sides of the action area, where they are safely off screen. To hide lights placed deeper in the set, deploy units that can be clipped or taped to moldings, curtain rods, door frames or the tops

Calculating Power Draw

The formula says that the amperage (size) of a power load (in this case, a video light) is equal to the ***wattage*** of the load divided by the ***voltage*** of the circuit, or

amps = watts/volts

It can be difficult to mentally calculate the ***amps*** used by a video light (and hence, the ability of an electrical circuit to take the load). Although North American current is nominally 110 volts, the actual voltage in a particular circuit may range from 105 to 130 and typically runs around 115 - 125. Without troubling to test each circuit, you cannot tell what its true voltage may be. Also, mentally dividing by a number like, say, 117.5 volts is not easy.

To solve both problems, divide by an arbitrary 100 volts, simply by moving the decimal. For example, a 750 watt light would draw 750 watts divided by 100, or 7.5 amps.

This not only simplifies the head math; it also builds in an automatic safety factor, since the nominal amperage will always be lower than the actual. (At a true 110 volts, a 750 watt light really draws 6.8 amps, not 7.5.)

Figure 15-14.
Mounting lights. A—Small lights can be gaffer-taped to walls. B—Larger lights can be clamped to walls and doors. C—Small lights can be supported almost anywhere. (Lowel Corporation)

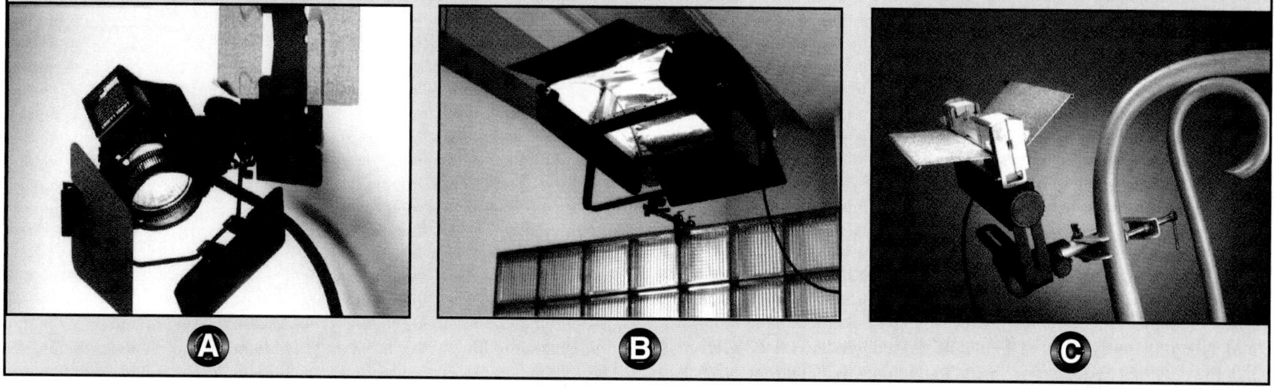

Managing Barn Doors

The edges of light beams look more realistic when they conform to natural features of the background. In this example, the spotlight is set to light a subject who will appear in the open doorway. The ***barn doors*** are set so that the beam edges are hidden by the sides and top of the door.

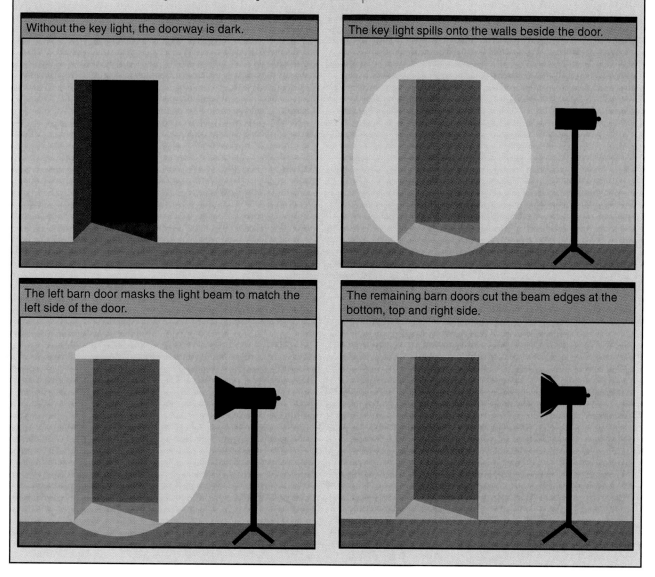

Without the key light, the doorway is dark.

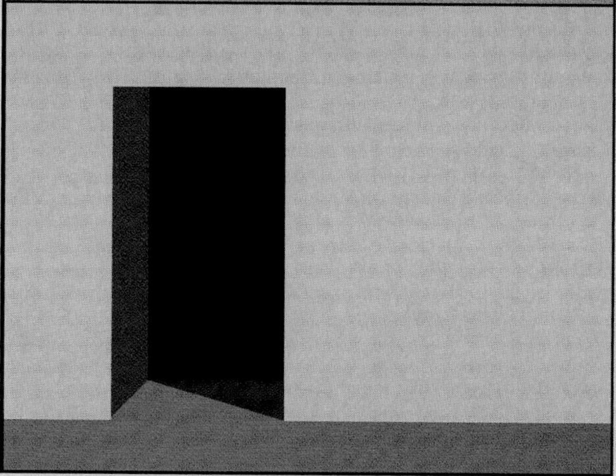

The key light spills onto the walls beside the door.

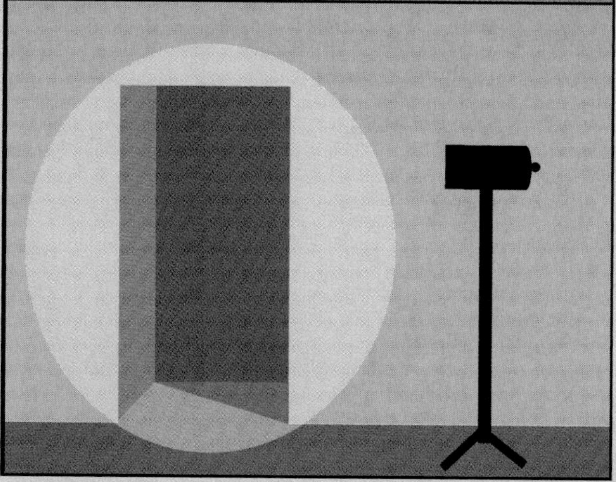

The left barn door masks the light beam to match the left side of the door.

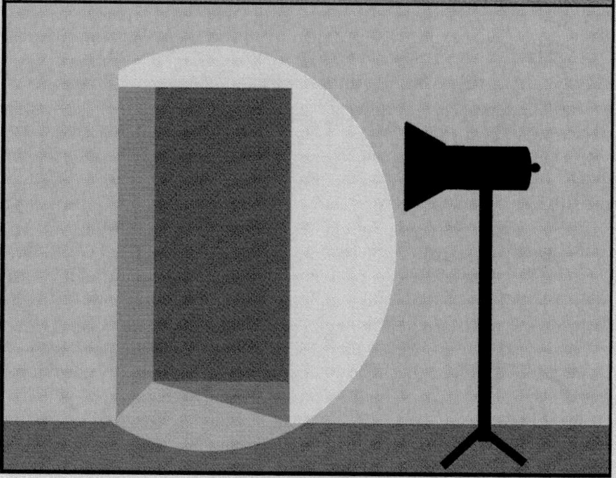

The remaining barn doors cut the beam edges at the bottom, top and right side.

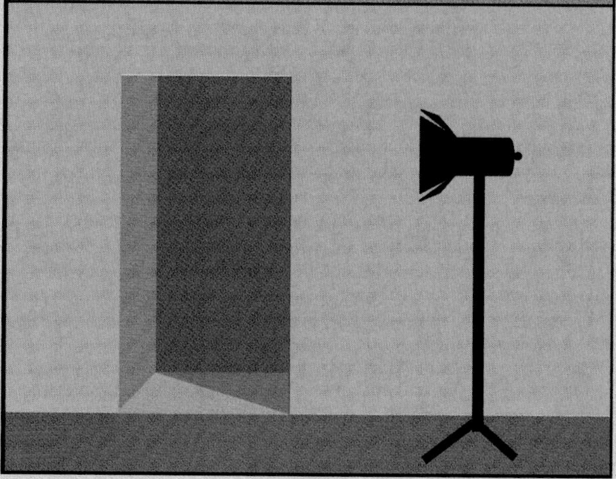

of open doors themselves. Fluorescent ceiling grids are also prime locations for clipping small lamps, **Figure 15-14**.

Some lighting instruments have built-in clips; others mount on posts fitted with clips or flat surfaces for taping.

Large Interiors

Large interiors are more comfortable to work in, but they present problems of their own, **Figure 15-15**. An area, say, 50 feet square, cannot be fully lit with the instruments in a typical small production kit. To solve the problem, you need to employ a two-part strategy.

Figure 15-15.
Too large to be lit with a small lighting kit, this room is lit only by its ceiling fluorescent fixtures—with unsatisfactory results.

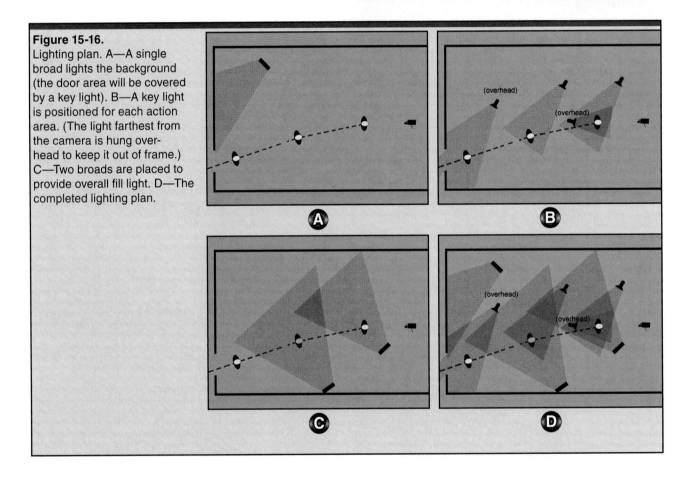

Figure 15-16.
Lighting plan. A—A single broad lights the background (the door area will be covered by a key light). B—A key light is positioned for each action area. (The light farthest from the camera is hung overhead to keep it out of frame.) C—Two broads are placed to provide overall fill light. D—The completed lighting plan.

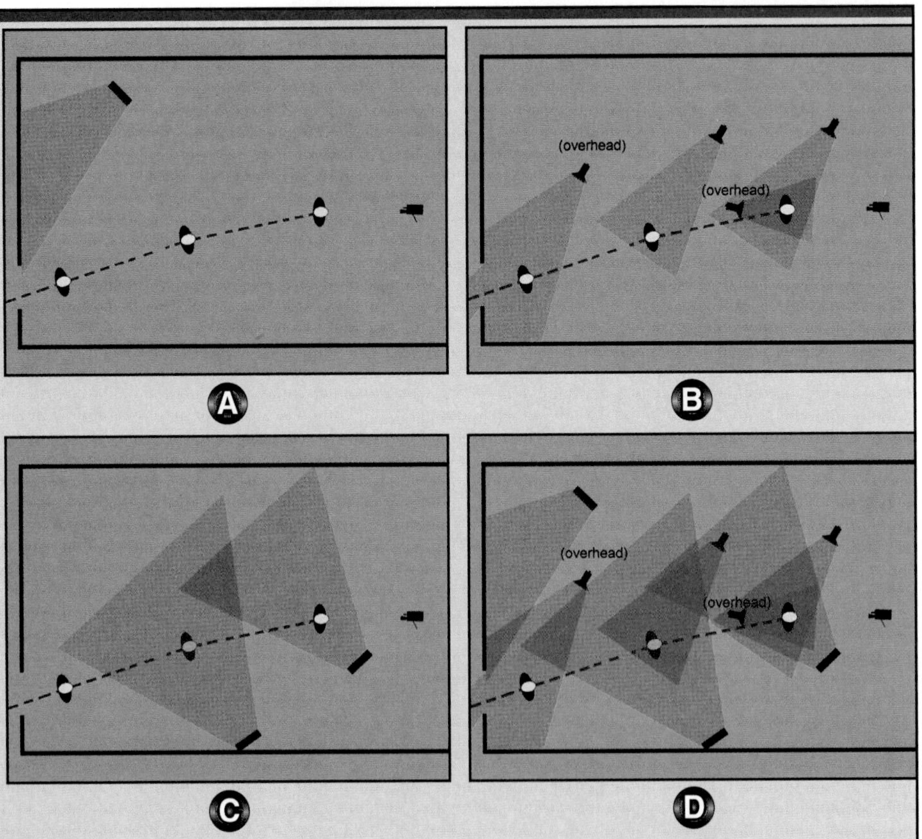

- Light the subjects, not the space, as discussed in Chapter 14.
- Break the action into parts; then light and shoot each part separately.

Here is how to implement this strategy. In this example, we will use four spotlights and three broads, **Figure 15-16**.

Start with the wide shots

When the camera will see most of the area, use your lights as follows:

Light the background fully. Since the walls or other backings will fill much of the frame, ensure that they are fully lit. Add lights to bring up furnishings or other contents of the area.

Light the action areas to be included in wide shots. Choose important places within the area for additional key lights. The closest action area also gets a rim light.

Add general fill light. This will bring up the overall light level. Ceiling bounce light works well, although direct fill from broads (as in Figure 15-16) is easier to control. Check floor areas carefully to make sure they get enough light.

The idea is to plan camera setups so that each major action area will also be covered in closer angles.

Light the close shots

After the wide shots have been recorded, you will re-light for each action area in turn:

Light the subject(s). Typically, this means fine-tuning the key and fill lights and adding some rim light for separation.

Light the background. You can take some instruments away from the background lighting by lighting only the parts that will appear in the close shots.

Take care that the key, fill, and background lights match the appearance of the wide shot lighting.

Lighting moving subjects

If there is subject movement from one action area to another, you will need to light all of them. Typically, the setup used for the wide shots will work well, **Figure 15-17**.

Figure 15-17.
To fully light each acting area, one fill light (1) has been moved closer, an extra fill light (2) now provides all-over fill, and two rim lights (3) and (4) accent the first two acting areas.

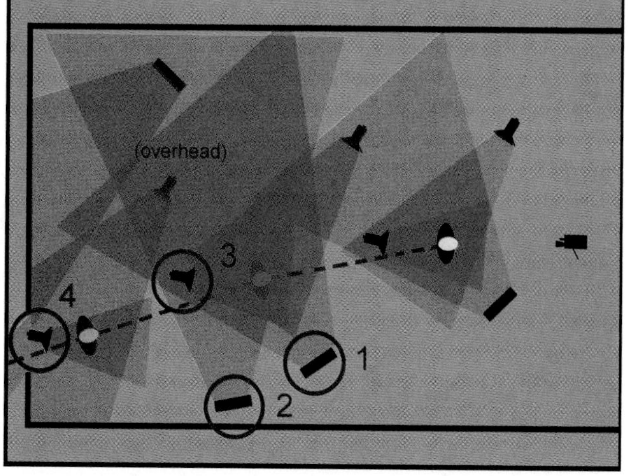

Exteriors

Chapters 13 and 14 cover, in passing, many outdoor lighting procedures. Following are some additional suggestions for professional looking results.

Figure 15-18.
Reflectors, screens, and silks are ineffective on sunless days. (Sue Stinson)

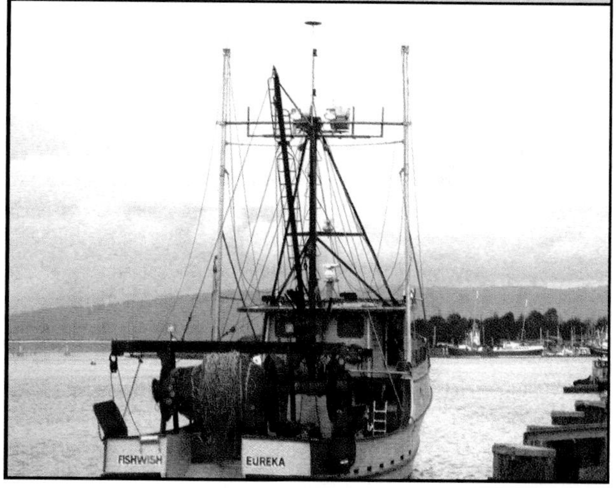

If the sky is overcast or you are shooting entirely in the shade, you cannot do much in the way of lighting, **Figure 15-18**.

In sunny weather, however, you can use your outdoor resources.

Choosing reflectors

Except for rim lighting, a rule of thumb for reflectors is "the softer the better," for evenness of coverage and subject comfort.

Aluminum. Choose aluminum reflectors when a long throw is needed, for backgrounds

Figure 15-19.
Matching lighting for wide and close shots. A—The wide shot is filled with a reflector at the 8 o'clock position, placed 25 feet away. B—The closeup lighting matches the wide shot. C—The reflector is 25 feet away.

or wide shots. Avoid using them for key lights (except when placed well-back) to keep them out of subject's eyes.

White. White reflectors are excellent for fill. In closeups, subjects can even hold them themselves, below the frame line.

Matching wide and close shots

Lighting wide shots is easier outdoors because subjects do not need as much modeling and because aluminum reflectors can throw effective fill light up to 50 feet or more, **Figure 15-19**.

Changing light and weather

Reflectors must be tended constantly, especially aluminum units that throw narrow beams. Between the time when a setup is begun and the moment when the shot is recorded, the sun can shift enough to misdirect the reflector light.

Reflectors vs. screening

Sometimes, you may prefer a screen or even a ***silk*** to a reflector. On the one hand, framed screens or silks cannot be used in wide shots, so matching the close shot lighting is more

Lighting Moving Subjects Outdoors

The technique for lighting outdoor movement is the same as for interiors, except that your lighting instruments are reflectors. Hard aluminum surfaces work well because they throw light a long distance.

Also, after traveling 50 feet or so, the light beam pattern is broad enough to cover a larger area and is reduced to a more manageable intensity.

Multiple reflectors fill a moving shot.

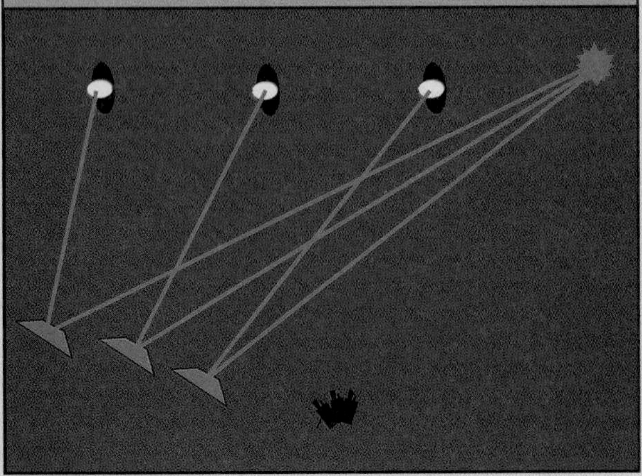

The longer reflector throw provides a wider, less intense beam.

Figure 15-20.
A silk on the left side, reflector on the right. (Photoflex)

Figure 15-21.
Narrow a wide face by highlighting the center and shadowing the sides.

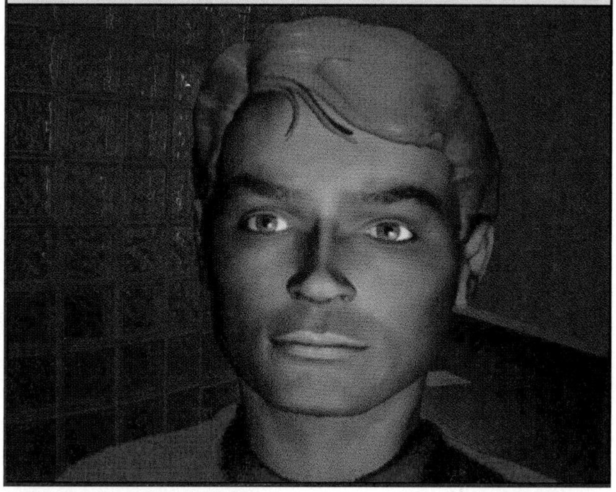

difficult. On the other hand, screens preserve the natural light patterns better than reflectors and silks can also replace white reflectors when used vertically, **Figure 15-20**.

Figure 15-22.
When soft-lighting a wide face, place the key light well to one side.

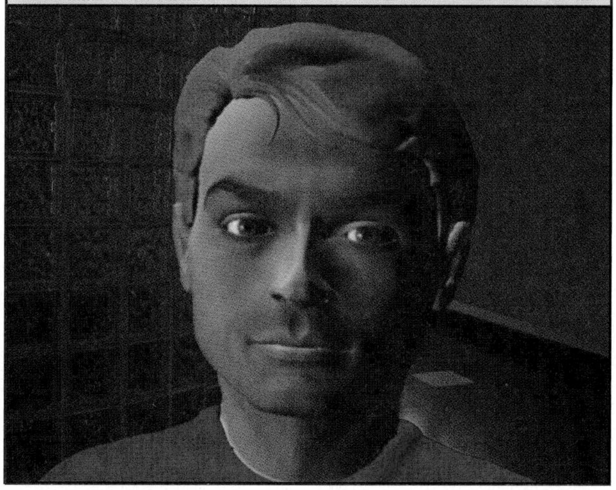

Lighting Problems

All lighting situations have problems, but some are especially common. One of these is subjects who are hard to light pleasingly; another involves specialized light sources.

Subject Problems

Unless you are lighting characters in story videos, you generally want to make your subjects look as good as possible. Here are four of the most common subject problems, with suggestions for solving them.

Wide faces

You can use lighting to make wide faces look slimmer. How you do this depends on your lighting method.

Classic three-light: When keying with a spotlight, the trick is to highlight the center of the subject's face, using the vertical barn doors, or ***flags*** if necessary. The key should be at nearly 6 o'clock, with *two* broads for fill: one at 3 o'clock and the other at 9 o'clock. On the more important side of the face, add moderate fill. On the other, use the lowest level fill that will reveal details in the shadows, **Figure 15-21**.

You may find that spill light alone is enough to fill the less important side of the face.

Natural light: With softlights, you have somewhat less control. Here, place the key light well to the side (8:30 or even 9 o'clock) and the fill light directly opposite. Check the amount of fill carefully. You may be able to use a reflector instead, or omit fill entirely. Although the "half moon" look is not as effective as a strongly modeled face in classic lighting, the slimming effect will still be considerable, **Figure 15-22**.

Darker complexions

Darker facial tones are beautiful when well lit, but they can offer ***contrast*** problems, especially in wedding videography, when the bride's

Figure 15-23.

White clothing with darker complexions create contrast problems.

Reducing the light to the lower part of the frame improves contrast.

skin tones are contrasted with a brilliant white wedding dress. The trick here is to get more light on the face than the dress, **Figure 15-23**.

Many ***on-camera lights*** can be fitted with barn doors that partially block light from the dress. With stand-mounted lights, half-screens can be added to the barn doors to further moderate the lower part of the light beam.

If your camera light will accept filters, you can buy or make a half screen to reduce the bottom part of the light beam.

Bald subjects

Balding heads are best handled by the makeup department: a little neutral powder will kill reflections and be quite invisible to the camera. If the subject refuses powder (as men sometimes do) try moving the lights up and then, if necessary, farther to one side, to reduce the reflections.

Subjects wearing glasses

For reflections on eyeglasses, the solution is similar. In general, small-source spotlights are easier to move out of the incidence/reflection path (as explained in the sidebar). However, reflections from the softlights often used for interviews can be tolerable, because viewers know that the subject has been lit for video.

For brief shots, it is often enough for the subject to lift the spectacle earpieces slightly off the ears, tilting the lenses downward and deflecting the reflection. When not overdone, this adjustment is generally invisible to the camera.

Specialized Light Sources

Practicals (such as table lamps) and environmental light sources (streetlights, signs, shop

Incidence Equals Reflection

In dealing with bald heads, spectacles, and other reflection problems, remember that light bounces off surfaces at the same angle, but in the opposite direction. So if a reflection is hitting the camcorder lens, the light is probably too close to the camera position, whether horizontally, vertically, or both. That is why raising a light or moving it sideways will often remove or at least lessen a reflection.

To minimize reflections, place lights at a 45° angle to the camera axis.

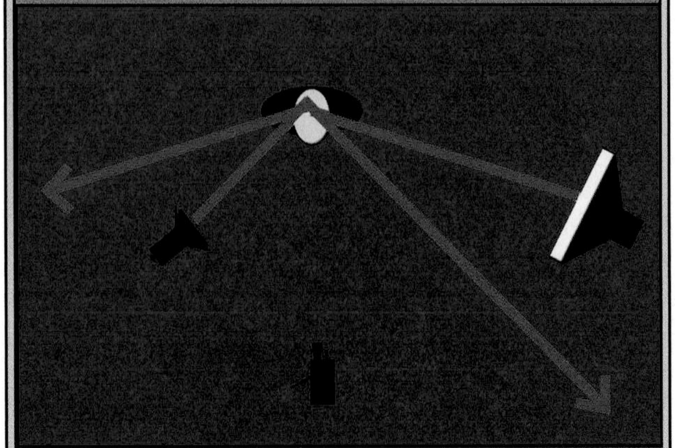

windows) create problems because they are rarely in the right place and/or operating at the right intensity. To solve these problems, it is usually necessary to establish the light source by showing it on camera, and then replacing it with a more controllable light.

Practicals

How you handle the lights that appear in the video frame depends on the lighting style you have chosen.

Naturalistic or expressionistic. Oddly, the opposite extremes in lighting styles can use the same technique: replace bulbs in practicals with screw-base ***halogen lamps*** and use them for actual video lighting. The resulting light will be contrasty, but excess contrast is acceptable in these styles, **Figure 15-24**.

Realism or pictorial realism. With these more common styles, you may want to establish the practicals and then use camera setups that exclude them, while simulating their light with video lights.

Adjusting intensity. To balance visible light sources with the rest of your lighting, try fitting larger or smaller lamp bulbs, as needed. Halogen replacement lamps can be dimmed somewhat to reduce intensity, but ordinary lights are already too orange to permit much further color shift through dimming.

On the other hand, if you light a scene entirely with household bulbs, you can simply set camera white balance manually, to match their 2700K–2800K color temperature.

Figure 15-24.
You can light with practicals if the result fits the style you are using.

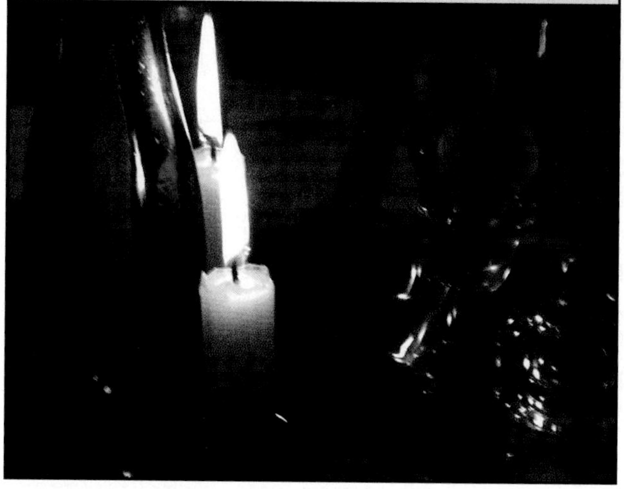

Moving light sources

Subjects often carry light sources such as flashlights or lanterns. When the practical light is not on-screen, you can simulate its light, for better control.

Flashlight. A small spotlight with a handle makes a good simulated flashlight. Focus the beam in the spot position to create a hard edge, **Figure 15-25**.

Lantern. To make a convincing "lantern," clip sheets of spun glass (for diffusion) and orange filter material (for "candle light") to a broad.

Candle. For a candle effect, omit the spun glass for a harder light. Wearing a leather glove, wave your hand and moving fingers slowly in front of the light to create a flickering effect.

Figure 15-25.
The practical flashlight is established...

and then its "light" is provided off-screen by a hand-held spot. (Lowel Corporation)

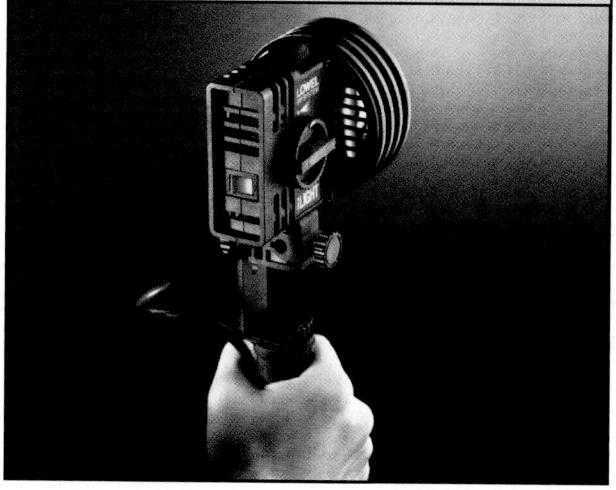

Firelight

Simulated firelight is created in somewhat the same way. First, place a broad very low, where a fireplace or campfire would be. Gel it with an orange filter sheet. Then staple a square of heavy cloth to a stick (denim works well) and slit the cloth at one-inch intervals to create a "grass skirt" effect, **Figure 15-26**.

Waving this device slowly in front of the "fire" light source will add a convincing flicker effect.

As with most such effects, you can enhance the realism with sound effects, in this case, a crackling wood fire.

Figure 15-26.
A "grass skirt" cookie can simulate the flickering of a fire.

Signs

In some night interiors, colored signs and other neon sources tint parts of the subject. If the sign is steady, simply gel a light with an appropriate color. If the sign turns on and off, have an assistant move a flag rhythmically in and out of the light path.

Electronics

Radar screens, computer monitors, and scientific instruments often bathe the faces of the subjects looking at them. With some units, the actual screen light may be bright enough, especially if the rest of the lighting is low key. If you cannot show the screens, however (say, because they are supposedly futuristic displays in a spaceship control cabin), place small lights low (at "screen" height) and gel them pale blue or green.

Lighting Night Scenes

Lighting scenes shot at night is challenging because there is little available light to help out. This section suggests some ways to create nighttime designs with relatively few instruments.

Interior Scenes

You can light indoor night scenes by using a few standard techniques.

Use low-key mode

First, create a low-key look, in which dark and medium values dominate in the background, with brighter accents and a well-lit subject.

Establish practicals

Since room lights are lit at night, establish practicals in the frame, and then mimic their light with video lights.

Control window light

If you can hang heavy screen or neutral density material inside or outside a window, you may be able to reduce its light to a "nighttime" level (the bluish color temperature will look like moonlight). If you do not have the resources to do this, exclude windows and their light from the frame.

Light for the highlights

In low-key lighting, you naturally use less fill light, so that shadows are deeper and show less detail.

Fake the darkness

It is common to show a subject in bed, turning off the bedside light and going to sleep. Shot in actual light, the scene would be very contrasty, and then would go black when the light went out. Here is a procedure for lighting this scene more effectively.

Establish the light level. Fit a halogen lamp in the bedside light and use it to key the scene. Use soft fill from the other side, with a crew member at the light. At this point, set and *lock* the camcorder exposure setting.

Establish "night." Next, turn off the key and fill lights and set up a very general overall fill, possibly with a pair of fluorescent pan lights.

Figure 15-27.
Faking darkness. A—The scene, as fully lit with practical, fill, and pan fill lights. B—After the subject and crew member have simultaneously switched off the practical and fill lights, the pan fill lights provide a very low level of light the viewer will accept as "darkness."

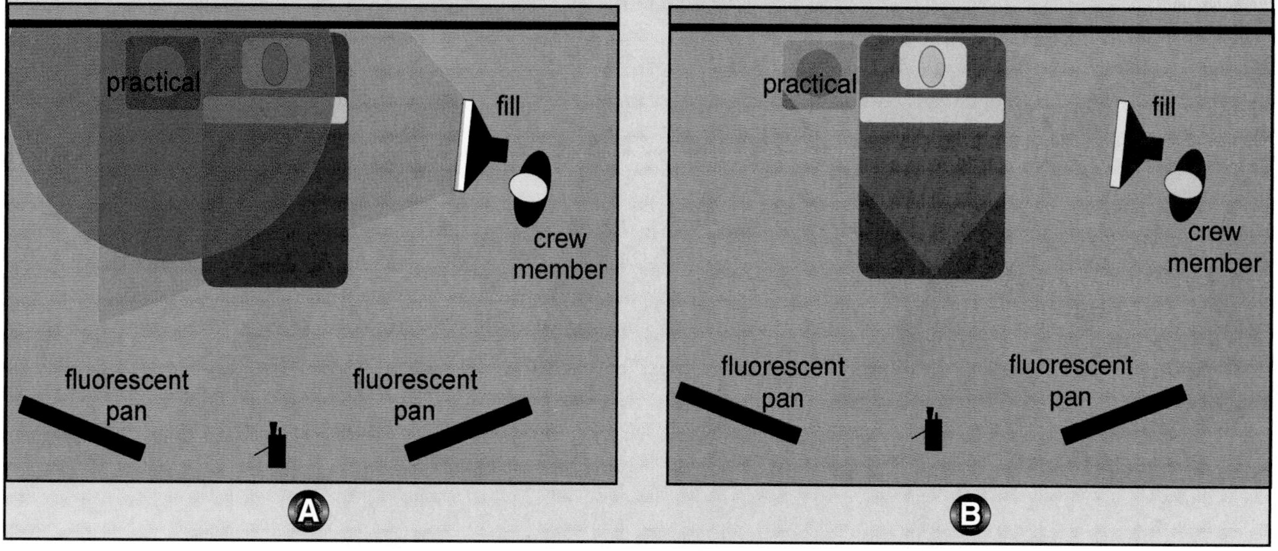

Without changing the camera aperture, adjust this fill light until the subject and bed are visible, though dark.

As an alternative, gel a spot or broad pale blue and place it at room-window height for a moonlight look. A window-frame *cookie* in front of the light will enhance the effect.

Synchronize the scene lights. Rehearse the shot until the crew member at the fill light switch can turn the light off exactly when the subject turns off the bed light, so that the two light sources look like a single light, **Figure 15-27**.

Although the remaining light (from the pan light fill) will be somewhat too bright for perfect realism, viewers will readily accept it as "dark," especially because of the bluish "moonlight" cast of its fluorescent lamps.

Exterior Scenes

It is impossible to light the whole outdoors for night scenes, but a few tricks will yield very satisfying results. First, light only the important action areas, as described previously. Then try these suggestions.

Use back-cross lighting

Except where the action must be clearly seen, place lights to the side and behind subjects (9 to 12 o'clock) to edge them with rims of light that will separate them from the background, **Figure 15-28A**.

Figure 15-28.
Lighting an exterior scene. A—Lights placed high and close to the building provide back lighting. B—This large castle background is lit by just three lights. C—Rain provides effective atmosphere. (Sue Stinson)

Lighting at "Magic Hour"

Magic hour is the brief period before sundown. On a sunny day, magic hour provides light qualities that look especially attractive on screen. Shadows from the low sun are long, molding objects and enhancing the impression of depth. The moisture in the air is often low, so everything appears exceptionally sharp and clear. The color temperature is warmer, lending a golden tone until near sundown, and then a distinctive sunset-orange tint.

McKinley at magic hour.

Lighting for magic hour is simple because the low sun makes reflector placement easy. The problem lies in capturing all the footage required in the relatively brief time before the sun actually sets. For this reason, you may wish to preset and rehearse several different camera setups, so that you can move quickly from one to the next as you shoot.

Another problem with magic hour is white balance, because the color temperature drops continuously as the sun goes down. One way to solve the problem is by compensating for the color shift while shooting. To do this, manually reset the white balance frequently. That way, you will capture the long shadows and clear light of magic hour, but all your original camera footage will have the same neutral color balance.

After you have edited a magic hour sequence, you can apply sunset tint to taste, or even warm the images up progressively as the sequence unfolds, to simulate an actual sunset. Be aware, however, that a digitally applied sunset color can have a mechanical, too-uniform quality. If you have the skill to capture it, there is no substitute for real "magic hour" light.

The characteristic long shadows and warm light of magic hour. (Adobe)

Rake the background

Night scenes are supposed to be dark, so use just a few lights to pick out features of the background. As usual with background lighting, place the lights to hit the background at oblique angles. See Figure 15-28B.

Look for motivation

Outside of urban centers, there is often little actual light. To motivate video lighting, simulate car headlights and house or shop windows with lights placed low and shooting horizontally. For streetlights, move rim lighting directly over subjects, to create eye socket shadows.

Atmosphere

Outside night scenes in movies are often wet because rain or fog (real or fake) picks up and scatters light rays. See figure 15-28C.

Finally, do not worry if some of your lights do not have enough motivation. This lighting problem is so common, even in big-budget productions, that viewers have come to accept night exteriors full of unexplained light sources.

Day-for-Night Lighting

Although you can use a camcorder in very low light levels, there are good reasons for shooting exterior night scenes during the day, using ***day-for-night*** lighting, **Figure 15-29**.

Daytime shooting is more convenient for everyone. Daytime light levels are high enough for optimal imaging. Fewer lighting instruments and accessories are required. Though electrical power is helpful, in some cases you can shoot without it.

Figure 15-30.
The setup for the shot.

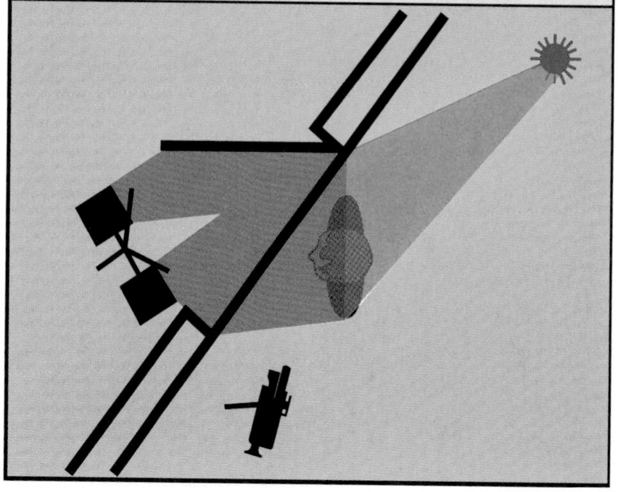

In video, making daytime shots look like night is easy if you follow a few simple guidelines (the setup is diagramed in **Figure 15-30**).

Set white balance for incandescent

Using an indoor white balance setting outdoors, **Figure 15-31**, will lend an overall "moonlight" bluish cast to the footage, while rendering any incandescent lights as true white. You may also want to set exposure so that the natural shadows fill in as deep black. If you will edit digitally (and no incandescent lighting is used) you can create the color, exposure, and contrast of

Figure 15-29.
A—An exterior shot recorded normally. B—The completed day-for-night version.

A B

Figure 15-31.
White balance set for incandescent (indoor) light.

"night" in postproduction. In this situation it is often safer to get conventionally balanced and exposed footage and then alter it later.

Use back-cross lighting

Instead of directing the main light source at subjects from the front or part-way to one side, position your subjects and reflectors as needed to splash the brightest light from the rear, onto hair and shoulders, while the faces remain somewhat darker, **Figure 15-32**.

Figure 15-32.
Back-cross lighting from the sun creates a rim light of "moonlight."

Frame off the sky

Unless you can darken the sky with a polarizing filter, use neutral or high angles to aim the camcorder away from the sky, which will appear much too light for a convincing night effect, **Figure 15-33**.

Figure 15-33.
Including the daytime sky reveals the trick.

In some lighting conditions, a polarizing filter on the camcorder can turn a blue sky dark enough to pass for night, if it is not allowed to remain on screen too long.

Include incandescent light

Finally, try to include some incandescent or halogen lighting in the shot, such as headlights, a street light, or light streaming out of a window or open door. Since the indoor white balance setting will render this light as "white" it will contrast convincingly with the "moonlight" cast of the overall scene, **Figure 15-34**.

Figure 15-34.
Incandescent light completes the illusion.

Lighting Assignments

In addition to the general lighting situations covered so far, there are a few specific assignments that come up frequently enough to deserve special attention. These include interviews, "standup" reports, compositing, very small areas, and graphic materials.

Interviews

Interviews are among the most common lighting assignments. Typically, they involve two subjects and a moderate amount of background.

One-person interviews

The current style of video interview has the subject on-screen all the time.

The interviewer's questions are posed as topics to be responded to, so that they can be omitted in editing. In the finished interview, the subject appears to be discussing the subject spontaneously.

Softlights are frequently used because they look natural, they generally light the background as well as the subject, and they are quick and easy to use.

Since interview subjects are rarely media professionals, lighting should be moderate and kept out of their eyes. Because interviews rarely use angles wider than medium (waist) shots, a reflector can be used opposite the soft key, for fill. Even with "natural" lighting, a small spot rigged as a rim light can add modeling and separate the subject from the background, **Figure 15-35**.

Two-person interviews

If your interviewer will appear on screen, you must light her or him as well, **Figure 15-36**. In establishing shots, the back view of the narrator in the foreground can often be lit by the lights on the interview subject. The same lights provide rim light to separate subject and interviewer from the background.

Figure 15-35.
One-person interview. A—A two-light interview setup on the subject only. B—Closeup of the subject.

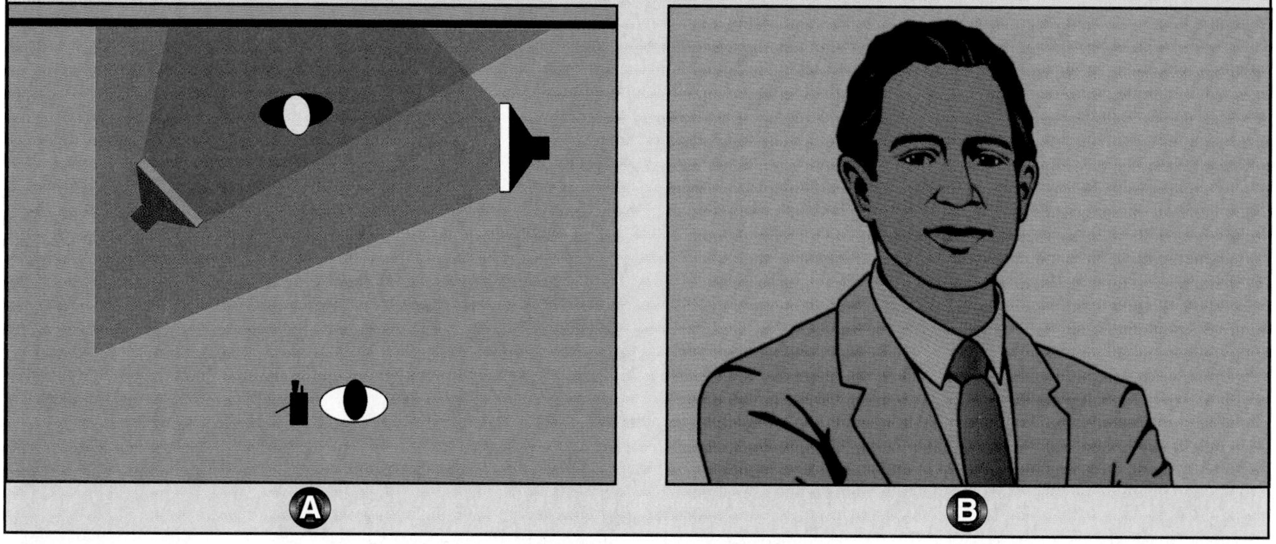

Figure 15-36.
Two-person interview. A—A three-light interview setup. B—Shooting over the interviewer's shoulder. C—A reverse setup features the interviewer.

When Cheating Is Legal

Cheating is the common practice of moving a subject between camera setups, usually to increase working room, separate subject and background, or find a better-looking background altogether.

Careful cheating is invisible to viewers, because the lack of true depth in video makes subject-to-background distances hard to judge, and because the background "behind" a subject may not appear in earlier shots.

A master two-shot of subject and interviewer, avoiding the windows.

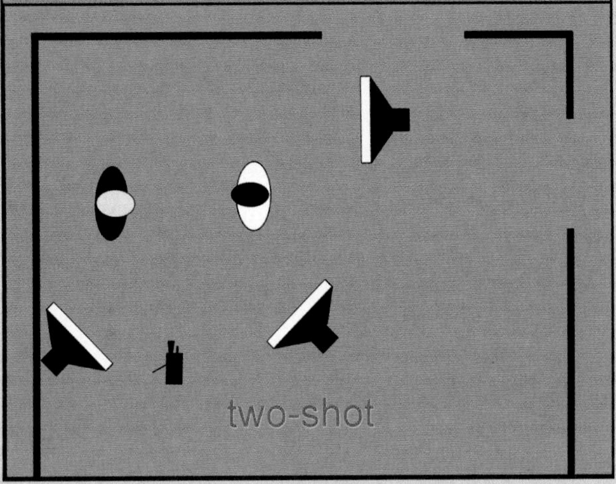

In this example, a subject and an interviewer are placed in the upper-left corner of a room, to keep back and side windows out of the frame for the master shot.

Lights, camera, and performers are all cheated to the right.

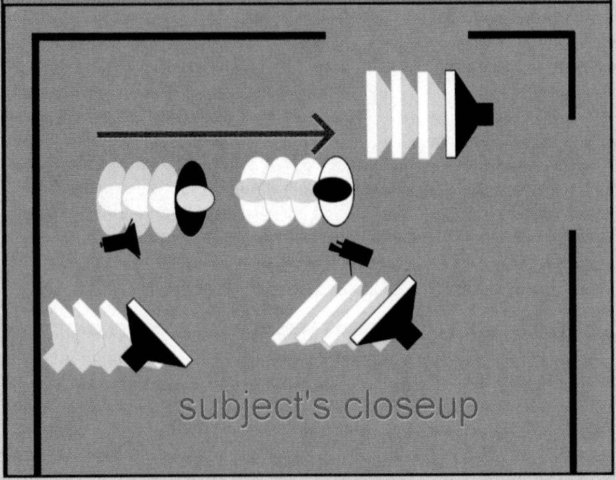

Because there is no room behind the subject to add a rim light for his closeup, the entire setup is "cheated" three feet out from the left wall. From the camera's new position, the move is undetectable.

Cheating a subject to light the background separately.

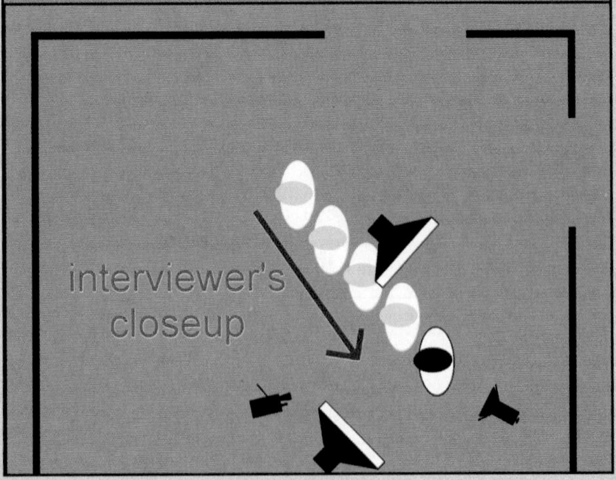

When it is time for the interviewer's closeup, she is cheated back and to the right (to avoid the right-wall window), and completely re-lit.

Because viewers do not see the right wall in the other shots, this cheat, too, is undetectable.

Figure 15-37.
Because viewers cannot see the backgrounds of both subject and interviewer, you can move the interviewer to a different location, if necessary.

For single shots of the interviewer asking questions or listening to answers, you have two different options. One is to light the interviewer separately for real-time recording with a second camera. To do this, position the reporter's camera to shoot over the interviewee's shoulder, so that both people are included in wider shots. Place the lights for both people so that they are outside the frame in both setups.

If you are using only one camera, or if some of the reporter's questions and reactions need to be re-shot, you can re-light and reset the camera after the main interview to pick up this material for later editing into the sequence.

If you shoot the interviewer separately, you can place him or her in front of whatever background looks best – even in another location, if necessary. The shift will be invisible to the viewer, Figure 15-37.

Standup Reports

Segments by reporters standing at the scene of news, sports, or entertainment events are called "standups." Even at the broadcast TV level, reporters often go into the field with a single technician who must function as videographer, lighting director, and sound recordist.

Daytime reports

To simplify lighting, this one-person crew will often work with just two tools: an on-camera light and a stand-mounted reflector, **Figure 15-38**.

The camera-mounted light is typically variable in light output and powered by its own substantial battery. Even in daylight, a front fill light can highlight the reporter's face just enough to emphasize it against the background.

News camcorders are often fitted with batteries big enough to power a light as well.

In Figure 15-38, the supplementary light is a stand-mounted reflector. Because the camera person must leave the reflector unattended while shooting, it needs to be heavy and relatively small, to resist the wind

Nighttime reports

Night shooting requires a different deployment of lighting tools. The on-camera light continues to provide much of the illumination. A second, battery-powered light on a stand can fulfill the same function as a reflector, providing more modeling on the reporter's face. Backgrounds at night are generally dark; but look for lighted walls or windows to include in the shot, so that the image behind the reporter has some design to it, **Figure 15-39**.

Figure 15-38.
Daytime reports. A—Two-person news teams are typical. B—In this setup, the sun provides the rim light and a reflector fills from the left. C—An on-camera key light would have better highlighted the reporter's face.

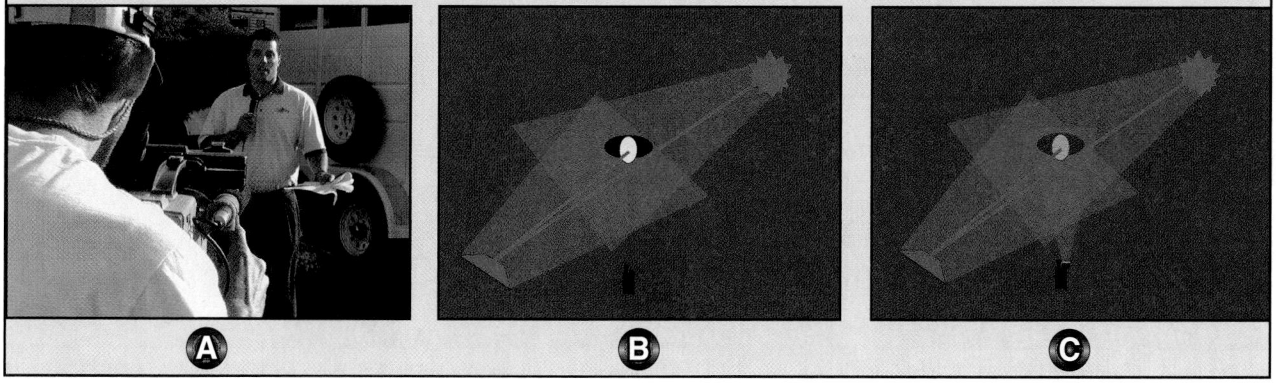

Figure 15-39.
Nighttime reports. A—A one-light setup. B—A two-light setup. C—Offsetting the camera light provides better modeling.

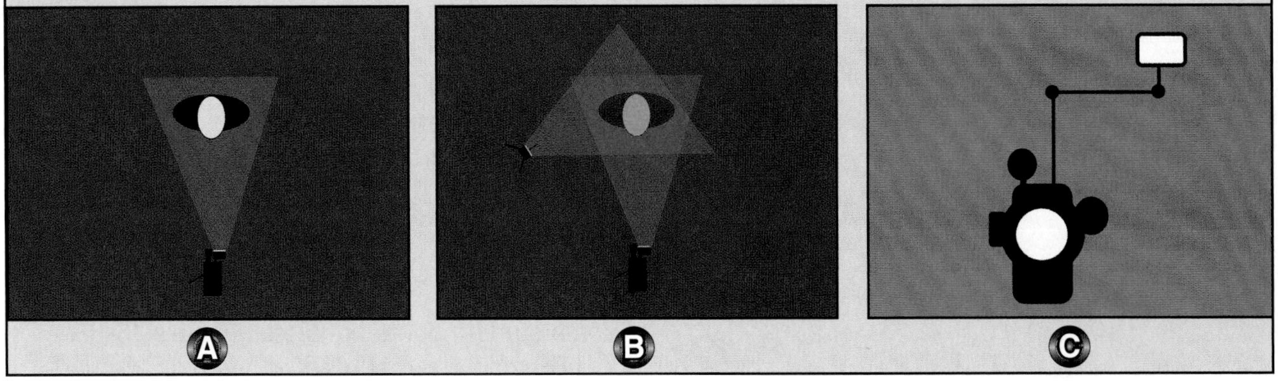

Lighting for Compositing

With the power of today's postproduction software, compositing is used increasingly.

Compositing is the process of videotaping subjects against flat, single-color backgrounds and then digitally replacing the background with other visuals during postproduction.

The more even the background color, the more perfectly it can be replaced; so the challenge is to keep the background lighting absolutely uniform. To do this, you try to light background and subjects separately, **Figure 15-40**.

Lighting the background

You can achieve very even background lighting with a pair of softlights (or spots with heavy diffusion) placed one at each side. Experiment with light positions and throws to create a near-perfect wash of light.

Lighting the foreground

Place the subjects as far forward of the background as possible. Classic spotlight lighting is easiest to use because the light paths are controllable. To keep foreground light off the background screen, position the lights relatively high and to the side (as far as 8:00 or 8:30) to keep hot spots and shadows below and to the sides of the frame .

Be sure to look at the background footage to determine the quality and direction of its own lighting. If your setup allows, key this footage into the image in place of the composite screen. Adjust subject lighting until it matches the composited background.

Small Objects and Areas

Tabletop videography involves shooting small objects and/or activities on a table,

Figure 15-40.
Lighting for compositing. A—Spotlights shooting through large silks wash the background evenly. B—The subject is far enough forward so that the key, fill, and overhead rim lights do not spill onto the background. C—Using softlighting for both subject and background eliminates shadows, but allows no subject modeling. (Bogen Manfrotto)

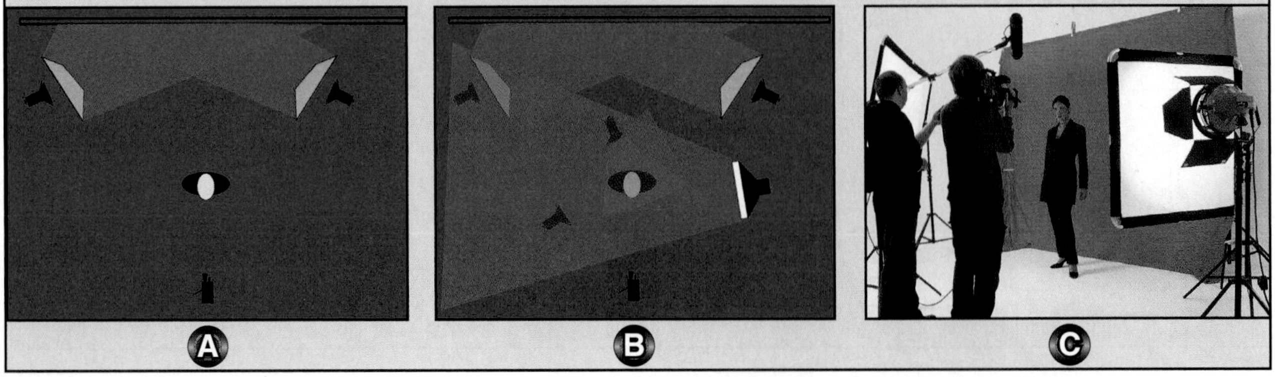

Figure 15-41.
A tabletop shot. (Photoflex)

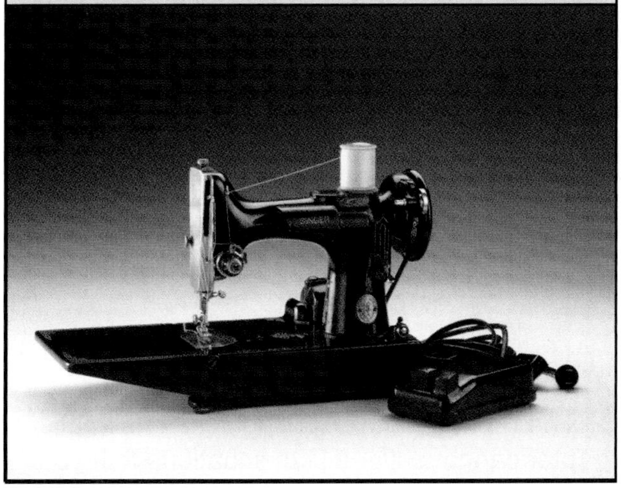

counter, or special photo stand. It is a common procedure for science experiments, product demonstrations, and how-to training sequence, **Figure 15-41**.

In shooting small subjects, camera and lighting problems generally arise from two causes. First, the short camera-to-subject distances (and/or telephoto lens settings) create very shallow depth of field. This makes small objects difficult to keep in sharp focus. Second, the hands, arms, and head of the demonstrator tend to get into the picture.

To help solve these questions, you need lights that are both very bright (to force smaller lens openings and thereby increase depth of field) and very soft (to eliminate or at least minimize shadows in the picture). Every situation has unique lighting requirements, of course, but here is a solution for a typical tabletop setup, **Figure 15-42**.

Figure 15-42.
Tabletop setups use very large, soft lights.

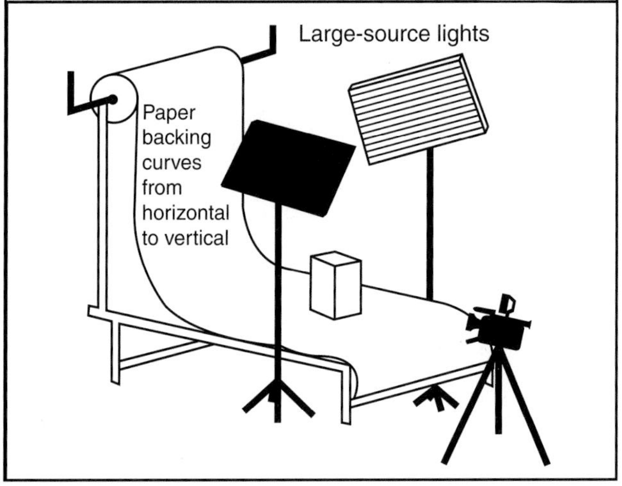

The subject and tabletop are bracketed by very large, bright fluorescent softlights. These instruments are excellent for this application because:

- Their four-foot square shape provides an extremely large source, to reduce shadows.
- Their multiple 40-watt tubes provide a bright light, permitting smaller F-stops to create greater depth of field.
- Their power requirements are low enough for use in most locations.
- Their output is cool enough for subject comfort, and for delicate applications such as food demonstrations and biology experiments.

Where you want completely shadowless lighting (and do not require great depth of field) you can use ***tent*** lighting. By hanging a white

Figure 15-43.
Tent lighting. A—A simple tent setup outdoors, using a bed sheet and clotheslines. B—A professional tent for small object videography. (Photoflex)

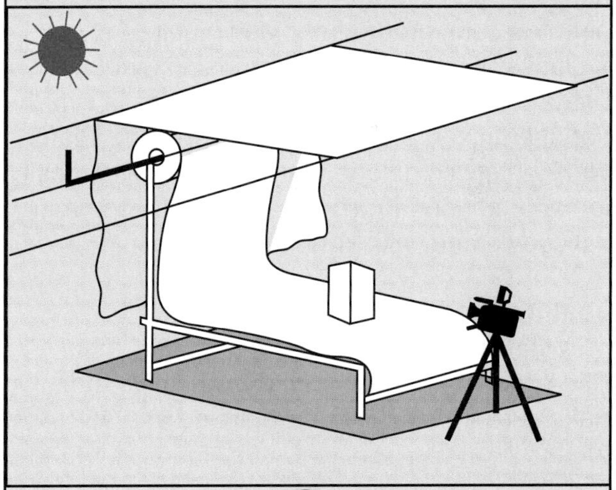

A

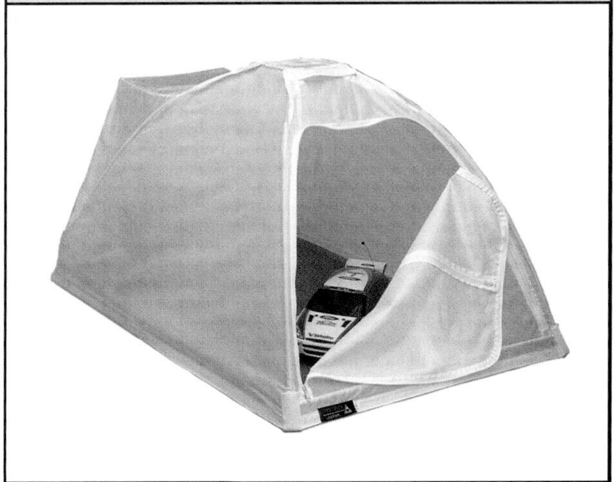

B

sheet over and around your subject and aiming lights through it, you can create the softest possible lighting effects. Tenting techniques are also useful for lighting products in commercials.

Tenting works very well outdoors, with a sheet hung over a pair of lines, **Figure 15-43**.

Graphic Materials

Many programs include two-dimensional subjects—photos, paintings, graphics, letters, book pages—in place of moving subjects. With today's quality equipment, it is often easiest to record subjects with flat bed scanners and import them into video programs during postproduction. If the flat material is larger than about eight by ten inches, however, scanning is often impractical. You will need to videotape these larger subjects directly, using a special lighting setup.

Organizing a setup

Most often, you will work with the graphic material on a flat surface and the tripod-mounted camcorder aimed down at it. A sheet of glass will help hold the material flat, but may create reflection problems. The lighting is simple: one unit on each side of the artwork. Clamp work lights with halogen lamps are easy to position and adjust. **Figure 15-44** shows a commercial copy stand, with light diffusion.

Dealing with reflections

With or without a glass cover plate, light reflections are often a problem. To solve them, make sure that the lights are aimed at a 45° angle.

Figure 15-44.
A professional copy stand. (Bogen Manfrotto)

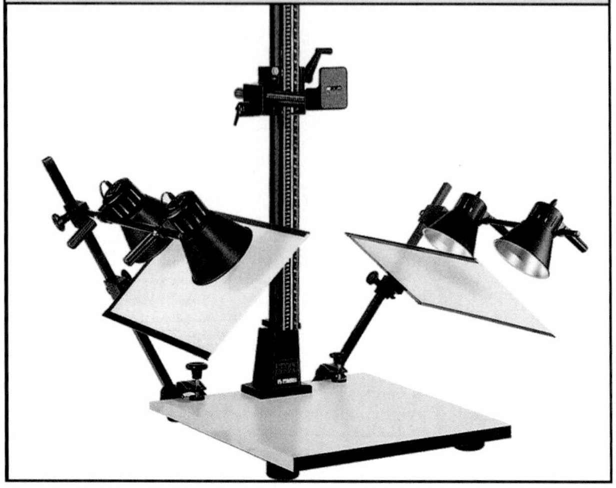

Working vertically

In many cases, posters, paintings, charts, and other large subjects are best handled vertically. Make sure that the camcorder is centered horizontally and vertically, at a true 90° angle to the artwork. Position the lights far enough back to wash the subject evenly, and keep them at a 45° or less angle from the wall, **Figure 15-45**.

Placing the camcorder far back, with a telephoto lens setting will improve the quality of the image recorded.

Other Lighting Applications

The representative lighting solutions in this chapter cannot cover all the situations that you may encounter, but they do demonstrate how to use the basic ideas behind all lighting designs to analyze each situation and create video lighting with style.

Figure 15-45.
A vertical lighting setup for copying. To avoid reflections, place lights at an angle of 45° or less to the wall.

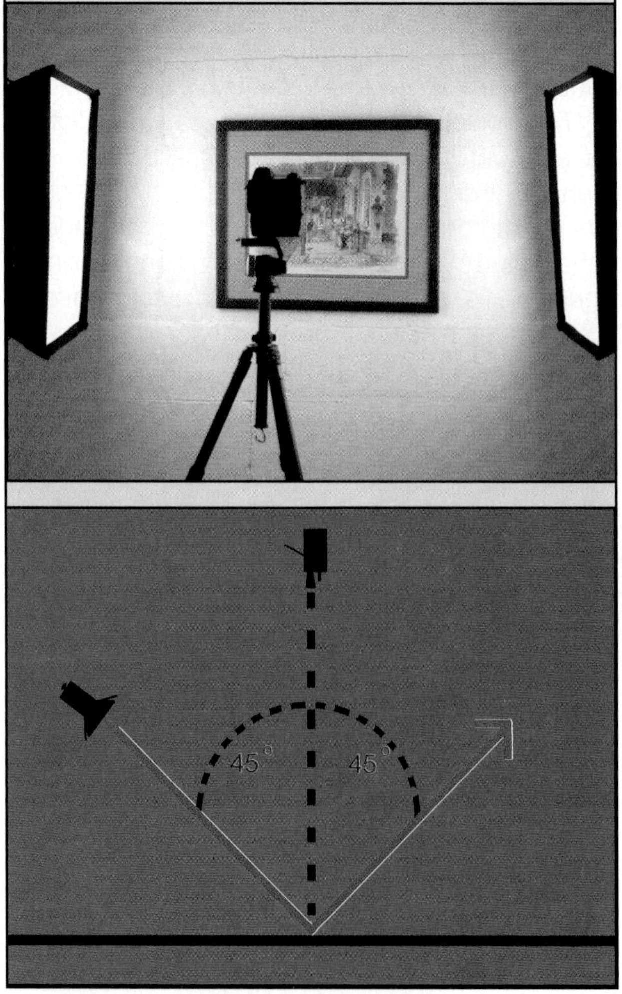

Chapter Review

Answer the following questions on a separate piece of paper. Do not write in this book.

1. Classic three-point lighting uses key, ______, and rim (back) lights on the subject, plus a background light.
2. ______ reflectors have the longest "throw" for use in outdoor scenes.
3. *True or False?* Because of their size, spots are the preferred lighting instrument for use in cramped quarters.
4. Using an "indoor" white balance outdoors will create a blue-tinted ______ lighting effect.
5. To produce completely shadowless lighting, a white fabric ______ can be used.

Technical Terms

For easier reference, some definitions are repeated from Chapters 13 and 14.

Amp (ampere): In lighting, the amount of power being drawn by a light.

Available light: The natural and/or artificial light that already exists at a location.

Background light: A light splashed on a wall or other backing to lighten it and add visual interest.

Barn doors: Metal flaps in sets of two or four, attached to the fronts of spotlights to control the edges of the beam.

Contrast: The difference between the lightest and darkest parts of an image, expressed as a ratio (e.g. "four-to-one").

Cookie: A sheet cut into a specific pattern and placed in the beam of a light to throw distinctive shadows such as leaves or blinds.

Day-for-night: A method of shooting daylight footage so that it appears to have been taken at night.

Diffusion: White spun glass or plastic sheeting placed in the light path to soften and disperse it.

Fill light: The light that lightens shadows created by the main (key) light.

Filter: In lighting, a sheet of colored or gray-tinted plastic placed over lights or windows to modify their light.

Flag: A flat piece of opaque metal, wood, or foam core placed to mask off part of a light beam.

Fluorescent lamp: A lamp that emits light from the electrically charged gasses inside it.

Glamorous lighting: Lighting that emphasizes a subject's attractive aspects and de-emphasizes defects.

Halogen lamp: A lamp with a filament and halogen gas enclosed in an envelope of transparent quartz.

Incandescent lamp: A lamp with a filament enclosed, in a near-vacuum, in a glass envelope ("bulb").

Instrument: A unit of lighting hardware such as a spotlight or floodlight.

Key light: The principal light on a subject.

Lamp: The actual bulb in a lighting instrument.

LED: Light Emitting Diode: an electronic light source for special applications.

Magic hour: The period, of up to two hours before sunset, characterized by long shadows, clear air, and warm light.

Neutral density filter: In lighting, a gray sheet filter placed over windows to reduce the intensity of the light coming through them.

On-camera light: A small light mounted on the camera to provide foreground fill.

(continued)

Practical: An instrument that is included in shots and may be operated by the actors.

Reflector: A large silver, white, or colored surface used to bounce light onto a subject or scene.

Rim light: A light placed high and behind a subject to create a rim of light on head and shoulders, to help separate subject and background.

Rugged lighting: Lighting that emphasizes three-dimensional qualities and surface characteristics of a subject.

Screen: A mesh material that reduces light intensity without markedly changing its character.

Silk: A fabric material that both reduces light intensity and directionality, producing a soft, directionless illumination.

Softlight: A lamp or small light enclosed in a large fabric box, which greatly diffuses the light.

Spotlight: A small-source lighting instrument that produces a narrow, hard-edged light pattern.

Tabletop: Videography of small subjects and activities on a table or counter.

Tent: A white fabric draped all around a subject to diffuse lighting completely for a completely shadowless effect.

Three-point lighting: So-called "classic" subject lighting, consisting of key, fill, rim, and background lights.

Throw: The distance between a light and the subject lit; also, the maximum useful distance between those points.

Voltage: The electrical potential or "pressure" in a system, nominally 110 volts in North America.

Wattage: In lighting, the power rating of a lighting instrument. 500, 750, and 1,000 watt lamps are common.

Location lighting can require large amounts of equipment. Chris Hayes of Haze Lighting and Grip, New York, displays his lighting equiment. (C. Abbiss)

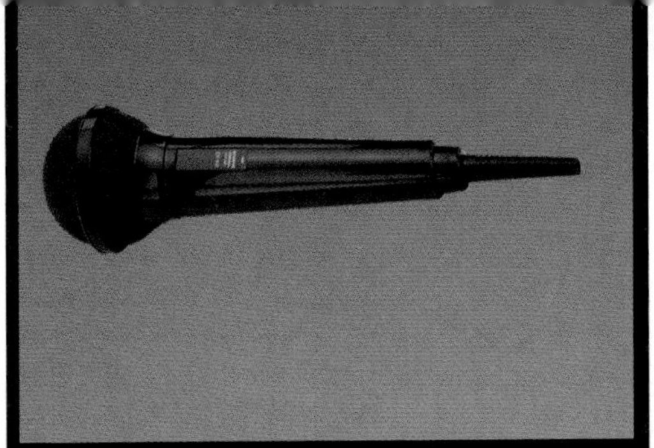

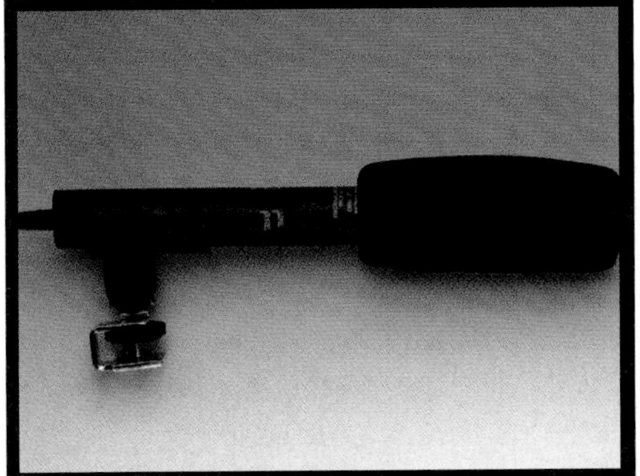

16 Recording Audio

Objectives

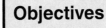

After studying this chapter, you will be able to:

- Distinguish the different types and purposes of recording.
- Understand audio equipment and select it appropriately.
- Record quality audio tracks of all the major types.
- Solve the problems associated with common recording situations.

About Recording Audio

Audio, as we use the word here, is the recording and reproduction of sound in support of video. Chapter 7 explains how the sound track affects viewers and why it is such an important part of every video program. With this information in mind, we are ready to consider the techniques used to create that sound track: the techniques of professional audio recording.

The sound "track" gets its name from the film medium, where the audio, recorded along the edge of the film, appears as a visible track.

Professional camcorders allow manual control of audio recording. Consumer camcorders, however, record sound automatically. Automatic recording is simpler, but if you have heard the often mediocre results, you can see why obtaining quality audio requires effort and skill.

Audio recording must always deal with two problems. First, recording captures everything, whether it is wanted or not. Second, on playback, recorded sounds may seem distorted or even quite unlike the originals. Before learning about specific audio equipment and techniques, you need to understand how these fundamental problems affect audio recording.

Problem: Audio Records Everything

Like your own ears, microphones pick up every sound within their range. Sound from an outdoor scene might be a mix of spoken dialogue, background traffic noise, wind, shuffling feet, and even sounds made by the camcorder itself. See **Figure 16-1.**

Figure 16-1.
A noisy background can degrade audio quality.

The problem arises because of the way in which people process the sounds they hear. You may hear sounds with your ears, but you *listen* with your brain. Without noticing it, you suppress everything in the incoming sound mix except for the parts that interest you. For example, you may focus on the dialogue, while blocking out the background noises. Then, if you should notice trees waving in the wind, you might shift your mental focus to the wind sound, to the point where you no longer hear what is being said.

For reasons that are not perfectly understood, this mental sound filtration is more difficult to do with recordings than with live sounds. As a result, the dialogue in a video scene may be ruined by wind or other noises that you would not even notice in the real world.

Problem: Audio Distorts Sounds

The second problem is that recording a sound does not guarantee that it will be convincing or even recognizable on playback. Many factors affect the quality and character of recorded sound; it takes skill to listen to what the microphone is picking up and then adjust recording conditions to make the resulting signal sound natural.

Even when the sound is not actually distorted, an audio recording can still sound unnatural, especially if it does not match the visual environment. For instance, sound recorded in postproduction and then synchronized with a video scene in a metal aircraft hangar may sound muffled and "dead," when it should have a metallic echoing quality.

The most common problem with sound quality is inconsistency from one shot to the next. The sound character of a wide shot may not match that of close-ups made of the same sequence because the camera's broad field of view has forced the microphone farther away from the performers in order to keep it out of the shot. See **Figure 16-2.**

In real-world listening, we tend to ignore the difference, but in an edited video sequence that alternates between wide and close angles, the constantly changing sound quality will be obvious and annoying.

Figure 16-2.
A wide shot may force the microphone far back from the subjects.

Figure 16-3.
Digital editing can handle dozens of audio tracks. (JVC Corporation)

In the real world, of course, we do not abruptly change position every few seconds, as the camera does from one setup to the next.

Overcoming Audio Problems

To address these problems, professional audio recordists employ a well-proven strategy:

- Identify the separate types of sound that are wanted for a sequence (such as dialogue, sound effects, and background noises).
- Isolate each sound component and record it separately.
- Select, position, and adjust microphones to obtain consistent, high-quality, natural-sounding audio, regardless of the visual perspective.

With clean, separate audio tracks, a sound editor can place and balance audio components to duplicate the function of the listener's brain, calling attention to certain sounds and minimizing others, as required, **Figure 16-3**. A well-mixed sound track directs the ear, eliminates audio clutter, and enhances the realism of the program.

Sound track mixing is covered in the chapters on editing.

Providing the editor with clean, separate components is the sound recordist's fundamental job. To do this job, you need to know the basics of audio recording, the equipment used to do it, and the techniques required to obtain high-quality audio.

Types of Audio Recording

In recording raw sound for later editing, you must deal with several different types of track, each requiring its own set of recording techniques. At different times you may need to record production sound, background sound, sound effects, voice tracks, and music. The most common type of recording is production sound.

Production Sound

Production sound is, of course, the sound that is recorded with the video. While most amateur sound is recorded by the camcorder's built-in microphone, professional productions generally use separate mikes. Because the camera position is rarely the optimal location for miking the shot, separate video and audio recording devices allow each to be placed in the best spot, **Figure 16-4**.

Production sound calls for special care because mistakes may require another take of the whole shot. As an extreme example, if your first attempt to record the actual sound of a train derailing is unsatisfactory, you cannot re-record the effect without derailing another train.

Because actual recordings are often impractical, train derailments and similar sounds are usually simulated and recorded separately as sound effects and added to the video in postproduction.

Figure 16-4.
Fishpole mike supports are sometimes used in studio production. (Sennheiser)

Background Sounds

In the real world, individual sounds are heard against a background of more-or-less continuous noise. This noise may come from wind, waves, birds, traffic, heating and air conditioning, or people talking or doing things like coughing in a theater or rattling dishes in a restaurant. See **Figure 16-5**.

Even without any obvious background noise, no location is truly silent. If the editor cuts unwanted pieces out of the production track, the "emptiness" of the resulting soundless passages is clearly audible. To prevent this, you record "room tone," which sounds like nothing by itself, but is used to fill the holes left by deletions from the production track. See **Figure 16-6**.

Though in real life, we filter out most background noise, a movie sound track does this job for us. To minimize background noise:

- Turn off controllable sounds like heating ducts or background conversations during takes.
- Minimize uncontrollable noises like traffic or wind as much as possible during dialogue or effects recording.
- Make lengthy (up to five minutes long) recordings of ambient (background) sound alone.

In editing, the continuous ambient noise is then laid in under the dialogue and effects tracks, where it helps conceal any background noise that makes its way onto the dialogue tracks, and it minimizes the differences in quality between production tracks made from different mike positions. Professional sound libraries include prerecorded background tracks of a wide variety of locales. These tracks are often substituted for background sound recordings made during a shoot.

"Mit Out Sound!"

Unlike video, film production usually requires two separate machines for recording picture and sound. For this reason, shots in which audio would be difficult to record adequately or microphones would interfere with the visuals are often made with the film camera only, omitting the sound recorder. Shots filmed without sound are slated M.O.S.

Tradition holds that the acronym "M.O.S." stands for "Mit Out Sound," in the imperfect pronunciation of a famous European director who came to Hollywood at the beginning of the sound era and hated the new technology so much that he did everything he could to shoot "mit out" it.

In video, by contrast, picture and sound are captured together, so sound of one sort or another is almost always recorded. Even if you think you will be unable to use the production audio in the finished video, you should still record it, at least with the camcorder's built-in mike. This will create a reference track, also called a "scratch" or "cue" track. In postproduction, you use a scratch track for several purposes: to synchronize replacement ("looped") dialogue with the original, to indicate the exact positions of sounds to be added from separately recorded sound effects, and to provide a record of minor dialogue changes, etc., that occurred during the shooting. For these reasons, it is always important to record the best quality production audio that is practical under the circumstances.

A waveform display simplifies identifying and editing effects.

Figure 16-5.
Heating and air conditioning are common sources of background noise.

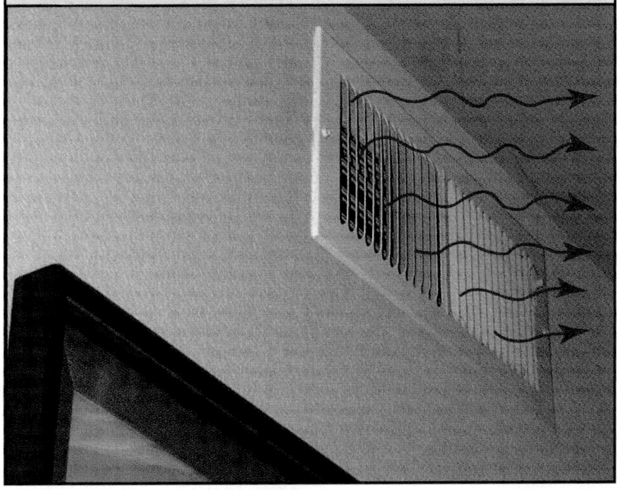

Figure 16-6.
Room tone is used to fill gaps in the production track.

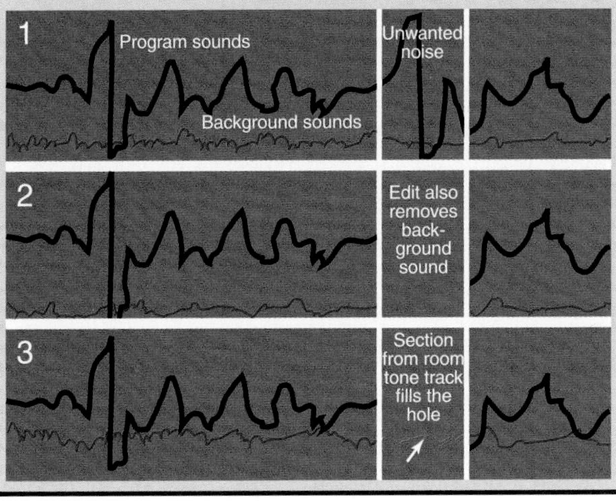

An ambient track is used in long pieces to blend other tracks. A room tone recording (also called a presence track) is used in very short pieces, to fill gaps where the editor has removed parts of the production track.

Sound Effects

Sound effects are often recorded "***wild,***" meaning separately from the production track. This usually happens when using that track is impractical, often because it is impossible to place a microphone for good quality sound or because the actual sound would be disruptive (like gunshots on a busy street).

Audio was originally recorded directly on film in a machine separate from the camera. To synchronize the two machines, their motors were electrically "slaved" to a "master" motor. Thus, a machine recording sound effects independently of a camera was running "wild," rather than "slaved."

Sound effects that would be hard to get under any circumstances are available in sound effects libraries, some of which contain thousands of different noises. See **Figure 16-7**.

In every recording situation, deciding whether to use production track or wild sound effects depend on how many there are and how closely they are synchronized to the picture. For example, the sound of a car starting and driving away could easily be recorded as a wild effect and post-synchronized with the picture. On the other hand, the on-screen feet of a man and two women crossing a marble floor would be extremely tedious to record and lay individually one person, foot, and footstep at a time.

For situations like this, professionals use what is called a ***Foley studio***—an environment in which technicians can make sounds while watching a shot on the screen and synchronizing the effects as they are recorded.

Voice Recording

Like sound effects, voices can be recorded separately from the production track. The most

Figure 16-7.
Continuous sounds like racing car engines are easy to record and use in postproduction. (Corel)

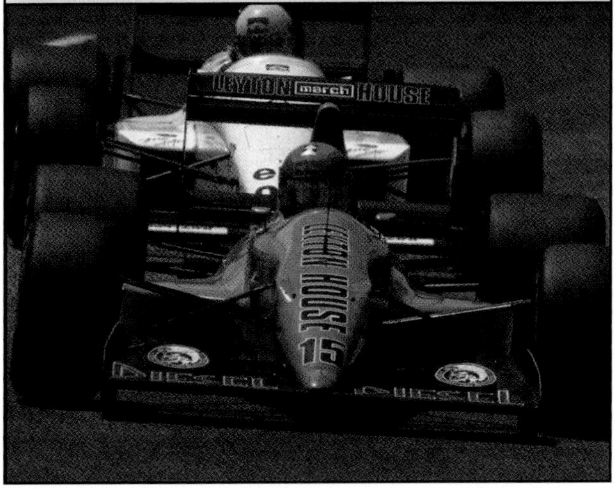

common application for this is narration recording. In a sound booth or small radio studio, the performer records all the narration in a program, leaving gaps between sections that allow the editor to select the best takes and lay each one in at its proper spot.

Dialogue replacement is also a common procedure. It works like Foley recording, only with voices rather than sound effects. One or more actors record their lines in a sound studio, while watching a dialogue scene on screen and listening to a reference production track through headphones.

Dialogue replacement is often still called "looping" because short lengths of movie film were once spliced end-to-end in loops for continuous projection while the watching performers rehearsed synchronizing the new dialogue.

Music

Music is seldom recorded with production sound. Instead, it is captured separately in studios designed to optimize musical sound, and then added to the sound track in editing. Music recording is a complex process that requires sophisticated microphones, mike placement, and electronic mixing and equalizing. As a result, almost all professional music is recorded by studio technicians rather than by production sound recordists.

However, production people do get involved in ***playback,*** the process of reproducing previously recorded vocal music while singers synchronize their simulated on-camera performance with the high-quality recording that they (or others) have made earlier, **Figure 16-8**. In playback, the camcorder will typically lay down a rough recording of the vocal lines sung by the performers on the set for later use as a reference track. In post-production, the editor will match the prerecorded track to the playback version of it, then remove the rough production mixture from the finished track, leaving only the high-quality original.

Figure 16-8.
For vocals especially, performers often lip-synch previously recorded audio. The process is called "playback."

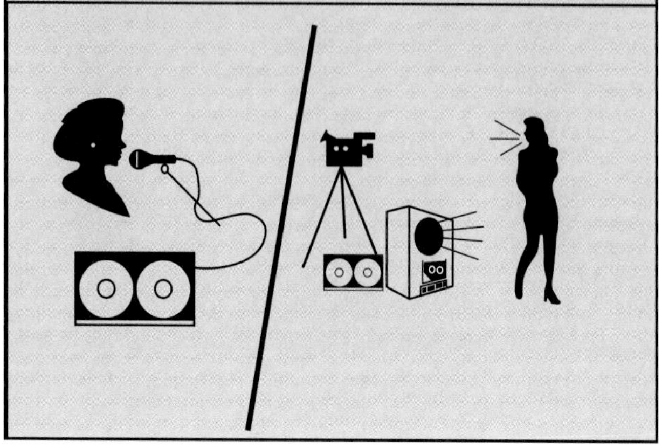

Microphones

All sound recording must begin with a ***microphone,*** a device that translates the original noise, created by a pattern of moving air, into an electrical signal that closely imitates that pattern. In order to select and use microphones effectively, you need to know the design basics of the principal types. Those design characteristics include physical form, transmission method, and pickup pattern.

Microphone Transducers

The ***transducer*** in a microphone is the electromechanical system that converts sound energy into electrical energy. Of the several transducer designs developed over the years, the most common are *dynamic* and *condenser.*

Dynamic transducers

A dynamic microphone transducer is a taut, flat diaphragm like a drum skin that is vibrated by incoming sound waves. Dynamic microphones are useful because they are tough and uncomplicated and the signal they

A dynamic transducer.

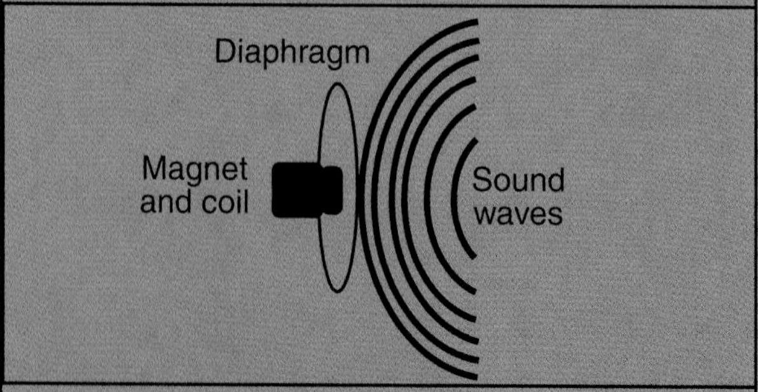

A dynamic microphone. (Sennheiser)

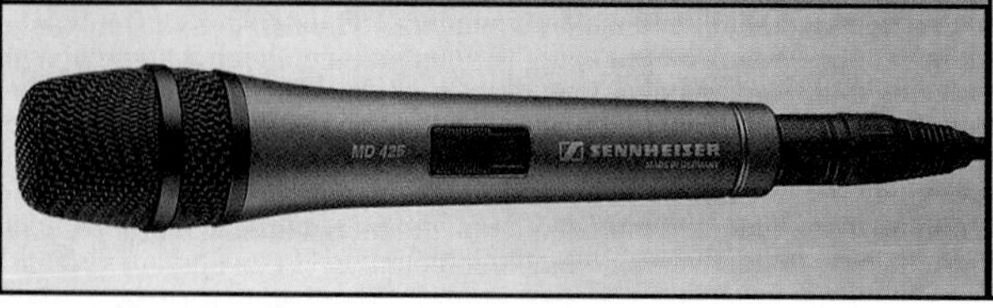

Physical Form

Physically, mikes are designed either to stand independently or to be worn by performers.

Independent mikes

These microphones are engineered to be held by the camera, by a boom or stand, or by the performer. Sometimes, independent mikes are suspended above the sound source, especially for recordings made in auditoriums.

User-worn mikes

Certain very small microphones are designed to be clipped onto the performer's clothing, **Figure 16-9**. Their advantage is that they do not have to be aimed or moved by a technician. Their disadvantage is that they are visible to the camera (unless specially concealed) and their cables or radio signals sometimes cause problems.

Transmission Method

Cable and radio transmission are the two methods by which microphones send their signals to the camcorder or stand-alone audio recorder.

Cabled mikes

Traditional microphones communicate via special cables. Because they are shielded against outside electrical interference, these cables provide the cleanest, most reliable signal. On the negative side, cables are difficult to conceal and to keep out of the way of technicians and wheeled equipment, and they require that the recorder be relatively close to the microphone.

generate is strong enough that it needs no amplification. The handheld mikes used in ENG (Electronic News Gathering) are often dynamic designs.

Dynamic mikes cannot always maintain a wide and even frequency response, and their sound is often described as "tight" or "less open" than that of the other common transducer type, the condenser.

Condenser transducers

A condenser transducer is a pair of electrically charged plates with air acting as an insulator between them. The distance between the two plates affects the voltage produced. One plate is rigidly fixed; the other is very sensitive to sound vibrations, which cause it to move toward and away from the fixed plate. The voltage fluctuations that result precisely imitate the sound waves that cause the flexible plate to vibrate.

Condenser microphones have wide, even frequency ranges and preserve high frequencies better than dynamic transducers. On the other hand, they are more delicate and the battery they usually require can fail in the middle of a recording task. Most condenser microphones need batteries in order to apply the electrical charge to the condenser plates and to drive a built-in preamplifier. The preamplifier is necessary because the signal strength of the output from the plates is quite low.

A condenser transducer.

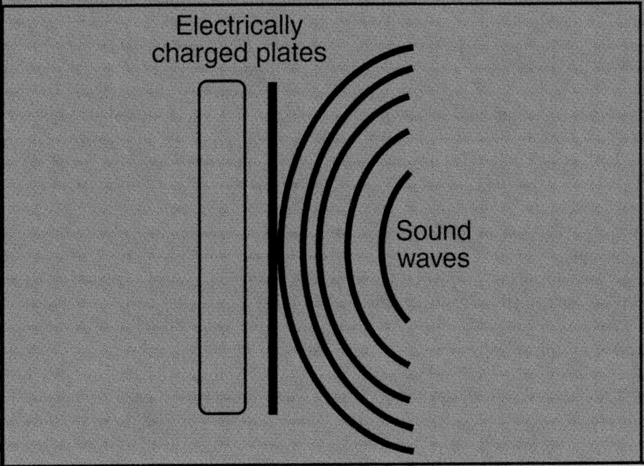

A stereo condenser microphone. (Sennheiser)

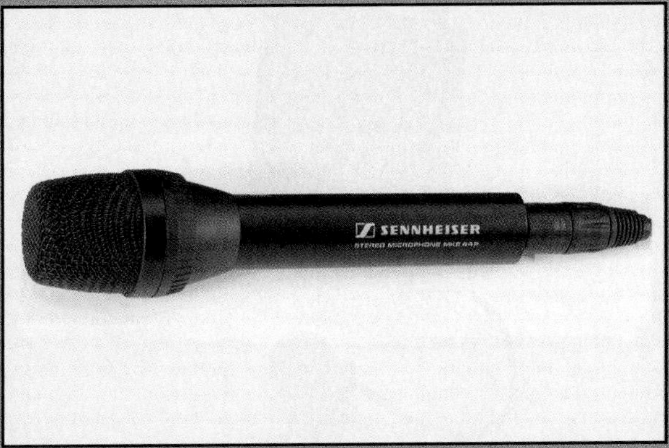

Figure 16-9.
A user-worn mike. (Sennheiser)

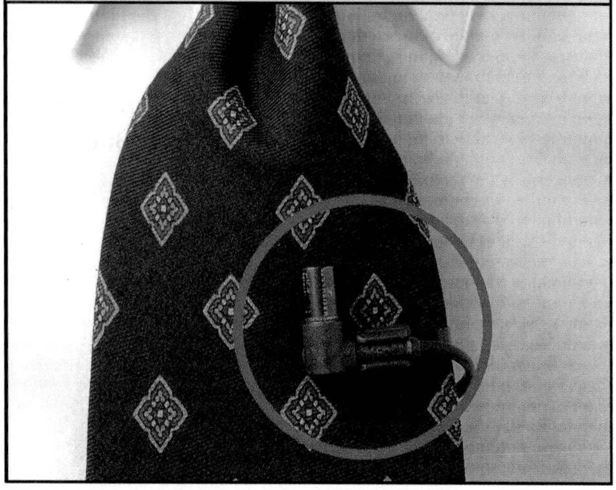

Wireless mikes

Wireless microphones, **Figure 16-10**, send their signals on specialized radio frequencies to receivers cabled to the audio recording system. Inexpensive wireless microphones are prone to interference from unwanted outside signals. High-quality professional types, however, capture sounds extremely accurately, transmit on frequencies free from most interference, and can be set to different frequencies. That way, several performers at once can transmit wireless signals to separate receivers (or channels on one receiver). These separate signals can be individually mixed and balanced for greater control.

Many wireless microphone/transmitter/receiver systems run on 120 volt power, for use in auditoriums and studios. Battery-powered units are available for use on location.

Figure 16-10.
A wireless microphone and transmitter. (Azden)

Level and Impedance

Level and ***impedance*** are electrical characteristics that affect the quality of recordings.

Level

Audio products from camcorders to VCRs to stereo receivers to computer sound cards have input and output jacks labeled "line" or "aux," "speaker," "headphones," "mike" or "mic," and sometimes other things. It is important to understand that each type of jack handles electrical signals of a particular strength, or "level."

In particular, line jacks carry high-level signals that average around one volt in strength. Microphone jacks, by contrast, carry low-level signals measured in millivolts. Microphones plugged into line level jacks will produce only feeble signals, so it is important to use only dedicated mike inputs.

Impedance

Impedance is resistance to the flow of electricity. Most inexpensive microphones are high-impedance types, while most professional units are low impedance. Low impedance translates into better signal transmission and better fidelity, especially in the high frequencies.

For best results, the microphones, cables, and equipment jacks should all be the same impedance. Cables fitted with XLR plugs almost always indicate a low-impedance system.

The impedance on these XLR camcorder jacks can be switched from high to low. (JVC)

To compensate for impedance differences in various system components, you can buy impedance-matching transformers.

Pickup Pattern

The most important characteristic of a microphone is its ***pickup pattern***—the directions in which it is most sensitive to sound. In order to select one or more microphones for particular recording tasks, you need to understand their pickup patterns. For example:

- A sound effect like footsteps needs a mike with a narrow pickup pattern, to exclude unwanted background sound.
- A musical performance needs a broad but directional pattern to capture all the music but exclude coughs and shuffles in the audience.
- Newscasters' lapel mikes need omnidirectional pickup patterns to capture their voices, no matter which way they turn their heads.

These are the most popular of the many available microphone pickup designs.

Narrow pickup

So-called "shotgun" mikes (named for their long, thin barrels) are highly directional, with most of the pickup areas directly in front. See **Figure 16-11**. However, the secondary pickup areas of some shotgun mikes will also capture sounds off to the sides; be sure to keep crew members at the camcorder quiet during takes. Shotgun mikes may be mounted on a camcorder, a boom, or a "fishpole." They are especially useful for distant sound sources, for recording in places with competing background noise, and for taping sound effects as cleanly as possible.

Broad, directional pickup

Most suspended, stand-mounted, and handheld performer mikes have a fairly wide but still directional pickup pattern. This compromise pattern is less narrowly directional than shotgun pickup, but also less inclusive than an omnidirectional pattern. Because their pickup pattern is shaped more or less like a valentine heart, these are called ***"cardioid"*** microphones, **Figure 16-12**.

Figure 16-11.
The sensitivity pattern of a shotgun microphone.

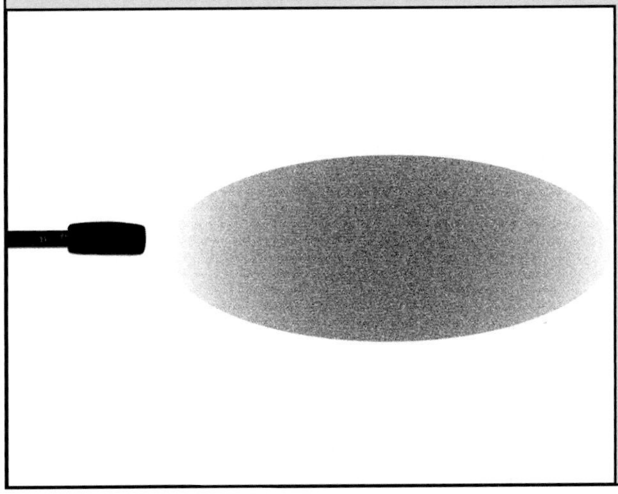

A shotgun microphone (Sennheiser)

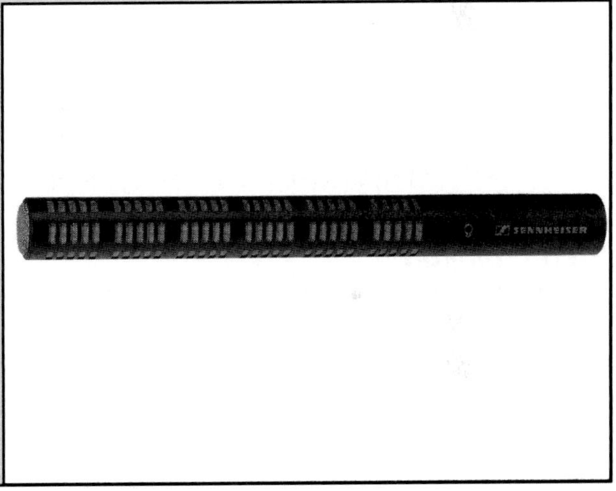

Figure 16-12.
The sensitivity pattern of a cardioid microphone.

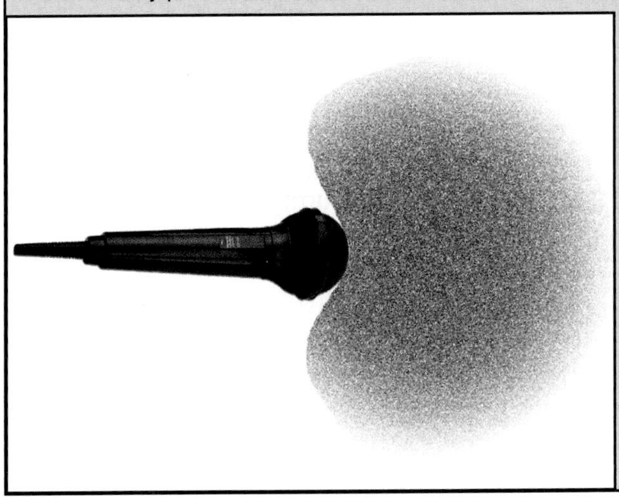

Cutaway view of a cardioid microphone. (Sennheiser)

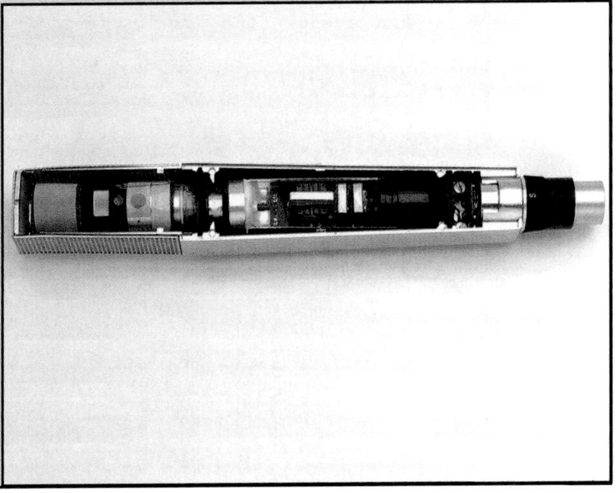

Microphone pickup patterns are actually much larger and more complex than the areas shown in Figures 16-11, 16-12, and 16-13.

Typically, a camcorder's built-in microphone is a compromise between a narrow shotgun and a broad cardioid pattern. Some camcorder mikes can be switched between wider and narrower pickup patterns.

Omnidirectional pickup

These microphones are sensitive in a spherical pattern stretching in all directions. The most common models, often called ***lavalieres*** or just "lavs," are worn directly on the body. In placing a lav on a shirt front, the goal is to position it high enough to get close to the performer's mouth but low enough to clear the "sound shadow" created by the chin. Because lavaliere mikes are balanced to compensate for their closeness to the wearer's chest, their relative insensitivity to lower sound frequencies makes them less suitable in some other applications.

In live television broadcasts, you will often see two lavalier mikes worn side-by-side. One is a backup mike in case the other fails in a situation where the performer must continue speaking uninterruptedly.

Stereo pickup

Stereo microphones have two separate pickup areas, each feeding its own audio channel. Stereo mikes can add presence and realism to almost any recording, and they are especially effective in studio and auditorium application, where the sound originates from a variety of directions. In studio situations, two mikes are often used, as shown in **Figure 16-13**. Camcorder stereo mikes use two pickups in a single housing.

Figure 16-13.
Stereo recording using two monophonic mikes.

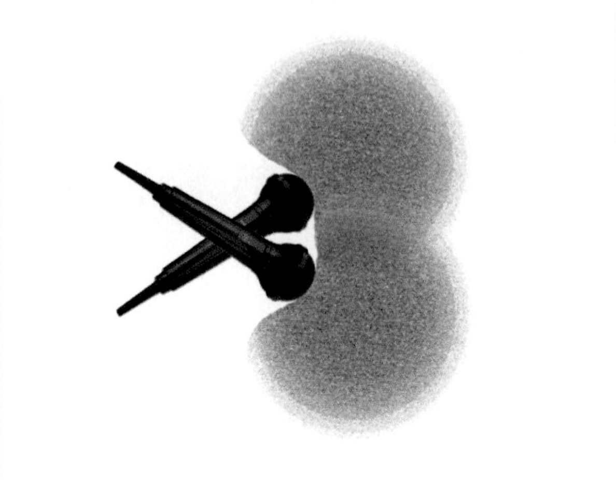

Other Recording Equipment

To complete our survey of audio recording equipment, we can follow the path of the audio signal downstream from the mike on its microphone *support,* along a *cable* (or via *transmitter/ receiver*), through a production *mixer,* to its destination at a *recorder.*

Microphone Supports

Microphones can be mounted anywhere and supported by anything that will hold a clip or gaffer tape.

The first task of any support system is to isolate the microphone from bumps and vibrations that are transmitted as unwanted noises. For this reason, boom, fishpole, and some stand mikes are typically suspended within frameworks by arrangements of shock cords. Microphones mounted directly on the performer cannot be shockproofed, and must be positioned so that they do not pick up the rustle of clothing.

Booms

Mike ***booms*** are typically used in studios, where their weight and bulk are not inconvenient. Booms are useful because the operators do not have to hold the weight of the microphone and its support. From one place, they can follow a moving performer by pivoting the boom arm, telescoping it in and out, and rotating the mike to point in the right direction.

Fishpoles

Fishpole booms must be held, moved, and aimed manually, but they are far lighter and more compact than booms. For this reason, fishpoles are preferred on many location shoots. See **Figure 16-16**.

Stands

Microphone stands are indispensable for recording narration, musical performance, and any other application in which the subject is off camera and remains in one place. Although

Using Wireless Microphones

Like all radio transmissions, the signals from wireless microphones are at the mercy of their environment. Electrical noise, other radio signals, and signals bounced around by the journey from transmitter to receiver—these and other problems can prevent you from obtaining quality audio recordings. Here are some tips for eliminating these problems.

- Always make the distance from transmitter to receiver as short as possible. If your equipment permits, place the receiver just outside the camera frame and then cable it back to the camera.
- Experiment with receiver position and receiver antenna. Radio mike signals are so sensitive to distance, direction, and reflection, that even small adjustments at the receiver end can improve quality dramatically.
- Have backup cabled mikes. Even the finest wireless mikes will not always work, so be ready to fall back on cabled lavalieres or a fishpole setup if necessary.

Finally, before recording audio in a location, scout it for problems, just as you would for video. Experiment with wireless miking to anticipate problems or, if all else fails, settle on an alternative solution using cabled mikes. Remember that production people may be tolerant of camera, lighting, and performance problems, but they (unfairly) grow impatient if the sound department holds up the shoot.

When the wireless receiver is placed near the camera, the distance to the transmitter may degrade quality. To improve quality, conceal the receiver as near the transmitter as possible and cable it to the sound recorder.

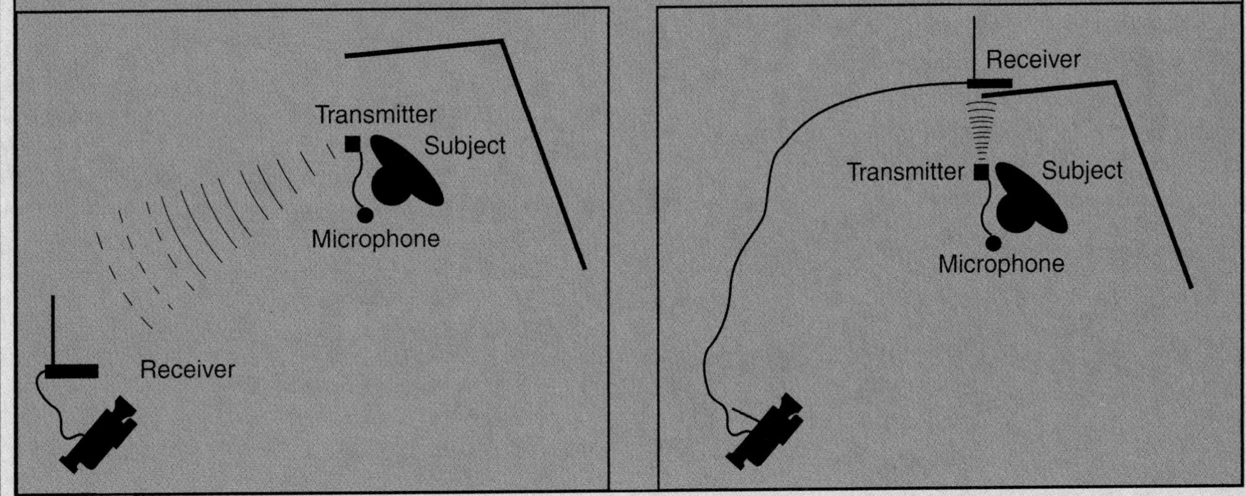

Figure 16-16.
A fishpole boom in use. (Gitzo/Bogen)

short table stands are often used for voice recording, low floor stands with short horizontal booms permit more flexible mike positioning, and they typically accept shock mounts.

Cables

Microphone cables contain wires that are both electrically insulated and shielded from stray noise and even audible signals that unshielded lines can pick up like radio antennas. For this reason, microphone connections should always be made with shielded cables manufactured expressly for that purpose. Cables should also be as short as possible, in order to minimize signal loss.

Though the plugs may match, never use a headphone extension cable for a microphone, as these lines are unshielded.

Line balance

Interference and signal loss are affected not only by the length of a microphone cable, but by its ***"line balance"*** as well. Cables with two electrical lines are unbalanced; those with three lines are balanced. See **Figure 16-15**.

Unbalanced cables are typically supplied with inexpensive microphones and fitted with phone- or mini-plug jacks. Though shielded from outside interference, unbalanced cables still suffer from this problem. They should be used for short runs only, usually less than 20 feet.

Balanced cables with three lines are quite secure from outside interference, and can be safely run for longer distances. Balanced and unbalanced lines are seldom identified, but you can usually tell them by their plugs. ***Balanced lines*** will have large XLR plugs, while ***unbalanced lines*** have phone plug connectors. A microphone fitted with an unbalanced line can be plugged to a balanced extension cable via a converter plug.

Figure 16-15.
Unbalanced and balanced audio cables.

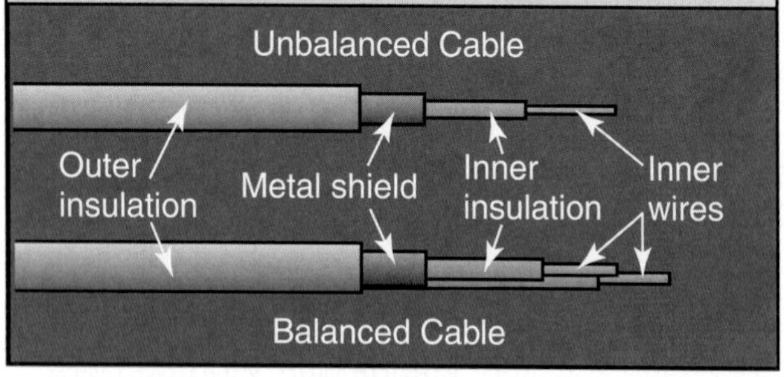

Connectors

Inexpensive microphone cables are typically fitted with phone plug connectors, either full-size or mini-plugs. Mono plugs have a single black band insulating the two connectors, while stereo plugs use a pair of insulators for their three connectors. See **Figure 16-16**.

Phone plugs, especially the mini-plug style, are unreliable because they easily lose their electrical connections and stop transmitting the audio signal. This is doubly dangerous if the recordist is not monitoring the input through headphones. When a plug from an external mike is inserted into the camcorder's "audio in" jack, it automatically disables

Figure 16-16.
Mono and stereo mini-plugs.

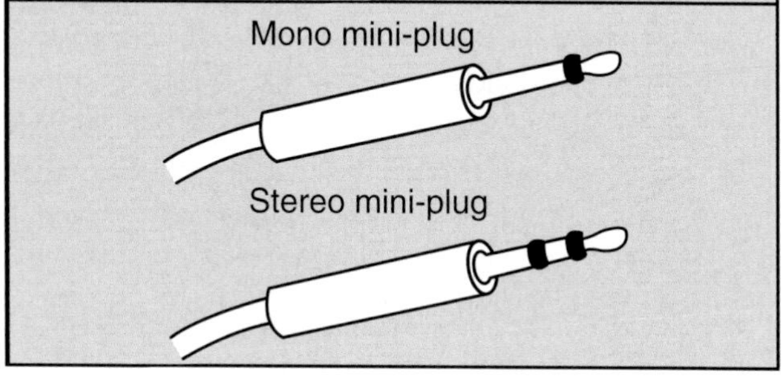

the unit's built-in mike. If the external plug becomes electrically disconnected but still remains in the camcorder jack, neither microphone will transmit an audio signal and the camcorder will capture no sound whatever.

Professional audio setups use XLR connectors that latch together for a reliable connection. You can obtain small, camera-mounted conversion units for some camcorders that allow you to use XLR plugs (with their balanced line cables) on camcorders fitted with mini-plug audio jacks.

The jacks for XLR plugs are labeled in Figure 16-17, and in the photo in the sidebar *Level and Impedance.*

Receivers

Wireless microphones incorporate radio transmitters that send the signals to matching receivers, **Figure 16-17**. The receivers, in turn, are cabled to the camcorder or separate audio recorder. Inexpensive transmitter/receiver systems use radio frequencies that can be subject to interference, while professional-grade models use much higher frequencies. Typically, high quality systems can transmit over longer distances with better signal quality and less interference.

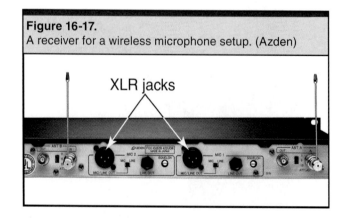

Figure 16-17.
A receiver for a wireless microphone setup. (Azden)

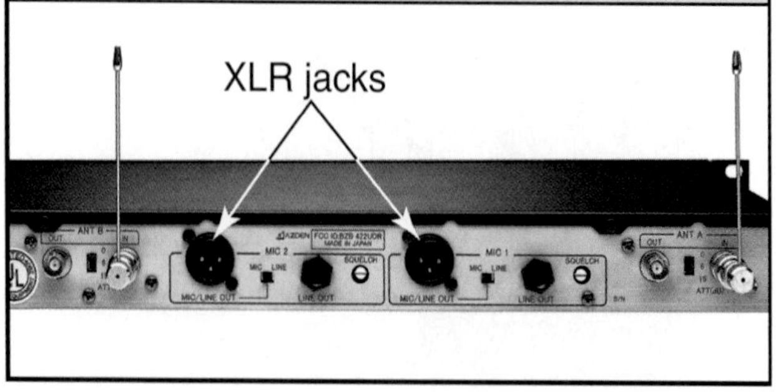

Some receivers can accept signals from more than one microphone, each on its own separate frequency. Some types feature switchable frequencies. If one channel is degraded by interference, a different one can be selected.

Mixers and Equalizers

Whether your microphones are cabled or wireless, their signals must be blended for recording as a combined audio track.

Mixer

To do this, you use a production ***mixer,*** **Figure 16-18,** which is basically a volume control for each incoming signal and circuitry for combining them. A production mixer is useful for just one microphone too, adjusting the signal to the optimal strength for transmission to the recorder. Almost all professional productions use a production mixer.

Equalizer

An ***equalizer*** improves overall sound quality by adjusting the relative strengths of selected sound frequencies. In production sound recording, it is most commonly used to minimize undesirable sounds like the rumble of distant traffic or the wheeze of an air duct. An equalizer should not be overused. For example, reducing *all* the sound frequencies that contribute to traffic noise can distort actors' voices so much that they sound like cartoon characters.

Recorders

Production sound is recorded by the audio system built into the camcorder. Backgrounds, sound effects, voice, and music recordings can also be made by other sound recorders such as VCRs, digital audio recorders, and analog audiocassette recorders.

In TV studios, the audio/video recorders are typically separate from the cameras. In small operations, however, camcorders may be used as floor cameras.

Camcorder audio recording

In most cases, a camcorder is the best choice, even when the video part of the recording will not be used in the program. Camcorder audio recording offers many advantages:

- Because it is in the same format as the program material, it can be edited or transferred to computer without connecting special-purpose source decks.
- It offers extremely high quality sound. Mini-DV audio can actually be better than CD music audio.
- Even if you will not use the picture recorded with the sound, it serves as a handy visual locator of audio clips, especially if you have identified them with written slates. See **Figure 16-19**.
- Because it is compact and battery-powered, a camcorder is easy to use in the field.

Figure 16-18.
A small production mixer. (Azden)

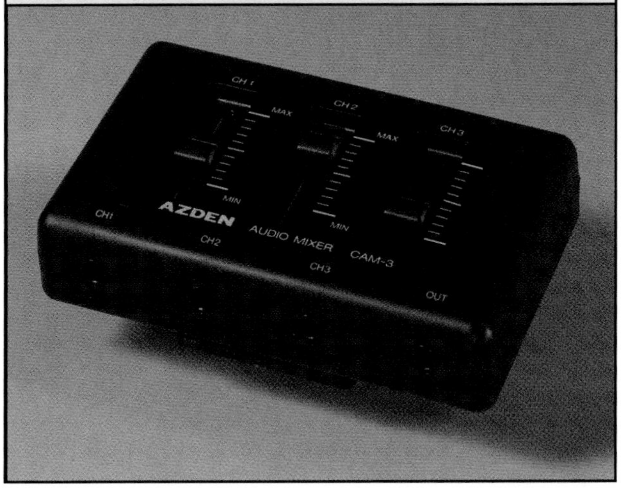

Figure 16-19.
During field recording, letter and record simple slates.

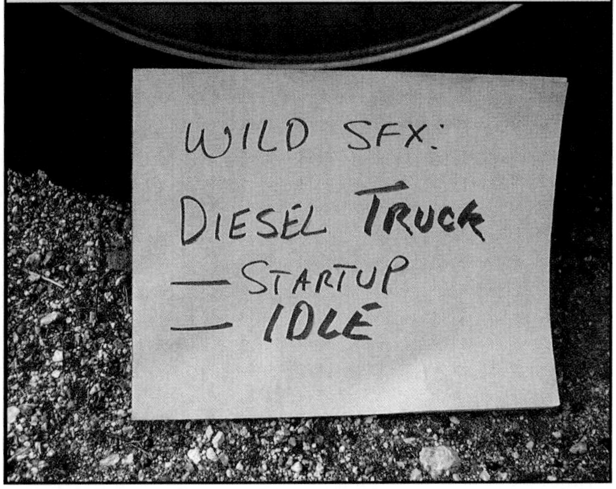

For many field recording applications, then, a stereo-quality camcorder will do the job as well as a special-purpose recorder.

Computer sound card

For studio audio recording, miking directly to the sound card in the editing computer offers all the advantages of a VCR, because the separate operation of inputting the audio from a source is eliminated. Computer audio cards are typically sold with sophisticated software for manual level control, equalization, and special effects like reverberation. Finally, the wave form display of most audio software allows you to verify that recording levels are optimal for the system.

Audio recorder

A good digital audio recorder is useful if you do not want to use a camcorder. For simple effects such as automobile sounds, explosions, etc., you can also use an analog device like the audiocassette recorders found on portable radios.

Accessories

A professional sound recordist will carry a broad range of specialized accessories. Of these, the following are perhaps the most widely used.

Headphones

Headphones are essential for quality sound recording. Like the viewfinder for video, headphones let you preview what you are recording to check the quality (or simply to verify that sound is being properly recorded). In choosing headphones, select the full-ear type that blocks out extraneous sounds, **Figure 16-20**. Ideally, you want to hear nothing but the sound picked up by the mike.

Windscreens

Wind noise is a common problem in exterior recording. A foam windscreen (sometimes called a sock or a doughnut) reduces wind noise by deflecting the air rushing past the microphone, **Figure 16-21**. A windscreen is also useful indoors

Automatic Level Control

All consumer camcorders and VCR decks are fitted with automatic level controls: circuits that adjust volume automatically to compensate for different levels of incoming sound. In general, automatic level control works well. Problems arise, however, when a distinct sound is preceded by a long silence. If there is a near-silent period before a distinct noise, the automatic level control may turn the record volume up, in search of a signal. When a normal-level sound begins, the circuits cannot turn down the record level fast enough to prevent the start of that sound from distorting.

Automatic level controls can also fail when the sound level rises or falls suddenly and dramatically. Here again, the circuits cannot anticipate the change and react fast enough to accommodate it. The solution is manual level control, a feature built into virtually all professional-level camcorders and decks. With manual control, the sound recordist can maintain a consistent level through quiet parts and anticipate sudden volume shifts by adjusting gain just before they happen.

Automatic level control can fail to keep up with sudden volume changes.

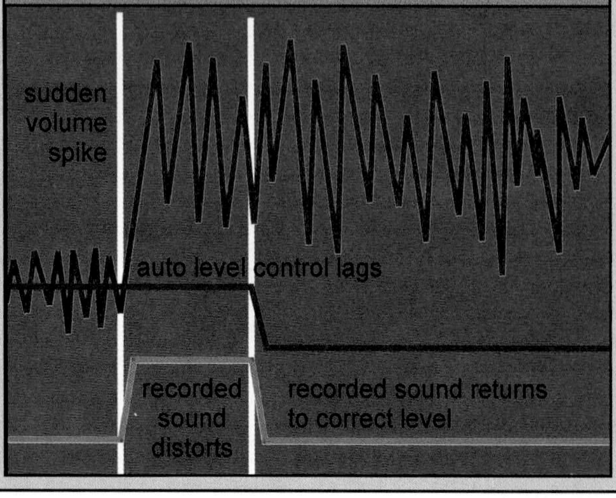

Defeating automatic level control

When you cannot disable the automatic level control, you can sometimes work around it. If a loud noise occurs in the middle of a shot, use an external mike, handheld or on a fishpole. That way, you can turn the mike away from the source of the loud sound, just before it happens. If you can, use a production mixer. By optimizing sound levels manually, before they reach the auto circuits, you can reduce unwanted variations. When recording voice for narration, follow each long pause with a voice slate. The slate, which will be removed in editing, gives the automatic system a sound on which to set a level, so that the narration is all recorded consistently.

Figure 16-20.
Full-ear headphones. (Sennheiser)

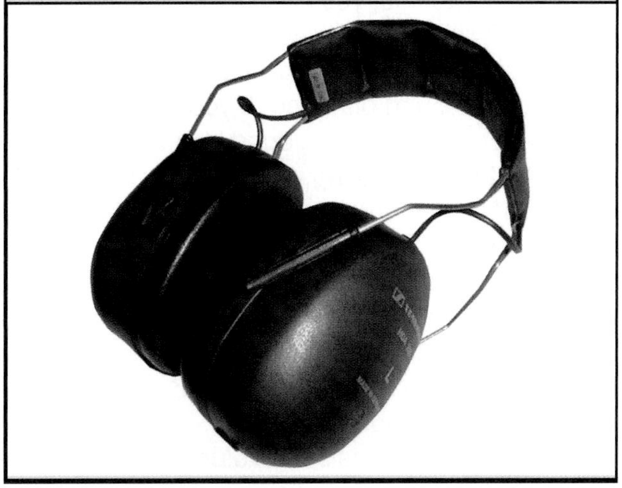

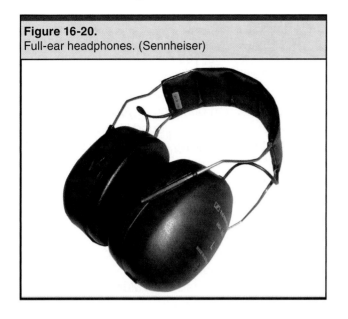

for narration or dialogue recording, where performers' breath can create wind noise across an unprotected mike.

The built-in windscreens on (or in) the small mikes of miniature camcorders may not be visible. Some camcorders filter wind noise electronically.

Furniture pads

To prevent unwanted echoes and boomy-sounding audio, drape blankets at strategic points or deploy the furniture pads used by moving companies. You can obtain these pads inexpensively at do-it-yourself moving rental companies.

Plug and cable adapters

If you put together your own audio equipment, everything will be cabled and plugged compatibly. But if you must connect to other equipment, such as a public address system or an auditorium sound setup, you will need a variety of plug adapters.

Slate

When recording audio with a camcorder, use a simple tablet and marker to create identifying ***slates***. Visible slates eliminate two major annoyances in audio editing. Unlike pictures, sounds cannot usually be scanned in fast-forward or reverse, so finding material can be time-consuming. Even at normal speed, some quite accurate, realistic sounds can be difficult to identify without visual clues. By visually identifying sound effects, slates will save a great deal of postproduction time and effort.

A complete slate should include the information shown in Figure 16-19. For simplicity, you can slate the production and date once, at the start of the session, then slate just the particular effect, recording the other information verbally.

Figure 16-21.
An inexpensive shotgun mike in a foam windscreen. (Azden)

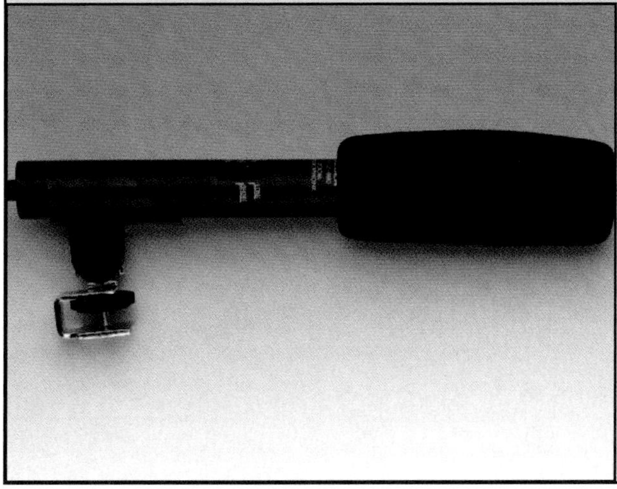

A foam windscreen can be added to most handheld mikes. (Sennheiser)

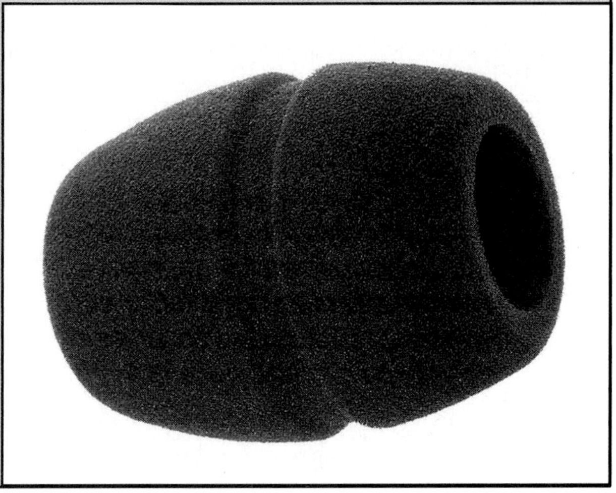

Audio Recording Techniques

Production, background, effects, voice, and music tracks all require their own recording techniques.

Production Tracks

The audio recorded along with the video is the most important sound track. A good production track is distinguished by its quality, consistency, and freedom from unwanted noise.

Quality

The keys to good audio quality are distance and direction. Place the mike as close to the subject as you can, without getting it into the picture. Then, aim the mike directly at the sound source. If two people are talking, point the mike at each one when he or she is speaking. Mikes worn by the subject(s) take care of distance and direction automatically.

If you are using a fishpole boom, begin by deciding whether to place the microphone above or below the frame line. Then orient the mike on the fishpole as shown in **Figure 16-22**. That way, you can swing it back and forth between speakers by simply twisting the pole.

If you need to use the camcorder's built-in mike, place the camera as close to the subjects as practical.

Remember that the wide angle lens settings required at very short distances can distort subjects' features unpleasantly.

Consistency

The next task is to keep all the audio sounding alike. To do this, you need to confine the volume level within a relatively narrow range, so that soft passages are well-recorded and sudden loud parts do not sound distorted. If you have manual volume control, maintain sound levels by rehearsing the action and noting where softer or louder sounds occur. During the shooting, anticipate the volume changes by adjusting the sound level just a fraction of a second before they happen. If you have only automatic level control, monitor the sound during rehearsals and restage the action (and/or move the microphone) if possible, to minimize excessive volume changes.

In addition to consistent volume, you want to maintain the same audio character, so that the shots recorded from different setups sound alike. If you are using the camcorder's built-in mike, try to keep the camera roughly the same distance from the subjects in every shot, changing image size by zooming instead. With a fishpole mike, experiment with different angles and positions while listening to the effect.

For this reason, it is important for the boom operator as well as the recordist to wear headphones.

Freedom from noise

Outdoor production recording is a constant struggle to keep wind noise and background sounds out of the dialog track. Wind noise can often be tamed by a microphone screen. Background noise is more difficult. If possible, move the subjects so that noisy traffic or pedestrians

Figure 16-22.
Mike boom positioned beside the camera (boom operator omitted for clarity).

By twisting the boom, the operator can aim the mike at each subject in turn.

Figure 16-23.
Avoid placing the microphone where it will pick up reflected background sounds.

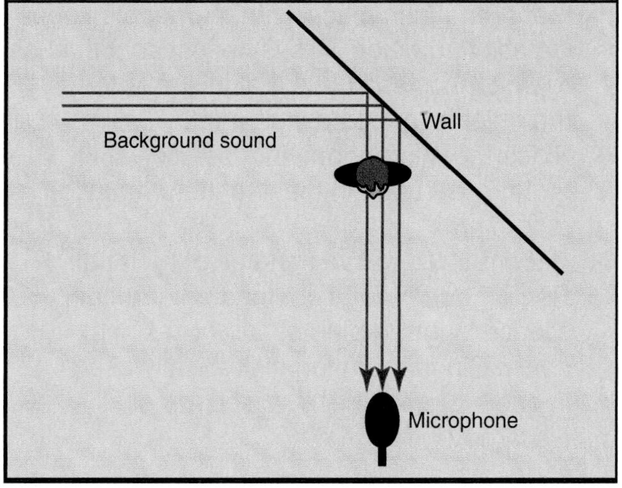

or ocean waves are not behind them. This avoids aiming the microphone directly at the background noise. Watch out for reflective surfaces like masonry walls, which can bounce noises originating behind the camera back into the microphone, **Figure 16-23**. When practical, place the mike above the subjects, pointed downward. This will reduce its sensitivity to unwanted background sounds.

Background Tracks

The key to good background tracks is consistency—a uniform sound unmarred by distinctive noises that call attention to themselves. For example, background traffic sounds should not include horns, brakes screeching, or the

Recording Stage Performances

Recording plays, musicals, and concerts in auditoriums can be complicated by several factors. Audio quality is degraded by the distance between performers and your microphone. The width of the stage offers a sound field too broad for the pickup pattern of your mike. Audience noise is difficult to keep off the production track.

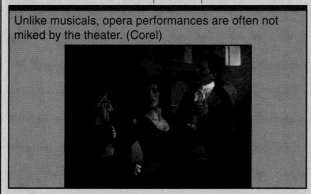

Unlike musicals, opera performances are often not miked by the theater. (Corel)

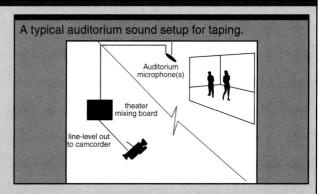

A typical auditorium sound setup for taping.

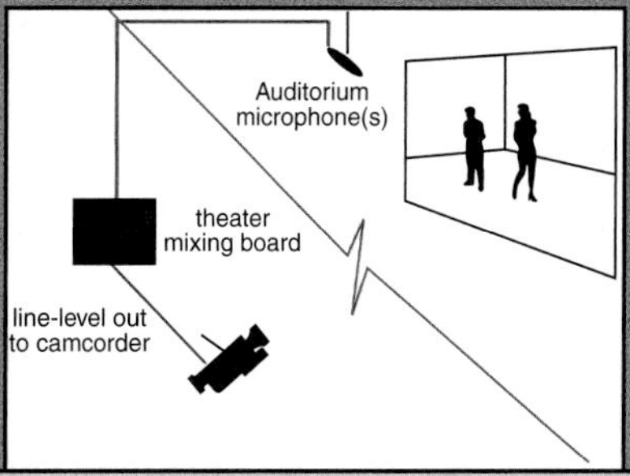

Tapping the source

If the performance is already miked by the auditorium's own sound system, the best solution may be to plug directly into it. Not only does this take advantage of microphones that are well placed for sound recording, it also eliminates sound from the auditorium loudspeakers.

If you do plug into the auditorium system, it will probably be to a line-level circuit, far too strong for the mike input on your camcorder. To match levels, you need to place a line-to-mike level converter in the signal path, or else cable the auditorium sound to the line level input on a VCR, as shown in the diagram.

Following the action

For a play or other performance where the sound sources move, a camcorder-mounted shotgun mike may offer the best reproduction. Assuming that its camcorder is constantly re-aimed to capture the most important action, the mike will be automatically reoriented with it.

Riding gain

Automatic gain control can be difficult to handle with the widely varying sound levels encountered in plays. During quiet moments, the volume circuits will increase the recording level to the point where audience shuffles and incidental stage noises may grow unnaturally loud. If you have a VCR with manual recording capability, feed picture and sound from the camcorder to a separate videotape recording in the deck. If your VCR lacks manual gain control, you can use an audio mixer between microphone and deck to smooth out extreme swings in level. For protection, you can record a second tape in the camcorder as well.

Recording on a VCR.

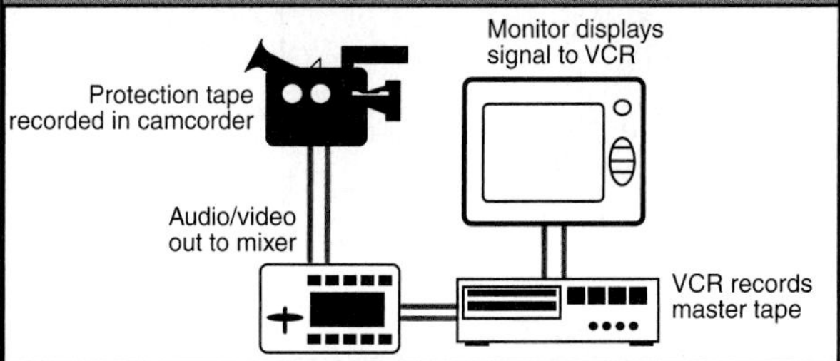

footsteps of occasional pedestrians because audiences will pick them out of the background and think they are part of the program. Background tracks fall into two types: background noise and room tone.

Background noise

Traffic or restaurant chatter or ocean surf is fairly easy to record: simply position the mike where the noises sound realistic and record about five minutes worth of material. You want a long recording so that it can be laid under an entire sequence without obvious repetitions or edits.

It is possible to edit out unwanted parts if the background is fairly uniform. See Figure 16-24. Digital sound editing allows cross-fades between different pieces of background track.

Remember to record at a good strong volume. It is easy to decrease sound levels in editing but difficult to increase them without obvious quality loss.

Room tone

As explained previously, room tone is the characteristic sound quality of an environment that is audible (though usually unnoticed) even when there are no overt noises. It is only when an editor cuts pieces out of a production track that the resulting dead silence is obvious. By replacing the dead spots with pieces of room tone, the editor preserves the ambient sound of the environment. Here is the procedure for recording room tone:

- Have the actors, crew members, and equipment remain in place, with the people keeping still and quiet. This is necessary because room tone is so subtle that it is affected by the objects in the environment.
- Place the microphone in a position similar to those used in making the production recordings.
- Record three to five minutes of material. Remember: it does not matter if the near silence is interrupted by an occasional noise. Since room tone is typically used in short pieces, these noises are easily discarded.

Keep in mind that audio tracks can be reused as required. For this reason, it is not necessary to record every minute of background sound that you might possibly need.

Figure 16-24.
A factory will often provide fairly uniform background noise. (Western Recreational Vehicles)

Sound Effects

The keys to good sound effects are isolation and realism. To isolate effects from background noises, position the mike as close as possible to the sound source. Use a highly directional shotgun mike if possible. See **Figure 16-25**. If you record wild effects on videotape, voice-slate each take so that the effect is identified even when the camera is too close to the sound source to record a recognizable visual.

To obtain realistic sound effects, there is no substitute for listening carefully through good headphones and changing mike angle and distance until you are satisfied with the sound. As always, record at a good, strong volume. If the script calls for, say, a car driving off outside in the distance, the editor can reduce the volume and adjust the sound quality in postproduction, as needed.

Foley Effects

The same rules apply to the recording of Foley effects, which are synchronized to video playback. Prepare for recording by making a duplicate copy of the footage that needs sound effects so that you can record them on the copy. Later, it will be synchronized with the master video.

Figure 16-25.
Because of the distance required, recording animal sound effects is especially difficult. (Corel)

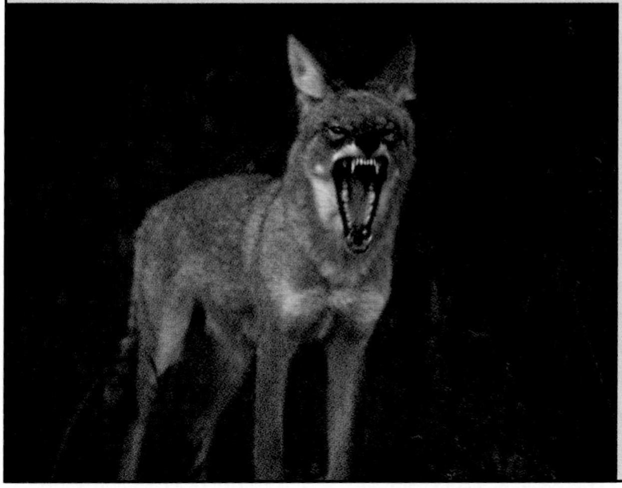

Highly sensitive directional mikes are used for capturing nature sound effects. (Sennheiser)

Voice Recording

As in all recording, the goal in taping narration or other voiceover material is a clean, quality track. The keys to success are studio acoustics, microphone selection, and mike placement.

Studio acoustics

Voice tracks should be completely free of reverberation and background noise. If you are not recording in a sound studio, select an area with carpet, draperies, upholstered furniture, and other sound-absorbing materials. Using good quality headphones, listen to the "room tone" of the selected recording area for background noises and sounds like ventilation system hum. If necessary, change the mike orientation, sound absorption material, or even the location itself, to eliminate unwanted background sound.

Microphone selection

Use a cardioid mike, if possible, because its pickup pattern allows for some movement on the part of the performer.

Shotgun mikes with narrow pickup patterns are usually unsuitable for the close distances typical of studio mike placement.

Even inexpensive units will perform adequately; but a better-grade model adds a professional-sounding crispness that makes voice-over speech stand out.

Mike placement

Poorly-placed studio mikes create two problems. First, they cause *plosive* sounds like "puh," (as in "*p*oorly *p*laced") to overload the mike and "pop" on the sound track. Second, their stands can interfere with the scripts that are usually used in voice recording. To correct both problems at once, place the microphone at a slight angle to the performer. Positioning a table stand to one side makes room for a script. Placing a boom mike overhead is even better.

If you are recording two or more performers at the same time, it is preferable to provide two mikes, separated by as much space as the location allows, **Figure 16-26**. That way, neither mike will pick up the sounds of the other mike's performer. Use a production mixer to balance the levels between the two mikes.

Figure 16-26.
Each microphone picks up only one performer.

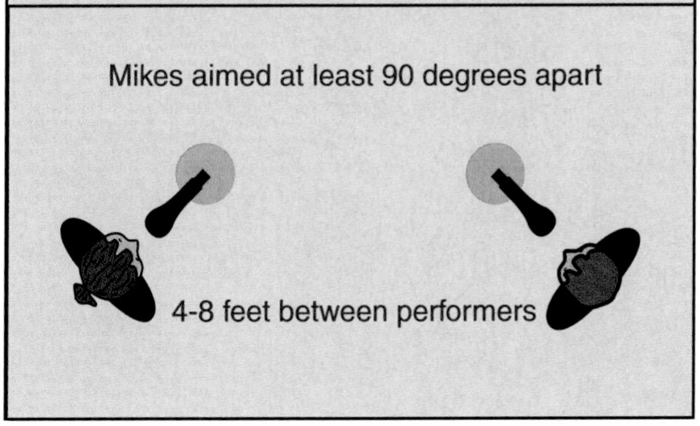

Managing Studio Voice Recording

In addition to placing the microphone and controlling record levels, you need to minimize script noises and identify parts of the recording for later editing.

Script rustle

To eliminate the noise of paper shuffling during narration recording, cover the table or music stand supporting the script with a scrap of carpeting. Place a blanket or rug on the floor beside the performer. Remove any staples or paper clips so that the script pages are loose. To use this setup, the narrator feathers the script pages as they rest on the carpet scrap. As each page is finished, it is picked up and dropped onto the carpet beside the reader.

Handling script pages silently.

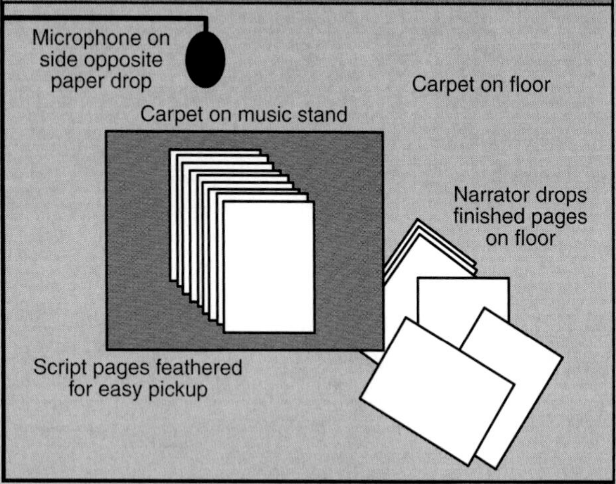

Slating voice recordings

Normally, a narrator reads through a script in sequence, until making a mistake. When this happens, have the performer repeat the line immediately and then resume reading the next line. To document this repeat for the editor, have the narrator voice slate the retake (for example, "Line 21, take 2"). Place a vertical line (I) in the margin of your own script to stand for the first take. Add a line for each additional take (II, III, etc.). If, say, line 21 has three lines in the margin beside it, then the editor knows that it took three takes to get the line right and captures only the last one. Or, if you should decide that the second of the three takes was better, circling the second line will tell this to the editor.

Dialogue Replacement

Dialogue replacement, **Figure 16-27**, is just a special form of voice recording, and all the suggestions presented earlier apply to this task as well. The major difference, of course, is the monitor, which is placed so that the performers can synchronize their dialogue with the lip movements on screen. Position the monitor so that the actors can easily look back and forth between their scripts and the screen.

Figure 16-27.
Scenes in noisy environments often require dialogue replacement.

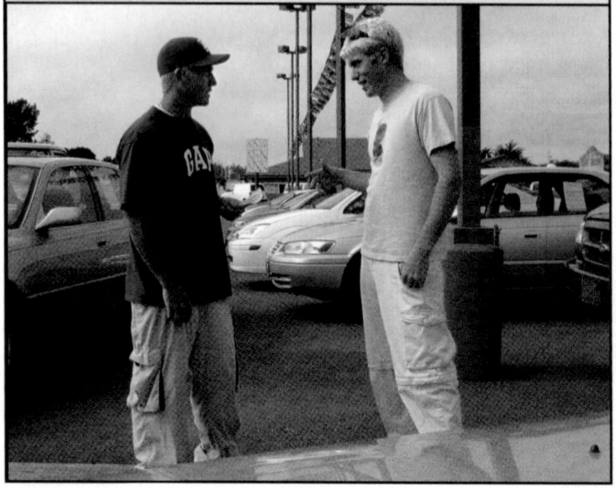

Here is a simple procedure for replacing dialogue. Edit the original footage, including the poor quality dialogue; then make a duplicate tape of the completed edit. Record replacement dialogue on the duplicate, using the audio dub function. That way, the new audio track will not interfere with the video track being used for reference. Feed the original stereo dialogue to the actors through headphones, so that it will not be picked up by their microphones. When the dialogue has been rerecorded satisfactorily, sync the new dialogue track with the original footage.

You may find it easier to follow this procedure even if you are doing digital editing. It is sometimes easier to export an edited sequence to tape, do your looping in linear mode, and then digitize the completed track. Using a timeline, you can easily sync the new dialogue to the edited digital picture.

Music Recording

Professional music recording typically involves many microphones, each balanced and

Figure 16-28.
In studio music recording, instruments are often miked individually. (Sennheiser).

Figure 16-29.
Place the mike far enough away to pick up the whole group.

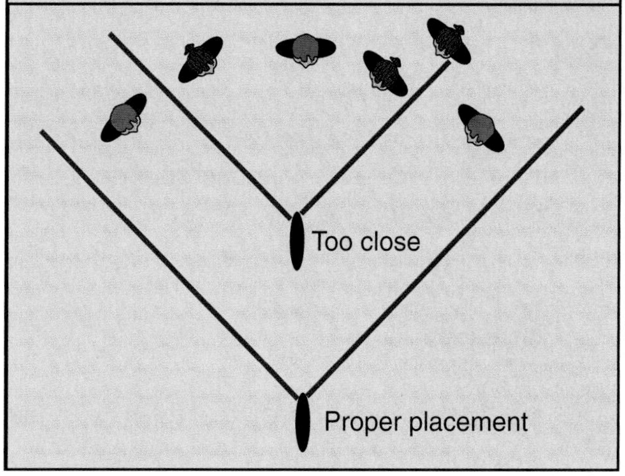

equalized by a large studio mixer and feeding a separate recording track. See **Figure 16-28**.

Production music recording is a simpler process. Taping a rock group, string quartet, or other small ensemble, you can obtain excellent results with a single well-placed microphone. As always, the keys to success are microphone selection and placement. For music recording, use a good quality stereo microphone of condenser design. Place the mike as close as possible to the group, while still keeping the performers inside the primary pickup pattern, **Figure 16-29**.

Camcorders with built-in stereo mikes are engineered to achieve separation automatically.

Without multiple microphones and recording channels, you can still equalize volume levels by moving the performers around so that the quieter instruments are nearest the mike and the louder ones are farthest away. Listen to the signal during rehearsals and experiment until you achieve a good balance.

The Rewards of Audio Recording

If your primary interest is in the video side of production, you may begin with the idea that audio recording is simply a technical process that has to be learned in order to provide the professional sound track that you need. But as you get into the creative aspects of sound, you may well find that audio is a complex craft that can be rewarding in its own right.

The optically recorded film audio from which we get the term "sound*track*".

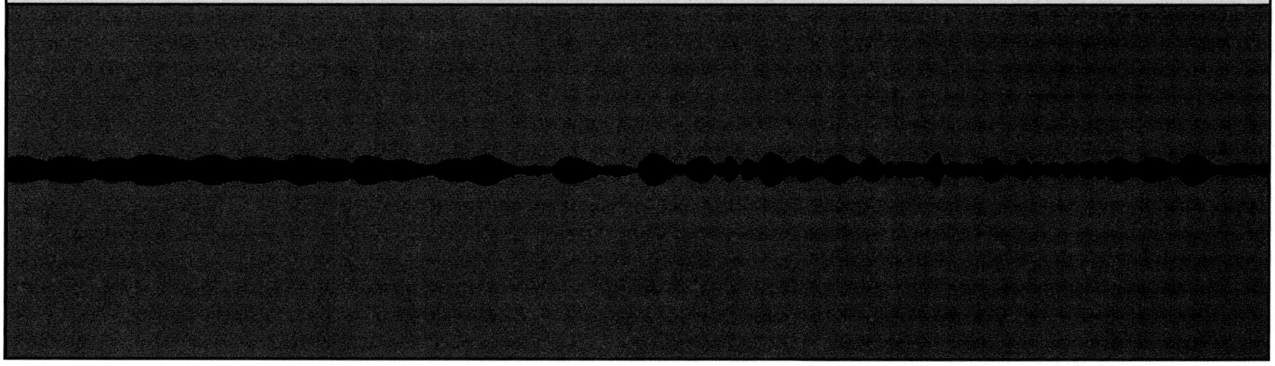

Chapter Review

Answer the following questions on a separate piece of paper. Do not write in this book.

1. Distortion is one of _____ problems in recording audio.
2. *True or False?* Professional productions use separate mikes for recording audio, rather than the camcorder's built-in microphone.
3. To fill gaps in edited audio tracks, a _____ or "presence" track is used.
4. Sound effects are created and synchronized with video in a _____ studio.
   A. Fogarty
   B. Farley
   C. Foley
5. Lavaliere microphones have a(n) _____ pickup pattern.

Technical Terms

Audio: Sound as electronically recorded and reproduced.

Balanced line: A three-wire microphone cable designed to minimize electrical interference.

Boom: A studio microphone support consisting of a rolling pedestal and a horizontal arm.

Cardioid: A spatial pattern of microphone sensitivity, named for its resemblance to a Valentine heart.

Equalizer: A device for adjusting the relative strengths of different audio frequencies.

Fishpole: A location microphone support consisting of a handheld telescoping arm.

Foley studio: An area set up for recording real-time sound effects synchronously with video playback.

Lavaliere: A very small microphone clipped to the subject's clothing, close to the mouth.

Looping: Replacing dialogue in real time by recording it synchronously with video playback.

Microphone (mike): A device that converts sound waves into electrical modulations, for recording.

Mixer: A device that balances the input strengths of signals from two or more sources, especially microphones.

Pickup pattern: The directions (in three dimensions) in which a microphone is most sensitive to sounds.

Playback: (1) Previously recorded video and/or audio reproduced so that actors or technicians can add to or replace parts of it synchronously, in real time. (2) Studio-quality music recording reproduced so that performers can synchronize lip movements with it while videotaping.

Slate: A written video and/or spoken audio identification of a component such as a shot, a line of narration, or a sound effect.

Transducer: The component of a microphone that converts changing air pressure ("sound") into an electrical signal ("audio").

Unbalanced line: A two-wire microphone cable subject to electrical interference, but less bulky and expensive, for use in amateur applications.

17 Directing for Content

Objectives

After studying this chapter, you will be able to:

- Deliver information on video completely and effectively.
- Communicate emotional effects to your audience.
- Guide on-screen talent in shaping their performances.
- Solve production problems arising from performance.
- Discharge typical directing assignments.

About Directing for Content

When you direct a video, your job is to determine exactly what is shot and how the result looks and sounds. That makes you the final decision-maker—the person in charge. Other people such as the writer, the videographer, and the editor make decisions in their own areas of responsibility. In addition, they often make recommendations that a wise director will consider with seriousness and respect. But even though production is a collaboration among many people, the final authority generally rests with the director.

Video directing requires the ability to juggle several different tasks at once, including:

- *Communication:* presenting the program content clearly to the audience.
- *Performance:* ensuring that the people in the video are effective on screen.
- *Editing:* providing the raw materials from which the editor can construct a smoothly assembled program.
- *Camera:* choreographing the way in which the action is recorded.

This chapter is devoted to *communication* and *performance*. Chapter 18 covers *editing* and *camera*.

Since the realities of production demand constant adjustments and improvisations while shooting, the director must evaluate each new idea to see how it might affect these four tasks. In addition, a good director also imparts an overall style to the material, giving it a distinctive "look and feel." It has been said that a good director combines the talents of a painter, a dramatist, a

Taking Charge

Even in professional productions, the powers and responsibilities of the director are not always automatically clear, especially when the director must share authority with the producer, the client, the financial backers and sometimes others. In order to be in charge, you must actively *take charge* by defining responsibilities, assuming the authority to discharge them, and ensuring that you have the power to enforce your decisions.

Where the director rules

Though directors often share overall responsibilities with the producer and others, three major areas should belong to the director alone:

- *Shaping the performances* by blocking the action, guiding the actors, and pacing the scenes.
- *Choosing the setups* by determining where to place the camcorder, where and when to move it, and how to select and change lens focal lengths.
- *Designing the coverage* by choosing how much action to include in each shot, how much overlap to allow between shots, and how many ways to re-cover the action with alternate setups.

You cannot accept the responsibility for directing a production if you lack the final authority over these areas.

Advice and Consent

In three other areas, the director should have the right to "advise and consent" (like the U.S. Senate approving judges). These areas are:

- *Videography and lighting:* the aesthetic and technical quality of the video images.
- *Art direction*: the overall design of sets, locations, costumes, etc.
- *Script*: the structure and language of the script that provides the blueprint for the production.

Advice and consent means the right to give input when these elements are being designed and veto power over the results. In other words, you can explain what you would like to achieve in these areas and you can express dissatisfaction when the results do not satisfy you, but you should not dictate the ways in which those results are to be achieved. That responsibility and authority should belong to the videographer, the art director, and the writer.

Real-World Compromise

In practice, every real-world professional relationship is different. In many cases, the director trusts the videographer to spot the camera setups as well as light the shots. In general, the more you can trust your co-creators, the more exclusively you can focus on the core tasks of directing.

diplomat, a psychiatrist, an accountant, and an air traffic controller. It is no wonder that it takes time and practice to master all the skills required for successful video directing.

Directing for Communication

Communicating with the audience means delivering information: selecting, organizing, shaping, and presenting it in a way that helps viewers absorb and respond to it. It is not enough to simply display information and expect the audience to pay attention to it. In directing, you must determine the way in which information is delivered by controlling *emphasis* and *effect*. Before you can shape information, however, you must see to it that the information actually gets into the program to begin with. That task is not as simple and obvious as it seems.

Information

Perhaps the most common error that inexperienced directors make is to omit important information. For example, **Figure 17-1** shows how our burglary might be staged by an inexperienced director.

Where did the burglar find a key to the cabinet? The inexperienced director has not shown us. Though this omission is not critically important, it will bother the audience. To satisfy their

A Demonstration Sequence

In order to contrast different solutions to directing problems, we need to apply all of them to the same situation. Throughout the following discussion, we will draw most of our examples from the same source: a sequence about a burglar and her hunt for a key to open a display cabinet.

Why not just break the glass in the cabinet doors? Let us assume that the noise of the shattering glass would arouse the sleeping household; or perhaps the burglar is too nervous to think of this obvious solution.

A burglar ... a locked cabinet full of valuables... a key to the locked cabinet.

Figure 17-1.
The burglar discovers that the cabinet is locked... and opens it with a key.

Figure 17-2.
The added shot explains the key.

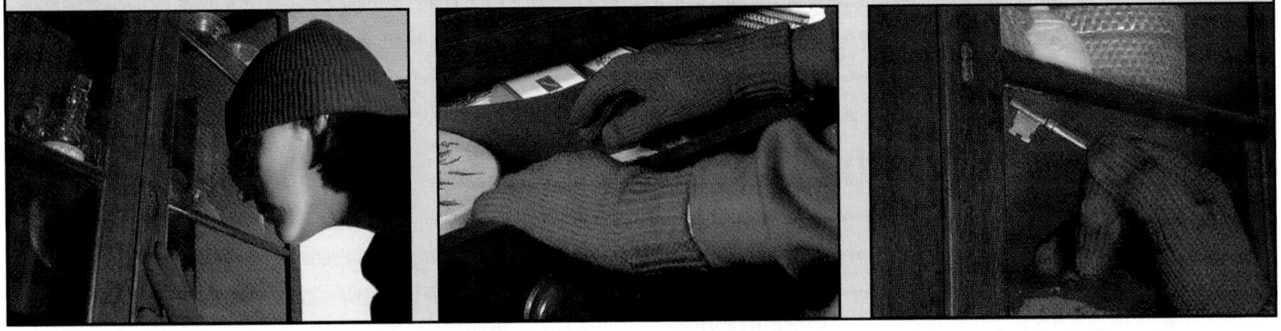

curiosity, we need to include the information about the key, as you can see in **Figure 17-2**.

In the added shot, we see the burglar open the top desk drawer and remove a key from it. Now the missing piece of information has been included in the scene. The director's very first job, then, is to identify the information that the audience needs to know and ensure that all of it is included.

Because the illustrations in this example are similar, it is important to study them carefully for small differences.

Emphasis

The added shot in Figure 17-2 is just another camera angle, with nothing special about it. Although we do see the discovery of the key, it is

Figure 17-3.

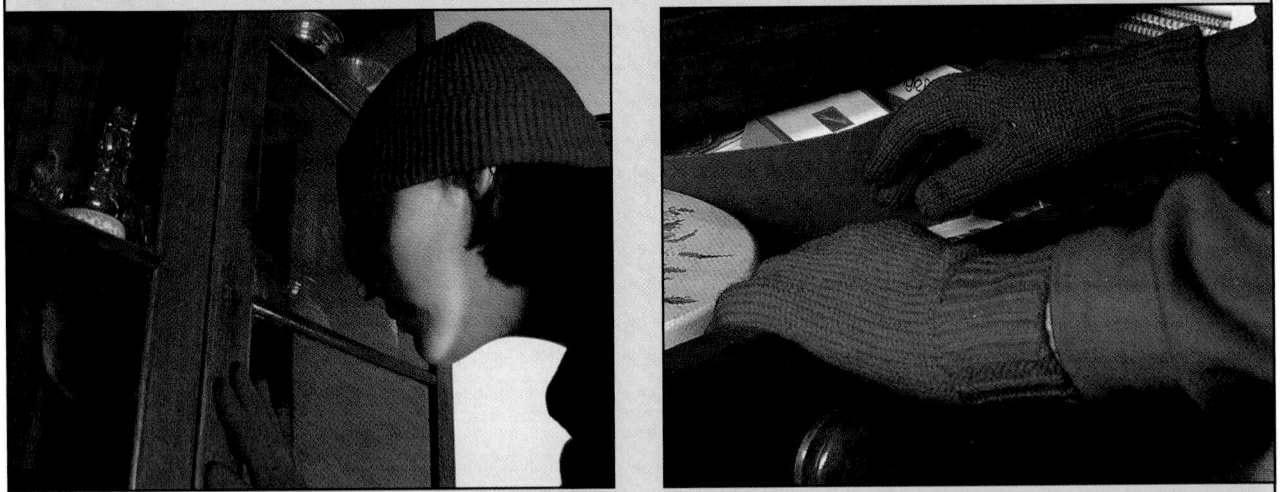

The insert emphasizes the key.

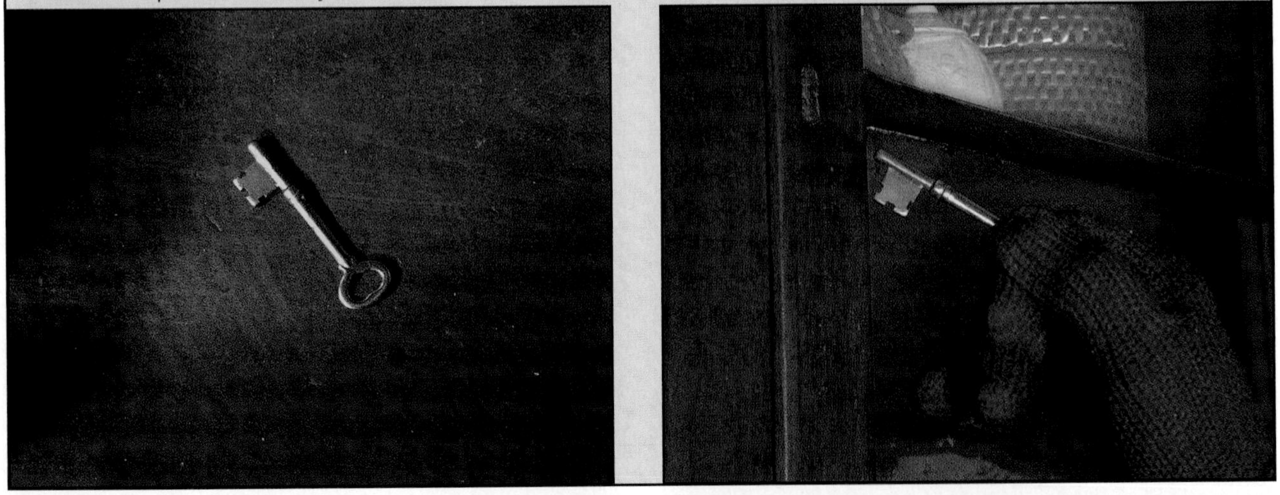

so unemphatically presented that we might very well miss it. To make sure that the audience gets the information, the director's second task is to emphasize it appropriately, as you can see from **Figure 17-3**. In this version, the audience sees the drawer being opened by the burglar and then the key inside, as her hand enters the shot and removes it. By giving the key its own closeup, the director adds emphasis to the information so that the audience will not overlook it.

In adding emphasis to information, you have several tools, including *image size, composition, camera angle*, and *shot length*.

Image size

The bigger the image, the more emphatic the information it contains. **Figure 17-4** shows two versions of the insert. Notice how the larger image presents the key more vividly.

Composition

You can call attention to the information you wish to emphasize by the way you organize the composition of the image. In **Figure 17-5**, notice that in the first shot, the key is hard to separate visually from the other clutter in the drawer. In the better-composed second shot, the key has been placed and lit to draw the viewer's eye to it.

Camera angle

A powerful tool for emphasizing information is camera angle: the perspective from which the audience sees the information. In **Figure 17-6**, both images are inserts of the key in the drawer.

Figure 17-4.
The closer shot adds more emphasis.

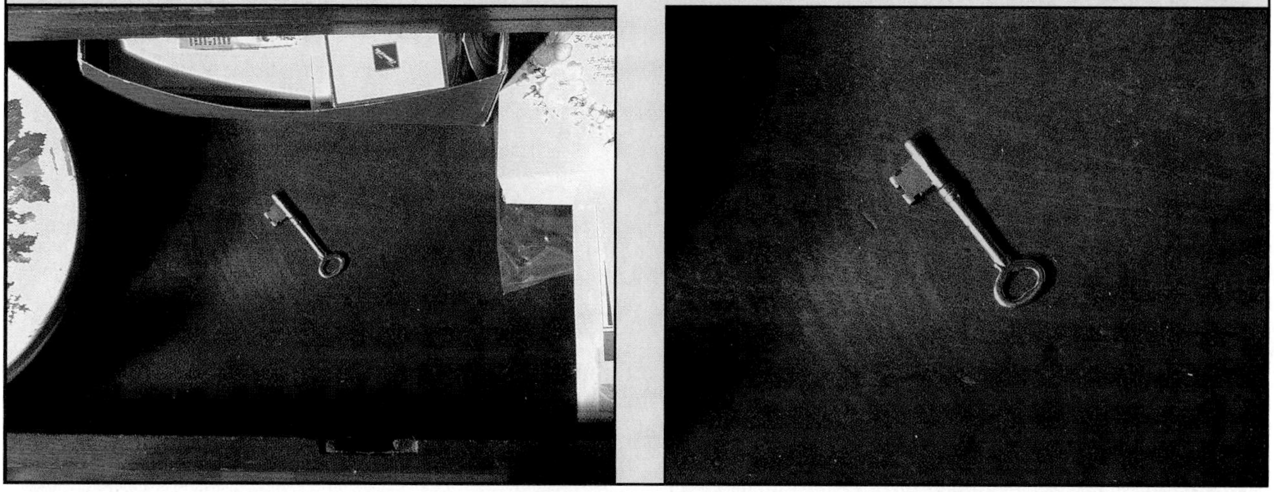

Figure 17-5.
The key is lost in the clutter. The key is emphasized by the composition.

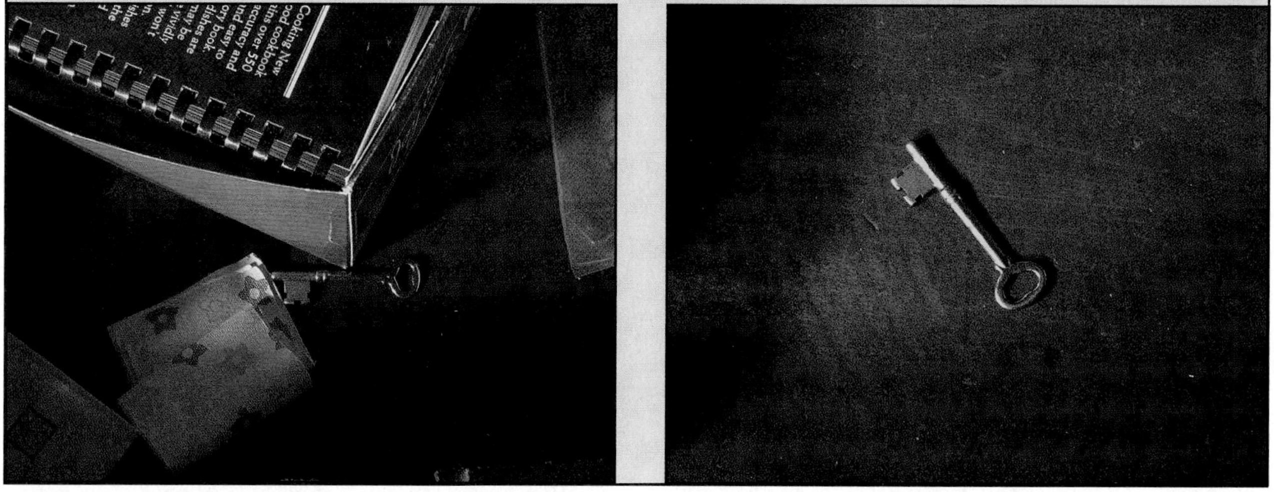

Figure 17-6.
An ***objective insert*** (shot from a neutral point of view).

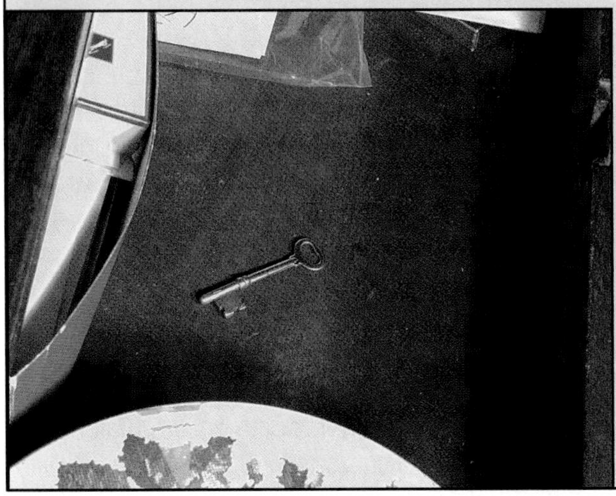

A ***subjective insert*** (shot from the burglar's point of view).

Other Forms of Emphasis

Accomplished directors have used a wide variety of tricks to emphasize elements in their shots.

Brightness and Color

In classic Hollywood lighting of the pictorial realist style, the center of interest was almost always emphasized by brighter lighting—perhaps a hotter key light or an arrow of background light that served as a pointer. In technicolor musicals, the stars typically wore costume colors that stood out from the overall color scheme.

Today, emphasis by light or color tends to be more subtle; but color especially, is still used to distinguish the most important element in a composition.

Though the third subject is small, his orange shirt attracts the viewer's attention.

Movement

Movement is an easy way to draw attention to a subject, especially when the rest of the frame is still. You can also reverse this procedure. In one of his many famous shots, Alfred Hitchcock showed a tennis match grandstand full of spectators whose heads swiveled left-right-left-right to follow a long volley on the court. Only the villain's head is not moving, because he is watching his prey rather than the match. Though he is small in the frame, the viewer's eye goes to him immediately.

Audio

Traditionally, audio had only a limited ability to emphasize a certain aspect of a shot. Today, however, most TV sets play stereo audio, and digital postproduction offers complete control over stereo imaging. This ability is especially useful for emphasizing important elements that are off-screen. The opening door, the car driving off in the distance, the incoming mortar shell—these sounds are greatly enhanced by stereo imaging. Surround sound systems take dimensional audio even further.

Although audio emphasis is ultimately created in editing, the director must anticipate it in production. To use the example of the off-screen door, the director may want a performer to react to the sound, even though it is not heard on the set.

The first is a so-called "objective" insert, meaning that the shot is from no particular point of view. The second is a "subjective" insert, showing the drawer as if the audience were looking at it through the burglar's eyes. Notice that the subjective camera angle is more emphatic than the objective (neutral) alternative.

Shot duration

You can control emphasis simply by how long you display a piece of information on the screen. If the audience sees only the single second during which the burglar's hand grabs the key, they may miss this information. But if they see the drawer opened to reveal a prominent key for a moment before the hand removes it, they have more time to notice and absorb the information.

Feeling

In the previous examples, the insert shot of the key in the drawer is added purely to convey information. But a good director presents that information in a way that involves the audience emotionally. By utilizing camera angles, image sizes, shot lengths, lens perspectives, camera movement, and other tools, the director creates *feelings.* In short, a merely competent director communicates only information; a good director communicates both information and feelings.

To see how feelings can be created, say that the effect you want for the burglary scene is suspense. Imagine that the burglar knows she has tripped the silent alarm and has only moments to get that cabinet open. But it is locked! Can the cabinet be opened and how long will it take? Your goal is to make the audience want to say, "Hurry, hurry!"

The two sequences of camera setups presented here show contrasting versions of the sequence. **Figure 17-7** includes the key only to convey the information that it exists.

Figure 17-8, however, breaks the action down into different camera angles and image sizes to emphasize the increasingly desperate search for that key. The resulting sequence adds the crucial effect of suspense.

As you study the second sequence, you will see that the suspense is developed by two principal techniques: image size and sequence length. These are the same directorial techniques that you use to emphasize information, only now they are employed to create emotional effect.

Image size

Notice that the camera stays close to its subjects: the burglar's face, her hands, the key. Generally, the bigger the image size, the more emotional impact it carries.

Sequence length

As you can see from the example, the information-only version requires just three shots, while the suspenseful alternative takes ten. By stretching the length of the sequence, the director increases the feeling of urgency.

Composition

Look again at the burglar's closeup, **Figure 17-9**. In the first version, the camera is not level (that is, not parallel to the floor). This compositional effect (often called ***dutch***) imparts a sense of imbalance and insecurity. Compare the shot with a level version of the

Figure 17-7.
Finding the cabinet locked, the burglar opens the top desk drawer.
Her hand locates, then snatches the key.
She jams the key into the keyhole of the door.

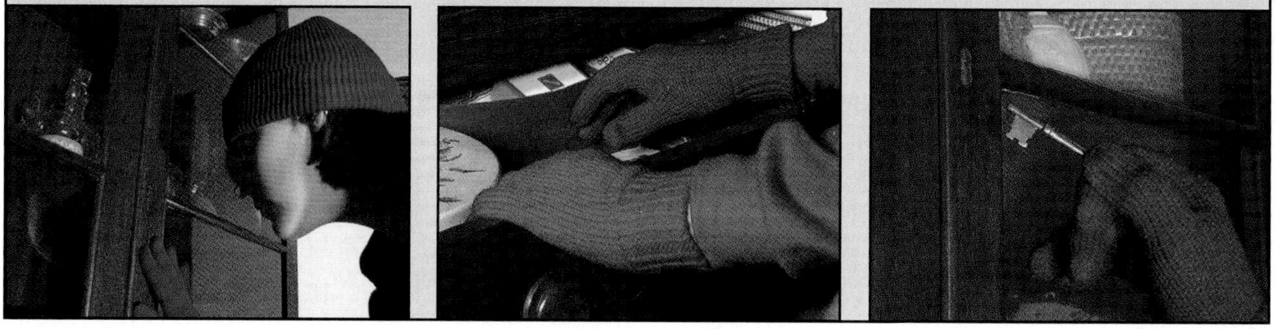

Figure 17-8.

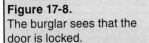

The burglar sees that the door is locked.

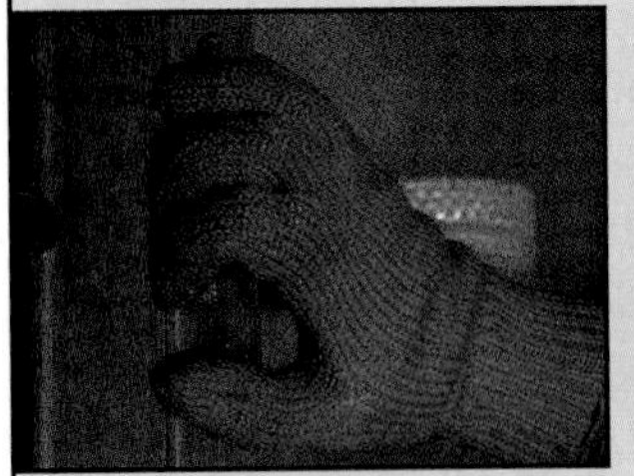

She realizes that she must find the key.

She opens the lower desk drawers.

They are empty.

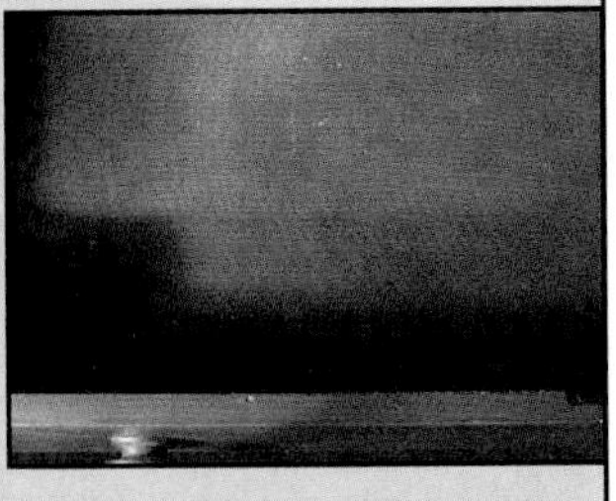

She is acutely aware that....

the motion sensor has picked her up.

The last drawer.

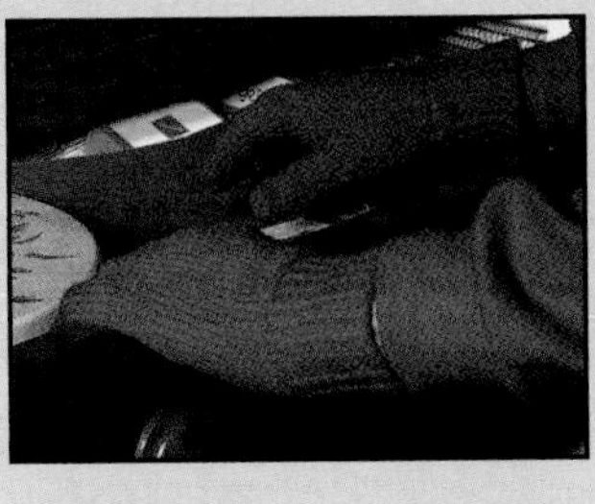

Her hand, offscreen, rummages the drawer.

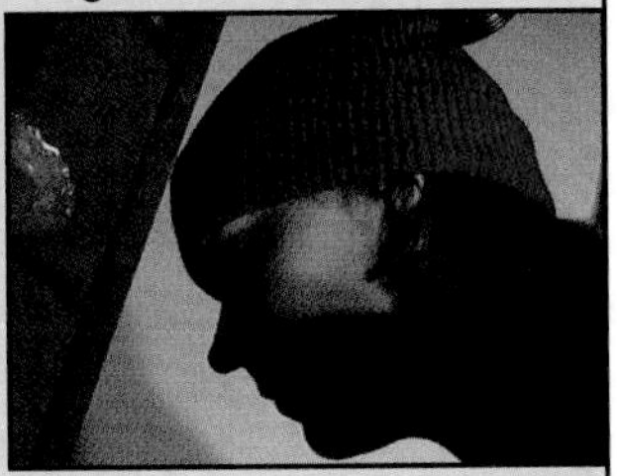

Her hand snatches the key.

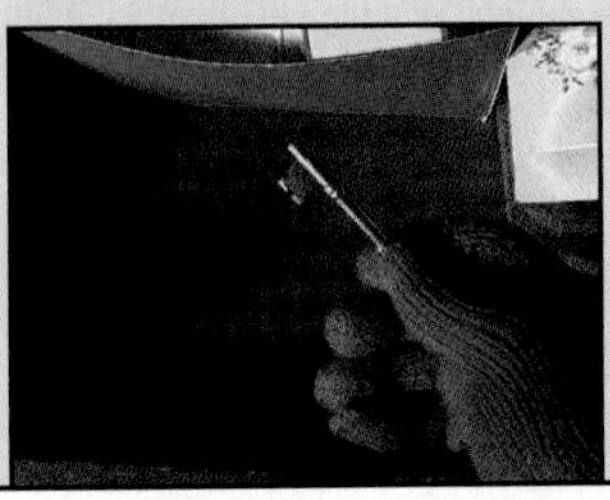

She jams the key into the keyhole.

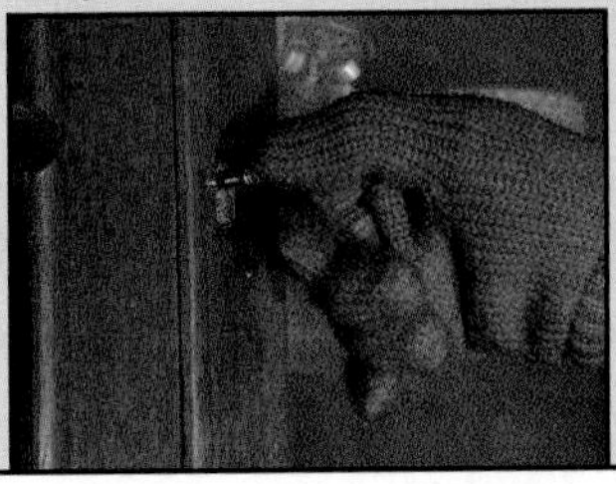

Figure 17-9.

Composition with off-level background.

Composition with level background.

same angle (the second shot) to see how tilting the horizon unsettles the composition.

The term "dutch," an informal version of Deutsch ("German") was first applied to off-level angles because they were favored by European directors who brought the expressionistic style of German cinema to Hollywood in the 1920s and '30s.

Camera angle

Both image size and composition are created by camera placement; and camera angles, in general, can contribute to the overall emotional effect. In **Figure 17-10**, you can see that high angles often convey a feeling of detachment, while low angles emphasize the power of the image (and the powerlessness of the

Figure 17-10.
A high-angle shot feels more detached.... than a low-angle setup. A formal, frontal angle is less dynamic than.... a strong diagonal composition.

Figure 17-11.
Wide angle lenses expand apparent depth... while telephoto lenses compress it.

viewer). Head-on angles feel formal and calm, while diagonal setups convey movement and energy.

In practice, image size, composition, and camera angles are all treated together, as aspects of the same thing.

Lens perspective

The "feel" of an image is strongly affected by the lens used to record it, **Figure 17-11**. Wide angle lenses exaggerate depth and motion toward or away from the camera. The effect they convey is dynamic and energetic. Telephoto lenses, by contrast, suppress depth and motion, keeping the viewer at a distance and conveying a sense of formality.

Camera movement

Camera movement can add significantly to the emotional cast of a scene. A stationary camera, panning and tilting from one position, conveys a sense of detached observation. A moving camera, especially when dollying in and out, becomes part of the action. Moreover, a constantly moving camera imparts a feeling of restless energy.

Shot length

To create an effect, a director can control the length of a sequence by shooting it in a greater or smaller number of separate shots. In the same way, the individual shots themselves can be made shorter or longer, depending on the effect desired. In general, short shots with quick cuts convey a sense of dynamic energy, while lengthy shots feel calm and deliberate. That is why fights are typically constructed from many brief shots, while romantic scenes are often assembled from fewer but lengthier angles.

The Other Half of Content

Directing to communicate with your audience means managing information, emphasis, and effect. Together, these elements make up the content of your program. However, communicating content is only half the job. Unless you are making a specialized program like a nature documentary, that content is delivered largely by the people in your video—by the performers.

Directing for Performance

Everyone knows that directors work with actors, but few people understand what they really *do* with them. This section introduces the craft of drawing effective performances from the talent in your video programs, **Figure 17-12**.

"Talent" is the collective term for the people who appear in films and videos.

What follows is not an extensive treatment of the art of directing professional actors in fiction films. That would require a book in itself,

Fixing Problems in Rehearsal

In working with nonprofessional actors, directors must deal with difficulties that come up again and again. Here are some of these common problems, with some suggestions for solving them in rehearsal.

Context

Videos consist of short individual shots that are often recorded out of order. This choppy, seemingly arbitrary production can leave performers confused about where they are and what they are supposed to be doing. An earlier chapter uses the example of two train station sequences that are shot back-to-back but are five years apart in story time. In another case, two screen lovers may be asked to enact their first meeting after playing many scenes in which they already have a romantic relationship.

To avoid problems with this shooting method, begin rehearsing each scene by explaining where it fits in the finished program and how it relates to scenes that the performers have already completed.

Memorizing

This chapter offers ways to help actors master their ***lines***, but movement and activities can be hard to remember too, especially since they must often be repeated precisely from one setup to the next. (A performer's movement within the scene is called ***blocking*** and his or her activities are termed ***business***.)

To control the amount of memorizing required of the actors, rehearse only a single sequence at a time. If the sequence is a long one, try to break it down further into ***beats***, short subsections of perhaps two minutes of screen time.

To obtain consistent performances from one setup to the next, consider letting actors refresh their memories by watching playback of previous setups that cover the same material.

Expressing emotions

As noted elsewhere, amateurs can have problems in expressing emotions convincingly. Some tend to overact, others have trouble making transitions between different feelings, still others cannot handle complex, mixed reactions.

Rehearsal is the place to spot these problems and adjust accordingly. Step one is often to help the actor with the problem. If this does not work, you will want to plan a strategy for shooting around the difficulty. Develop solutions before you start shooting, whether this involves simplifying the performance requirements, breaking the recording into shorter shots, or thinking up cutaways to help edit out mistakes.

Tuning performances

In some cases, the quickest way to get the performance you require is by demonstration: you act out the role yourself to show how you want the blocking and business handled, and how you want the lines of dialogue to sound. Professional actors often object to being given these ***line readings***, as they are called (though they seldom mind having their movements demonstrated); but nonprofessionals are often grateful for the help.

In modeling movement or speech for an actor, do just enough to get the idea across. If you go through the whole thing, you risk getting performances that are just mechanical imitations. (Charlie Chaplin, however, was famous for acting out every second of every role and then expecting his cast members to imitate him exactly.)

A 12-volt monitor allows playback on location.

Figure 17-12.
The "burglar" is, of course, a performer.

and several are available. Instead, this section focuses on the craft of guiding people through all kinds of videos from stories to training programs to promotional pieces. Some of your performers may be professionals, but others may be amateurs like corporate spokespersons or workers demonstrating job skills or simply members of your family or community.

Actors and Performance

Directing performers like these means making them appear believable and effective on screen. To be believable, the performers should not appear to be acting, but simply living and behaving normally while the camera just happens to record them. Though a skilled actor is practicing a complex craft, the result looks simple and natural on screen because that craft is invisible to the audience.

To be effective, the performance must be more than simply believable. It must also engage viewers' attention and make an impression on them. Some professional actors achieve this by sheer force of personality, but most performers can do it if they get enough help from the director.

Many people think that performing means creating a character in a fictional program, and that is often the case. But the people who appear in nonfiction programs are also performing. As a director, you may be coaching a spokesperson for an organization, a person being interviewed on camera, a subject of a documentary program, or a person who demonstrates the skills or techniques being taught in a training video. Even the most seasoned professional actors will benefit from good direction because the director can show them where they fit in the program as a whole and how best to do their part in it.

Some professional directors feel that 90% of directing actors lies in casting exactly the right people for their roles and then getting out of the way and letting them work.

Outside of Hollywood-style production, most performers are not full-time professionals. Working with less-seasoned talent, directors spend considerable time solving performance problems—problems like self-consciousness, difficulty in speaking lines, ignorance of production techniques, and trouble expressing emotions. The most widespread problem with actors is a general insecurity: a fear that they look ridiculous, that they are inadequate, that they are failing.

Actors and Insecurity

Camera fright is a form of stage fright: a fear of subjecting yourself to public attention, of placing yourself under a spotlight for critical inspection by the audience, **Figure 17-13**. Even professionals can suffer from it and almost all actors may occasionally feel ill at ease about performing.

Fortunately, camera fright is easier to combat than stage fright, because the performer is *not* actually appearing in public. With reluctant actors (in fact, with *all* actors) it helps to remind them that their mistakes are completely private and will never be seen by anyone except the

Figure 17-13.
Amateur performers are often insecure.

cast and crew. Shooting multiple takes lets performers practice until they get it right. Assembling the program from separate shots allows the editor to cut out all the bad parts and show only the best. Finally, reassure the actors that your job is to make them look good on screen, and that is exactly what you are doing.

Helping Actors Overcome Self-consciousness

Self-consciousness is a painful awareness of face and body that makes the performer behave stiffly and awkwardly. It also makes faces look tense and uncomfortable. There are several directorial tactics for reducing self-consciousness. First, allow plenty of time for rehearsal. The more familiar the performers are with the material, the more comfortable they feel about it. Often amateurs will perform more naturally if they think they are not being recorded, so it may help to disable the ***tally light*** on the camcorder (the button that glows red to indicate that the camera is recording). That way, you can record rehearsals without alerting the performers.

If you cannot turn off the tally light, cover it with a small piece of opaque tape.

To reduce physical awkwardness, keep actor movements simple and natural. It helps to check with performers after each rehearsal to see if anything feels uncomfortable to them. If so, correct the problem. In the first shot of **Figure 17-14**, the performer does not know what to do with his hands. The chair provides a natural place to rest them.

Finally, the best way to make performers look natural is by having them do things that they do in real life, so that they do not have to learn new skills. For instance, if you are making a training program on how to operate a fork lift, use an actual fork lift operator as your on-camera demonstrator.

Helping Actors Deliver Lines

All performers can have trouble memorizing speeches and amateurs have the additional problem of stiffness.

We have all seen local commercials in which the owner of a small business stands woodenly in front of the camera and intones something like "So-Come-On-Down-To-Billy-Bob's-House-Of-Barbecues-And-Fireplaces..."

If the performer cannot read aloud convincingly, the easiest solution is to allow him or her to improvise. For example, the director might say something like, "Okay, now tell the folks to come visit your store and be sure to say the whole name." Allowed to say his own words in his own way, the performer/client will probably sound more natural.

If the problem is remembering lines, you can minimize it by avoiding long speeches and breaking conversations down into several brief shots. Remind performers that it does not matter

Figure 17-14.
Simply standing still can be difficult for amateur performers. You can often use a prop to anchor them.

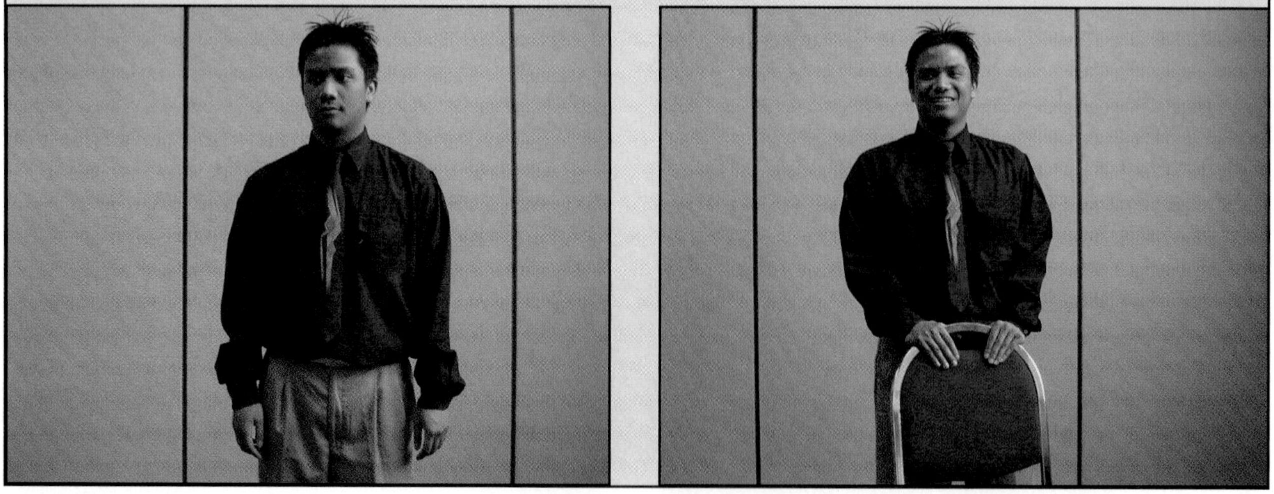

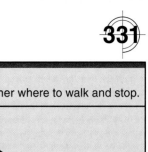

Figure 17-15.
A hand-lettered cue card.

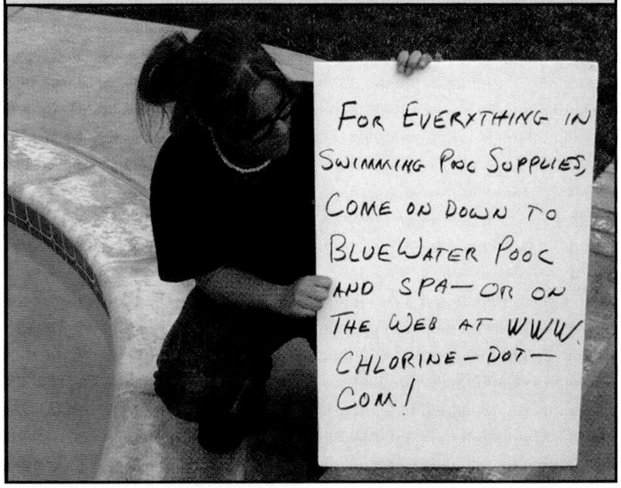

Figure 17-16.
Chalk marks show the performer where to walk and stop.

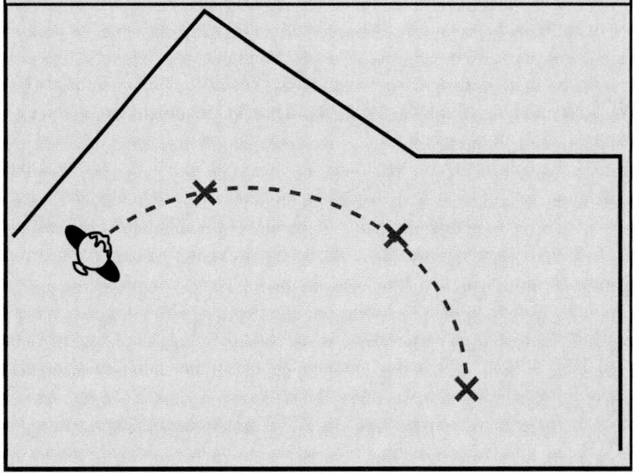

if they make a mistake or how many times they make it. In the finished program, only the correct version will appear.

Finally, be prepared to use cue cards so that performers can read their lines instead of memorizing them, **Figure 17-15**. Place the cue cards off-camera but in positions such that the performers' eyes appear to be looking in the right direction.

Do not transcribe all the lines to cue cards before shooting. Instead, have sheets of cardboard and broad markers available at the shoot and make cards only when performers are having real difficulties. Otherwise, your talent will expect cue cards for every scene and refuse to memorize their lines.

With a speech or lengthy remarks by an on-camera spokesperson, you will, of course, need to prepare cue cards in advance.

Helping Actors Master Production Techniques

Many times, amateur performers have problems because they do not understand the technical requirements of the video medium. Here are some of their most common mistakes, along with suggested remedies.

Hitting marks

Often a performer is asked to move through a shot and then stop at a predetermined spot (usually because that spot is specially lit for the performer and the camera is set to create a good composition there). Some amateurs have trouble stopping at the right place, or can do so only by searching obviously for their ***marks*** on the floor, **Figure 17-16**.

The simplest solution is to avoid the problem entirely. For example, instead of having the actor move into frame and then stop, have the actor stand off-screen while the camera moves to frame him or her.

Looking at the camera

It is natural for amateurs to look directly into the camera lens, especially after they think they have finished a shot. Their look asks, "Is the shot over?" or "Was that okay?" This can be a problem if you need extra footage at the end of the shot. Explain diplomatically that the shot was not usable because the actor looked at the camera; then retake the shot. If the actor cannot soon learn to avoid looking at the camera, this is a sure sign that you will encounter other problems and you had better replace this performer sooner rather than later.

Matching action

Once you explain the importance of doing (and saying) everything identically in each angle on a scene, most amateur actors will try to do this. To help them out, keep good continuity notes so that you can remind them of details before every shot. Use an on-set monitor to play back their previous angles so that they can see exactly what they said and did.

Overall, the best solution to technical problems with amateur actors is to remove

Improvised Teleprompters

For reading lengthy texts on camera, professionals use ***teleprompters***—special displays that scroll the copy. Commercial models can be expensive, but you can achieve the same results with a personal computer.

The system is more flexible on a laptop model because the computer screen is easier to position off-camera where the actor can read it comfortably. With a little ingenuity, however, you can also use the monitor of a desktop computer.

Preparing the text

The best applications for teleprompting are slide show programs such as Corel Presentations and Microsoft PowerPoint, because they can display slides full-screen, without surrounding menus. Also, slide programs are specifically designed to display text in large sizes.

If a slide program is not available, you can use a word processor by hiding all unnecessary workbars and menus, setting the copy in a large type size, and setting the viewing width so that the text fills the computer screen side-to-side.

The first step is to key (or import) the text into the program and format it in convenient sections. Set the typeface in a simple font like Arial or Swiss, at a very large size (try 32, 36, or 48 points).

Unlike true teleprompters, slide programs cannot scroll text continuously. Instead, they display one full page at a time. For this reason, inexperienced

Improvised teleprompter setup.

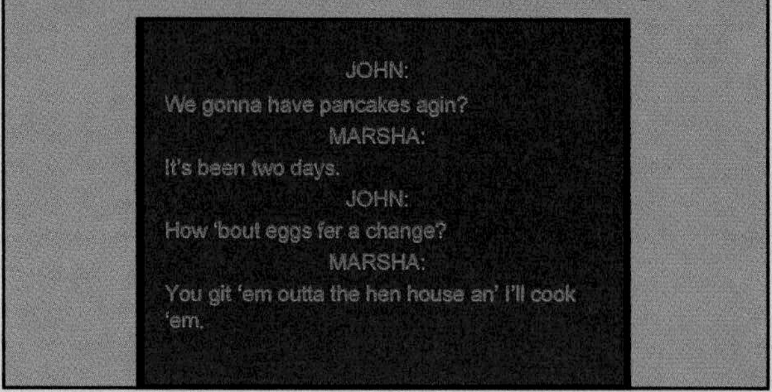
JOHN:
We gonna have pancakes agin?
MARSHA:
It's been two days.
JOHN:
How 'bout eggs fer a change?
MARSHA:
You git 'em outta the hen house an' I'll cook 'em.

A word processor display used as a teleprompting device.

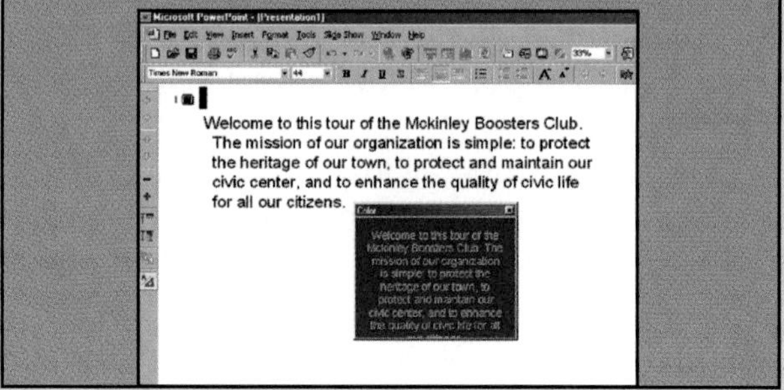
Welcome to this tour of the Mckinley Boosters Club.
The mission of our organization is simple: to protect the heritage of our town, to protect and maintain our civic center, and to enhance the quality of civic life for all our citizens.

A slide program work screen.

the problems rather than ask the performers to solve them. It is unlikely that you can turn your talent into professionals during a single shoot.

Helping Actors Express Emotions

Professional actors are skilled at adjusting the intensity of their performances to suit the director. Amateur performers often lack this ability.

This section applies mainly to fiction movies, since nonfiction programs seldom require performers to simulate feelings.

Overacting is often easy to correct. Simply show the offenders some closeups of their work on a monitor and they will realize that they need to rein themselves in. With persistent overactors, you may want to pull the camera back from them because the smaller the image, the less intense it appears to the audience. Be aware, however, that this technique has other consequences. If the wider shot of one actor is intercut with closer angles on another, the actor with the larger image may well dominate the scene.

With amateurs, a bigger problem is complex acting involving combinations of emotions. For example, if you say to an amateur performer, "You're delighted that your no-good brother-in-law is leaving, but unhappy that your mother-in-law's staying," the actor may well be unable to show these two opposite emotions at the same time and the result will be a failure, **Figure 17-17**.

To solve the problem, rework the scene so that the actor can react separately to the good news and the bad news. When the transition readers can be confused by rapidly changing text pages. To solve the problem, repeat the last sentence of each outgoing page at the top of the incoming page. On both pages, set the repeated text in a contrasting color. This makes it easy for the reader to follow continuing copy from one slide to the next.

Setting up the system

Position the computer screen facing the performer, and as close to the camera lens as possible. To conceal the fact that the person is looking to one side of the lens, use a telephoto setting and place the camera as far away as practical. (The setup must be close enough for the performer to read the computer display.)

First slide, with text to be repeated in white.

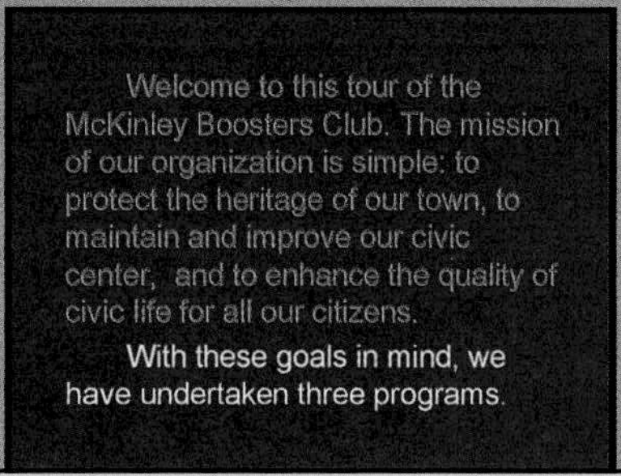
Welcome to this tour of the McKinley Boosters Club. The mission of our organization is simple: to protect the heritage of our town, to maintain and improve our civic center, and to enhance the quality of civic life for all our citizens.
With these goals in mind, we have undertaken three programs.

Using the system

To use the program as a set of cue cards, have the continuity supervisor follow the dialogue, advancing the slides to change the virtual "cards."

To minimize the angle between the lens and the display screen, set the camera as far from the performer as practical.

Too close to the camera, the subject is obviously looking to one side.

Farther from the camera, the subject's look is not obvious.

Poor placement
Better placement

Figure 17-17.
HUSBAND: My brother's leaving tomorrow, but Mom's staying six more weeks.

WIFE: (dismayed) What!?

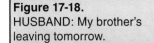

Figure 17-18.
HUSBAND: My brother's leaving tomorrow.
HUSBAND: But Mom's staying six more weeks.
WIFE: What!!?

Figure 17-19.
A low angle conveys a feeling of authority.
Setting the camera just below eye-level conveys the effect more subtly.

between emotions happens off screen, the actor has to play only one in each shot, **Figure 17-18**.

Since you shoot the wife's happy and unhappy closeups as separate shots, the performer can take all the time she needs between setups to change from happiness to anger.

Helping Actors Project Authority

Low angles tend to make a person seem more powerful, more commanding. High angles have the opposite effect. When shooting an executive or a spokesperson for a corporate video, make certain that the camera is at least slightly below the subject's eye level, **Figure 17-19**.

Directing Assignments

So far, we have focused on helping actors perform effectively. However there are situations where you will be directing the shooting of actual activities rather than rehearsed performances. Some of the more common types include interviews, documentaries, and "wallpaper."

As noted elsewhere, wallpaper is footage designed to fill screen time while the audio delivers the important information.

Directing Interviews

Today, most video interviews show only the subject speaking informally to someone beside the camera, **Figure 17-20**.

The interviewer is never seen or heard. Though the result looks spontaneous, it is really the result of patient questioning by the director and adroit cutting by the editor.

It is not unusual to shoot two hours of raw footage to obtain 10 minutes of edited screen time.

Setting up the interview

First, set up the interview in an interior with a suitable background. (Some production teams carry draperies and century stands to create neutral backgrounds when needed.) Test for

Figure 17-20.
Most interviews show only the subject. (Corel)

Figure 17-21.
Plan of a typical interview.

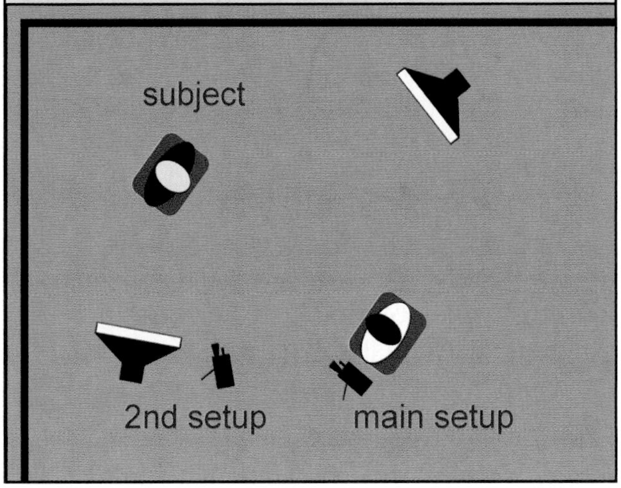

good acoustics and lack of background noise, because amateurs will be less tolerant of interruptions and extra takes due to sound problems.

Position seats so that the subject is facing the interviewer, who sits close to the right side of the camcorder. Use a low-backed chair to keep the chair back out of the shot, **Figure 17-21**.

By convention, the subject usually sits slightly screen-left, looking screen right. However, when presenting opposing viewpoints in separate interviews, have the opposition sit screen-right and facing left. Cut together, the opposing compositions will give the illusion of a discussion between the two subjects, **Figure 17-22**.

Making the subject comfortable

To reduce camera anxiety, place the camcorder back from the subject—as much as eight feet away (location permitting). If possible, light with two or at most three softlights, to simplify the lighting and reduce heat, as shown in Figure 17-21. Fluorescent video lighting is especially useful for interviews.

Framing questions and answers

The key to successful interviewing is good questioning. The objective is to get the subject talking about a topic rather than answering a question. (In the examples that follow, a video director is interviewing Mr. Winesap for the Discovery Channel. Mr. Winesap, President of the McKinley Boosters Club, tends the flowers in the McKinley Plaza.) Here is the wrong way:

Figure 17-22.
Mirror-image compositions give the impression that the subjects are debating.

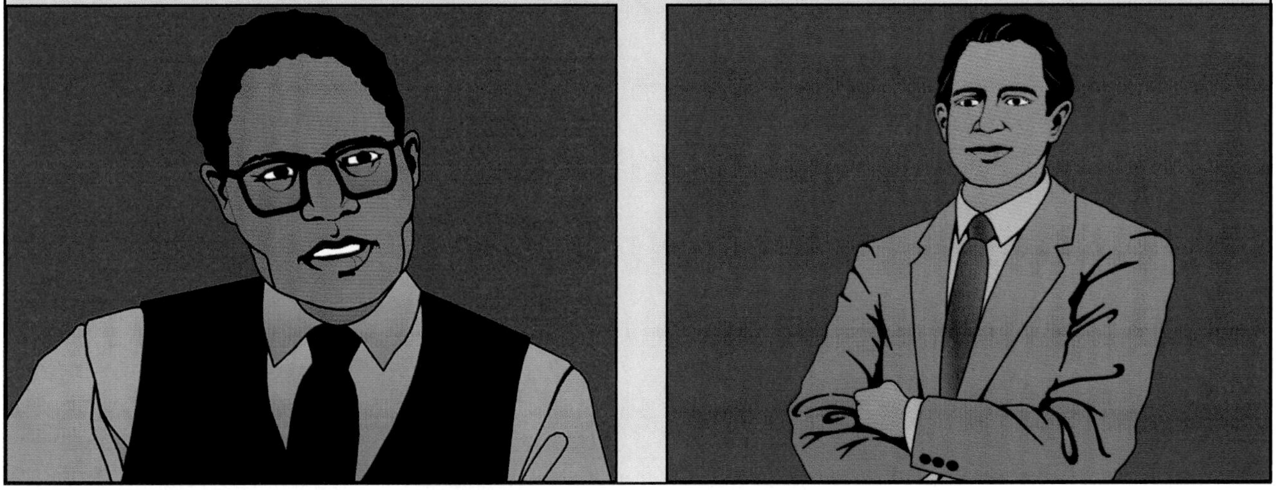

Figure 17-23.
Shooting at different angles or image sizes adds variety to the edited interview. (Corel)

INTERVIEWER: How long have you cared for the plaza flowers, Mr. Winesap?

SUBJECT: I've been running the show for 22 years!

When the question is removed, the answer, "I've been *running the show* for 22 years" will not make sense to viewers. Instead, ask the subject to talk about a subject. The right way:

INTERVIEWER: Mr. Winesap, talk to us about tending the flowers, including where and how long.

SUBJECT: I've filled the McKinley town square with flowers for 22 years."

Now the editor has an answer that makes sense without a question.

Warming up

Most subjects will need to practice this skill, so develop some "throwaway" questions to get the subject comfortable.

INTERVIEWER: Tell us how you feel about McKinley.

SUBJECT: I'm very fond of it.

INTERVIEWER: Great! Now tell us again and this time, include the town's name.

SUBJECT: Oh, I see. (Ahem) I've always been fond of McKinley.

INTERVIEWER: That was perfect. You have a knack for this!

Listening carefully, the interviewer decides that the throat-clearing can be cut from the sound track, so the shot (if used) can begin after the "Oh, I see."

You may also need to gently coach the subject to avoid looking at the lens. Explain in advance that the subject should look only at you. Ask for repeats if he or she forgets and looks into the lens.

Finally, explain that any mistakes or fumbles will be cut out and only the subject's best answers will be used. As always, remind subjects that your job is to make them look good.

When you feel that the subject is ready, move on to the actual interview topics.

It is common to record during the warmup (which may produce some usable material), and then simply continue with the actual interview.

Camera angles

If you have the luxury of two cameras, you can obtain two different angles on the entire interview, to cut back and forth between them when editing. With just a single camera, you may want to pause partway through and move the camera to another setup, to add variety. Make sure that the two setups show a significant difference in both horizontal angle and subject size, **Figure 17-23**.

Directing Documentaries

Documentary programs about events that happen in real time are a special challenge. The director (who is often camcorder operator as well) must anticipate important shots

Figure 17-24.
Taping a famous draft-horse team.

because the action usually cannot be repeated, **Figure 17-24**.

Talented documentary makers develop a sixth sense that enables them to frame the right subjects at the right moment. Here are some suggestions for obtaining good documentary footage.

See "around" the camcorder

If your camcorder has an external LCD viewing screen use it. If you must use the viewfinder, keep your other eye open. Either way, you are using your peripheral vision to keep tabs on the area outside the video frame. That way, you can spot interesting activities to record**, Figure 17-25**.

Since the LCD monitor is on the left side of the camera, it may help to reverse the usual interview positions, placing the director/operator behind the left side of the camcorder, and having the subject sit right and look left.

Anticipate the next shot

If a softball batter gets a hit, the action will move from home plate to first base. Knowing this, you can re-frame and zoom in on first, before the runner gets there. The idea is to predict the results of the activity you are taping and switch to them in time to tape them as they happen.

Stage pickup shots when possible

Even in real-time shooting, it may be possible to pick up crucial action that you were unable to capture. For example, taping the team of horses as they pass by, you spot a Dalmatian dog riding on the wagon–but too late to get a closeup. So when the team stops for review, you frame a closeup, **Figure 17-26**. The closeup ***pickup shot*** can then be cut into the action, **Figure 17-27**.

If you pan across the dog from left to right, you can simulate the movement of a rolling wagon (since there's nothing in the background to reveal the trick).

Get cutaways

In any documentary situation, there are moments when nothing important is happening. Those are the times to get ***cutaways***—shots

Figure 17-25.
Keep your other eye open... to see details outside your frame.

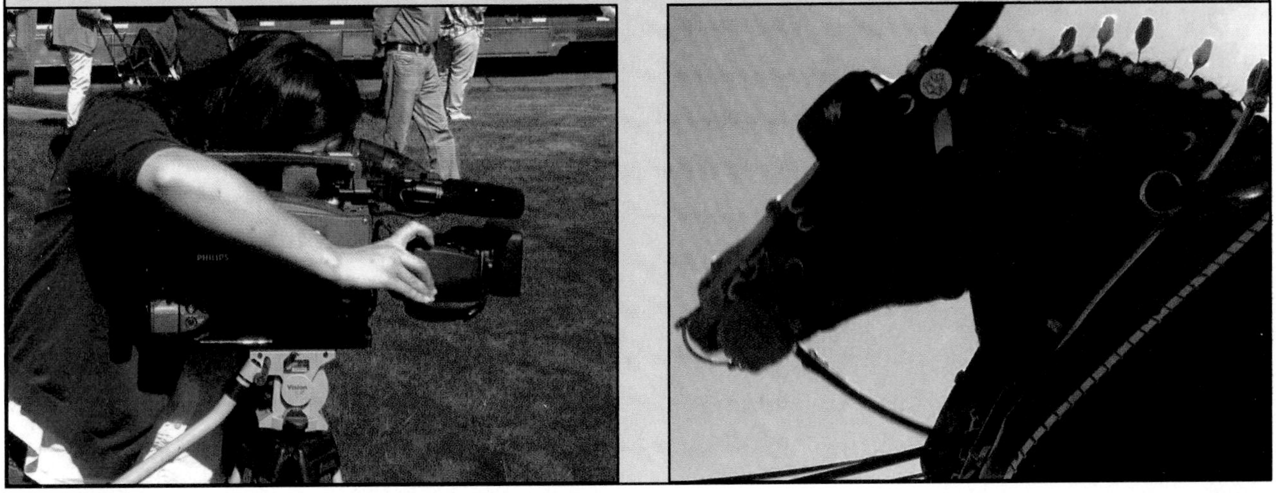

Figure 17-26.
Missing a closeup of the dog the first time around... you get a pickup shot later.

Figure 17-27.
The pickup shot is inserted into the edited sequence.

other than the main action. By using these cutaway shots, the editor can make widely separated actions look continuous.

At the Leibniz County Fair Jersey heifer judging, there are long, uninteresting stretches while the judge walks around looking at the entries. By using cutaways of the spectators, the editor can invisibly cut out these passages, **Figure 17-28**.

Good documentary makers are supreme opportunists–ready to think and move fast to record all the pieces that the editor will need.

Cutaway shots are covered more fully in the next chapter.

Directing "Wallpaper"

The sidebar on wallpaper in Chapter 9 explained that abstract ideas are impossible to visualize. Instead, you must fill the screen with images that seem to fit the words on the sound track, even if the relationship is not very close. This filler footage is often called ***wallpaper***.

Figure 17-28.
While the judge (blue shirt) inspects the entries... the editor can shorten the process with a cutaway. (Sue Stinson)

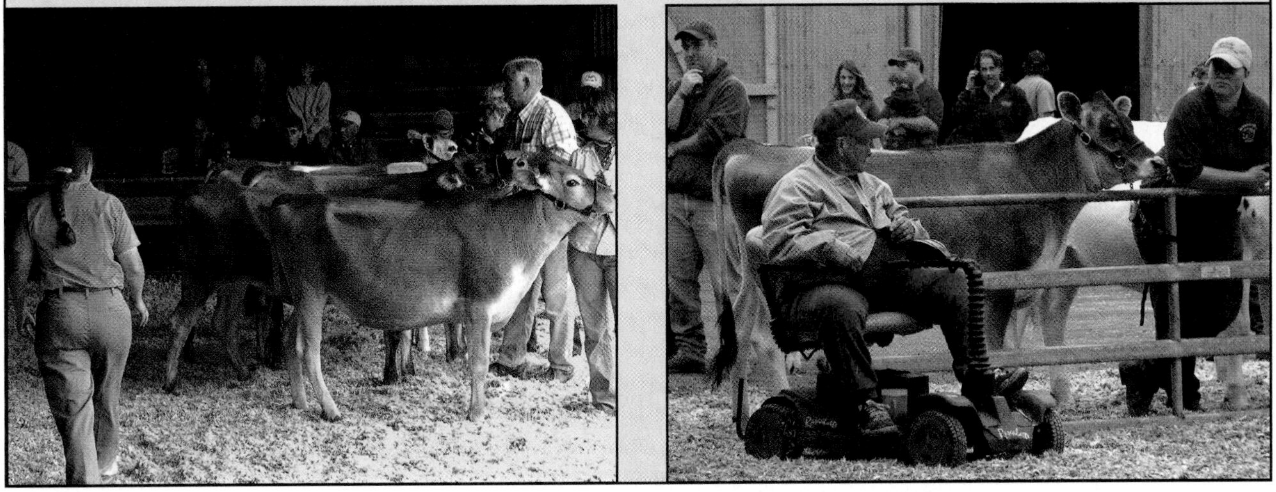

Figure 17-29.
NARRATOR: Established two decades ago by the McKinley arts community... the great annual kinetic sculpture race... has become an internationally famous event... with imitators around the world.

Figure 17-30.
REPORTER: Meanwhile, at City Hall... the mayor refused to comment on the allegations... saying only that "appropriate administrative interventions... are ongoing at this point in time."

Some wallpaper can be prescripted, but there are times when you will have to improvise with what you find on the scene. Once again, the trick is to look for cutaways and color shots to fill the screen during necessary but nonvisual narration.

In **Figure 17-29**, the narrator's introduction to the annual McKinley kinetic sculpture race is covered by establishing shots of the opening parade.

When confronted with static material, you can compensate somewhat by panning and/or zooming—anything to get movement into the wallpaper, **Figure 17-30**.

Other Assignments

Pickup shots, inserts, cutaways, color shots —collectively these are known as "***cover***," the extra content a good director captures to provide the editor with full *cover*age of the action. These and other types of shots are the subject of the next chapter, *Directing for Form.*

Chapter Review

Answer the following questions on a separate piece of paper. Do not write in this book.

1. Inexperienced directors often omit important _____.
2. *True or False?* The bigger the image, the more emphatic the information it contains.
3. To help amateur performers feel less self conscious, you can _____ the tally light on the camcorder.
4. Using a _____ camera angle makes a person appear more powerful.
5. A _____ is a shot other than, but related to, the main action.

Technical Terms

Beat: A short unit of action in a program, often, though not always, corresponding to a scene.

Blocking: A performer's movement within a shot. It must be rehearsed and memorized so that the actor hits the correct spots with respect to camera and lighting, and so that the movements can be repeated exactly for other angles on the same action.

Business: Activities performed during a shot, such as writing a letter or filling a vase with flowers.

Cover: Additional angles of the main subject, or shots like inserts and cutaways, recorded to the material needed for smooth editing.

Cutaway: A shot other than, but related to, the main action.

Dutch: Referring to an off-level camera. The "put dutch on a shot" is to purposely tilt the composition.

Line reading: A vocal interpretation of a line that includes its speed, emphases, and intonations. In simple terms, "You came back!" is one reading of a line, and "You came *back*?" is a different reading.

Lines: Scripted speech to be spoken by performers.

Marks: Places within the shot where the performer is to pause, stop, turn, etc. These spots are identified by marks made of tape or chalk lines.

Objective insert: A detail of the action presented from a neutral point of view.

Pickup: A shot obtained later to record action that was either missed or inadequately covered previously.

Subjective insert: A detail of the action presented from a character's point of view.

Tally light: A small light on a camera that glows to show that the unit is recording.

Teleprompter: A machine that displays text progressively as a performer reads it on-camera.

Wallpaper: Footage designed to fill screen time while the audio delivers the important information.

18 Directing for Form

Objectives

After studying this chapter, you will be able to:

- Deliver footage that can be edited smoothly and flexibly.
- Create a three-dimensional world on the two-dimensional screen.
- Choreograph effective camera movement.

About Directing for Form

Complementing Chapter 17 on directing for content (communication and performance), this chapter covers directing for form (the techniques of the medium). Mastering the video form requires technical knowledge because of the special nature of the medium. Every program consists of hundreds of very short pieces (shots) that are often recorded out of order. When rearranged by the editor, these pieces must fit together seamlessly, even though each is made quite independently of the others. Also, because space and time on the two-dimensional screen follow their own special rules, action must be carefully staged to appear as if it is happening in the real world.

The world on the screen is summarized in Chapter 8.

For these reasons, a video director needs a thorough understanding of the formal aspects of the craft, beginning with the fundamental principle that *absolutely everything recorded, both visually and aurally, is nothing more than raw material for future editing.*

Covering the Action

As you direct a production, you need to remember that you are not making a video program, but only its separate component parts. The actual program will be assembled from your shots at a later time by the video editor. What you are doing is providing the building blocks, the component materials from which the editor will construct the final program.

For clarity, we will treat the editor and director as two different people, though directors often edit their own programs.

To help ensure a successful result, the components you supply must provide the editor with everything that is needed, in a form that the editor can use. The process of shooting the material to anticipate the editor's requirements is called *directing to edit.* To begin with, directing to edit means delivering full coverage of the action so that the editor has enough material to work with.

Coverage

Coverage involves more than just taping all the action. It also means ensuring that the action is recorded and then rerecorded in multiple shots that repeat all or at least part of the content. Good coverage typically requires *repetition, overlap, variety, protection*, and ***cutaways***.

Repetition

Repetition involves recording the same action more than once by shooting it from more than one point of view. For instance, as shown in **Figure 18-1**, you might record the entire scene from three different camera setups: a two-shot, a closeup favoring one actor, and another closeup favoring the second actor.

Multiple angles allow the editor to cut between one shot and another at any time to improve performance, adjust timing, and control style.

Overlap

Even when you do not rerecord the entire action in each of several setups, it is usually

A Demonstration Sequence

The discussion of coverage uses a simple sequence of a couple picnicking on a wooded bluff above the ocean. The basic actions in the sequence are summarized in the following illustrations.

In actuality, this short sequence might be covered by just three or four setups; but for demonstration purposes, we will use many more angles than usual.

The couple picnicking.

He takes a drink of soda.

He points out a storm over the ocean.

She doubts that it will reach them.

Figure 18-1.
A neutral two-shot covers the entire sequence.
Her closeup repeats the entire sequence.
His closeup repeats the entire sequence.

important to overlap action from shot to shot. This is illustrated in **Figure 18-2.** Overlapping means beginning a new shot by repeating the last part of the action in the previous shot.

By overlapping action, you allow the editor leeway in setting the edit point between the two shots. Even if you think you already know exactly where you want the edit to be, protect yourself and the editor by providing extra footage.

In addition to overlapping parts of the action, you should begin recording at least five to ten seconds before any action begins and continue shooting for the same length of time afterward as well. These extra seconds allow the editor more leeway in choosing edit points.

Variety and protection

To offer the editor a broad range of options, you should provide coverage from a variety of angles. Varying angles helps make invisible edits and it also tends to hide small mismatches between different takes (recordings) of the same action, **Figure 18-3**.

Sometimes mismatches or other mistakes are too large to hide, so the editor cannot use the shots in which they appear. That is another reason for covering the same action from different angles. In fact, if you are not sure about something in a shot, obtain an extra shot from another angle.

Such an alternative setup is sometimes referred to, literally, as a "protection shot."

Cutaways

Cutaways are shots that show information other than the main action. Typical cutaways, **Figure 18-4**, include:

- ***Inserts:*** closeups of details.
- ***Reaction shots:*** the responses of subjects other than the one who is currently the center of interest.

Figure 18-2.
He picks up his soda at the end of the first shot.
He picks up his soda at the start of the second shot, drinks, then puts it down again at the end.

Figure 18-3.
His action is covered in three different styles of closeup.

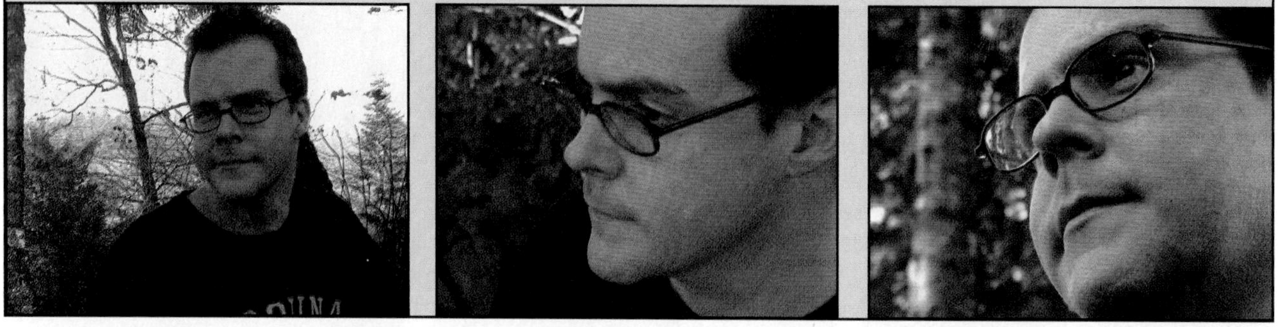

Figure 18-4.
Cutaways. A—Insert of the soda being set down. B—Reaction shot of her while he talks offscreen. C—Color shot of the ocean view. (Sue Stinson)

A B C

- ***Object shots:*** the second half of a "glance-object pair," an object shot shows what a subject is looking at, from his or her point of view.
- ***Color shots:*** shots that show expressive details of the environment in which the scene is unfolding.

Nowadays, using a shot for color alone is considered somewhat old-fashioned. It is better to relate color cutaways to the main action.

In addition to delivering extra information, cutaways are extremely useful for interrupting other shots so that they can be invisibly shortened or so that other changes can be made. For example, suppose that the two-shot of the couple includes more dialogue than necessary, **Figure 18-5**.

The trouble is that the basket is already packed. Discovering this potential problem, the director adds a cutaway shot of the ocean. By inserting this cutaway into the closeup, **Figure 18-6**, the editor can snip out the line about packing without leaving a hole in the original shot.

Of course, a reaction shot of the woman could be used as a cutaway instead.

Figure 18-5.
HE: I don't like the looks of that squall coming in. I'll start packing the picnic basket. You'd better finish your soda.

Maintaining Continuity

Continuity is the complex process of ensuring that all the small pieces of a video (the shots and the details recorded in them) add up to a smooth, consistent whole. A program with good continuity appears to be a single *continuous*

Figure 18-6.
HE: I don't like the looks of that squall coming in. You'd better finish your soda.

presentation that has no mismatches in information, action, or screen direction.

On larger productions, the script supervisor keeps track of these elements, but the director still needs to be keenly aware of continuity.

The script supervisor's job is not to create and maintain continuity, but only to monitor it.

Achieving good continuity is the director's job. In smaller productions, script supervision may be only a part-time responsibility of a production manager or an assistant. In such a situation, the director is the only person who can ensure consistency.

Continuity of Information

In shots that include duplicate or overlapping action, the details should all be the same. For example, dialogue should be identical. Do not allow an actor to say "Hello!" in the two-shot and "Yo!" in his closeup. Though an editor can cut mistakes and trim pauses (if given enough coverage), it helps if the actor's lines are consistent from shot to shot.

Physical details should match. If the actor is holding a glass in the two-shot, then in the closeup, he or she should continue to hold the glass, it should be in the same hand, and the liquid in the glass should remain at the same level.

This may seem obvious, but remember that shots appearing back-to-back in the program may be recorded days or even weeks apart. For example, an actor may park a car in one location and then enter a building hours later and miles away, **Figure 18-7**. Between the two shooting sessions, the clothing carried on hangers has changed.

Continuity of Action

You also need to ensure that action is repeated accurately from one shot to the next. In our demonstration picnic, for instance, the man puts down his soda bottle, points off screen, and talks about a squall over the ocean.

Now suppose the editor wants to use a wider shot of his warning about the squall. The problem is that, in the performer's first shot, he says the line *after* he puts down the bottle. But when he

Figure 18-7.
Since the first shot shows the light shirt on the outside... the same shirt must be outside in the next shot.

Figure 18-8.
He sets down the bottle before he says his line.

He gestures with the bottle as he says the line.

Figure 18-9.
The bottle is already down when the next shot begins.

repeats the action for his next angle, he points off screen with his soda while he says the line and never puts the bottle down, **Figure 18-8**.

The problem arises because the next shot starts with the bottle already on the bench, **Figure 18-9**. If the editor cuts from the closeup to the two-shot, the bottle will jump from the man's hand to the bench.

But since the alert director has shot an insert of the bottle being set down on the bench, **Figure 18-10**, the editor can use it to cover the mistake.

Now we have seen the cutaway of the bottle used for two different reasons: first to cover the omission of the line about the squall and then to conceal the action mismatch between two other shots.

Continuity of Direction

The third and final consistency is continuity of direction. It is called ***screen direction*** because actors' looks and movements are measured by reference to the borders of the screen. Although direction is just one of three principal types of continuity (along with information and action), it is such a large, complex, and important subject that it deserves a more extensive discussion.

Figure 18-10.
HE: I don't like the looks of that squall coming in.

You'd better finish up your soda.

Styles of Coverage

Although most directors have their own working methods, general approaches to ensuring full coverage fall into three broad types: classical, contemporary, and personal.

Classical coverage

In Hollywood before about 1960, it was standard practice to record a scene like this:

Notice several things about this classical approach to coverage. Starting with a wide view, it gradually moves closer and closer to the characters. The complete scene is recorded at least five times, offering the editor a wide choice of shots. Finally, the pairs of over-the-shoulder two-shots and closeups are symmetrical, in that they are almost mirror images of one another.

Classical coverage. A— Establishing shot. B— Master shot. C— Over the shoulder CU of her. D— Over the shoulder CU of him. E— CU of her. F— CU of him.

1. Establishing shot, **A.** Usually a wide shot including much of the location (restaurant, bedroom, campfire, etc.) to orient the audience and establish the positions of the performers.
2. Master shot, **B.** Usually somewhat closer than the establishing shot, but still including all the principal performers and all the action and dialogue.
3. Over-the-shoulder two-shot favoring one character, **C.** This angle moves closer but still holds more than one person.
4. Over-the-shoulder two-shot favoring the other character, **D.** This angle is the reverse of the other one.
5. Closeup of one character, **E.** Closer now, and only one person on screen.
6. Closeup of the other character, **F.** The reverse of its predecessor.

Keep in mind that this example demonstrates classical coverage in its fullest and most conventional form. In actuality, many Hollywood directors omitted some of these angles and covered only parts of the action from each of the angles that they did use. Some directors, like John Ford and Alfred Hitchcock, shot very little extra material, to force editors to cut the film the way the director envisioned it. Others, like George Stevens, often shot many more angles than they could use, to allow themselves the greatest possible freedom in editing.

Contemporary coverage

Today, the establishing and master shots are sometimes omitted, or at least combined. In these wide shots, the actors may begin the scenes with the opening lines, then jump directly to the closing ones, knowing that the remainder will be captured by

Styles of Coverage (continued)

Contemporary coverage.

other camera setups. The photos above present an approximation of the same scene shot in contemporary style. Notice that the angles are not symmetrically opposed and that the coverage tends toward tighter angles for the smaller video screen.

Personal coverage

In the personal style typical of commercials and music videos, each part of the action is treated separately and composed without apparent regard for a formal plan. (Notice that screen direction is reversed in the third and fifth shots.)

Even with the most personal style, however, an astute director will give the editor options by shooting at least some material from multiple angles and by overlapping action from shot to shot.

Personal-style coverage will often not show a formal pattern. (Sue Stinson)

Managing Screen Direction

Screen direction is the concept that people and objects on *screen* should usually point in a consistent *direction*. This means that they should aim more or less at the same edge of the frame (usually left or right) from one shot to the next, until either the sequence (scene or continuous action) ends, or the director indicates a purposeful change in screen direction. **Figure 18-11** illustrates a mistake in screen direction.

In **Figure 18-12**, the screen direction is consistent.

The Importance of Consistent Direction

It is important to maintain screen direction for three reasons. First, whenever you present a new subject, your viewers respond by establishing their basic point of view: where they "are" in relation to the scene. They do this automatically and without thinking about it.

As long as people and objects on screen continue to face approximately the same direction, viewers do not have to make the subconscious effort to reorient themselves. But if someone suddenly switches screen direction, viewers are forced to mentally relocate themselves to match. In Figure 18-11, for example, the viewpoint suddenly jumps to the opposite side of the bench. Though viewers can easily figure out that the second woman has not really moved, the brief effort required to do so diverts their attention from the content of the program and reminds them that it is only a video, rather than reality.

Another reason for maintaining screen direction is to enhance the illusion that an action is unfolding continuously in real time, instead of in numerous separate shots.

Finally, assigning different screen directions to different actions allows you to alternate between them without confusing the audience.

Keep in mind that screen direction is only a convention, and that it can be successfully broken when you have a specific need to do so.

The Power of the Frame

Screen direction can be confusing because you must establish and maintain it in the real world, where you are shooting your program. But screen direction is not determined by events in

Figure 18-11.
A two-shot establishes the screen direction. The woman on the left looks to the right in her closeup, but the second woman also looks screen right, when she should look left.

Figure 18-12.
Now the second woman's closeup matches her direction in the two-shot.

Figure 18-13.
In the real world, the subject walks in three different directions.

the actual world. *It exists only in relation to the frame around the image.* To see how this works, study Figure 18-14, in which a subject walks around three sides of a house.

Notice that in the real world, the subject walks in three different directions: south, then east, then north (as shown in the composite image, **Figure 18-13**).

In the video world, however, she never changes direction, always moving from screen-left to screen-right. See **Figure 18-14**.

But if the sequence includes an angle from the point of view of the man on the front stoop, the screen direction must be temporarily reversed, as shown in **Figure 18-15**, even though the subject follows exactly the same route in the real world.

In short, screen direction has nothing to do with the action itself and everything to do with the viewpoint of the camera.

There are three types of screen direction: look, movement, and convention.

- *Look* means that actors remaining in one place face in the same direction from shot to shot.
- *Movement* means that, from shot to shot, moving actors and objects like vehicles always go across the screen in more or less the same direction.
- *Convention* means that subjects face and/or move either toward the left or toward the right in accordance with the conventions established by maps.

Figure 18-14.
In all camera setups, the screen direction remains left to right.

Figure 18-15.
Basic left-right screen direction. Shot from the man's point of view reverses screen direction. The basic screen direction resumes.

Screen Direction: Look

When two people are speaking to each other, good screen direction calls for one person to face one side of the screen and the other person to face the opposite side. **Figure 18-16** illustrates correct screen direction looks.

Note that screen direction does not have to be exactly identical from shot to shot, as long as the subject is oriented more-or-less toward the same screen edge. Note, too, that the subject is positioned toward the screen edge opposite the direction of the person's look.

Screen direction is often preserved even when the subjects are not together. In a telephone sequence, for example, it is customary for one caller to face screen right and the other screen left. As you cut back and forth between the two, the opposing screen directions reinforce the idea that the two people are talking to each other**, Figure 18-17**.

It is also common to divide the screen with a partial wipe and show both callers at once, **Figure 18-18**.

Figure 18-16.
From almost full front to 3/4 rear, all these angles maintain the same screen direction.

Figure 18-17.
Opposing screen directions... of subjects speaking to each other.

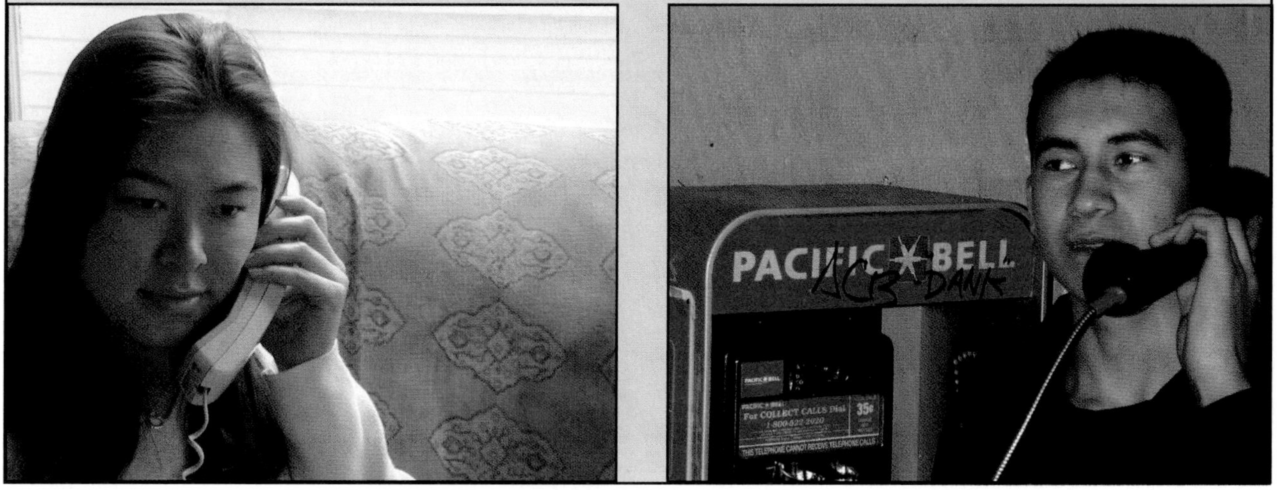

Figure 18-18.
Split-screen conversation between two subjects.

Figure 18-19.
A split screen showing two different phone conversations.

You can also use screen direction to indicate that subjects are *not* related. For example, suppose you show a disk jockey offering prizes for the first person who calls in. The following shots show two different callers responding, as in **Figure 18-19**.

Notice that this split screen contains exactly the same shots as the previous example. But because the callers are facing in opposite directions, it appears that they are not talking to each other.

Screen Direction: Movement

When movement of a subject continues through several shots, the screen direction should be essentially the same in every shot, **Figure 18-20**.

Frequently, you will need to manage the screen directions of more than one subject at a time. In doing so, you help the audience remember who is who in a sequence. You can also use screen direction to show what type of action you are presenting.

Parallel movement

When two subjects start at the same point and head for a common destination, you keep the screen directions of both subjects the same, **Figure 18-21**.

Opposing movement

When subjects are moving simultaneously, but toward different destinations, their screen directions are opposite, **Figure 18-22**.

Divergence

You can also use opposing screen directions to indicate that two people are starting from the same point and heading for different destinations.

Figure 18-20.
Shot-to-shot screen direction maintained.

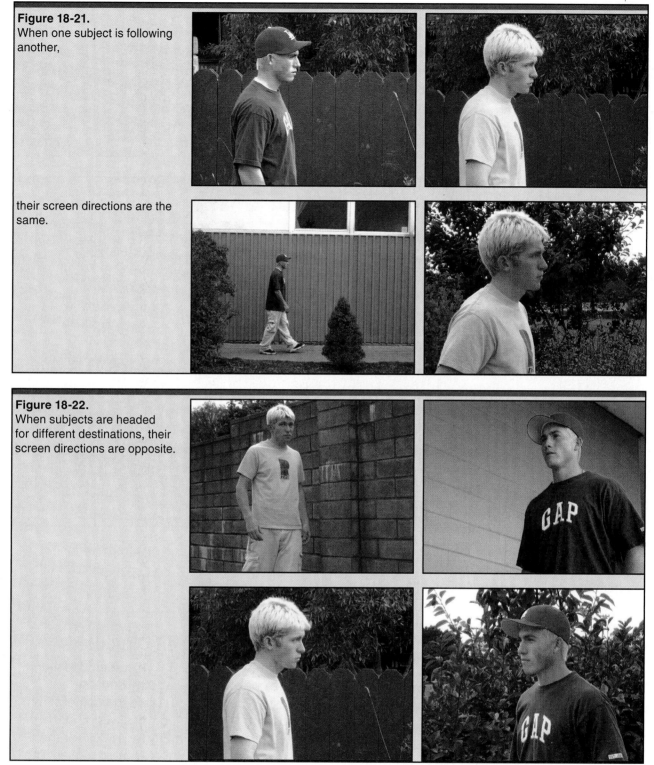

Figure 18-21.
When one subject is following another,

their screen directions are the same.

Figure 18-22.
When subjects are headed for different destinations, their screen directions are opposite.

Random movement

You can use screen direction to indicate that a succession of shots is not a continuous action. To do this, you purposely vary screen direction so it shows no pattern at all. **Figure 18-23** shows the subject headed in different directions, as if to present an anthology of her activities.

It helps, of course, that the backgrounds and costumes are different, as well.

Screen Direction: Convention

Finally, some screen directions are set by convention. As noted in the sidebar, *Scenes in*

Figure 18-23.
A complete lack of consistent screen direction says clearly that the shots are not continuous.

Scenes in Automobiles

Shots made inside automobiles follow special screen direction conventions because movie makers have no choice. If you shoot the driver from the passenger side, the performer will face screen right. The passenger, shot from the driver's point of view, will face screen left.

If you establish that a car is headed from New York to California, convention dictates that it travel from screen right ("east") to screen left ("west"). By the conventions of normal screen direction, that would make the driver and the passenger travel in opposite screen directions and the driver would be moving in the opposite screen direction from the car. However, since viewers have ridden in cars, they recognize the passenger's viewpoint of the driver, and are able to ignore the apparent violations of screen direction.

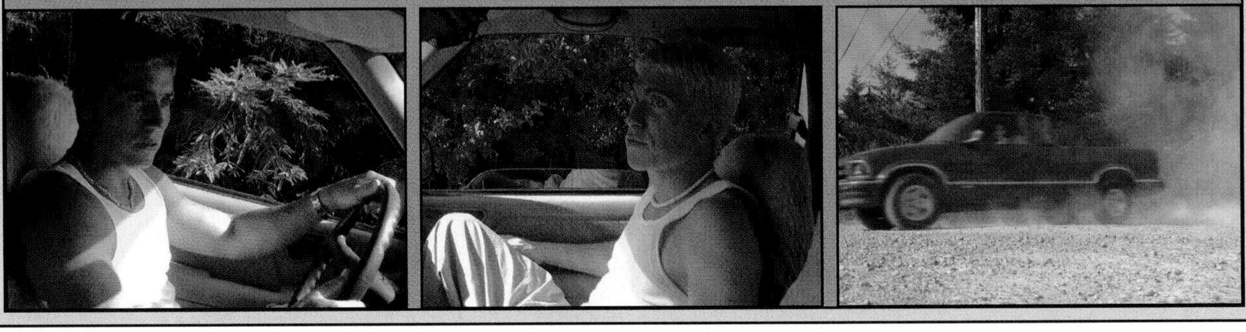

Automobiles, a car traveling from New York to California is established as moving from right to left because on a map, convention places New York to the right of California. See **Figure 18-24**.

Figure 18-24.
Convention dictates some screen directions: a car traveling westward should move from right to left.

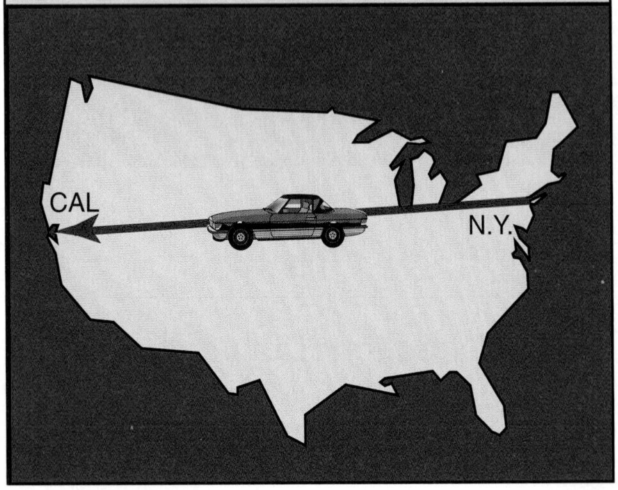

For the same reason, a person watching a sunset will often face screen left because left is the "west" side of the screen.

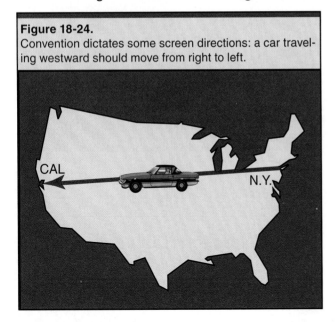

Controlling Screen Direction

To create and maintain a screen direction, use the concept of the ***action line***, an imaginary line drawn to divide the camera from the subject(s). To see how this works, study the camera ground plan in **Figure 18-25**, for the walk around the cottage presented earlier. The action line is represented as an arc between the subjects and camera setups A, B, and C.

As long as subjects and camera remain on opposite sides of the action line, the screen direction will remain consistent, **Figure 18-26**.

But if the camera crosses the action line, screen direction will be reversed. From camera setups D and E, the subject reverses screen direction, **Figure 18-27**.

Of course, there are situations in which you *want* to reverse screen direction. As noted earlier, the man watching from the front stoop of the house would see the walker from the other side of the action line. For this reason, a shot from his point of view would correctly reverse screen direction, **Figure 18-28**.

Figure 18-25.
Ground plan showing camera setups and the action line.

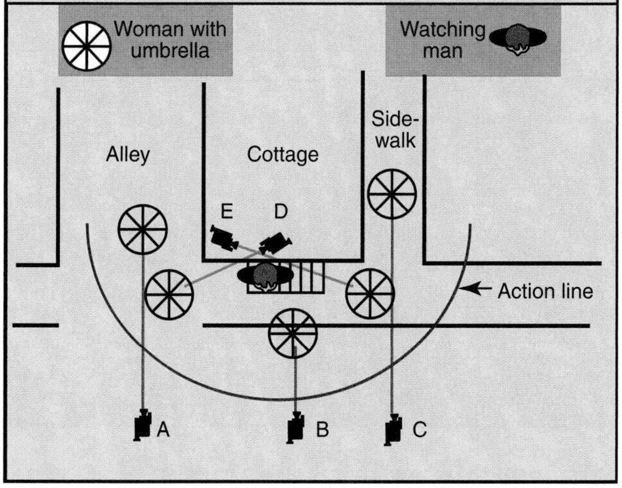

Changing Screen Direction

It is neither possible nor desirable to continue a screen direction indefinitely. When a sequence ends, the next sequence will naturally start with its own new screen direction; but it may also be necessary to change direction within a single sequence. When you need to do this, you have four options.

On-screen change

One way is to show your subject changing direction in the middle of a shot, **Figure 18-29**. That establishes a new screen direction.

Neutral direction

A second way to soften a change in screen direction is by using a neutral screen direction (either toward or away from the camera), as you can see from **Figure 18-30**.

Figure 18-26.
Consistent screen direction:
Shot A. Shot B. Shot C.

Figure 18-27.
Reversed screen directions.
Shot D. Shot E.

Figure 18-28.
Setup A sets the screen direction.
Setup B establishes the man looking.
Setup E represents his point of view.
Setup C resumes the original screen direction.

Figure 18-29.
The audience watches as the subject reverses screen direction.

Cutaway buffer

You can conceal a direction change by inserting a cutaway shot of something else between shots with different screen directions, **Figure 18-31**.

Empty frame

Finally, you can have the subject change direction while off the screen entirely. In the first shot, the subject leaves the frame. The camera holds on the empty screen for a moment, then we cut to a second shot in which the subject is moving in a different screen direction, **Figure 18-32**. This technique works equally well in reverse: cutting to an empty second shot, which the subject then enters.

Figure 18-30.
A neutral screen direction softens the change from one direction to another.

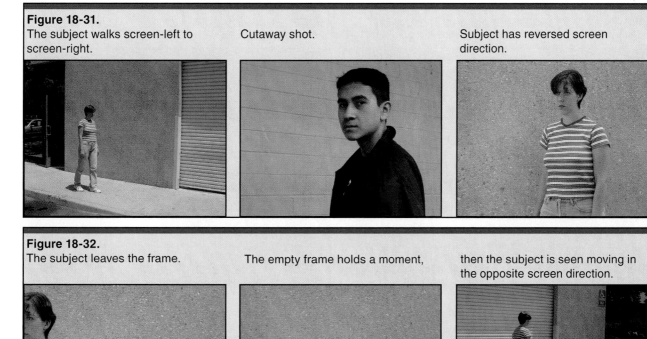

Figure 18-31.
The subject walks screen-left to screen-right. Cutaway shot. Subject has reversed screen direction.

Figure 18-32.
The subject leaves the frame. The empty frame holds a moment, then the subject is seen moving in the opposite screen direction.

For the smoothest possible transition, use both techniques at once: have the subject exit shot one, hold the empty shot, cut to empty shot two, have the subject enter the shot.

In general, strict observance of screen direction is considered less important today; some sequences —or even whole programs—pay little attention to it.

Staging for the Screen

Maintaining screen direction is just one part of a larger task: staging action for the screen. It has been said that half a director's job is to put the camera in exactly the right place at the right time with the right lens pointed in the right direction—in short, to select the ideal camera setup for each shot. What is the most effective setup for each shot and how does the director know it? Partly by instinct—an artist's gift—and partly by practice.

Some would say that there is no such thing as an "ideal" or "right" camera angle. They argue that the "right" angle is whichever one seems to work the best.

The craft of staging action for the camera is grounded in four basic principles concerning the special world on the screen:

1. The actual world is nothing more than raw material for the creation of the screen world.
2. The only unchangeable part of the screen world is the frame around it.
3. The dimension of depth in the screen world is controlled by the director.
4. The window on the screen world is small and the view is somewhat indistinct.

In short, the director does not record the real world, but creates an artificial world on a relatively small, two-dimensional screen that is surrounded by a frame. No matter what you are shooting, your selection of camera angles should be guided by these four principles.

Creating a Screen Geography

1. The actual world is nothing more than raw material for the creation of the screen world. Whether you are shooting in a tiny room, on a large soundstage, or on location in an endless landscape, your actual surroundings are nothing more

Figure 18-33.
Creating screen geography. A—On location at an actual farmer's market. B—Closer view of a vendor's stall.

A

B

Figure 18-34.
By establishing the church in the first shot, the editor makes the graveyard shot seem to be in the same location. (Sue Stinson)

than raw material that you use to create screen images. In the real world, you can see where everything is and how it relates to everything else, but the audience watching the world on the screen cannot do that without your help. So your first job is to create a screen world geography that shows viewers where the action is taking place.

Depending on your needs, you may wish to show the whole environment, or you may want to reveal only enough of the locale to let your audience know where they are. Since **Figure 18-33A** was shot at an actual farmer's market, it can include a considerable amount of real-world space. But since the stall in **Figure 18-33B** is actually a small set erected on a quiet street, we see only enough to tell us what we are looking at.

If needed, you can combine two or more separate real-world places into a single screen environment. For example, the graveyard in **Figure 18-34** is actually many miles away from the church.

As these examples suggest, you create a screen world environment by controlling what you show (and how much of it) and how you combine multiple real-world environments into single screen locales.

Working within the Frame

2. The only unchangeable part of the screen world is the frame around it. The frame is by far the most powerful tool that a video director has to work with. It can hide anything that you do not

Figure 18-35.
The power of the frame to conceal. A—This shot is from the main burglar sequence shown in chapter 17. B—Inserts of the burglar's hand... C—were shot later with a different performer.

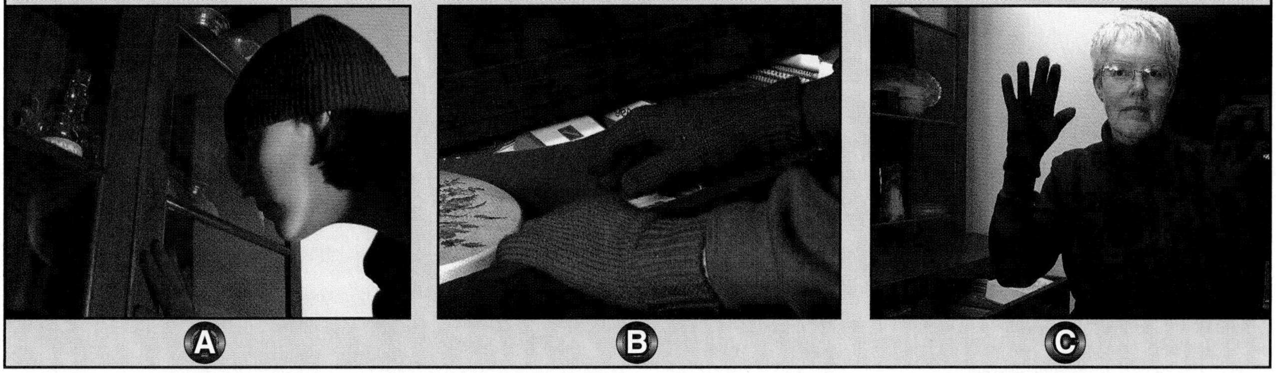

Figure 18-36.
Once the actual farmer's market has been established... the close shots of the set seem to be part of it.

want viewers to see. For example, **Figure 18-35B** seems to be the gloved hand of the burglar shown in **Figure 18-35A**. But as **Figure 18-35C** reveals, the sequence's inserts were actually shot with a different performer. The frame around shot B, with the gloved hand, hides the deception by framing off the second actor.

By hiding the edges of sets or unwanted details of locations, the frame can also suggest that the world displayed on the screen continues indefinitely beyond the borders of the image, **Figure 18-36**. This, of course, is what makes the farmer's market stall set seem convincing.

Managing Depth

3. The dimension of depth in the screen world is controlled by the director. We have repeatedly emphasized that depth in the world of the screen is only an illusion. Controlling this illusory third dimension is an important part of staging for the screen. Sometimes, you may wish to suppress the feeling of depth; but more often, you are trying to enhance it.

Enhancing depth

To strengthen the feeling of three-dimensional space, a director can use several strategies:

- Stage the action to move toward and away from the camera, rather than from side to side. That way, your subjects appear to move through the depth of the scene.
- Set up shots so that the camera looks diagonally at walls, roads, and other surfaces with parallel edges.
- Look for multiple planes parallel to the camera.
- Stage movement against stationary backgrounds, for contrast.
- Use wide-angle lenses. In conjunction with action staged to flow toward or away from the camera, wide lenses exaggerate apparent depth and enhance the sense of dynamic movement, **Figure 18-37**.

Figure 18-37.
To increase the sense of depth, the second image was taken closer to the sign, using a wider angle lens.

Suppressing depth

There are times when you want to reduce apparent depth instead of enlarging it; when you want to turn the action into a design painted on the flat surface of the screen, **Figure 18-38**. To do this:

- reverse all the previous suggestions for enhancing depth.
- keep action parallel to the camera.
- avoid diagonals and converging lines.
- omit foreground objects that would reveal the scale of the action.
- use telephoto lens settings—the longer the better. They suppress the sense of apparent depth and emphasize that what the viewer is really looking at is a video painting with the screen as its canvas.

Figure 18-38.
The telephoto lens squeezes the image into a design on the picture plane.

Scaling for the Video Screen

4. The window on the screen world is relatively small and the view is somewhat indistinct. In staging action for the camera remember that video is a low-resolution medium that cannot capture fine details. Even digital high definition TV images are able to resolve only a limited level of detail. Video is also intended for relatively small screens, most of which are 60 inches or less, measured diagonally. (Video projection systems merely enlarge the images without improving them.)

There are strategies for combating the effects of low resolution and small screen size. To compensate for low resolution, work close to your subjects in order to make details more visible. To offset the loss of dynamic impact on a small screen, use lenses and stage action to emphasize depth. See **Figure 18-39**.

Remember that video's smaller scale is not always a drawback. Well-handled by a director who understands it, video has an intimate quality that welcomes viewers and makes them feel comfortable. This intimacy and comfort is part of the reason for the enormous psychological power of TV.

Moving the Camera

In considering staging for the screen, we have assumed that the camera remains essentially in one place for each shot. In actual video directing, that is often not the case, because the camera is almost always in motion to one extent or another. Sometimes the movement is as obvious as a dramatic crane shot that swoops down

Figure 18-39.
As soon as you have set the scene, work close to your subjects. Use wide angle lenses to enhance impact.

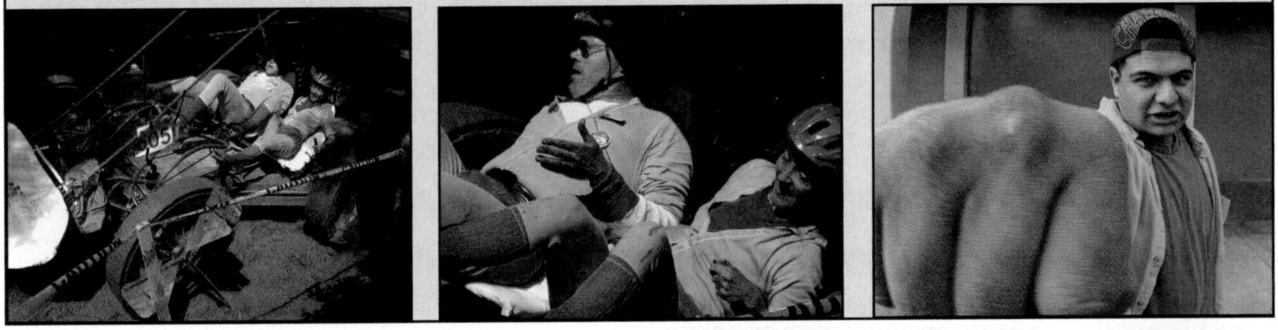

from high above the action. At other times, the movement may be confined to subtle panning and tilting. But even when the camera appears to be still, the operator is usually making small adjustments to keep the shot well-composed. (This technique is often called ***correcting***.)

To make these continuous small corrections, a good camera operator never locks the movement controls on a tripod, even if the shot is supposed to be a fixed one.

Camera movement is one of the director's most powerful and versatile tools. Used well, it can greatly enhance your programs. Used badly, it can consume valuable production time and resources while contributing little or nothing.

Moving shots, especially shots using a dolly, crane, or stabilizer, can be expensive and time- consuming to set up, rehearse, and execute.

Types of Camera Movement

There are several different types of camera movement, and each type creates a distinctive visual effect.

Panning and Tilting

With rotation, the camera support (tripod or dolly) remains in one place while the camera is pivoted on it, **Figure 18-40**. Horizontal rotation is called ***panning*** and vertical rotation is ***tilting***.

Figure 18-40.
Panning and tilting.

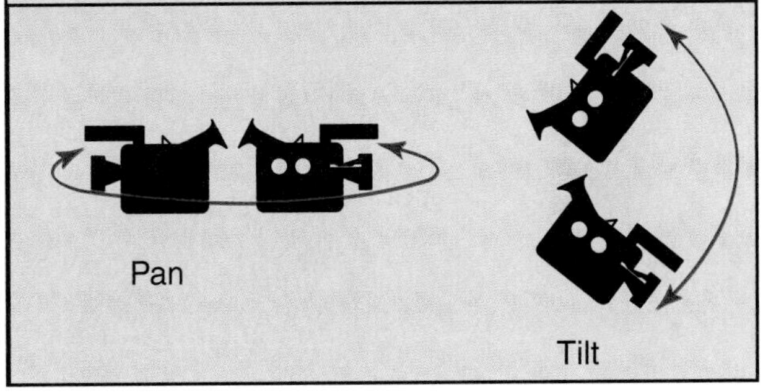

Dollying

Dollying shifts both the camera and its support from one spot to another during the shot. Dollying in and out moves the camera into the action, while dollying laterally moves the camera beside the action. Moving the camera up or down in an arc is called ***booming***, while raising or lowering it on a central column is termed ***pedestaling***. See **Figure 18-41**.

Except for "pan" and "tilt," there are no standard terms for camera movement. You will see "truck," "track," and "dolly" used for horizontal movement and "boom," "crane," and "pedestal" applied to vertical shifts.

Figure 18-41.
Booming and dollying.

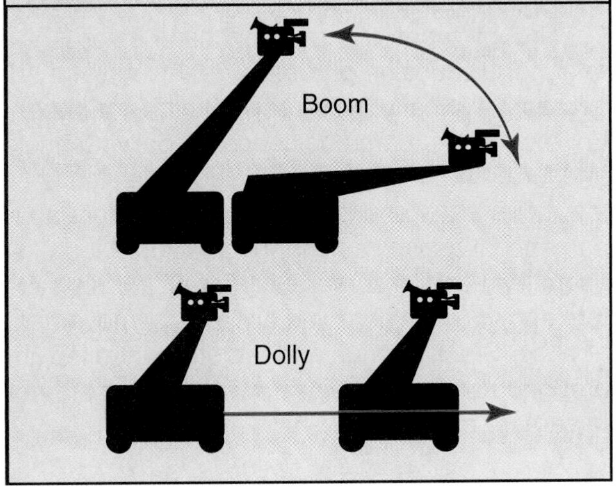

Zooming

The term ***zooming*** refers to changing the lens focal length during the shot to show a wider angle with smaller subjects or a narrower angle with larger ones. Because a zoom out reveals more of the scene, it somewhat resembles a dolly out, and the opposite is true of a zoom in. But a zoom is not a true movement. Because zooms do not involve changing the position of the camera, they do not shift the point of view of the lens. Instead, they merely reduce or enlarge the same image, **Figure 18-42**.

Figure 18-42.
Zooming in progressively enlarges a portion of the original image to fill the frame. (Sue Stinson)

The distinctions between rotations, moves, and zooms are not merely theoretical, because each type of movement produces a different visual effect. In professional production, dollying is often preferred to zooming, though zoom lenses are, of course, indispensable for real-time coverage of news and sports.

Composite movement

Though types of movement can be separated for discussion purposes, real-world videography typically mixes several types at once. For example, a crane designed for dollying back and forth and booming up and down is also fitted with a pan head so that the camcorder can pan and tilt as well as change position. Mount a zoom lens on the camera supported by this system and you can combine all types of movement at once. Stabilizer rigs and remote-controlled jib arm booms also combine zooming with all forms of camera movement.

Today, directors tend to think less in terms of panning, tilting, and dollying. Instead, they simply want the camera to flow freely with the action. For clarity, we will continue to organize camera moves into separate types (rotate, dolly, and zoom), but keep in mind that real-world videography is moving toward a fully integrated form of composite motion.

Reasons for Moving the Camera

Why move the camera at all? Why not compose the action in a succession of fixed shots and then edit them together? That approach is simpler, faster, and cheaper. There are several good reasons for camera movement. Though some people use it just because "it feels right for this shot," thoughtful directors understand what a moving camera can do and why they want to use it.

Traveling shots take extra time to set up and rehearse; they also tend to require more takes because they are more difficult to execute properly.

Moving to follow action

The most obvious reason for moving the camera is to follow action that is also in motion. For example, to keep a walking subject in a full shot, the camera has to dolly along beside her, **Figure 18-43**.

Moving to reveal information

Because the audience cannot see anything that is outside the frame, moving that frame, by moving the camera, delivers new information. For example, **Figure 18-44** shows two successive shots revealing that what looks like a statue is actually alive.

As an alternative, imagine that the camera frames the lower robes of the "statue" and then slowly tilts up to reveal the living eyes in its white face, **Figure 18-45**.

Although both approaches deliver the same information, the slow revelation by the moving camera is subtler and perhaps more effective.

In this example, the camera tilts up to reveal the human eyes in the "statue's" face. When the movement is a dolly or crane shot, the new information may be about spatial relationships, which the camera reveals by rolling through the acting area.

Figure 18-43.
By dollying along with a moving subject, the camera maintains the same perspective.

Figure 18-44.
Shot 1 establishes the "statue." (Sue Stinson) Shot 2 shows her moving.

Figure 18-45.
The beginning of the tilt. The end of the tilt.

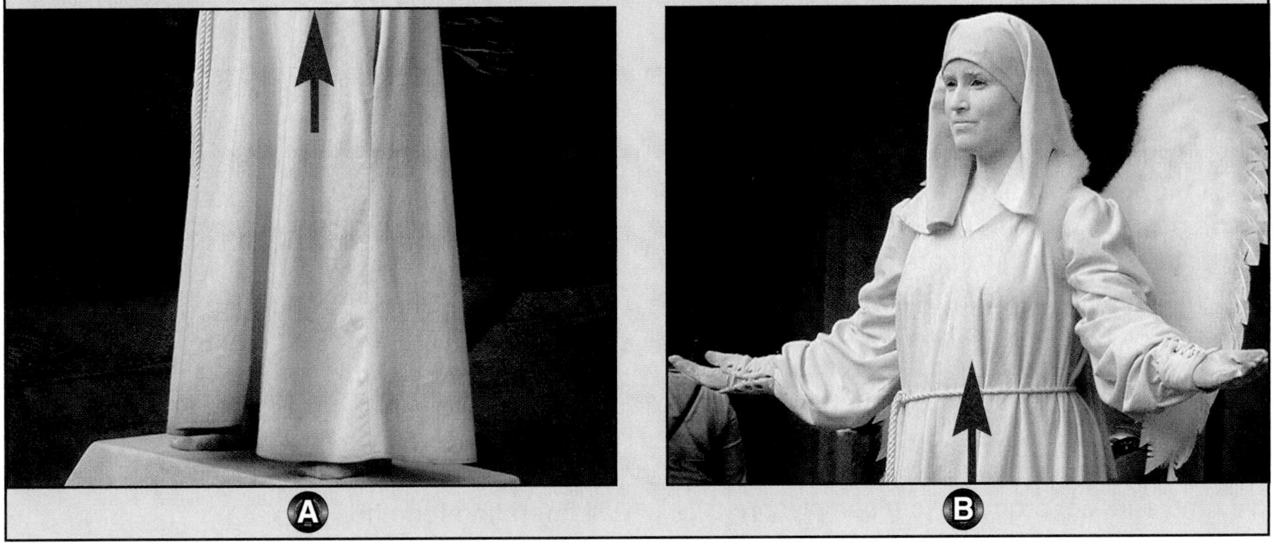

Moving to emphasize depth

By dollying into or out of the action, the camera strengthens the feeling of a third dimension in the scene. As explained in Chapter 5, *Video Composition,* moving sideways instead of in or out can also enhance the sense of depth by changing the relationships of different elements within the image.

Moving to involve the viewer

Since the camera functions as the viewer's "eyes," it carries them with it into the center of the action, involving them kinetically in the story.

Moving to increase energy

By its very nature, movement is more dynamic than stillness. A good director knows when to keep the camera still and let the action move across and through it, and when to move with the action so that the background changes behind it.

Moving to enhance continuity

Above all, a moving shot strengthens the feeling that events are unfolding right before the audience's eyes. No matter how smoothly separate shots are cut together, they are still unconsciously perceived as assembled pieces. A continuous moving shot, however, feels like "the real thing."

Techniques for Moving the Camera

Like other aspects of directing and camera work, good camera moves are carefully planned and executed. The audience may be unaware of a well-designed move and yet unconsciously pleased and satisfied by it.

In professional productions, responsibility for camera moves may be shared among the director, the videographer, the camera operator, and often the chief of the dolly crew.

In planning camera movements, you need to consider composition, motivation, and speed.

Composition

A moving shot consists of three parts: before, during, and after the actual motion. Of these, the start and finish are especially important. In composing them, remember that both the beginning and ending frames should be pleasing compositions. When practical, the ending composition should be the stronger of the two. By following these suggestions, you give viewers a satisfying feeling of a beginning, middle, and end—of progress to a completion.

Motivation

Earlier, we identified a number of reasons (motivations) for moving the camera. Here, we use the word motivation in a special sense: the visible justification for making the move. The most common justification is subject movement. When performers walk across the screen and the camera starts moving to keep them in the frame, the audience takes that motion for granted. When the camera dollies forward to get into the center of the action, the viewer instinctively understands and accepts the move. In the red truck example on page 365, the camera move brings the viewer forward to greet the truck and its driver. The audience will often accept a move that lacks clear motivation, but the best moves justify themselves to the viewer.

Speed

A well-executed camera move accelerates and decelerates like a car that goes from 0 to 60 and then slows to a stop again. (Obviously, this does not apply to moves intended to parallel moving subjects.)

Finally, keep in mind that the best camera moves are unnoticed by the audience. They may be aware that their viewing position has changed, but they do not notice all the video craft required to change it.

The exceptions occur in programs like commercials and music videos, where dynamic camera moves may be used purely for dramatic effect.

A Reminder

This chapter has concentrated on the "cinematic" part of the video director's job: the design and execution of images and shots that succeed individually while providing suitable material for editing. As you think about these visual aspects of directing, remember the other half of the job, covered in Chapter 17: directing to communicate with the audience and to draw effective performances from the cast. Directing at the professional level requires all these skills, all operating together.

A Typical Composite Movement

Here is a single shot that includes several different camera movements. As a truck pulls off a forest road and into a campground, the camera:

1. Dollies to the right to hold the truck in frame.
2. Dollies forward while booming downward.
3. Makes a right angle and continues forward and downward to the side of the truck.
4. Booms down to a low angle shot.
5. Dollies back to hold the subject as she gets out of the truck.
6. Pans with her as she walks away.

Notice how the foreground tree acts as a fixed reference that helps enhance the feeling of forward and downward movement.

The truck and camera movements. A—As the truck appears, the camera dollies right. B—At the end of the dolly, the tree appears. C—The camera dollies and booms down. D—The dolly/boom continues. E—The foreground tree enhances the sense of movement. F—The truck stops as the dolly/boom down continues... G—and ends framing the driver. H—The camera dollies right to hold the driver... I—and booms down to a worm's-eye angle. J—The camera dollies back while panning to hold the subject.

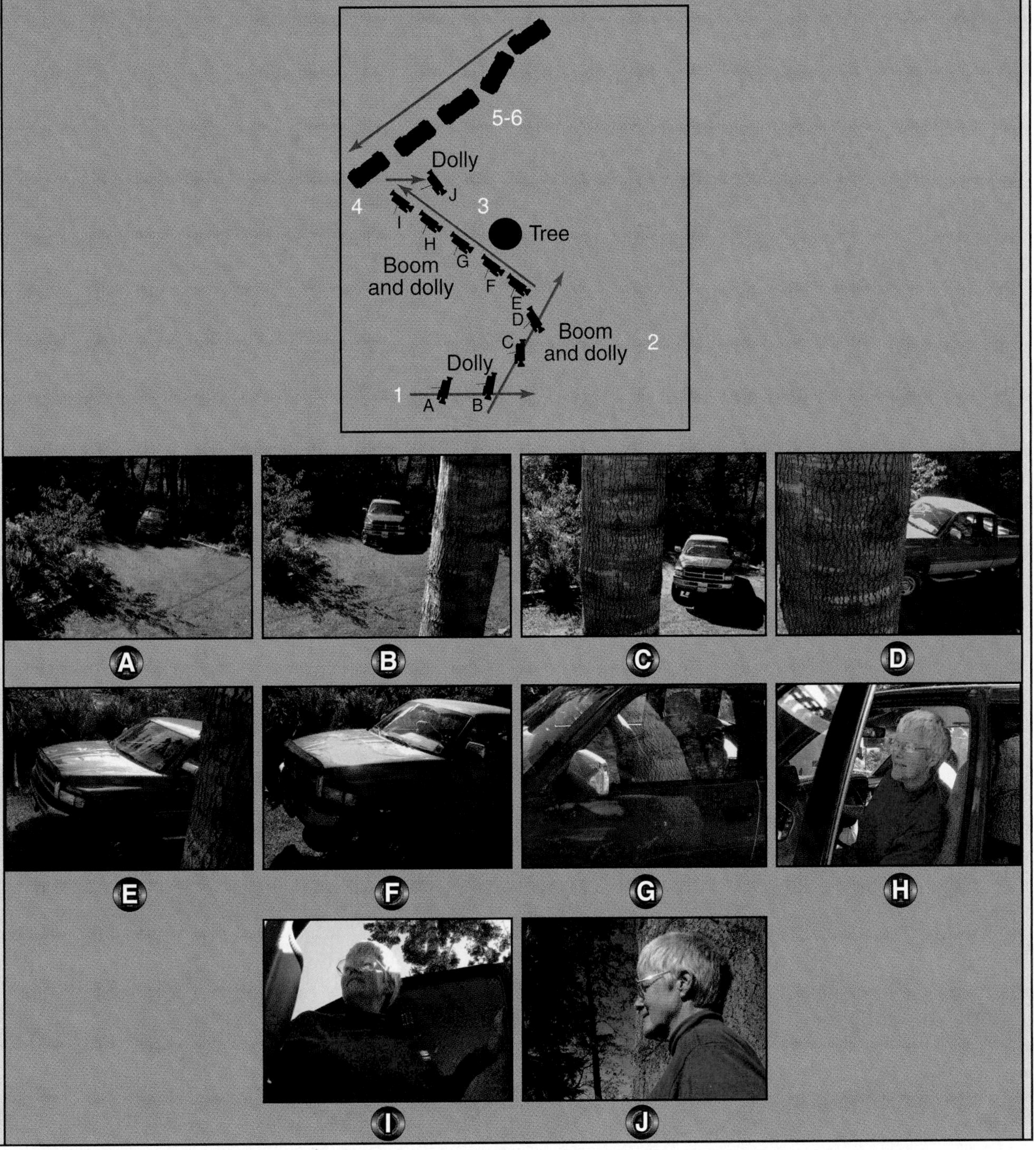

Improvised Dollies

Almost any moving (or movable) object can be pressed into service as a camera dolly. Here are some suggestions.

Chairs

A thrift-shop wheelchair makes an excellent dolly because it is light, collapsible, and fitted with very large wheels. With larger cameras, the system is limited to low angles because the operator shoots sitting down. With an external LCD screen, however, you can hold the camera above your head for a more normal height.

Office chairs also make smooth-rolling dollies, especially on the hard, even floors of commercial buildings. The sound recorded during moving shots may not be usable, however, because the small wheels are noisy.

For this shot, the videographer was pulled along in a child's wagon!

People movers

When possible, take advantage of systems that do your moving for you. Airport beltways allow the camera to keep pace with subjects moving along side. Escalators are naturals for improvised crane shots. You may wish to follow a subject on the escalator, or else shoot down on a subject walking on a lower floor.

Some hotels, casinos, and even parking structures have glass-sided elevators that allow dramatic booming up or down. It is difficult to conceal the camera's means of transport, however, because the passing floors interrupt the view.

An available-light moving shot in a mall.

Vehicles

Vehicles make good camera platforms. Big-budget productions use camera cars and trucks, as well as special suction-cup camera mounts; but you can obtain excellent shots out car windows by framing shots in an external viewfinder and using your arms as shock absorbers. To reduce vibrations that shake the image, make sure that the camcorder does not touch any part of the car.

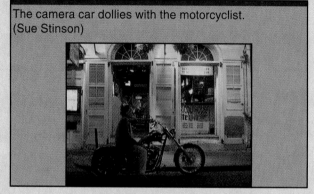

The camera car dollies with the motorcyclist. (Sue Stinson)

Hand-held crane shots

Camera stabilizers permit hand-held moving shots that closely imitate the movements of large and expensive studio cranes.

For example, the previously described moving shot of a truck entering a campground was actually made using the steps of a forest cabin.

Remember: in the screen world, if it is not in the frame, it does not exist!

The "crane shot" was made with a hand-held camcorder.

Chapter Review

Answer the following questions on a separate piece of paper. Do not write in this book.

1. Providing coverage from a variety of angles gives the editor ______ shots.
2. *True or False?* Screen direction exists only in relation to the frame around the image.
3. The ______ is an imaginary line drawn to separate the camera from the subject(s).
4. Horizontal rotation of the camera is called ______; vertical rotation is ______.
5. A ______ makes a good improvised dolly because of its large wheels.

Technical Terms

Action line: An imaginary line separating camera and subject. Keeping the camera on its side of the line maintains screen direction.

Booming: Moving the entire camera up or down through a vertical arc (also *craning*).

Color shot: A view of the scene, or a detail of it, not directly part of the action.

Correcting: Making small continuous framing adjustments to maintain a good composition.

Cutaway: A shot other than, but related to, the main action.

Dollying: Moving the entire camera horizontally (also *trucking* and *tracking*).

Insert: A close shot of a detail of the action, often shot after the wider angles, for later *insertion* by the editor.

Panning: Pivoting the camera horizontally in place.

Pedestaling: Moving the camera up or down on its central support.

Protection shot: A shot taken to help fix potential problems with other shots.

Screen direction: The subjects' orientation (usually left or right) with respect to the borders of the screen.

Tilting: Pivoting the camera vertically in place.

Zooming: Changing the lens' angle of view to fill the frame with a smaller area (zooming in) or a larger area (zooming out). The effect is to apparently reduce or expand the size of subjects in the frame.

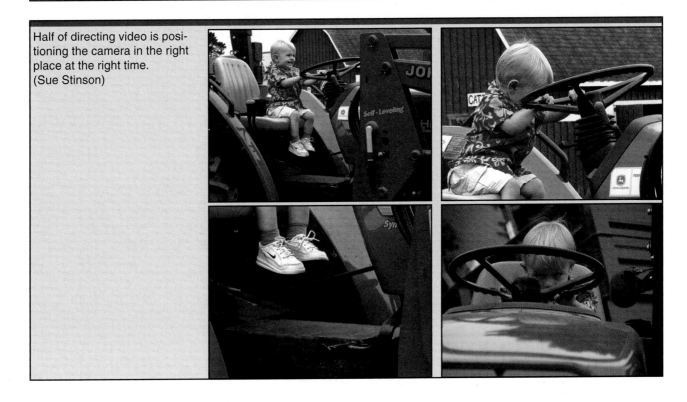

Half of directing video is positioning the camera in the right place at the right time. (Sue Stinson)

Camcorder Stabilizers

Today, perhaps a majority of moving shots is made with a Steadicam, the industry standard in camera stabilizing systems. (Glidecam and other designs are also available.)

In this photo, Melanie Motiska, a professional Steadicam operator, demonstrates a large model designed for heavyweights like the Panaflex camera mounted on it. Steadicam models are also available for video cameras ranging from HD broadcast units down to very small mini-DV camcorders.

19

Editing Operations

Objectives

After studying this chapter, you will be able to:

- Describe the craft of editing.
- Demonstrate the fundamental operations of editing.

About Editing Operations

Editing is the task of taking the materials recorded during production and transforming them into a finished audiovisual program. Like videography, simple editing can be performed by a beginner with almost no instruction; while at the professional level, editing is a rich and satisfying art that requires mastery of the many tasks, skills, and techniques of video postproduction. These tasks and techniques are the subject of this chapter.

Video Postproduction

This is the first of six chapters on video *postproduction*, the general term for the fascinating process of turning raw video footage into a finished program ready to play on a movie screen, a TV set, or a website.

The terms *editing* and *postproduction* are more or less interchangeable. We prefer the shorter "editing."

Editing deserves extensive coverage for three reasons. First, it is a complex process made up of many different operations that require explaining. Next, editing is performed almost entirely with computer programs — powerful software that can be daunting to approach and time-consuming to master, **Figure 19-1**. Most importantly, editing is the production phase in which everything finally comes together to create the very different "video world" that has preoccupied us throughout this book.

Creating the Video World

Chapter 9, *Program Development*, shows how the writer designs a blueprint for the video world to be constructed for a program. Chapter 18, *Directing for Form,* explains how the director and production staff build the pieces that will be assembled to make this world. Now, Chapters 19–24 on postproduction reveal how the editor assembles the video world from the raw materials designed by the writer and constructed by the director.

This three-stage process is *additive,* as explained in the text and sidebar that follow.

An edited program consists of many layers, all working together and all transparent to viewers (who notice only the program content). These layered components, **Figure 19-2**, typically include:

- Production video and audio.
- Audio effects and background tracks.
- Music.
- Transitions.
- DGEs (Digital Graphic Effects).
- Titles and other graphics.

Historically, different editing tasks have been performed by different specialists, particularly music and sound effects editors, audio mixers, special effects companies, title houses, and

Figure 19-1.
The Video Studio work screen. (Ulead)

Figure 19-2.
A program contains many elements. (Ulead)

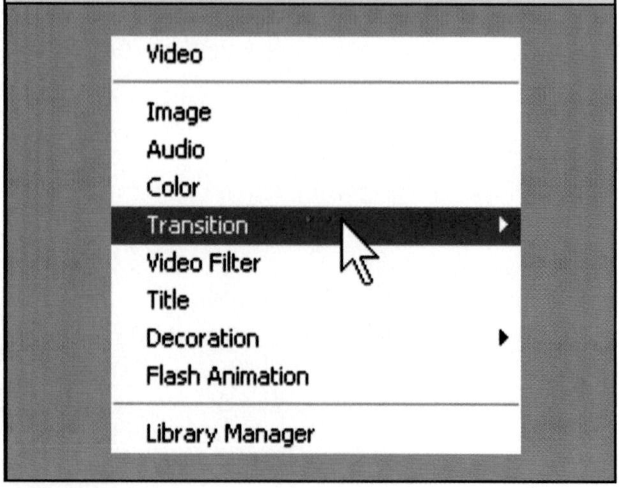

The Convergence of Film and Video

As noted in Chapter 1, the technologies of video and film have been growing closer together; this is especially true in postproduction.

Today, most films are copied onto digital video for editing. This edited video "workprint" eventually serves as a pattern by which the original film negative can be cut to match the tape. Release prints are then made from the cut negative.

Most transitions, titles, and special effects are now created by making high-quality digital duplicates, which are ultimately copied back to film.

In fact, many films are edited as "digital intermediates." Instead of low-quality digital workprints, ultra-high-quality video copies are made of the films. When postproduction is complete, the digital intermediates are copied to fresh film, to make masters for film release prints. (In this system, the original film negative is not used again.) In some theaters, the programs are even projected as high-quality videos.

Film is likely to remain a preferred recording medium for movies because many directors and cinematographers like its visual characteristics. Some high-budget movies (like the second *Star Wars* trilogy) are recorded as film-quality digital video, and low-budget features can look very good indeed on HD (High Definition) Video.

In all but lighting and cinematography, then, video professionals can feel equally comfortable in film production, because so much of the process is now the same in both media.

laboratories. Today, however, every one of these specialties can be performed right on the desktop — and a single editor is expected to master all of them.

The Challenge of Digital Post

If digital editing is challenging because it demands mastery of so many specialized operations, it is made doubly difficult by the peculiarities of editing software.

"Post" is widely used shorthand for the longer "postproduction."

To begin with, software program components simply *must* be organized, or else no one could use them, or even find them. The problem is that the organization scheme is always somebody else's and never your own. You are forced to follow thinking and adopt work methods that may not be anything like yours.

Secondly, features must be labeled, often with unfamiliar names and/or tiny pictures (icons) that fail to illustrate their functions. Learning new software is often like learning a foreign language.

In addition, many of these functions are buried in menus or sub- and sub-sub menus, where their relationship to the main menu is not intuitive, **Figure 19-3**.

Moreover, editing software is complicated by its own versatility. To fill a variety of needs, the programs offer more features than most individual editors will use; and to accommodate different work styles, the software may include several different ways to activate each function. In many cases, you can perform an action via a pull-down menu, a screen icon, a keyboard shortcut – or even more than one of each!

Finally, no two editing programs look or work exactly alike, and some are highly individual. Others are not really unique, but their quirky appearance (sometimes called a "skin") conceals their similarities to other programs. See **Figure 19-4**.

As a result, learning editing software can take hours of exploration and frustration; sometimes,

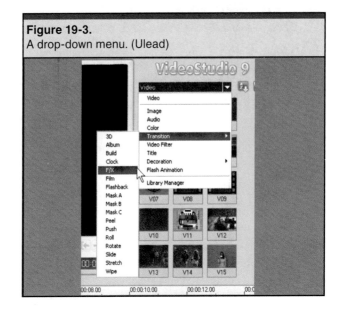

Figure 19-3.
A drop-down menu. (Ulead)

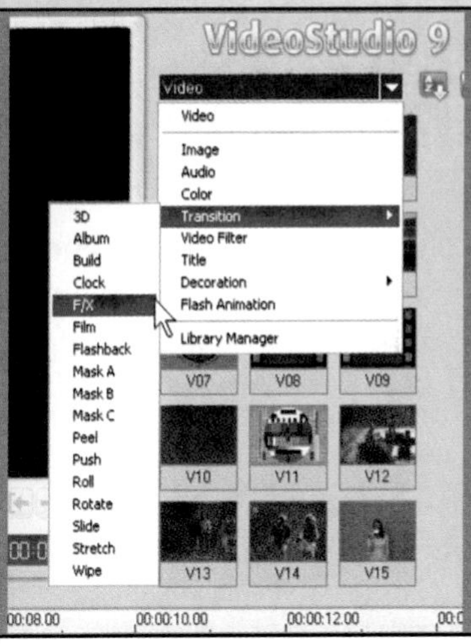

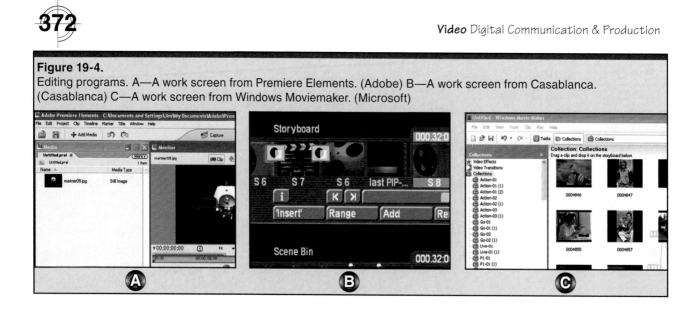

Figure 19-4.
Editing programs. A—A work screen from Premiere Elements. (Adobe) B—A work screen from Casablanca. (Casablanca) C—A work screen from Windows Moviemaker. (Microsoft)

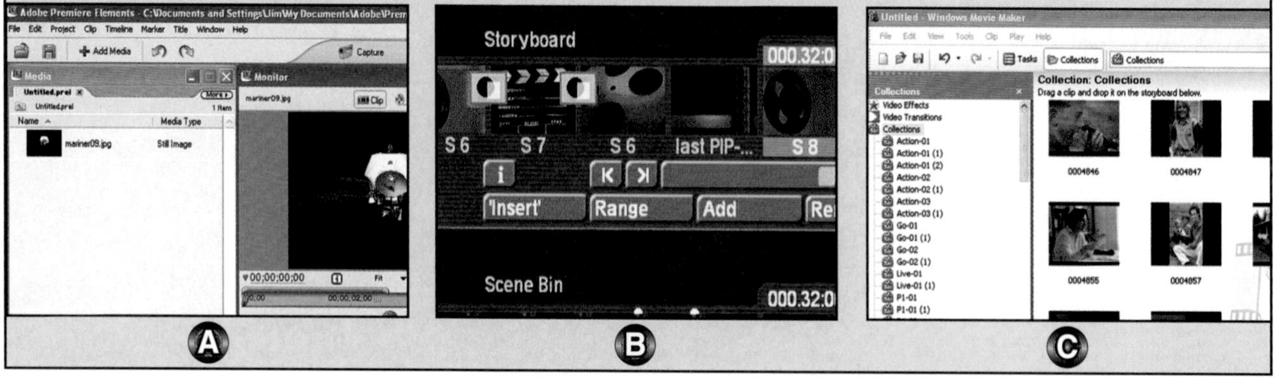

The Importance of Experience

Carnegie Hall is a New York auditorium in which the world's finest musicians perform, often in solo concerts. In a famous old story, a New York pedestrian flags down a taxi; but instead of getting in, he just leans in the window.

PEDESTRIAN: Hey, Cabbie: how do I get to Carnegie Hall?

CAB DRIVER (*disgusted*): Practice! Practice!!

The same good advice applies to editing software.

high-end professional packages cannot be mastered completely, even after months of full-time effort.

These peculiarities are common to almost all complex and powerful software applications, whether designed for video, graphics, word processing, or management applications.

The Underlying Concepts

With all these obstacles, how do editors learn a software application—and how do they transfer their hard-earned skills to a different editing program when needed?

They do it by grasping the concepts that lie under the program skins, behind the menus, below the icons. As we will see in later chapters, all editing programs share the same basic workflow, from starting a new project to burning DVDs. All of them use one (or both) of two graphic metaphors for assembling footage: either a slide/storyboard or a timeline. The tabs and buttons and icons address the same features and the various working windows display the same kinds of information, no matter how different everything may first appear, **Figure 19-5**.

Once you have learned a program, you should have little trouble in using another one of similar complexity. That will free you to move up to a more powerful and versatile software package.

The Range of Editing Software

It is probably easier to start with entry-level software and then progress to more sophisticated applications. To accommodate this growth path, editing programs are available for several different kinds of users:

- Casual amateurs are often satisfied with the simple applications bundled with their computer operating systems, such as Windows Moviemaker.

Figure 19-5.
Video Studio 9 organizes the workflow. (Ulead)

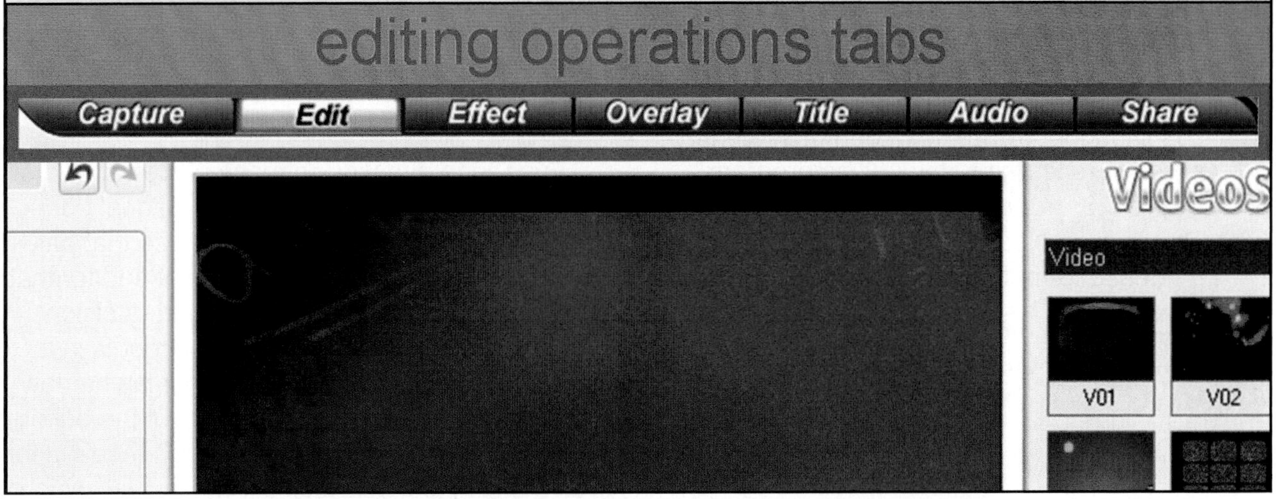

- Hobbyists may want easy-to-use software with features for more professional-looking programs. Adobe, Sony, Ulead, and others make popular programs addressed to this level.
- "Prosumers" (advanced amateurs and beginning professionals) can master the more sophisticated features of these same programs, or move up to professional grade software from the same companies.
- Professionals typically use one of several powerful programs such as Final Cut Pro, Avid Express, Sony Vegas, Adobe Premiere Pro, or Ulead Media Studio Pro.

Some editors prefer stand-alone systems consisting of software designed for and built into special-purpose hardware that is used only for editing. Systems from Applied Magic and Casablanca are examples of these "black box" products, Figure 19-6.

Figure 19-6.
A stand-alone editing system. (Casablanca)

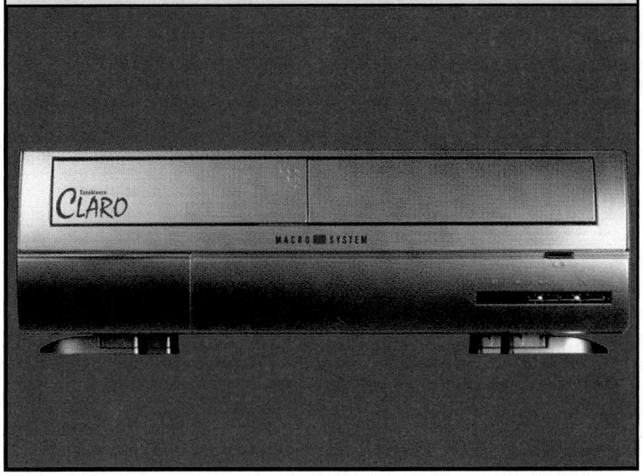

It would be confusing to illustrate postproduction operations with bits and pieces of all the fine editing programs available. To provide consistent examples, these chapters use VideoStudio from Ulead Systems, Inc. This software has several advantages:

- It covers most of the range of editing programs by allowing users to select any of three different (though related) applications of varying sophistication and complexity.
- It uses the overall approach favored by most amateur and professional applications.
- Its tabbed screen system reflects the basic steps in editing programs.
- It shares its more sophisticated capabilities with professional products like Adobe Premiere Pro and Apple Final Cut Pro.
- Its clean, attractive screen design makes it easy to reproduce on the printed page.

In short, VideoStudio is not the only excellent postproduction software; but it is well-suited to the needs of this book.

The Design of These Editing Chapters

The five chapters on digital postproduction are designed to move from the general to the specific:

- Chapter 19, *Editing Operations* (this chapter), lays out the basic processes of all media editing, whether digital, analog, or even film.
- Chapter 20, *Editing Principles,* shows how to build the special world of video programs

from the materials created during preproduction and production.

- Chapter 21, *Digital Editing,* explains the concepts behind digital editing activities that are common to all software programs.
- Chapter 22, *Mastering Digital Software,* uses Ulead VideoStudio to demonstrate an approach to learning any editing program.
- Chapter 23, *Authoring DVDs,* explains how to produce copies of programs in this popular release format.

The book concludes with *Chapter 24, Analog Editing,* which can be useful for occasional projects by beginning editors.

The Craft of Editing

At its simplest, video editing means placing one shot after another to create an organization that makes sense while it presents information or a story. At the next level of complexity, editing achieves that organization and presentation invisibly. The audience perceives the program as a simple continuous flow, without noticing that it is carefully built up, one piece at a time, of many separate units of picture and sound. At its most sophisticated, editing does more than organize information and present it invisibly. It does those jobs with *style*, with an emotional character that touches viewers' feelings as well as their minds.

A two-hour program is a mosaic of hundreds or even thousands of individual pieces of video and audio.

Every one of a video's many pieces appears in the program as the result of an editorial decision. The editor has determined to use that particular element at that particular point and for that length of time. Multiply this three-part decision by the hundreds or thousands of elements in a professional program, and you can imagine how much artistic control is in the hands of the editor.

To illustrate the number of choices required for even the simplest edit, look at **Figure 19-7**, which shows the raw materials for a typical sequence.

Should the editor begin with an establishing shot, then move to a two-shot, as in **Figure 19-8**?

Perhaps the first shot should establish Subject A before revealing the whole scene, as in **Figure 19-9**.

Which choice is right? Either one, depending on the program's needs at that point. In fact, there are 20 different ways to select and sequence the first two shots in this sequence. How do you decide, then? Ultimately, you do it

Figure 19-7.
Master shot. Two-shot favoring A. Two-shot favoring B.

Closeup A. Closeup B.

Figure 19-8.
Master shot. Two-shot favoring A.

Figure 19-9.
Closeup A. Master shot.

by an instinct for the right thing at the right time. At its highest level, editing is an art that cannot be fully described.

Like all arts, however, it is created through craft. The craft of editing consists of the *operations* of postproduction and the *principles* that guide them. The editing operations discussed in this chapter are the things that you do to create a program. Editing principles (covered in Chapter 20) are your *reasons* for doing those things. If you master these editing operations and principles you will command the craft, and once you have mastered the craft you are ready to practice the art.

Since editors are dependent on their directors for competent footage, we will occasionally look back at the director's job of recording appropriate raw material for editing.

In editing as in directing, many people evolve through three stages:

- Beginners often record material without planning the shooting and without modifying the raw footage. The result is called "home movies."
- Experienced amateurs record material from selected viewpoints and cut out the mistakes and repetitions in the footage.
- Professionals create and record material in short, well-planned pieces and assemble programs by selecting and sequencing those pieces.

To put it another way, amateur video editing tends to be *subtractive*, while professional editing is *additive*. In order to understand the nature of true editing, you need to recognize the fundamental difference between the subtractive and additive approaches.

Subtractive Editing

Subtractive editing means removing elements from a shooting session by copying the camera original while omitting the unwanted parts. Editing subtractively resembles the way you might organize a folder of still photos for mounting in an album: first, you discard the obvious mistakes; then you remove duplicate or near-duplicate shots. Removing the mistakes improves the look of the album page and cutting duplicate shots makes it more interesting. For the same reasons, videos created by subtractive editing are at least less boring than raw footage.

The problem is that a video is not an album of pictures that happen to move, but a presentation of actions and/or ideas with a beginning, middle, and end—in short, with a *continuity.* Since video shots are not always recorded in their intended sequence, showing them in their original order can deprive them of some of their meaning and much of their effect.

Additive Editing

Even when a video records real events like a vacation or family party, its component shots can often be taken out of shooting order and rearranged to make better sense or tell a more coherent story. Used in this way, shots become self-contained building blocks that can be selected and placed in any order to construct a program from the ground up. This is additive editing.

With ***additive editing***, a video program starts as an absolute blank, like a canvas before an artist puts paint on it or a foundation slab before a house is constructed, **Figure 19-10**. On that empty canvas (a blank computer timeline) the editor places pictures, sounds, graphics, titles, and effects one piece at a time, building up a structure from nothing until it becomes a complete program.

In fabricating your program, you are limited only by the availability of material. You can ignore shots that you think unnecessary, arrange shots in whatever order you choose, repeat shots or pieces of them, and include material that was not originally shot for the program. In short, you can do anything you like, as long as you remember that you are not improving an existing program that was created in the camcorder. Instead, you are creating a *new* program, completely from scratch.

Without the key concept of additive editing, you can only make photo albums that move. But if you understand the basic idea that editing means creating something quite new, then you can master the five phases of postproduction.

Editing Phases

Here are the five major editing phases.

- **Organizing.** First, the video shots, sound components, and other raw materials are catalogued and filed so that each separate piece can be located for use.

Figure 19-10.
Like this empty work screen, an additive editing project begins with a perfect blank. (Ulead)

Figure 19-11.
Digital storage is preferable for archiving programs.

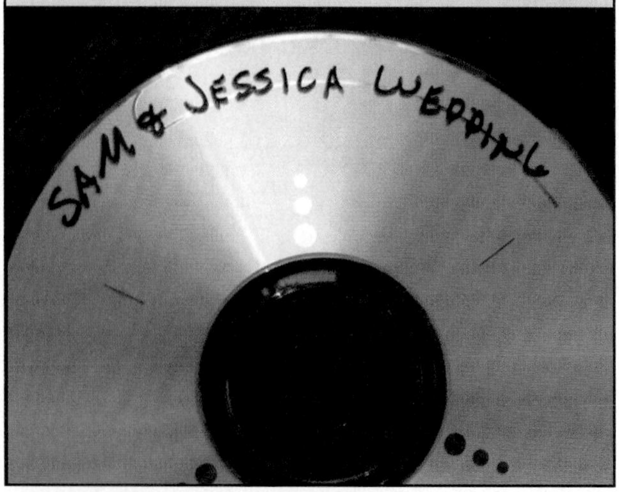

Subtractive and Additive Editing

Here are sequences from two videos editied from the footage shot when a famous team of Clydesdale draft horses visited the town of McKinley. Both programs were constructed from essentially the same raw footage.

Subtractive editing

The first program was edited subtractively. Although two near-duplicate angles have been cut, the remaining footage is presented in the order shot. Notice that the effect is repetitive, and arbitrary changes of screen direction give the program a choppy look.

Editing subtractively means cutting bad or repetitive shots.

Additive editing

The second video was created by additive editing. This opening sequence was built shot by shot to introduce the team and establish the locale. Notice that screen direction is changed by cutting in a head-on shot of two horses before reversing the orientation of the team in the final shots.

Editing additively means creating a sequence from scratch.

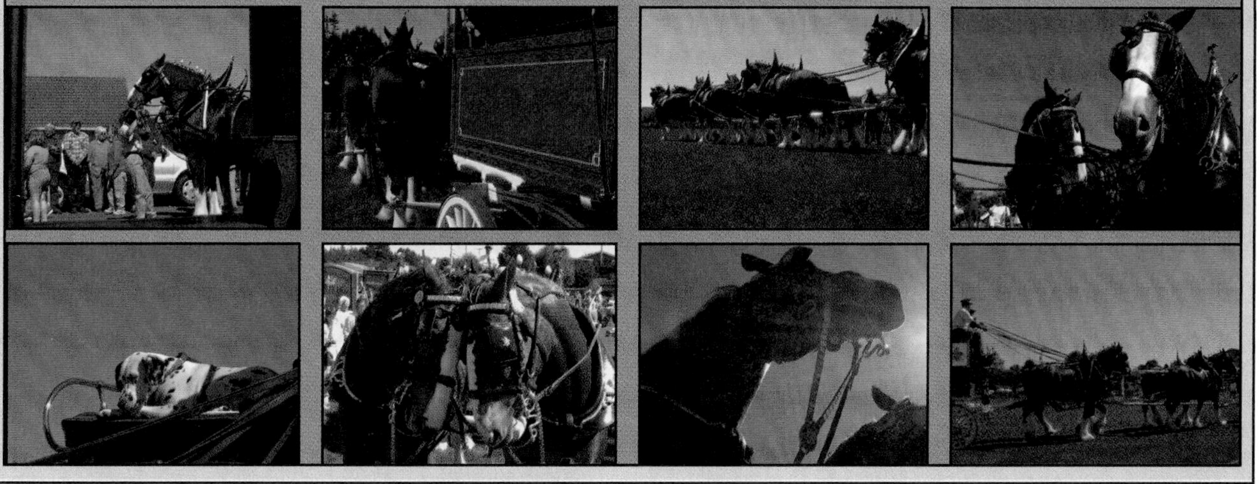

- **Assembling.** The pieces of the program are chosen and arranged in the proper order.
- **Enhancing.** The selected picture and sound elements are modified, where necessary, to improve their quality or adjust their effect.
- **Synthesizing.** The separate layers of material (for instance, live action and graphic visuals, plus production sound, sound effects, and music) are blended together and the shots and sequences are linked by transitions to create a single, seamless program.
- **Archiving.** Finally, the finished video is stored on tape or disk, **Figure 19-11**.

We will look at each operation in detail. Though we must examine them one at a time, keep in mind that a video editor typically works on several of these operations at once.

Remember, too, that "video" is the general term for electronic motion pictures, so that the phrase "video postproduction" covers sound as well as image editing.

Organizing

All video projects begin with a great deal of picture and sound material, and this material commonly suffers from several problems. Every shot is somewhere on a recording that may be an hour in length. The only way to reach a shot is by rolling its tape down to the place where that shot may be found. The fastest way to roll through a tape does not play visible picture or audible sound; so you cannot identify material by inspection as it goes by. When you do reach the right section of a tape, you often find various angles and takes of a shot that look quite similar. In short, materials to be edited can be time-consuming to locate and difficult to identify. For these reasons, it is useful to organize them systematically before you start to use them.

Organizing audiovisual materials involves *identifying*, *describing*, *locating*, and *managing* them.

The change to tapeless video recording at least allows random access to footage.

Identifying

In most professional productions, shots are identified by a ***slate*** recorded at the start of each take and noted in a master copy of the script, **Figure 19-12**.

Figure 19-12.
Slate information is often written with a felt marker.

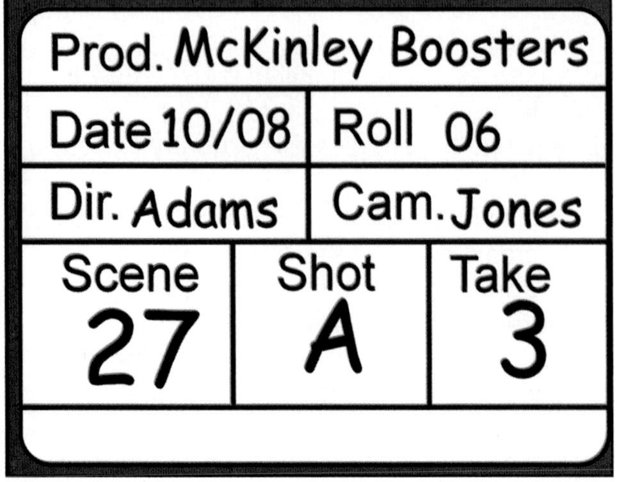
Prod. McKinley Boosters
Date 10/08
Roll 06
Dir. Adams
Cam. Jones
Scene 27
Shot A
Take 3

When a shot cannot be slated before it starts, an "end slate" is recorded afterward, with the slate held upside down to signal that it goes with the preceding shot instead of the following one.

Professional slates generally show the program title, director, and videographer—information that remains the same from shot to shot. The tape cassette in the camera is also identified, typically by a simple number. The individual shots are usually identified by code in which the scene is numbered, the shot is given a letter, and the take is numbered. (A *shot* is an uninterrupted length of audio/video footage from camera start to camera stop. A *take* is one attempt to record a shot, an attempt that is often repeated until the director is satisfied.) In this coding system, "27 A 3" translates, "scene 27, shot A, take 3." As you can see, a slate labels

Slating by Timecode

Since digital camcorders display timecode in the viewfinder, it is easy to make a slate log by noting the starting address of each take. Typically, the person handling continuity keeps a database on a laptop computer, filling in the information shot-by-shot. The timecode is read off by the camera operator or transcribed from a reference monitor.

Keep in mind, however, that timecode contains no information whatever about shot content. Classic slating, by contrast, identifies the scene, setup, and take of each separate piece of material. For this reason, timecode information is a useful addition to, but not a good substitute for, a descriptive slating system.

All digital camcorder viewfinders can display timecode.

each program element, no matter how small, so that it can be catalogued and tracked.

In professional news, sports, and documentary programs, as well as in most amateur videos, shot-by-shot slates are not generally recorded during production. Instead, the editor identifies each shot in postproduction. Even when applied by the editor, the production slating system of sequence/shot/take ("27 A 3") is highly effective because it describes the piece of material as well as naming it. If you adopt a simpler slating system, the minimum identification should include the tape and a shot number. In this case, "2-27" indicates the 27th individual piece of footage (whether a shot or one take of a shot) on roll 2.

Laptop editing in the field in real time requires novel ways to identify footage.

Describing

Typically, a shot description includes several items.

- **Angle:** The image size and camera position, as "neutral angle medium shot" (or in brief form, "Neutral MS").
- **Content:** What the shot depicts, as "Tammy takes sip of coffee."
- **Quality:** NG for no good, OK for acceptable, or perhaps B for best.
- **Notes:** It is often helpful to add notes about the shot, such as "she smiles in this take only."

Spending the time to describe every program component before starting to edit can save far more time later in locating shots.

Locating

Other aids in finding shots are tape roll or disc number, timecode address, and filename.

- **Roll number.** If you are working with more than one cassette, you need to know which materials are on which roll. The same applies to multiple discs.
- **Timecode address.** All digital recording formats label every single frame of video with a unique code, based on the time elapsed from the first coded frame on the tape, **Figure 19-13**. Thus, a ***timecode address*** of 01:23:16:12 labels the twelfth frame of the sixteenth second of the 23rd minute of the second hour on the tape. (The first hour would read "00.")
  In some recording formats that use 60-minute tapes, the videographer can set the timecode to start at any hour, to match the number of the cassette. That way, tape number 3 can start with the timecode 03:00:00:01. This system builds the tape number right into each timecode address.
- **Filename.** In digital editing, shot identification can often be made while importing the program materials to the computer. As each shot is stored on a hard drive, it is given a unique ***filename***. Though it is tempting to devise descriptive filenames (like "Tammy coffee") it is much better to file shots by slate number (like "27 A 3"). Names that seemed descriptive when first applied can grow very difficult to interpret days or weeks later. A slate number, however, refers you directly to the complete shot description in the shot list or database.

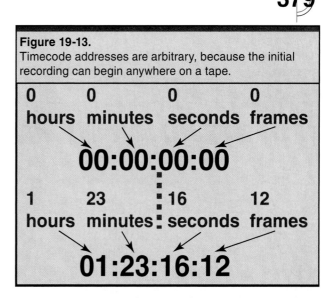

Figure 19-13.
Timecode addresses are arbitrary, because the initial recording can begin anywhere on a tape.

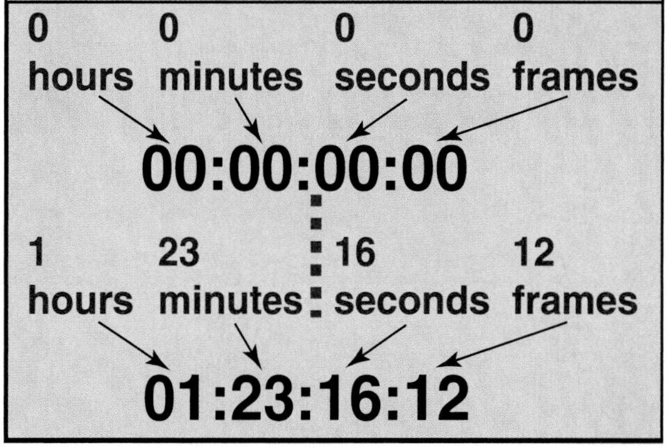

In digital editing, a piece of footage can be copied, changed, and filed many times, in many forms and in many different locations. For this reason, a logical filing system is doubly important.

Managing

For any program that utilizes more than perhaps 25 pieces of raw footage, all the shot identification, description, and location information can be logged in a database like the simple model shown in **Figure 19-14**.

Figure 19-14.
A simple shot log database.

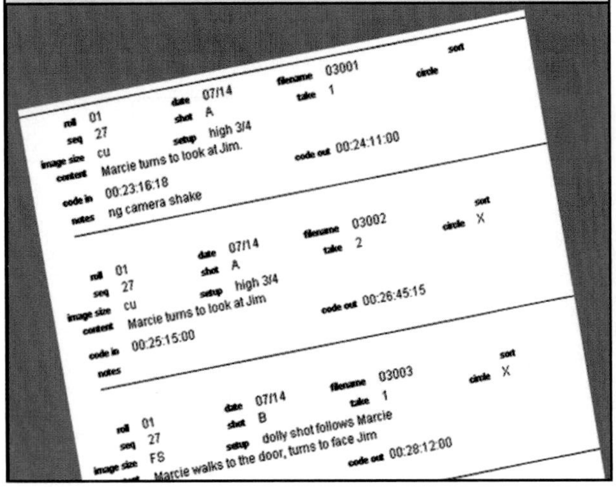

Since "footage" literally means "recorded feet of tape," perhaps we should start calling raw tapeless recording "timeage."

By working with a database, you can organize and search your program materials in any way you need. To create an inventory of each source tape, design a report form that sorts first by roll number and then by timecode address. To collect all the shots needed for a sequence, sort by sequence (or by slate number, if it begins with the sequence). If, for example, you are digitizing only selected takes for nonlinear editing, sort by tape roll and then use a database filter to display only the takes you wish to transfer.

Professional editing software has database features built in, so you can often log footage automatically as you import it to your computer. Most software, however, does not log every shot, only the ones selected for transfer. Since you may frequently need to return to your sources for footage that you did not originally capture, it helps to have a full log of untransferred footage, too.

Admittedly, the identifying, describing, locating, and managing of your material is not a very exciting process; and you may be tempted to skimp on it or even omit it. But after you have spent a tedious hour in a frustrating search for a shot that you know exists but cannot locate, you will appreciate the importance of systematically organizing your editing components.

When you have organized your raw footage, you are ready to work with it. You start by assembling the pieces of your program.

Assembling

Even when it is well organized, unedited video footage is still just a collection of separate shots, takes, and pieces of audio. At its simplest level, editing means putting together these pieces of raw material to create a finished program. This assembly operation includes *selecting*, *sequencing*, and *timing* each individual shot.

For simplicity, we will treat each step individually, as if you selected all the shots in a sequence at once, then put all of them in order at once, and so forth. In actual practice, you are likely to select, sequence, and time each shot in order before proceeding to the next shot.

Some software encourages you to select and sequence whole groups of shots at a time.

Selecting

Since all professional projects start with far more footage than will appear in the finished program, the editor begins by picking the shots to be included and selecting the best take of each. The sidebar *Raw Footage* shows all the unedited shots in a demonstration sequence. This simple action of pouring and offering a drink has been covered by five separate shots:

- *Slate 15 A 1* and *Slate 15 A 2.* A two-shot favoring Frank, the pourer (Takes 1 and 2).
- *Slate 15 B 1.* A reverse two-shot favoring Jack.
- *Slate 15 C 1.* A medium shot of Jack that zooms to a closeup.
- *Slate 15 D 1.* An extremely high ("birdseye") insert of the drink being poured.
- *Slate 15 E 1.* A somewhat lower alternative insert of the drink being poured.

In reviewing this raw material, the editor notes several things:

- *Shots A, B,* and *C* all cover the entire action, so any one of them alone could deliver all the information in the sequence.
- Frank's two-shot offers two slightly different takes to choose from.
- Jack is covered in two completely different angles.
- The pouring insert has been shot in two different versions.

Raw Footage

The discussion that follows refers to the shots in a single sequence that shows the pouring and offering of a drink. (The three images in each shot represent the beginning, middle, and end of the shot content.)

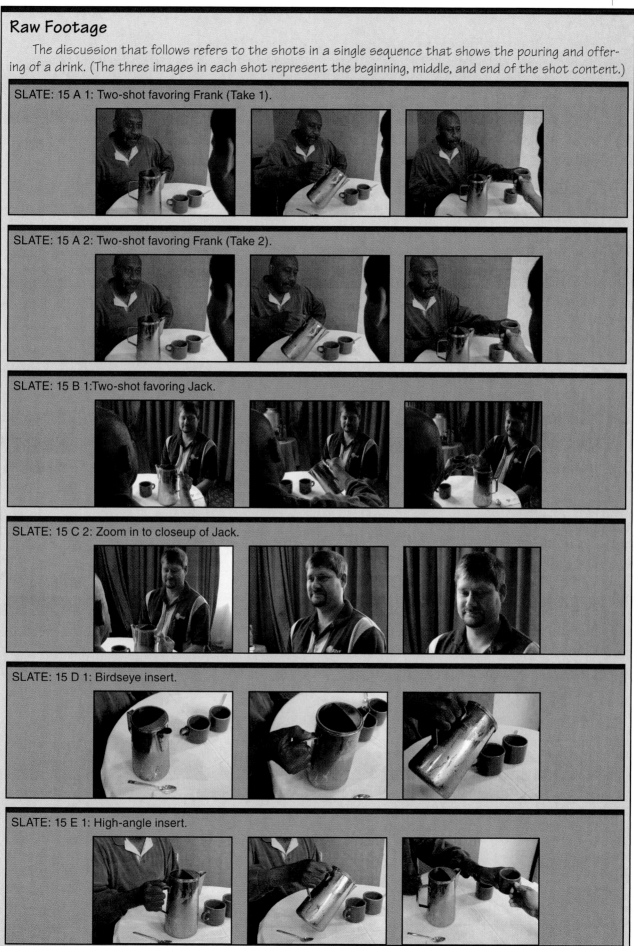

SLATE: 15 A 1: Two-shot favoring Frank (Take 1).

SLATE: 15 A 2: Two-shot favoring Frank (Take 2).

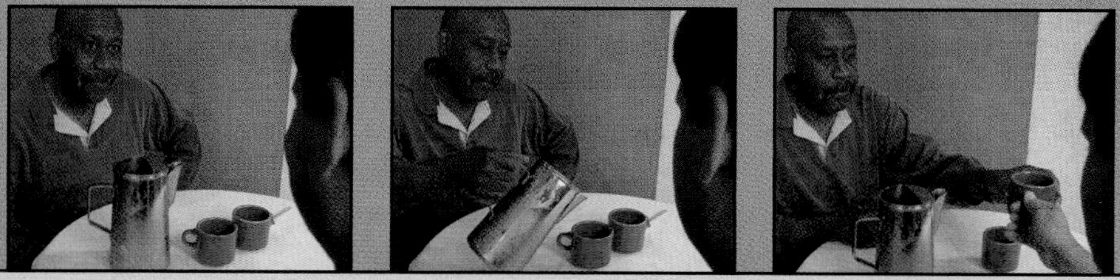

SLATE: 15 B 1:Two-shot favoring Jack.

SLATE: 15 C 2: Zoom in to closeup of Jack.

SLATE: 15 D 1: Birdseye insert.

SLATE: 15 E 1: High-angle insert.

Figure 19-15.
Shot 15 A Take 1 covers the action more completely.

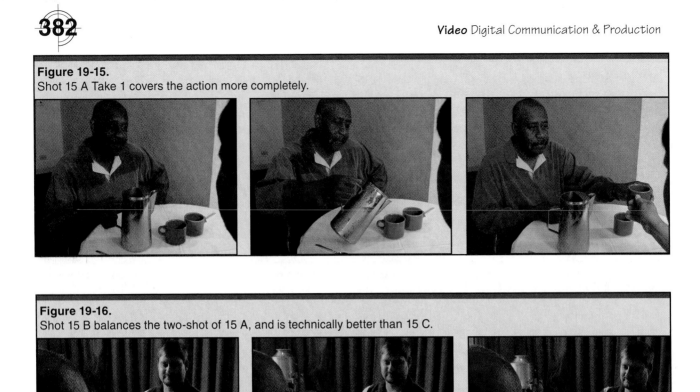

Figure 19-16.
Shot 15 B balances the two-shot of 15 A, and is technically better than 15 C.

Figure 19-17.
Shot 15 E is a less-extreme angle than 15 D, and it covers more of the action.

Figure 19-18.
The inserts were originally taped after the main shots.

Figure 19-19.
The insert is moved to its proper place in the action.

Accordingly, the first task is to decide which shots to use. In this case, the editor chooses *Shots 15 A 1, 15 B 1,* and *15 E 1.* See **Figures 19-15**, **16**, and **17**.

Sequencing

With the preferred shots and takes chosen, the next task is to put them in order. On the camera recording, the insert of the pitcher and cup follows the other shots, **Figure 19-18**.

Now the editor makes the decision to move the insert to its proper place in the action, **Figure 19-19**.

Timing

In our example, the action is repeated three times, so the next task is to time each shot by assigning it just one part of the action and discarding the overlapping parts. Each shot is trimmed to its usable section by setting its in-point (start point) and out-point (end point).

Enhancing

When you have assembled your images and sounds, you can enhance them to better suit the requirements of the program. Enhancing raw editing materials can mean *improving* them, *conforming* them, and/or *redesigning* them.

Improving

Even in high-level professional production, some footage may not measure up to the quality standards of the program.

Video. A shot may have luminance and chrominance problems that make it too dark or too light or degraded by an unwanted color cast, **Figure 19-20**. Its white balance may be off, due to a mistake or to a camera movement from, say,

Figure 19-20.
Correcting exposure. A—Original footage too dark. B—Exposure corrected digitally. (Sue Stinson)

Figure 19-21.
Eliminating a distraction. A—Microphone showing in upper right part of frame. B—Mike removed by digital cropping.

a window-lit subject to a scene lit by incandescent lighting. Contrast is often a problem, especially in outdoor locations. Finally, framing may need improvement to recompose a shot or eliminate a visible microphone, **Figure 19-21**. With digital manipulation you can often fix all of these problems completely.

Audio. Audio problems include incorrect volume levels, as well as poor sound equalization, presence, and mike-to-subject distance. With digital sound editing, you can make substantial enhancements.

Even the simple software included with Mac or Windows operating systems can make respectable improvements in sound quality.

Conforming

Even when various components are all equally acceptable for quality, they may not always match. Frequently, you need to adjust video and audio to eliminate differences between one piece and the next.

Video. Often, white balance is not consistent from shot to shot, especially when different setups are shot at different times of day, or even on separate days, **Figure 19-22**. When intercutting these shots, you tweak the color balance to conform their look.

Audio. Since every new camera angle demands a different microphone position,

Figure 19-22.
Late afternoon light can change color temperature between one setup and the next.

Speeding or Slowing Action

Digital processing allows you to change the speed of a shot, not only to create obvious fast or slow motion, but also to make subtle adjustments. You can usually alter speed as much as 10 percent without making the shift noticeable (though you need to try different modifications and judge the results on a case-by-case basis).

The simplest reason for changing speed is to adjust the length of a shot for timing purposes, usually to fit an available slot in the program.

Slowing an action slightly can also lend it added significance by giving it greater visual weight.

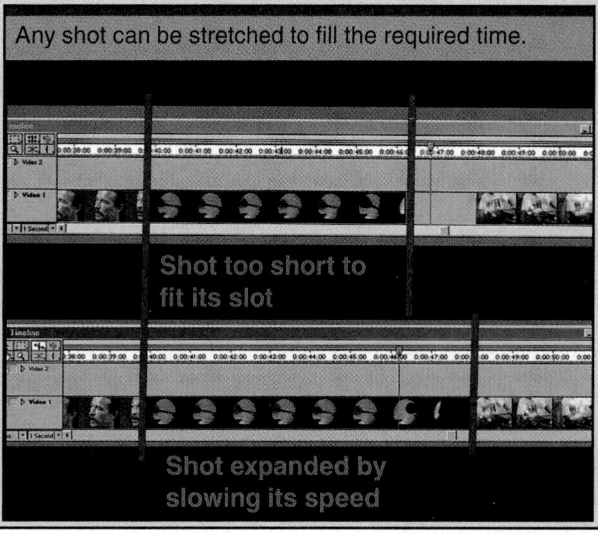

equalizing characteristics like distance, presence, and background noise is a constant chore.

Redesigning

Digital processing allows you to change picture and sound character as well as quality. Sometimes this involves computer-generated sounds and images; but more often it involves processing raw footage for editing.

Video. On the video side, you can adjust overall color to provide a sunset glow, create a nostalgic sepia monotone, or create a black and white program. You can apply slow- or fast-motion effects, or strobe. You can transform the image with computer graphics effects, and apply any of the dozens or even hundreds of image filters available for use in computer-based editing.

Audio. In the sound department, you can alter material to create any effect you choose.

Today, with digital editing, image and sound enhancement is so widespread that many editors process most of their shots and audio clips as a matter of routine.

Synthesizing

Synthesizing means fusing separate elements together to create a single, new product —in the editor's case, a video program.

To understand synthesizing, it helps to think of programs the way many computer editing applications represent them: as multilayered timelines.

A ***timeline*** is like a very long, thin matrix, with the program's shots represented in order on a horizontal axis. The program begins on the left and the timeline scrolls leftward as you move along it to the end, which appears on the right. The timeline is several levels high, usually with separate lines for two or more video streams, transitions, titles, and filters. The audio component may appear as a dozen or more separate tracks of production sound, background, effects, and music.

Integrating material horizontally, from moment to moment through the length of the program, may be called *connecting*; while blending the many levels of picture and sound may be thought of as *layering*.

Connecting

Creating a seamless horizontal flow from start to finish requires connecting shots to form sequences, sequences to form sections, and sections to form whole programs.

In short videos, the entire program may be a single section. Also, in fiction programs, sections are often called "acts," like the acts in theatrical dramas.

Typically, you select shots and choose their in- and out-points to make invisible edits, so that each sequence seems to the audience like a single presentation rather than a collection of individual pieces.

You connect sequences with transitions like fades, dissolves, wipes, and digital effects. The decisions you make in selecting and timing the transitions between sequences will determine the coherence of the program as a whole.

Layering

Where connecting joins elements end-to-end, layering melds them from top to bottom. Even a simple program may involve working with several visual layers at once, including:

Composites. Pieces of different images may be combined into a single picture through color matting or other digital techniques.

Superimpositions. Two or more images may be layered so that all are visible at once.

In compositing, pieces of one image completely replace pieces of another. In superimposing, each is displayed in its entirety, but at partial intensity, so that both pictures are visible at the same time. Compositing is discussed in Chapter 23.

Multiple images. The screen may be divided into separate areas to display either one or more small images in a larger one ("picture-in-picture") or to create a mosaic of images. See **Figure 19-23**.

Titles. Titles are often layered over live action, or even vice versa, **Figure 19-24**.

Audio layering is often even more elaborate. Usually, the visual track presents only one image at a time, but a sophisticated audio track is almost always a mix of many elements, including dialogue, background effects, synched sound effects, and music. To permit more precise control, these components are usually

Figure 19-23.
Multiple images. A—Picture-in-picture. B—Mosaic.

Figure 19-24.
Two versions of the same title. (Ulead)

divided into subcomponents. Often, the dialogue will include separate tracks for each speaker. Sound effects tracks may number two, four, or even more to permit multiple overlapping effects. Music may be distributed among multiple tracks to allow cross-fading between sections and stereo imaging.

Historically, video has been limited in its ability to synthesize program material out of multiple streams of picture and sound, but the development of digital nonlinear postproduction has offered editors almost limitless possibilities in integrating program components.

Archiving

The final task in editing a program is storing it in its finished form, so that it can be presented, duplicated, and preserved. In digital format, archiving is the essential last step in the postproduction process. Archiving involves *rendering, storing*, and *presenting.*

Rendering

In ***rendering*** a finished program, a computer takes the low-quality editing version that you have assembled and uses it as a blueprint for automatically creating a high-quality version. To save hard-drive space and editing time, some digital editing applications work with low-resolution images and crudely rendered effects and titles. The resulting "work print" contains all the instructions necessary to make a duplicate that is exactly the same, except that the images, sounds, and effects are all full quality. Rendering is the process of creating this finished-quality copy, **Figure 19-25**.

Other editing software, designed to run on fast computers with large memories, can render programs in real time, as each subcomponent is completed.

Storing

The rendered version of a digital program is automatically stored on a hard drive; but that storage is only temporary. Digital video requires so much space that programs can quickly fill even very large arrays of hard drives.

To overcome this problem, the rendered program is exported for permanent storage on a high-capacity medium, like a tape or disk. Because of the long potential life of the program signal, the program is truly archived.

No medium is absolutely permanent. Digital tape signals fade like their analog cousins and disks are eventually degraded by handling. But as long as a complete signal can be read from a digital tape or disk, a perfect duplicate can be made. For this reason, well-maintained digital video libraries are indeed archives.

Storing and Distributing

The last step in postproduction is the preparation of the program for presentation in a particular medium. Digital movies are transferred to film for making release prints, or set up for direct digital projection. Most other programs

Figure 19-25.
In rendering your completed project, you can specify its video and audio parameters.

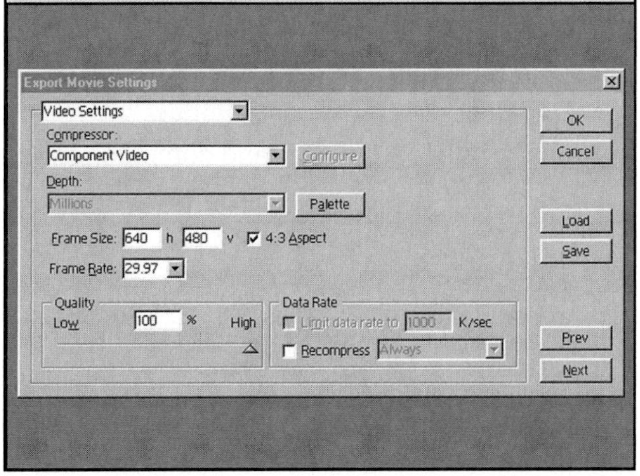

Figure 19-26.
A DVD menu. (Casablanca)

are prepared for viewing on DVD and/or the Internet, via streaming.

Creating ("authoring") a DVD means organizing it into chapters and providing a menu for accessing them, **Figure 19-26**. These menus often include live action or moving graphic backgrounds and elaborate audio. DVDs prepared for commercial release usually have multiple language tracks and/or subtitles, copy protection, and regional coding.

DVD authoring is covered in Chapter 23.

Operations and Principles

In this chapter on video editing operations, we have reviewed the fundamental editing procedures that are common to all programs. The next step is to see why we carry out these operations—what we are trying to achieve in editing. That is the subject of Chapter 20, *Editing Principles.*

Chapter Review

Answer the following questions on a separate piece of paper. Do not write in this book.

1. *True or False?* Postproduction and editing are separate operations.
2. Learning new software is often like learning a ______.
3. Professionals follow an ______ editing process in assembling a program; amateur editing tends to be ______.
4. The timecode provides a unique label for every ______ of recoded material.
   A. frame
   B. second
   C. minute
5. ______ is the process of making a finished high-quality version of a program.

Technical Terms

Additive editing: Creating a program from raw footage by starting with nothing and adding selected components.

Compositing: A digital editing process in which elements of one image replace elements of another image to create a combination of both.

Editing: Creating a video program from production footage and other raw materials.

Filename: The identification of an editing element as it is stored in a computer. Filenames may or may not be identical to timecode addresses or slate numbers.

Rendering: Creating a full-quality version of material previously edited in a lower quality form.

Slate: The identification assigned a take before shooting it.

Subtractive editing: Creating a program by removing redundant or poor-quality material from the original footage and leaving the remainder essentially as it was shot.

Timecode address: The unique identifying code number assigned to each frame (image) of video. Timecode is expressed in hours, minutes, seconds, and frames, counted from the point at which timecode recording is started.

Timeline: The graphic representation of a program as a matrix of video, audio, and computer-generated elements.

20

Editing Principles

Objectives

After studying this chapter, you will be able to:

- Connect shots with edits invisible to the audience.
- Adjust and improve actors' performances.
- Direct viewers' attention within a program.
- Present program materials at an appropriate pace.
- Elicit emotional responses from audiences.

About Editing Principles

Editing principles determine the qualities that you want in your finished programs. In effect, editing *operations* (covered in Chapter 19) are what you do; editing *principles* are what you want to achieve by doing it. Important editing principles include:

- *Continuity:* the information or story should be presented in a coherent order that the audience can follow.
- *Performance*: the people in the program should appear believable, and they should create the intended effect on the audience.
- *Emphasis:* information should be presented with an impact proportional to its importance, and the audience's attention should be directed to the most important aspects of the program.
- *Pace:* the program should flow briskly enough to maintain interest, but deliberately enough so that viewers can absorb the content.

While performance, emphasis, and pace are very important, the principle of continuity is absolutely essential: without it, your program cannot communicate adequately.

Continuity

Video is usually a linear form of communication. When telling a story, explaining an idea, teaching a skill, or presenting a person or place, you take the viewer through one thing after another, in a predetermined order from start to finish. The principle that governs the order of presentation is ***continuity***; the art of organizing and sequencing program content so that it makes sense to the audience.

We are concerned here with the "classical" continuity popularized by Hollywood and now more or less standard. There are, however, other forms of visual narrative.

Methods of Organization

Continuity can be achieved through different methods of organization, depending on the type of program. These methods include *narrative*, *argument*, *association,* and *subjective*.

Narrative

Stories are usually organized in chronological order. Documentaries and training programs —especially about events, tasks, or processes —are also chronological. The program illustrated in **Figure 20-1**, for instance, shows the building of an Alpenlite fifth-wheel trailer from chassis to finished RV.

Argument

Programs intended to persuade viewers are typically organized by logical argument. For example, an argument for recycling is expressed in **Figure 20-2**.

Association

Programs (or program segments) intended to convey an impression of people, places, or events may be organized by association, with similar

Figure 20-1.
A narrative (chronological) sequence on building a trailer. (Western Recreational Vehicles)

Figure 20-2.
A sequence organized by argument.

We produce a lot of trash.

Discarded trash degrades the environment.

A degraded environment harms wildlife.

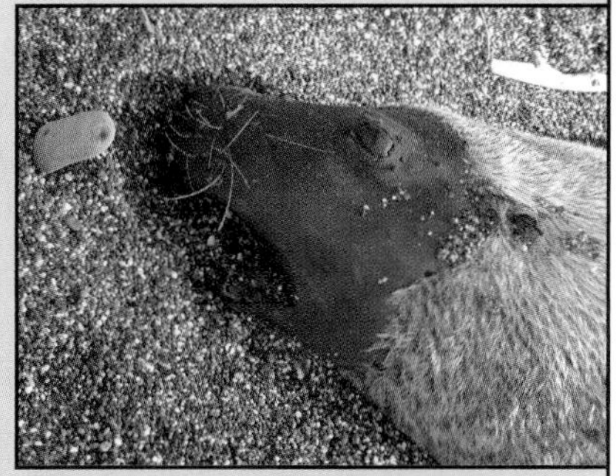

Disposing of trash is difficult.

We should recycle trash instead of disposing of it.

Recycled paper saves a lot of trees.

things grouped together. At the Liebniz County Kinetic Sculpture Race, for example, one sequence might associate the attempts, by several different extravagant vehicles, to climb the notorious "slippery slimy slope." See **Figure 20-3**.

Subjective

The most elusive form of thematic continuity is purely subjective: like a poet or a painter, the editor presents content in a structure that viewers can feel, even if they cannot clearly explain it.

Figure 20-3.
A sequence organized by association.

Subjective continuity is not illustrated here because it is impossible to suggest on the printed page.

In all its forms, continuity is evident at several levels of organization: the individual ***shot***, the ***sequence*** made up of shots, and the ***program*** made up of sequences. In addition, longer programs are often divided into *acts*.

Shot-to-shot Continuity

You establish basic continuity at the shot level by putting shots in order. Returning to an example introduced previously, **Figure 20-4** shows three shots in the order they were recorded on the camera tape.

Since the insert of pouring the drink is out of order, the editor establishes correct continuity by placing it between the pair of two-shots. See **Figure 20-5**.

Matching action

When inserting the closeup between the main shots, the editor has to match the movement in order to preserve the illusion that the audience is watching a single continuous view rather than three independent shots. In matching action, the editor often has three different options:

- *Start or end* off screen. The easiest method is to make the transition off screen, where the audience cannot tell whether or not the action matches. **Figure 20-6** shows the sequence starting closer on Frank, the pourer. Because the camera frames off the pitcher, viewers cannot see where it is when the insert begins.
- *Cut during a pause.* In **Figure 20-7**, the wide shot continues until the static moment when water is flowing into the cup, before the closeup begins. Matching is relatively easy because movement is minimal in both shots.
- *Cut during the movement.* Though this is the most difficult option, it is often the most effective, because the seemingly continuous motion helps sell the illusion that there is a single, uninterrupted view. See **Figure 20-8**.

Except in special circumstances, matching action is usually done with every shot in a continuous sequence.

It is not always necessary or even desirable to match action exactly, as we will see.

Figure 20-4.
The insert was originally shot out of order.

Figure 20-5.
The editor moves the insert to its proper place.

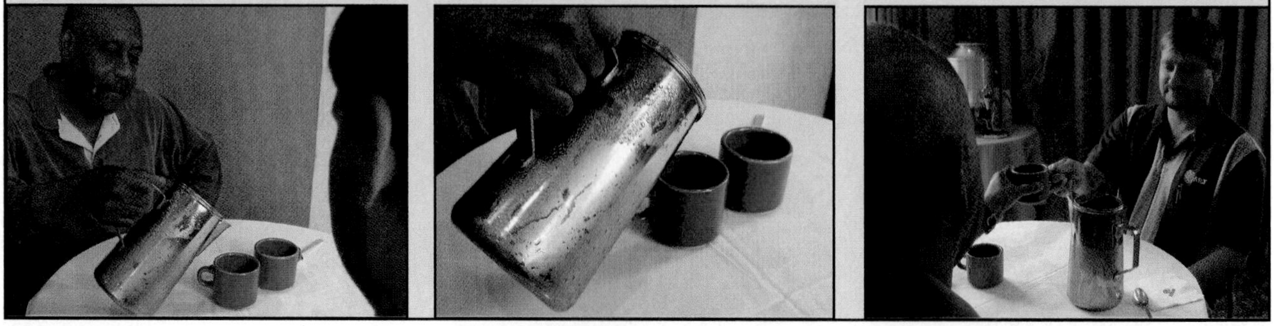

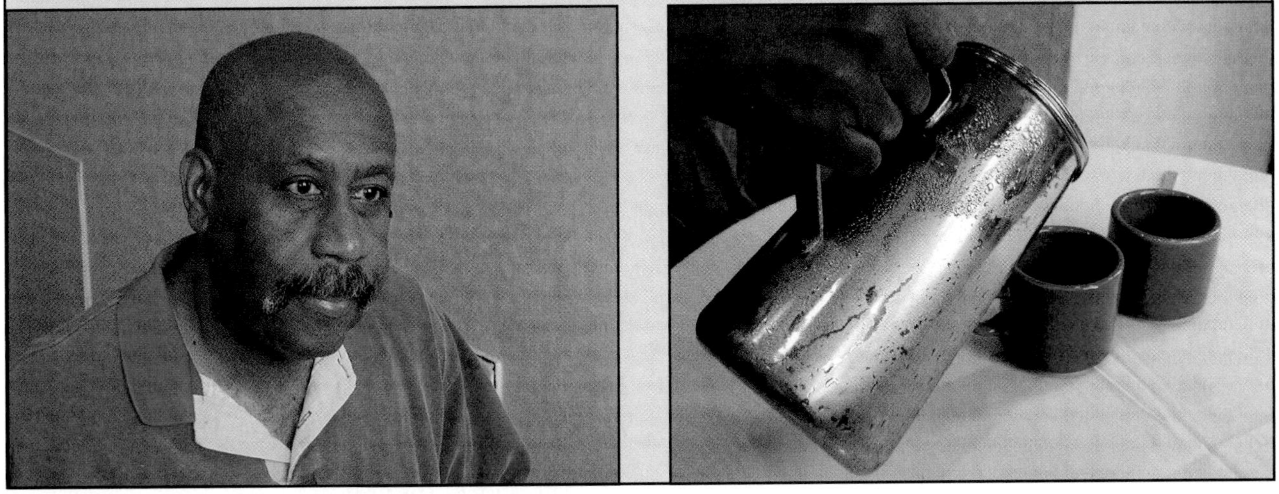

Figure 20-6.
Because the pitcher is not in the first shot... its movement need not be perfectly matched.

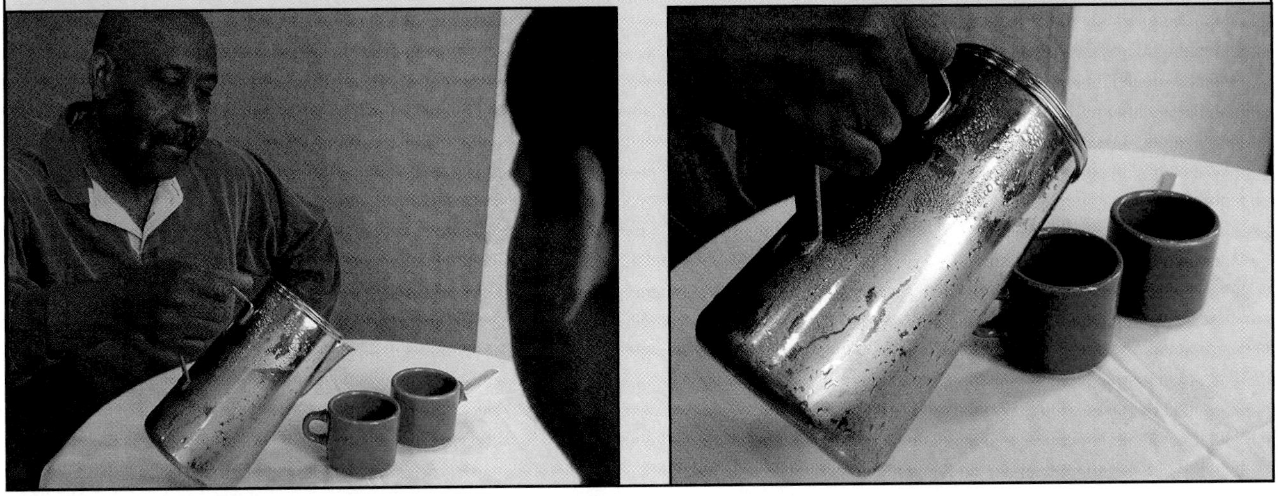

Figure 20-7.
At the end of the first shot, the pitcher has already been tipped... so the insert begins after the movement has stopped.

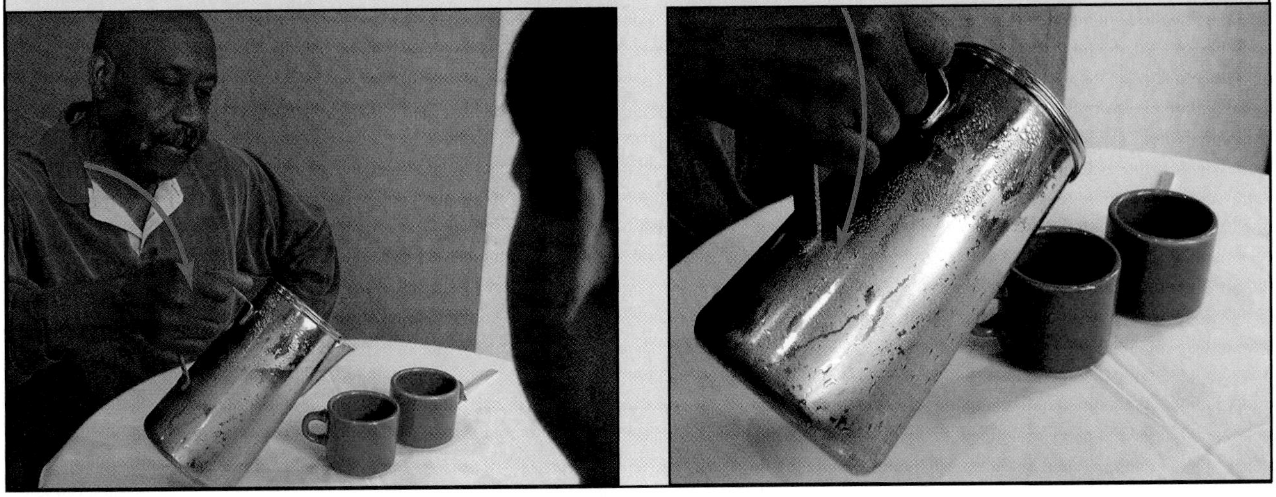

Figure 20-8.
The pitcher is descending as... the editor cuts to the insert.

Scenes and Sequences

The words sequence and scene can cause confusion because they are sometimes used interchangeably. Strictly speaking, though, sequences and scenes are not the same. A ***scene*** is a single action or process. A sequence is a single unit of fiction or nonfiction content. (The word "scene" itself can cause confusion because in stage plays, the word is used where movies use "sequence." Also, some discussions of films use "scene" as a synonym for "shot."

A scene

A scene usually has the following characteristics: it consists of a small number of individual shots (typically fewer than ten), the shots all cover (or at least relate to) one short piece of a larger action, and—though the shots may be made from different camera setups — they are all in a single area of the set or location.

A sequence

A sequence often consists of several separate scenes, all of them related so that they build a continuous action; a sequence will often (but not always) end with a transition (like a dissolve or wipe) to the new sequence that follows.

An example

To demonstrate the difference between scene and sequence, imagine a short training program titled, *How to Make Your Own Firewood.*

The sequences. This process consists of three major parts; cutting down a tree, cutting the felled tree into firewood, and building a wood pile. In the edited training video, these parts will become separate sequences.

The scenes. The first sequence, cutting down a tree, breaks down into individual scenes: preparing the chain saw, selecting the tree, and felling the tree. Each scene covers a small part of the action that unfolds continuously in a single place and at a single time.

The shots. The first of these scenes, preparing the chain saw, is made up, in turn, of several individual shots.

Production vs. editing

Another difference is that scenes are units of organization used during production, while sequences are created during the process of editing.

Note that camera slates carry scene numbers rather than sequence numbers. (Sequences are not identified by number or letter.) So, "27A3" on a slate identifies the third take of shot A in scene 27.

The shots in the scene "Preparing the chain saw."

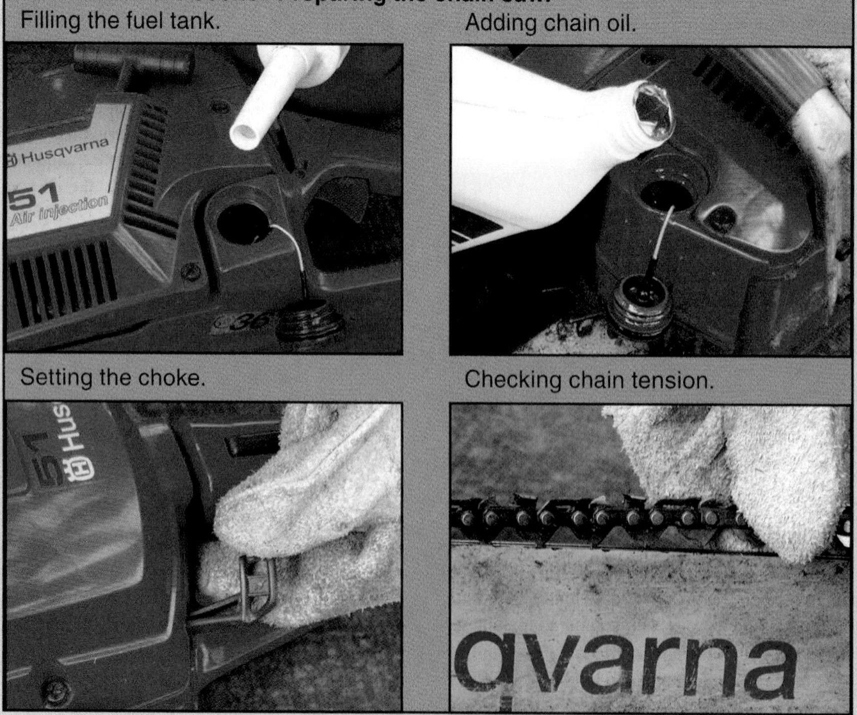

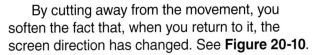

Maintaining screen direction

In addition to seeming uninterrupted, continuing action usually needs to maintain a consistent screen direction. If the director has done a good job, the raw shots in a sequence will have their screen direction already established.

Screen direction is covered in Chapter 18, *Directing for Form.*

If there are inconsistencies, you may want to remove them. To do this, you have three available methods:

- *Use only correct shots.* If you can cover all the action adequately without the incorrectly oriented shots, simply omit them from the sequence.
- *Insert cutaways.* If you must use an incorrect screen direction, precede it with a shot that does not show the action. **Figure 20-9** shows two mismatched shots.

  By cutting away from the movement, you soften the fact that, when you return to it, the screen direction has changed. See **Figure 20-10**.
- *Flop the incorrect shot.* If you lack material for a cutaway, you can sometimes reverse the shot. With digital processing, it is simple to correct screen direction by reversing the left/right orientation of an image, **Figure 20-11.** Of course, reversing the shot will not work if details in it reveal that it is backward.

Editors often use the term "flopping" for reversing an image because in film this was accomplished by turning a strip of film over and copying it through its transparent backing. Today, the process is usually digital and is often called "flipping" instead.

Intentional mismatches

In narrative and other linear types of continuity, matching action and direction is usually

Figure 20-9.
In the second shot, the screen direction is reversed (and the cup is in the wrong hand).

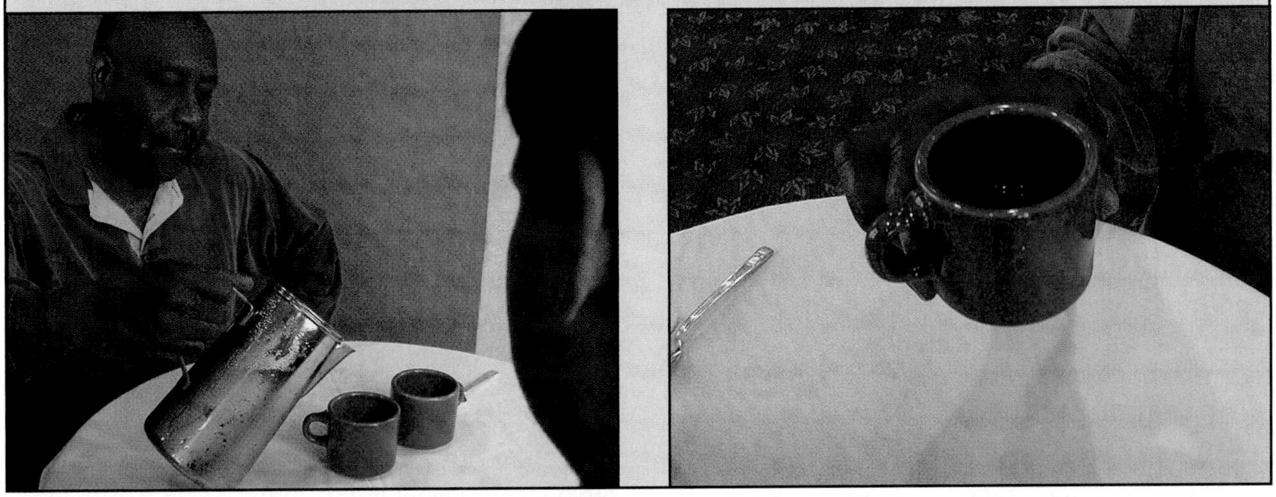

Figure 20-10.
The cutaway to the other subject... softens the mismatched screen direction.

Figure 20-11.
Flopping (reversing) the closeup of the cup.

important. But in ***associative continuity***, when shots are linked only by theme (as they are in the kinetic sculpture race example), there is no need to match action or screen direction. In fact, it is often better to juxtapose shots with clearly different actions and directions, to emphasize their independence.

Sequence Continuity

Sequence continuity is the organization of all the shots in a sequence.

What is a sequence? Though there is no precise definition, a sequence may be thought of as an assembly of shots organized to present a single action, process, or idea. The continuity of a sequence may be *linear, associative,* or *subjective.*

Linear sequences

The majority of sequences employ ***linear continuity***, with the shots lined up in temporal order for stories and training programs, and in logical order for editorial presentations.

In creating a linear sequence, the challenge is to decide how much detail to include. That is:

- How many sub-steps in the action do you need to select for showing?
- How much of each step should be shown?

The answers to these questions depend on the purpose of the sequence.

Timing each shot is addressed in the discussion of emphasis, later in this chapter.

Since the editorial decision process is easier to illustrate than to explain, consult the sidebar, *One Action, Three Sequences,* later in this chapter, for a demonstration.

Associative sequences

Sequences organized by theme can be more difficult to arrange because they lack the built-in constraints of chronological order. For this reason, it is helpful to discover some other organizing principle in the raw footage and use it to govern the sequence. The possibilities depend entirely on the material you have to work with, so here are some examples from a documentary about the weekly McKinley farmer's market:

- *Color.* Sequence the shots to begin with muted reds and gradually work up to bright, dramatic scarlet, **Figure 20-12**.
- *Subject.* Often, you can find several shots related by the fact that they contain similar subject matter — in this case, hats. See **Figure 20-13**.
- *Size.* An alternative approach is to begin with small subjects that grow in size as the sequence progresses, **Figure 20-14.**

As you can see, there is no "logical" reason to group shots by size or color or any other characteristic. The point is to give the audience a feeling, however unconscious, of coherence and order. Without consciously thinking about it, viewers generally want to feel that there is a purpose behind the program they are watching. A well-ordered sequence conveys that sense of purpose.

Subjective sequences

In specialized programs like certain music videos or very personal expressions, sequences may have no organizing principle other than the editor's purely subjective preferences. Though this is a perfectly legitimate approach to creating sequences, it can be difficult to handle because there are no organizational guidelines. The only test is, does the result succeed—that

Figure 20-12.
Shots associated by color.

is, does the intended audience find satisfaction in watching and listening to it? If viewers sense an underlying coherence in a sequence, however obscure, they feel reassured that they are not just watching random footage.

Program Continuity

We have noted that the basic building block of a video is the individual ***shot***; a number of shots that are related, ordered, and matched is a ***sequence***. Sequences, in turn, are organized into a complete ***program***.

Figure 20-13.
Shots associated by subject matter.

Figure 20-14.
Shots associated by size.

This three-level organization, shot/ sequence/ program, is often enough for short works. Longer videos usually play better when the sequences are first combined into larger groups that, collectively, make up the program as a whole.

These larger groups of sequences have different names in different program types. The scripts of feature films and TV shows are typically organized into ***acts***. Long nonfiction programs like documentaries, infomercials, and training videos are also constructed in multisequence groups that are referred to simply as parts or sections. Like individual shots in a sequence, these larger groups are also organized to create a coherent, effective continuity of presentation.

Parallel sequences

Often, separate sequences are presented concurrently, in alternating pieces. This technique is known as ***parallel cutting*** or cross cutting. Classic parallel cutting is as old as the movies:

- Our hero hangs from the roof railing of a tall building. CUT TO:
- Our heroine races up to the roof to rescue him. CUT TO:
- He looks down at … CUT TO:
- The ground far below. CUT TO:
- He reacts with dismay. CUT TO:
- Our heroine races onto the roof.

...and so on. In Chapter 4, *Video Time,* parallel cutting is illustrated by the shots repeated as **Figure 20-15**.

For simplicity, this example consists of single shots. However, each segment may actually consist of a short scene, say three or four shots of his struggle to climb up over the railing, then several more of her racing up flights of stairs to reach him, and so forth. Also, complex actions may involve intercutting several different sequences at once.

To choose a more complex example, the conclusion of *Star Wars Episode I* intercuts four separate actions as the princess, the boy, the Jedi knights, and the alien army lead separate attacks at the same time on their common enemy. Each visit to each of these actions consists of a full sequence made of many individual shots.

Acts

As noted previously, the sequences in longer programs are often grouped into larger units called acts. An ***act*** is a major division that is traditionally opened with a fade-in effect and closed with a fade-out.

In commercial television, act breaks are also punctuated by commercials.

In programs that are not constrained by TV time requirements, act divisions are flexible. Typically, an act will end for one or more reasons:

Figure 20-15.
Two sequences presented through parallel cutting.

- The preceding group of sequences creates a cumulative effect on the audience (like the suspenseful effect of an elaborate robbery or escape).
- The audience needs a moment to relax and/or absorb the information in the previous group of sequences.
- The material that will follow is different in style or tone.
- The material that will follow is different in content, whether a distinctly new part of a story or a major subject change in an informational program.

Transitions

The conventions of traditional continuity separate sequences and acts by transitions. A dissolve signals a change from one sequence to the next. A fade-out/fade-in marks the transition between acts. Today, however, it is common practice to speed up the pace by joining sequences with direct cuts. Act breaks, however, are usually still signaled by fades or dissolves, **Figure 20-16**.

With the availability of digital effects (DVEs), new transitions are beginning to acquire standardized meanings. A slow wipe, especially a horizontal one, can replace a fade-out/fade-in, creating a major pause without fully closing and reopening the program. (Of course, this alternative will not work if the program must stop for a commercial.)

The Star Wars movies are also notable for using soft-edged horizontal wipes for transitions.

Figure 20-16.
A fade-out. (Sue Stinson)

One Action; Three Sequences

Here is a single action edited into three very different sequences.

The action covers the process of connecting a travel trailer to its tow vehicle and driving off with it. The three sequences show how this action might be presented in three very different types of video program:

- *Instructional.* This is a training program on how to operate a trailer.
- *Promotional.* This video is designed to sell potential buyers on the delights of trailer travel and recreation. The objective is to show how easy it is to manage a fifth-wheel trailer (the kind that overlaps the pickup truck that tows it).
- *Dramatic.* This is the story of a woman who sets out to see the world on her own.

Instructional version

In this version, the objective is to show the steps in hitching up as clearly as possible. The shots detail mechanical and electrical hookup procedures.

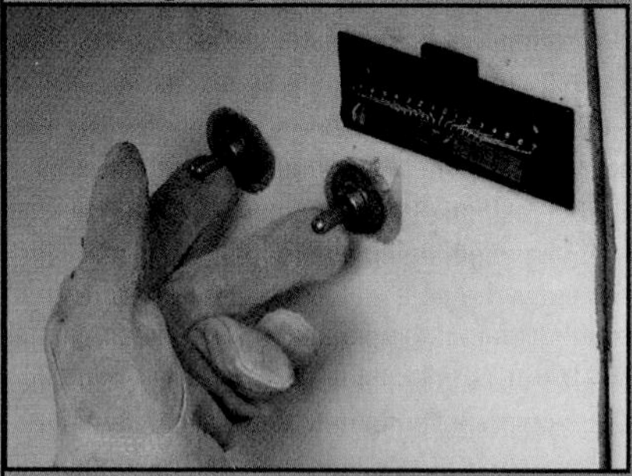
Operating the jacks.

Hitching the trailer.

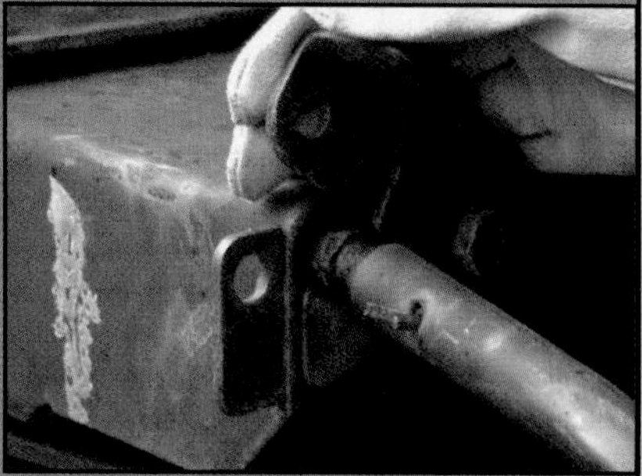
Latching the hitch.

Attaching the safety cable.

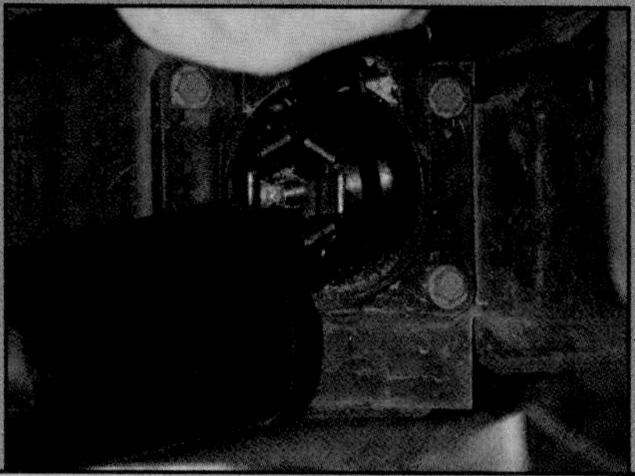
Plugging in the electrical cable.

Hand position for backing the rig.

Special DVEs like flips, fly-ins, or hundreds of other choices, can signal a change in the program without indicating whether the audience should consider it minor (like a dissolve) or major (like a pair of fades).

Choosing and timing transitions is a subjective process that depends greatly on the taste of the editor.

Sound and Continuity

In this discussion of continuity, we have repeatedly mentioned audio. Of all the tools for creating continuity at the shot, sequence, and program levels, sound is by far the most powerful. Audio can be flexible, subtle, evocative, easy to work with, and very inexpensive. For a more comprehensive treatment of audio's expressive capabilities, see Chapter 7, *Video Sound.*

Performance

After continuity, the next major editing principle involves performance, since almost all video programs involve people. Though you cannot

Promotional version

The purpose of this version is to suggest that hitching up is a quick and easy process.

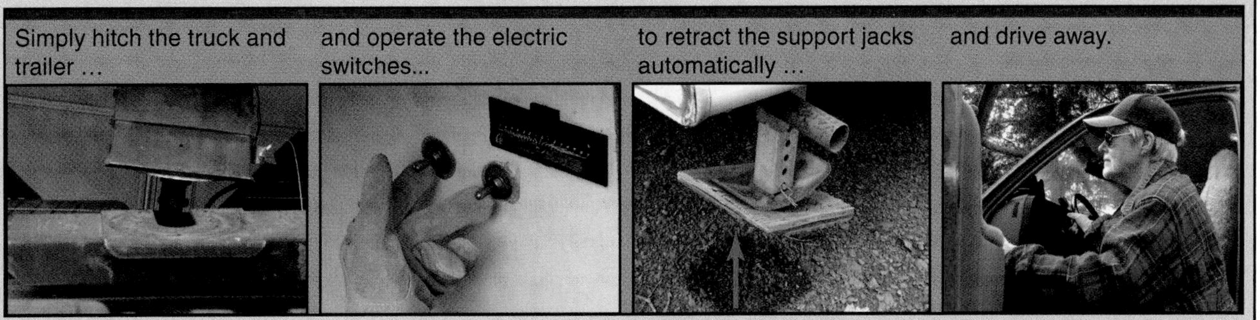
Simply hitch the truck and trailer ...
and operate the electric switches...
to retract the support jacks automatically ...
and drive away.

Dramatic version

The goal of this version is suspense: can she succeed in the tricky business of mating truck and trailer?

The bystanders add suspense to the sequence.

transform an awkward amateur into an award winner, you can often rescue a poor performance, improve an adequate one, and fine-tune even an accomplished job of screen acting. Editing performance involves four major tasks:

- *Selecting* the best parts of the best shots.
- *Adjusting* the timing of dialogue and action to create sequences that move effectively.
- *Enhancing* the content by underlining, shading, or even changing the actors' original meanings.
- *Directing* viewers to the most important aspects of the performances.

To understand how these four processes shape performance, imagine a sequence in which Bob is desperate to buy a car and Bill is reluctant to lend him the down payment, **Figure 20-17**.

Selecting the Best Performance

Typically, each moment of an actor's performance is recorded more than once. The same action is usually shot from two or more different camera setups, and each shot may be repeated to provide multiple takes. Since no two shots are

Figure 20-17.
Bill and Bob at the used car lot.

Figure 20-18.
BOB (Take 1): Please lend me that money. I need the money, and I need it now!

exactly identical, your first task is to select the best ones for your purposes.

The selection task is easy if one entire shot is better than the others, but this is not always the case. Sometimes, the best performance is in separate pieces that are distributed among several setups and multiple takes of each one. When this happens, your task is to cement the best sections together with joints that are invisible to the audience. To demonstrate, imagine that an actor misreads a line, as in **Figure 20-18**.

The problem is that the actor was supposed to say, "I need the *cash*," in order to avoid repeating the word *money.* To correct the error, the director shot a second take, **Figure 20-19**.

In Take 2, the actor misread the *other* half of the speech, saying "the" money instead of "that"

Figure 20-19.
BOB (Take 2): Please lend me the money. I need the cash, and I need it now!

Figure 20-20.
Butting two similar shots results in a jump cut.

money. To make the entire speech read correctly, you need to join the first line in Take 1 to the second line in Take 2. If you butted two takes of the same angle together, the result would be an unacceptable jump cut, **Figure 20-20**.

To avoid this problem, you have three options:

- Cut to a different angle of the performer.
- Cut to another actor in the scene.
- Provide an insert or a cutaway buffer shot between the two halves of the line.

Which option you choose depends on the materials available.

Changing angles

If the director has covered the actor from more then one angle, you can try for a matched cut, as in **Figure 20-21**.

Changing performers

If you lack another angle of the performer, you can sometimes substitute another actor in the scene, letting the performer's offscreen voice carry over the shot, as in **Figure 20-22**.

Using buffer shots

Another way to hide the join is by adding a ***buffer shot***, an image of something completely different. There are two types of buffer shots: inserts and cutaways. An ***insert*** is a small detail of the action that excludes the actor, **Figure 20-23**.

A ***cutaway*** is an image outside the action, **Figure 20-24**.

All of these methods of selecting pieces of a performance require cutting from the first shot to another angle of the same performer, a shot of a different performer, or an insert or cutaway. These same techniques are also employed to adjust actors' performances and to control the pacing of sequences.

Controlling Performance Pace

With judicious editing, you can change the pace of an actor's performance or even the

Figure 20-21.
BOB (Take 1): Please lend me that money.
BOB: (Take 2) I need the cash and I need it now!

Figure 20-22.
BOB (Take 1): Please lend me that money.
BOB (Voice over, from Take 2): I need the cash…
BOB (Take 1, cont.) and I need it now!

Figure 20-23.
BOB (Take 1): Please lend me that money. | BOB (Voice over, from Take 2): I need the cash... | BOB (Take 1, cont.) and I need it now!

Figure 20-24.
BOB (Take 1): Please lend me that money. | BOB (Voice over, from Take 2): I need the cash... | BOB (Take 1, cont.) and I need it now!

speed of a whole sequence. Usually the goal is to shorten it.

You can do this if the available footage includes individual shots of each performer and/or over-the-shoulder two shots in which one actor's mouth is invisible to the camera.

Some actors tend to pause before and during their lines or business (actions) to make their responses appear more deeply felt; but these pauses slow down the sequence. If you can cut away from the pause, you can often shorten or omit it and speed up the scene. The techniques are the same as the ones demonstrated earlier.

In other situations, you may want to stretch a scene out rather than tighten it up. To do this, you could leave in the performers' pauses; but too much of this device can seem artificial. To lengthen a sequence more naturally, try adding performers' reaction shots or cutaways or both. As you can see from **Figure 20-25**, lengthening the sequence heightens the suspense: will Bill lend Bob the money or not?

Enhancing or Adding Meanings

Figure 20-25 uses Bill's reaction shots to build suspense. Notice that they also make him seem reluctant to lend the money. We can imply all kinds of performer feelings by the way we time the edits.

To use a different example, we can make Bob seem as reluctant to ask for a loan as Bill is to give it, **Figure 20-26**. By repeating the word "please" and inserting significant pauses, we seem to drag the request out of the borrower.

Now, suppose the director did not shoot Bill's reaction, but we do have that insert of his hand holding the money. By timing that insert, we can make the hand act for us. In **Figure 20-27**, Bill has no problem with giving up his cash.

In **Figure 20-28**, however, he is reluctant to part with his cash.

As a general rule, showing pauses *ahead of* actors' reactions makes them appear to be thinking. Since viewers know what is taking place in the sequence, they will generally supply the thoughts.

Conferring Importance

In fiction programs especially, one performer may be more important, dramatically, than other players in a sequence. You can increase a performer's prominence by selecting image size, assigning screen time, and giving him or her the critical moments in the scene.

Figure 20-25.

BOB (Take 1): Please lend me that money.

BOB (Voice over, from Take 2): I need the cash…

BOB (Take 1, cont.) and I need it now!

Figure 20-26.

(A long pause as Bob looks at Bill.)

BOB: Please…

BOB (after another pause): Please lend me that money.

Figure 20-27.

BOB: Please lend me that money.

(Bill's hand thrusts the cash forward at once)

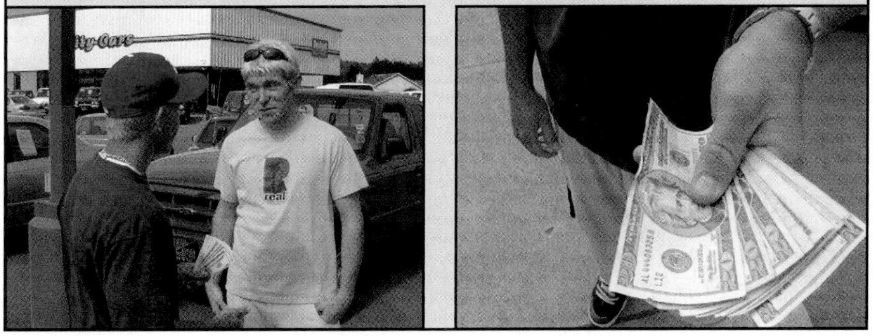

Figure 20-28.
BOB: Please lend me that money. After a long hesitation... Bill's hand thrusts the cash forward.

Selecting image size

In many cases, sequences will be covered by more than one camera angle, for example, a medium (waist) shot and a closeup (head) shot. From the standpoint of viewer psychology, the bigger the image, the more impact it has; you can make key players more important by intercutting closer angles of them with wider shots of others. **Figure 20-29** makes Bill bigger and, therefore, more important.

Apportioning screen time

A second way to enhance a character's importance is by giving him or her more time on-screen. For example, in Figure 20-29, Bob has two shots to Bill's one. To strengthen Bill's status as the more important character, we could reverse the shot ratio, as in **Figure 20-30**.

By giving Bill most of the screen time, we make him the focus of the sequence, even though he does not say a word.

Giving the moment

In addition to image size and screen time, performers also gain importance from *when* they are on screen. Showing a character at a critical point may be called giving him or her the moment. **Figure 20-31** focuses our attention on Bob's emotions at getting his loan.

On the other hand, if Bill is the more important character, you can give the moment to him

Figure 20-29
BOB: Please lend me that money. BOB (Offscreen): I need the cash. BOB: I need it right now!

Figure 20-30.
BOB: Please lend me that money. BOB: (offscreen): I need the cash. BOB: (offscreen): I need it right now!!

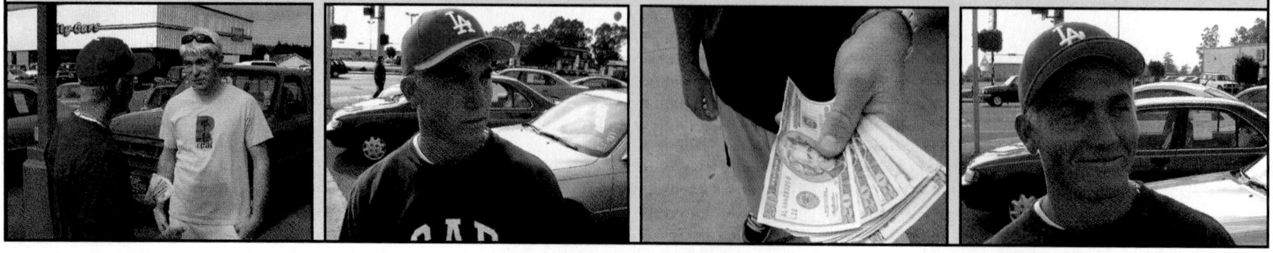

Figure 20-31.
BOB: I need the cash. I need it right now! (His expression turns to delight as he looks down at... Bill's hand holding the money out.)

Figure 20-32.
BOB: I need the cash. BOB (offscreen): I need it right now! (BILL smiles, giving in, and looks down at... his hand holding out the money.)

instead by putting him on screen as he makes up his mind, **Figure 20-32**.

Of the three methods of conferring importance on a performance, giving the screen to an actor at critical moments is perhaps the most powerful.

Emphasis

After continuity and performance, the next editing principle concerns emphasis. We have already seen how editors emphasize one aspect of video programs: actor performance. As a matter of fact, *all* aspects of movies are shaped by editorial emphasis. You emphasize something by making it more prominent, so that the audience pays more attention to it. In the hands of a skilled editor, controlling emphasis can be a sophisticated process; but the basic techniques are quite straightforward. Simply put, you highlight selected material by managing:

- *Content*: What you show.
- *Angle*: How you show it.
- *Timing*: How long you show it.
- *Shot order*: When you show it.
- *Reinforcement*: How you call attention to it with additional material.

In this discussion, we will assume that the director has supplied the editor with complete coverage of the sequence.

To illustrate these techniques for adding editorial emphasis, we will return to the sequence about hitching up a trailer. The essentials of the sequence are reviewed in **Figure 20-33**.

With the footage available, we can create several quite different sequences, simply by controlling the editorial emphasis. We do this by managing *content*, *angle*, *timing*, *shot order*, and *reinforcement*.

Content

To illustrate controlling content, imagine that we decide we want to emphasize the support of the female bystander, while downplaying the skepticism of the male. The result might resemble **Figure 20-34**.

We have emphasized the response of the female bystander by replacing a piece of content: the original two-shot is now a closeup of her reaction.

Angle

To emphasize the difficulty of the process, we can use our control of camera angles. Up

Figure 20-33.
Hitching the trailer.

to now, shots of the trailer and truck hitch components mating have been made from neutral angles. By showing this action from the driver's point of view, we show how hard it is to line up the hitch, **Figure 20-35**.

Timing

The sequence still lacks suspense, because it all seems too cut-and-dried. By lengthening the first half of the sequence, **Figure 20-36**, we can build suspense before showing the successful hitch.

By adjusting content, camera angle, and timing, we have edited a tight, suspenseful sequence.

Now, suppose we decide to change the story slightly by showing that the woman never doubts her ability to hitch up the trailer. Since we have no footage that specifically shows this, we will make the change by simply altering the shot order.

Figure 20-34.
The female bystander gets an extra closeup.

Figure 20-35.
Her worried closeup emphasizes the difficulty.

Figure 20-36.
Adding shots helps build suspense.

Shot order

In most programs, events happen in chronological order. That means that the sooner something appears, the earlier it is in story time. Though we have no footage designed specifically to reveal the woman's inner confidence, we do have a shot of her reacting to the success of her efforts.

To make her seem confident from the start, all we need to do is move her reaction closer to the beginning of the sequence, **Figure 20-37**.

By showing her grin *before* she hitches the trailer, we inform viewers that she is confident of success.

Figure 20-37.
Changing shot order.

She grins *before* the two halves mate.

Reinforcement

Up to this point we have controlled emphasis by *selecting, timing,* and *positioning* shots. To add the final touch to our sequence, we can use editorial reinforcement. Reinforcement can take visual form. For example, we could have the hitch components mate in slow motion, to lengthen and emphasize the success of the operation. To enhance the effect, we can add further reinforcement with sound, **Figure 20-38**.

As the hitch halves mate in slow motion, we lay in the low, reverberating *clang* of a great iron door slamming shut. Though this effect is not realistic, it will be accepted by viewers because it is subjectively "right" for the feeling of the image.

Pace

After continuity, performance, and emphasis, the last major editing principle involves pace. Pace is the sense of progress, of forward movement, that the audience gets from a program. "Pace" is not identical to "speed," although speed is one of its main determiners. In fact, a very fast-moving program can still feel boring if it lacks the other two components of pace: variety and rhythm.

Of course, if the content is boring, it cannot be made interesting by adjusting its pace.

Delivering Content

The easiest way to evaluate the pace of a program is by seeing how well it delivers its content to the audience. If viewers miss important details, the pace is too fast. If content points are belabored after the viewers have already absorbed them, the pace is too slow. If viewers feel that they are getting the material while moving briskly forward, then the pace is about right.

Editing *pace* should not be confused with content *density*. How rapidly a subject is presented also depends on how much detail is included. Suppose, for example, that you finish

Figure 20-38.
Reinforcing with sound.

Figure 20-39.
Good cooking programs deliver their content at a carefully measured pace.

a how-to video on creating a dinner party with a cherry pie dessert. See **Figure 20-39**. Depending on your needs, you could detail every step in baking the pie, or cover only the highlights of that process, or just show the finished pie being served.

Since content density is a function primarily of scripting, it lies outside a discussion of editing. Here, we are concerned with the pacing of the material that the editor is given to work with, whatever its density may be.

Because pacing is judged by the response of the viewers, you will want to consider your target audience in establishing a pace for a program. An elementary school class, a stockholders' meeting, and a classic car club respond to video programs in quite different ways.

The concept of the target audience is covered in Chapter 9, Program Development.

Content delivery is also determined by the type of program. Often, fiction programs are paced energetically so that audiences can follow the story but not catch all the enriching details on the first viewing. (The *Star Wars* episodes and the best Disney animated features are good examples.)

At the other extreme, training programs are paced quite deliberately to ensure that viewers absorb all the essential information. In the final analysis, evaluations of pace must be based on personal taste and experience. In effect, if it *feels* right, it *is* right. Though variety and rhythm contribute to the sense of pace, the most important determiner is speed.

Speed

Speed may be expressed as the average length, in seconds, of the shots in a sequence. See **Figure 20-40**. Though there are no fixed rules about appropriate speed, here is a rough guide:

- One-half second per shot (120/minute) is quite fast.
- One second per shot (60/minute) is fast.
- Two–three seconds per shot (30–20/minute) is brisk.
- Four–six seconds per shot (15–10/minute) is moderate.
- Ten–fifteen seconds per shot (6–4/minute) is slow.

In nonfiction videos, shots typically average between three and ten seconds apiece. It should be emphasized that these figures are not rules, but only examples.

It is possible to have 30 single-frame shots per second of NTSC video (1,800 per minute) and this ultra-high-speed cutting can be used quite effectively, in short bursts. At the other extreme, some documentary programs include shots lasting a minute or more.

In the movie Rope, Alfred Hitchcock made an entire film out of shots lasting several minutes apiece.

The average speed of the shots in a sequence typically depends on the nature of the subject matter. For instance, a fight or car chase

Figure 20-40.
Typical shot durations.

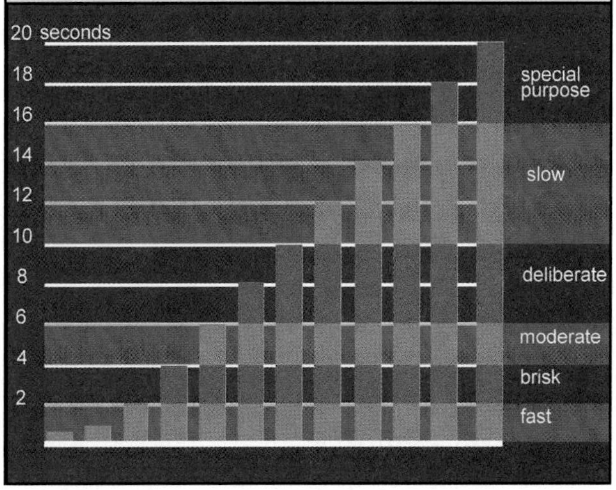

will typically be cut together from very short shots. A business meeting or dinner conversation might consist of moderately long takes, and a languorous love scene might be built from individual pieces many seconds long. Alternatively, you can time your shots *against* the subject matter, presenting a chase in long, deliberate pieces to build suspense or cutting a love scene quickly to mirror its emotional intensity.

Variety

Read these sentences. They are similar. They are the same length. They are short. They are boring.

Shots are like sentences: even short ones grow boring if they are too similar, and long ones can become agonizing if they go on and on without relief. The solution to the problem is variety: mixing shots of varying lengths to avoid falling into repetitive patterns. In most programs, good editors achieve variety instinctively, that is, they do not consciously think, "*The two preceding shots are six and eight seconds long, respectively, so I will follow them with a short shot in order to achieve variety.*" Instead, they rely on their feel for the rhythm of the sequence, while keeping the general goal of variety in mind as they work. In doing so, even experienced editors operate by a sort of successive approximation:

- Building a trial sequence.
- Evaluating the effect.
- Fine-tuning it if it seems to be working, *or*
- Discarding it and start over if it is not working.

The easiest way to achieve variety in shot lengths is by reviewing your work as you go and adjusting shot timing until you like the effect. When this style of cutting is successful, the result is like a path made of irregular flagstones: though it has no visible pattern, it feels like a coherent design.

Rhythm

But, sometimes, an editor wants a sequence to resemble a patterned brick walk instead of an informal path. Sometimes you want to time your shots to develop a clear and repetitive rhythm. See **Figure 20-41**.

Rhythm for contrast

Instead of varying shot length continually, you may want to group several long, slow shots,

Figure 20-41.
A sequence of rhythmic edits is like a patterned brick walkway.

punctuate them with a group of very brief shots, then resume the pattern of lengthy shots. This is the classic western gunfight rhythm: slow cutting while the adversaries stalk each other; a hail of gunfire handled in very quick shots; then another slow passage of cat-and-mouse stalking.

Rhythm for drive

Another classic editing rhythm might be called "marching the prisoner." Imagine that, flanked by Imperial Troopers, our Hero is escorted from the battle cruiser's landing bay to the command center of the Admiral, deep within the giant mother ship. To achieve this sequence, the director has had the actors march through five different sets at exactly the same pace. Using the rhythmic footsteps of the troopers, the editor cuts four out of five shots to exactly the same length:

1. Landing bay: TROMP! TROMP! TROMP! TROMP!
2. Curved corridor: TROMP! TROMP! TROMP! TROMP!
3. Straight corridor: TROMP! TROMP! TROMP! TROMP!
4. Antechamber: TROMP! TROMP! TROMP! TROMP!
5. Command room where the Admiral is waiting: TROMP! TROMP! TROMP! TROMP! TROMP! TROMP! TROMP!

Notice the rhythm of the final shot, which contains three additional steps. The extra paces contribute a sense of closure to the sequence. Also, it consists of an odd number of steps (7), where the other shots contain even numbers (4 each).

For psychological reasons that are not perfectly understood, a sequence often feels more powerful when it concludes on an odd number.

The two most common English verse forms are iambic pentameter with five beats per line:

Shall I compare thee to a Summer's day?
Thou art more lovely and more temperate.

and ballad measure with seven beats per line:

It looked extremely rocky for the Mudville nine that day.

The score stood four to two with but one inning left to play.

Rhythm for comedy

Comedy often relies on surprise, or the unexpected. Since rhythmic cutting sets up the expectation that it will be continued, an editor can help a joke by building rhythmically to the payoff and then suddenly breaking the pattern. For example, suppose that the storm trooper marching sequence was shot by a comic director like Mel Brooks. The edited result might resemble this:

1. TROMP! TROMP! TROMP! TROMP! Landing bay
2. TROMP! TROMP! TROMP! TROMP! Curved corridor
3. TROMP! TROMP! TROMP! TROMP! Straight corridor
4. TROMP! TROMP! TROMP... Whoa! Somehow, they are back in the landing bay.

Here, the editor sets up a strong, repetitive rhythm and then deliberately breaks it in order to reinforce the climax of the joke. If you study classic movie comedy, from Charlie Chaplin to the present, you will see how dependent it is on rhythmic editing.

Rhythm for montage

The sidebar called *The Art of Montage* explains how multiple images and sounds can be presented to dramatize events over time. Because ***montages*** are constructed in repetitive patterns, they are usually edited to strict rhythms set by musical underscoring. Occasionally, however, a montage may be purposely built without any discernible rhythm. For example, imagine a football game. If the story calls for one side to march triumphantly over the other, this might be expressed in a classic rhythmic montage. But if the game is fought to a messy, brutal scoreless tie, then the montage might be as formless and confusing as the game itself.

Cutting to music

Rhythmic cutting is widely used in matching visuals to music. This is usually done for music videos, and for sequences in other programs in which shots are linked by association rather than by chronological order. For instance, a succession of views of the Italian countryside might be edited to the lyrical rhythms of Vivaldi's *The Four Seasons.* A montage of New York City going about its complex daily business, **Figure 20-42**, could be cut to the rhythm of a modern jazz composition.

At its best, editing picture to music is like dancing: though you follow the basic beat, you do not match the music exactly, phrase for phrase and measure for measure. Instead, you maintain an organic collaboration between picture and sound. Handled this way, cutting to music can be the most satisfying kind of rhythmic editing.

The End Result

In this discussion of editing principles, we have considered continuity, performance, emphasis, and pace. There is more to the craft of editing, especially in subtle matters of tone, mood, and style. We have not addressed these matters directly because as an editor, you cannot do so either. Tone, mood, and style are not guiding principles but ultimate goals: they are what you will achieve through the way in which you handle *continuity, performance, emphasis,* and *pace.*

Figure 20-42.
A New York street suggests up-tempo jazz music. (Sue Stinson)

The Art of Montage

A montage is a quick succession of sounds and images designed to condense a lot of material into a brief amount of screen time. For example, a montage dramatizing a winning baseball season might be communicated in a rapid-fire succession of shots showing hitters, scoreboards, umpires signaling "safe!" and so on, intercut with repeated newspaper headlines announcing team wins, pages flying off a desk calendar, and shots of the team bus traveling to one game after another.

Dramatic effect

Most montages are not absolutely necessary for conveying information. After all, the content of a baseball montage could be fully communicated by a single insert of a newspaper headline, such as this one:

The essential information could be conveyed in a single shot.

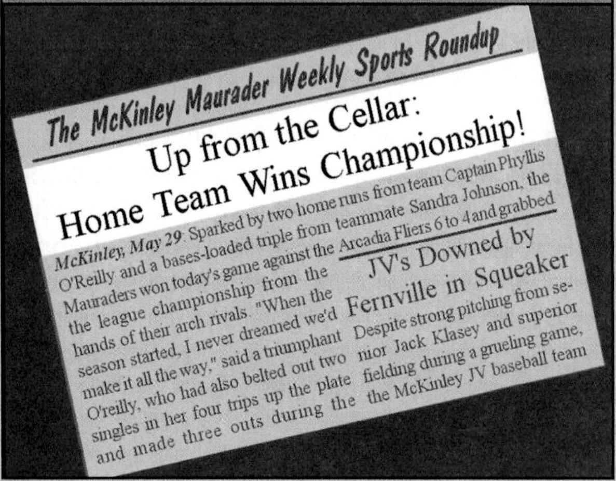

The McKinley Maurader Weekly Sports Roundup

Up from the Cellar:
Home Team Wins Championship!

McKinley, May 29: Sparked by two home runs from team Captain Phyllis O'Reilly and a bases-loaded triple from teammate Sandra Johnson, the Mauraders won today's game against the Arcadia Fliers 6 to 4 and grabbed the league championship from the hands of their arch rivals. "When the season started, I never dreamed we'd make it all the way," said a triumphant O'reilly, who had also belted out two singles in her four trips up the plate and made three outs during the

JV's Downed by Fernville in Squeaker

Despite strong pitching from senior Jack Klasey and superior fielding during a grueling game, the McKinley JV baseball team

The real purpose of a montage is drama: developing a sense of momentum and excitement from the kaleidoscope of images. By showing highlights of a story, a montage preserves the feeling of narrative, in highly condensed form.

Styles

The simplest montages are linear, with one quick shot after another filling the screen. More complex versions may display multiple images in several forms:

- Superimposed double or even triple exposures.
- Screen divided into several separate images.
- Changing background images composited behind a foreground character.

Superimposition and split screen are not usually employed together because the result might be too confusing to follow.

The role of sound

Sound is very important in montages for several reasons:

- Because audiences can process audio information very quickly, sound can be several layers deep.
- Compared to video, audio is easy and inexpensive (laying in the sound of a steam train is much easier than videotaping a real engine).
- Music, especially, can communicate nostalgia and many other feelings.

As a general strategy, you may wish to keep the visuals relatively simple, while mixing a complex and evocative composite sound track.

An example

Here is an example of a simple montage. For demonstration purposes, it progresses through only six stages, and contains just six elements. A real montage might run longer and contain many more layers of information. Our montage dramatizes the rise of the brilliant actor Laurence O'Livery from a callow youth to a revered icon of the theater. The montage contains six kinds of information:

1. A succession of theater programs that show O'Livery's rise from extra to superstar.
2. Calendar years changing constantly, to show the passage of time.
3. Titles of the plays our hero appears in.
4. Lines spoken by the great O'Livery,
5. Evocative music of each period.
6. Sound effects of applause.

Solicitor — Craig Sassoon
Landlady — Walla Warfield
First Policeman — Jason Quigley
Second Policeman — Harlan Stander
The Crowd
Angela Aames, William Bassett, Charles Dark, Melissa Evans, George Grimes, Cedric Haddon, Mary Jones, Gwyneth Kalem, **Laurence O'Livery,** Michael Porter, Daryl Ransom, Gale Starr, Herschel Tannen, Gregory Wilson.

Lyceum Theatre Programme/The Faceless Crowd

ACTION: Flipping calendar pages show the years advancing.
VOICEOVER: Crowd noise
EFFECTS: Scattered applause
MUSIC: Big band.

with
Lady Landsdowne — Bernice Duprey
Lord Landsdowne — Guy Bainbridge
Rev. Cheadle — Jack Crammell
A. Hobo — Launce Gobbo
Parlor Maid — Fanny Pryce
Butler — **Laurence O'Livery**

Presented by arrangement with his Majesty's Theatre, Hammersmith. Grateful acknowledgement is made to the estate of Lord Wolverhampton for provision of the suits of armour appearing in Act II.

Kinghtsbridge Theatre Programme

ACTION: Flipping calendar pages show the years advancing.
VOICEOVER: You rang, Sir?
EFFECTS: Polite applause
MUSIC: Pop song.

The Theatre Company, Ltd.
presents
Edward Maybridge and Vanessa Prawn
in
Weekend at Windemere
with
Cedrec Cubbins, George Brandt
and Laurence O'Livery as Freddie

ACTION: Flipping calendar pages show the years advancing.
VOICEOVER: Anyone for tennis?
EFFECTS: Strong applause
MUSIC: Do-wop pop song.

Selwyn and Selwyn
presents
A Love for the Ages
by
Hamish Korngold
starring
Laurency O'Livery Angela Bredhan
with
Percy Adline
Maurice Rample

ACTION: Flipping calendar pages show the years advancing.
VOICEOVER: I love you, my darling!
EFFECTS: Very loud applause
MUSIC: Beach Boys song

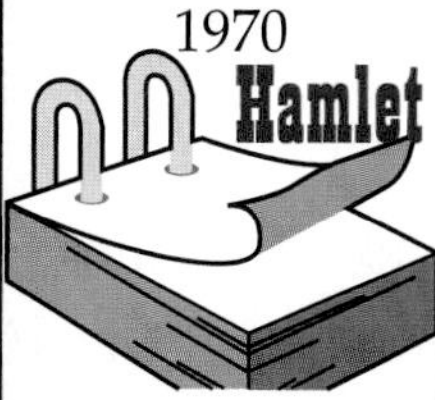

The Imperial Shakespeare Company
presents
Laurence O'Livery
in
Hamlet
Prince of Denmark
by
William Shakespeare

ACTION: Flipping calendar pages show the years advancing.
VOICEOVER: To be or not to be...
EFFECTS: Thunderous applause
MUSIC: Paul Simon song

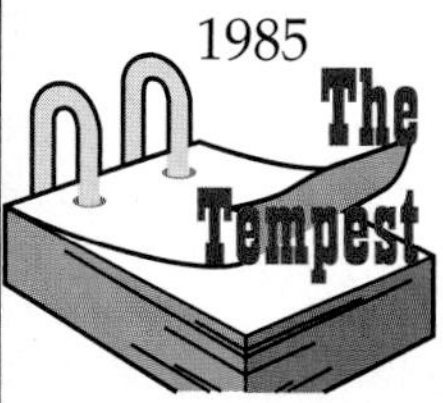

presents the
Laurence O'Livery
production of
The Tempest
starring
Laurence O'Livery
by
William Shakespeare

ACTION: Flipping calendar pages show the years advancing.
VOICEOVER: We are such stuff as dreams are made on...
EFFECTS: Distant, filtered applause
MUSIC: New Age/Celtic instrumental.

Notice how inexpensive this montage is to create. There is not a single shot of an actual play in progress.

Montage and video

Though popular up through the 1950s, montages almost disappeared for a few decades, partly because they went out of fashion and partly because the film laboratory work required was very expensive. But with digital editing, most software can create complex and effective montages for just the price of the editor's time.

Chapter Review

Answer the following questions on a separate piece of paper. Do not write in this book.

1. Programs intended to ______ viewers are typically organized by logical argument.
2. Reversing an image to correct screen direction is known as ______.
3. *True or False?* Editing to control performance pace usually involves shortening.
4. The principle of editing emphasis is accomplished by managing content, angle, timing, ______, and reinforcement.
5. Music videos are a good example of ______ or "cutting to music."

Technical Terms

Act: A major division of a program, typically ten or more minutes long, and divided from other acts by fade-out/fade-in transitions.

Associative continuity: A sequence of shots related by similarity of content.

Buffer shot: A shot placed between two others to conceal differences between them.

Continuity: The illusion that a video program is a single continuous piece.

Cutaway: A shot other than, but related to, the main action.

Insert: A closeup detail of the action.

Linear continuity: A sequence of shots organized chronologically and/or logically.

Montage: A brief, multilayer passage of video and audio elements designed to present a large amount of information in highly condensed form.

Parallel cutting: Presenting two or more sequences at once by showing parts of one, then another, then the first again (or a third) and so on.

Program: A complete video presentation.

Scene: A brief content component, usually recorded in several shots.

Sequence: A coherent assembly of shots.

Shot: A single recording, from camera start to stop.

Subjective continuity: A sequence organized according to the feelings of the director/editor, rather than by chronological or other objective criteria.

Although the screen direction is the same, the real world directions are exactly oposite.

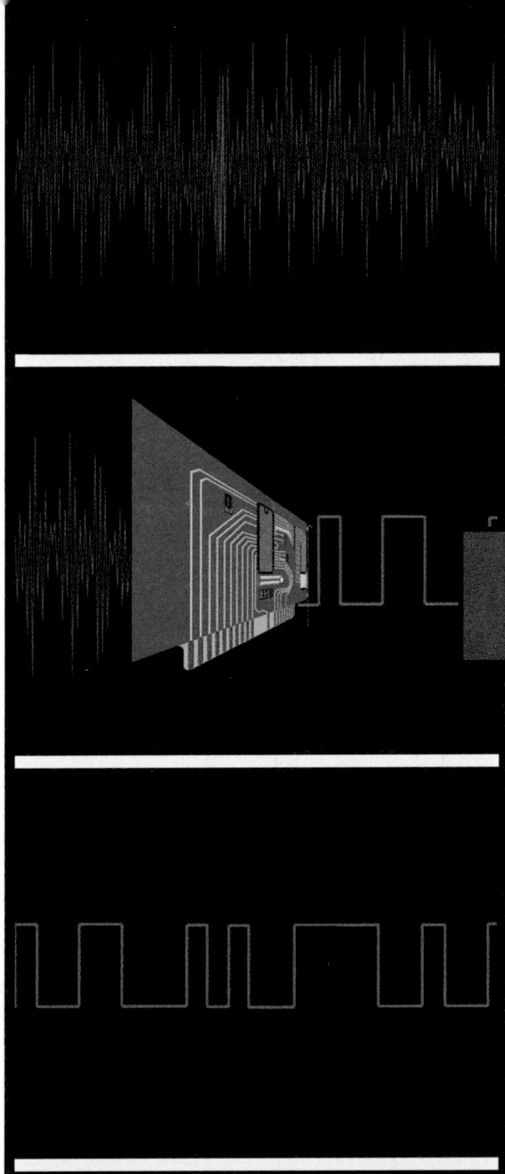

21 Digital Editing

Objectives

After studying this chapter, you will be able to:

- Explain the advantages of digital editing.
- Contrast digital and analog editing methods.
- List the basic steps in digital postproduction.
- Configure a new editing project.

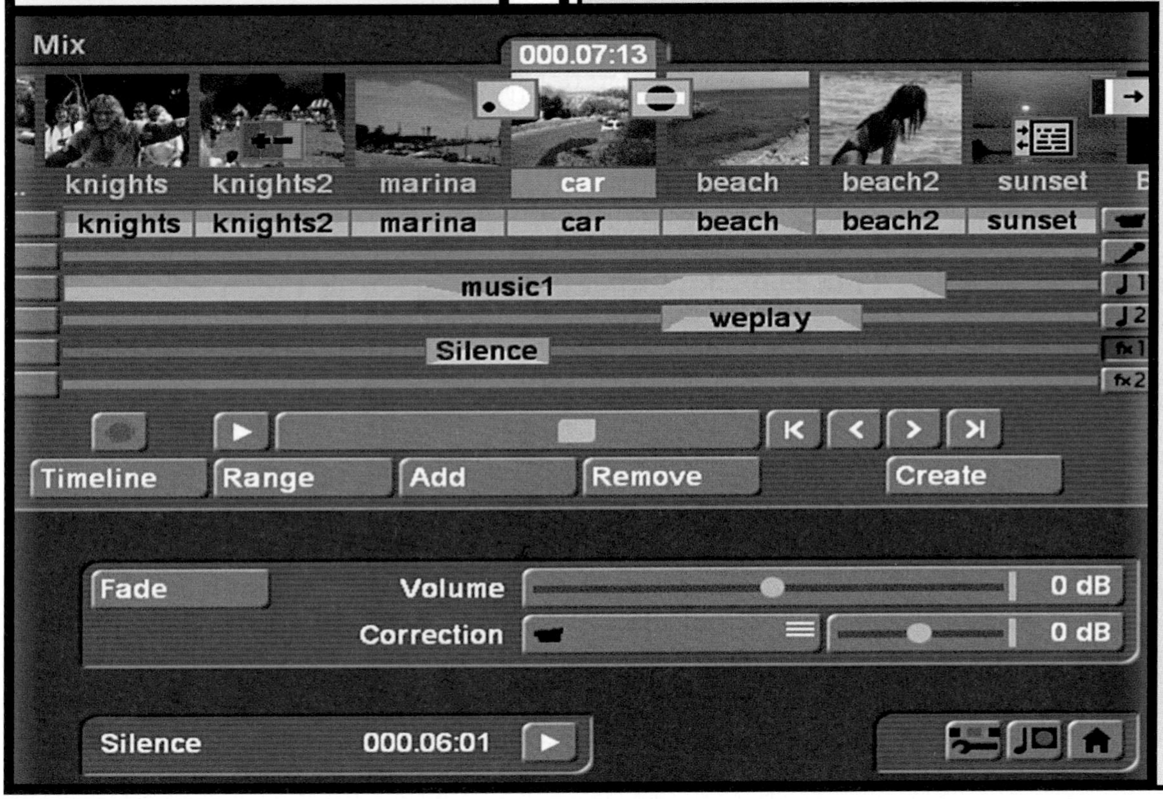

About Digital Editing

Now that we have covered editing operations and principles, we are ready to apply both of them to the actual procedures of digital postproduction. It is here that we first encounter the problem of editing programs that look different from each other and, to some extent, operate in different ways. To address the problem, this chapter focuses not on menus, screens, and icons, but on the digital operations *behind* them, operations that are common to all editing programs.

The term *digital editing* refers, of course, to postproduction using video and audio stored in digital form and processed on a computer.

Traditional "analog" editing uses video and audio stored as recorded copies of electrical signals. Analog editing is covered in Chapter 24.

Digital has taken over professional postproduction completely because it is easier to work with, it produces results of a higher technical quality, and it allows use of operations that are difficult or impossible with old-fashioned analog editing. Moreover, digital editing can be performed on a high-quality personal computer, so that all operations can be controlled by a single person in a single place.

See the sidebar on the advantages of digital postproduction.

The Advantages of Digital Postproduction

As noted, the obvious benefits of digital editing are the concentration of all postproduction on a single desktop and the ability to perform operations like titles and effects that would otherwise be difficult or impossible. But digital post offers advantages that are even more fundamental. To appreciate them, it helps to compare digital editing to the old analog (tape-based) editing system that it has replaced.

	Digital Editing	Analog Editing
Data format	Numerical encoding	Electrical signal
Storage medium	Hard drive or media disc	Tape
Access	*Parallel:* Because your material is stored on a drive, you can jump almost instantly to any point in it.	*Serial:* To reach any shot, you must roll through all the footage between your start point and your destination.
Work order	*Random:* You can work on any section at any time, and you can return to previously edited material and revise it without disturbing the rest of the program.	*Sequential:* You must assemble your edited program in strict chronological order. Once you have edited a sequence, you cannot revise it without re-editing everything that follows it on the tape.
Multiple tasks	*Additive:* You can work on one element at a time. The final multilayered program can be richer than anything you could achieve with tape-based postproduction.	*Concurrent:* You must perform all editing operations at the same time. This makes complex operations with multiple video and/or audio tracks difficult.

Nonlinear editing offers other advantages as well.

Speed: Random access, additive editing, and easy revision let you work more quickly.

Permanence: Although digital recordings are not themselves permanent, they can be copied through many generations without appreciable quality loss. (Analog recordings suffer visible decay each time they are copied.)

Flexibility: Digital recordings are easy to modify, so effects, composites, filters, titles, and other processes are simpler to work with and revise. Digital audio is equally flexible. Although analog video can be pro-

The typical work flow for digital editing involves several basic operations:

- Configuring the project.
- Capturing video and audio.
- Assembling the shots.
- Adding transitions, titles, and effects.
- Building the audio.
- Outputting the program for storage and distribution.

The rest of this chapter considers each of these operations in order, beginning with capturing video and audio materials.

Though configuring the project is the first operation, this rather technical process has been saved for last.

Capturing Materials

The first step in building a new program is to bring the raw materials for it into the computer. This process is called ***capturing***.

Capturing Camera Footage

Typically, the video files are recorded on tape, although some systems record source material directly on disc or other media. Several considerations apply to capturing source files.

Digital or analog

The way in which your computer captures external sources depends on whether those sources are analog or digital. ***Digital*** files are

cessed externally by digital switchers, compositors, and titlers, the process is easier when all the material is stored in a computer.

Economy: Because a single desktop computer can perform the operations of many different stand-alone editing components (switchers, titlers, audio mixers, timebase correctors, etc.) a disc-based digital editing setup is typically less expensive. When outside postproduction services are purchased for analog operations, the cost differential is even greater.

Universality: Broadcast video is recorded and displayed in several incompatible formats, including NTSC, PAL, and SECAM. In addition, high-definition video exists in a number of different standards. Internet streaming and computer video displays use still other protocols. As a result, converting analog video from one format to another has been a complex and imperfect process. In digital video, by contrast, you work in the digital format that your software prefers, and then output the result in whatever standard you need for storage and playback.

Time code: Digital video brings time code (formerly reserved for expensive professional postproduction) to even the simplest setup. With this ability to give every single frame a unique and permanent address, you can maintain precise control of all your program components.

An analog editing system.

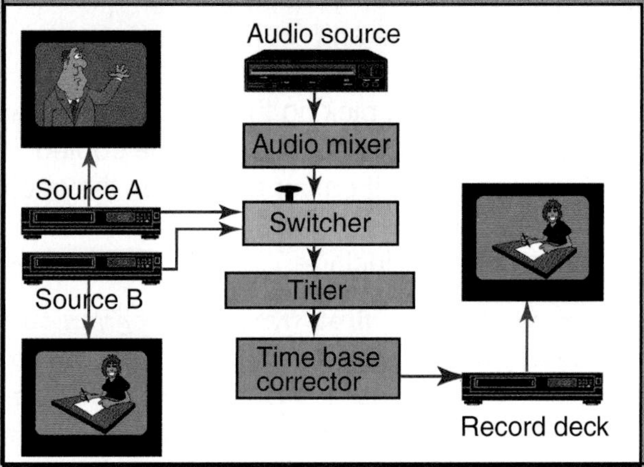

A digital editing system.

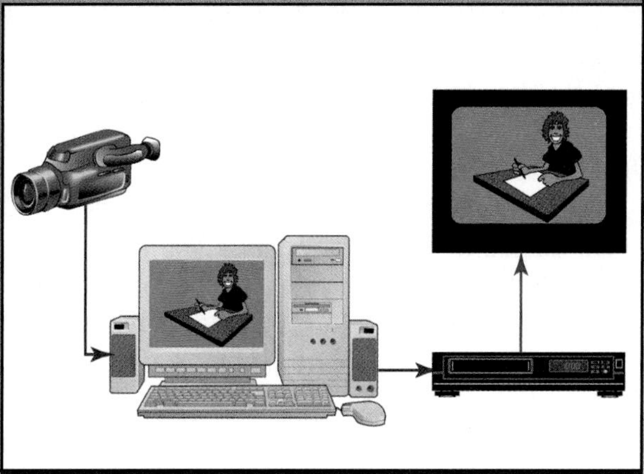

Figure 21-1.
Ready to capture video. (Ulead)

Figure 21-2.
Device control lets you operate your camcorder through your computer. (Ulead)

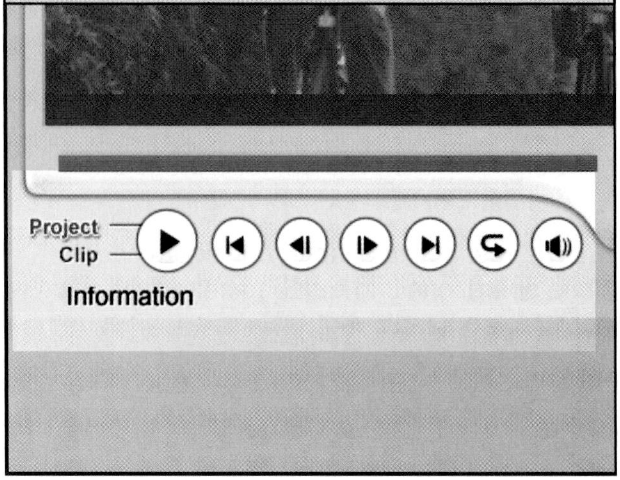

captured and stored in their native form, without need for conversion. See **Figure 21-1**. Usually, camcorder and computer are cabled together through "IEEE 1394" ports.

These ports are also known as "I-link" and "firewire."

To bring in the older ***analog*** format video, you will need a capture device: a piece of hardware that will convert the incoming analog signal to digital. Older computers typically use internal capture cards, but external boxes (many of which contain TV tuners too, for capturing broadcast or cable signals) are more versatile and usually easier to use.

Device control

Device control is the ability to operate your source camcorder from your computer, without touching the actual controls, **Figure 21-2**. Device control is a great convenience because you do not have to work with two separate pieces of hardware at a time.

Online or offline

In digital postproduction, you can work at two different levels of quality. *Online* editing creates the final program, so file capture is conducted at the highest quality level that your system can sustain. *Offline* editing creates a low-quality "work print," plus a frame-accurate record (called an edit decision list or ***EDL***) of every element in the program, for later use in making a high-quality final version at a professional postproduction service facility.

Offline editing is still used in many high-end productions. In recent years, however, the power of personal computers and the sophistication of editing software have grown so much that many highly professional programs are completed without ever leaving the desktop.

As used in editing, the term "online" has nothing to do with being connected to the Internet.

Individual or batch

Most editing software allows you to capture individual shots one at a time, or to identify all the shots you want captured and have them processed automatically as a group after you have finished selecting them. Both methods have advantages:

- Individual capture allows you to examine each shot, select in and out points, capture the shot, and verify that the copy is free of dropped frames and artifacts.
- Batch capture, **Figure 21-3**, allows you to concentrate on picking the footage you want, without waiting for each one to be copied. This can make it easier to keep all the shots in your head as you make decisions about which ones to include.

The *in-point* is the first frame of program to be captured and the *out-point* is the last frame. A *dropped frame* is a frame that was not copied, usually because some components of your computer could not keep up with the speed of data transfer from the camcorder or VCR. An *artifact* is a small blemish on an image.

Figure 21-3.
The batch capture "wizard" in Video Studio 9. (Ulead)

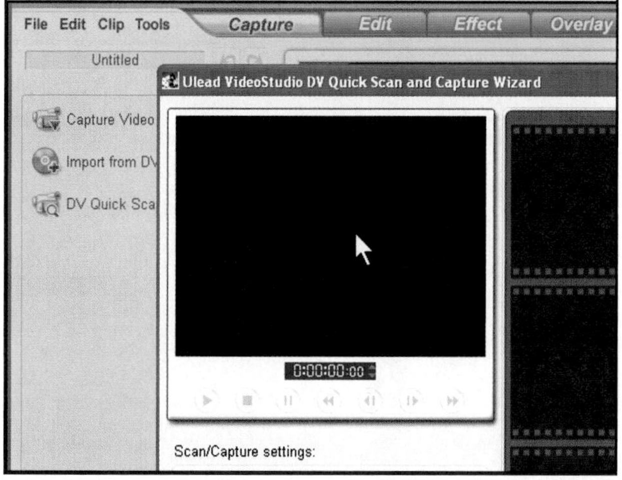

Many editors employ both methods, often using batch capture for initial copying and individual capture for picking up other shots during the course of editing.

Many editing programs take batch capture a step further. If you select the option, the software will watch for evidence of change from one shot to another (usually jumps in the time stamp created by the camcorder's clock) and capture the changed image as the first frame of a new shot. When the capture is complete, you can review the shots as transferred and make any needed adjustments.

Saving captured material

Because of their large file sizes, shots are saved to disc as part of the capturing process. To conserve disc space, it is possible to bring them up, trim unwanted footage, and re-save them under the same file names. But it is faster (and often safer) to do most of the selecting and trimming as you capture.

You can also save storage space by discarding unwanted components, such as the production audio track when you want only the video, **Figure 21-4**.

Also, some editing software allows you to add descriptive comments about each shot and many programs let you organize your shots in a searchable database. It is a good idea to annotate your footage while it is still fresh in your mind. This will save time later, when you need to hunt for a particular shot.

Importing Other Components

Many editing programs allow you to incorporate still photographs, graphics, and presentations from PowerPoint and similar applications. Digital photographs are uploaded from discs or camera storage devices. Photographic negatives and prints are digitized and uploaded by scanners.

These materials are said to be "imported" rather than "captured" because the editing software does not bring them into the computer, but only accesses them on a hard drive, **Figure 21-5**.

Once in the editing software, however, other graphics can be treated just like video shots. In fact, very effective programs can be made without any video sources at all.

Figure 21-4.
A clip is captured, trimmed, and filed as a video-only shot.

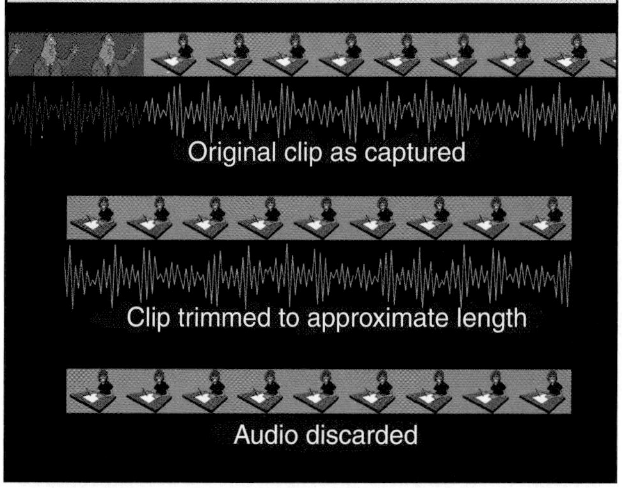

Figure 21-5.
Imported still images, ready for use. (Ulead)

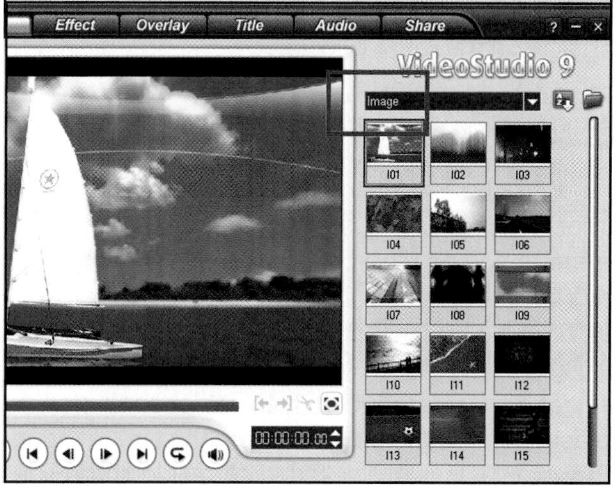

Online and Offline Editing

The terms *offline* and *online* were originally coined to label editing at different types of facilities: *online* editing meant postproduction performed on the high-quality equipment of a specialized facility informally referred to as a "***post house***." *Offline* editing meant assembling a work print of the program elsewhere, on inexpensive equipment.

The system of editing offline and then finishing online at a post house is still used when the quality of the source material and (especially) the final program is higher than a desktop editing system can conveniently handle. For example, footage shot in a professional digital format and intended for broadcast can be dubbed to a Mini-DV copy, edited offline, and then reassembled online and archived in the original professional format. Since postproduction houses rent their services at high hourly rates, producers try to do most of the editing offline, before bringing the program to a post house for completion online.

Offline editing

To do this, editors work with inexpensive copies of the source footage that include the original time code. In traditional offline editing, the time code is sometimes recorded twice: both electronically and also visually in a small rectangle "burnt in" at the bottom of every frame so it can be read right off a monitor screen. A work tape with visible time code is often called a "***window dub***." By means of a desktop computer, time code addresses of each element are recorded on an Edit Decision List (EDL) that will be used later to create the final program in online editing. In modern digital editing, a window dub is unnecessary because time code is automatically displayed by the editing software.

A frame from a window dub. (Sue Stinson)

Online editing

When offline editing is complete, the work print, the original source tapes, and the EDL are then taken to the postproduction house for online editing. There, a new, high-quality duplicate of the work tape is created by matching every shot on the work tape with the identical footage on the source tapes and copying the original to an edit master tape. The matching process is partially automated by loading the EDL into a computer at the post house. Titles, effects, transitions, and other processing are added during online editing. Sound tracks are mixed and "sweetened" to create a composite sound track.

Production audio is captured with video. Music and sound effects are imported like still graphics. Audio is covered separately, later in this chapter.

Building the Program

With the raw materials captured and/or imported, the next step is to assemble the program, piece by piece.

Assembly Editing

The simplest method is to select shots and place them on the storyboard or timeline in order, **Figure 21-6**. This is called assembly editing. The next job is to adjust each shot so that it contains exactly the right content and runs exactly the desired time.

Shots and clips

The term "shot" can be slightly misleading. Because a digital original can be copied and recopied as often as necessary, it helps to think of the original shots as no more than the raw materials of your program. The working units of digital editing are called "clips." A ***clip*** is a piece of a shot (or shots) selected and used in editing a program. You can create clips that use or re-use all or parts of shots; you can make clips out of pieces of several shots together; you can modify clips to change color, speed, direction, length, and other attributes.

Figure 21-6.
The work screen for editing. (Ulead)

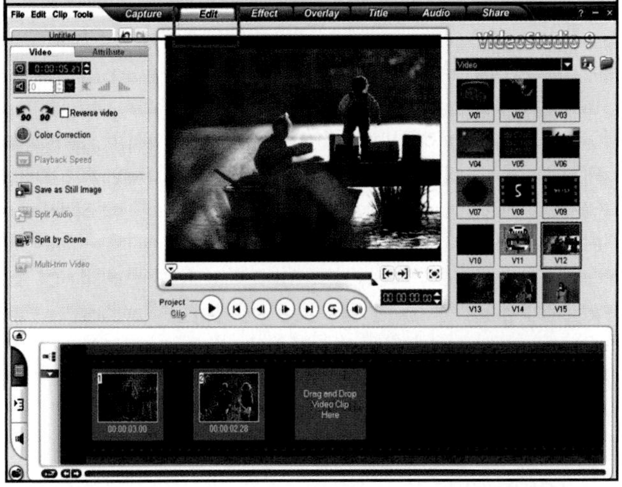

Figure 21-7.
Video Studio calls the shot folder the "library." (Ulead)

Most editing software allows you to name subclips and otherwise identify material that is divided, duplicated, or modified.

Even after you have placed clips in your program, you can divide them into multiple clips and continue to manipulate them as long as you like.

Creating clips

Most editing programs allow you to work with a folder or "bin" containing the shots you will use in creating and placing clips. In effect, a shot becomes a clip when you select it from the folder and begin placing and shaping it for your program, **Figure 21-7**.

Trimming clips

Since you will rarely use all of an original shot, your first task is to trim it. Trimming means deciding exactly where the action in the clip will start and stop, and/or how long the clip will remain on screen.

One way to trim a clip is to display it in a window of its own, complete with time code readout and movement controls (play/stop/single frame, etc.). By doing this, you can adjust length and content to your liking before placing the clip in the program. After trimming the clip, you place it on the storyboard or timeline.

Another way to trim a clip is by adding it to the assembly immediately and making adjustments in place. By trimming on the time line, you can see the clip in relation to the clip ahead of it. This is useful for establishing cut points and matching action. In practice, many editors use both methods at once, establishing the basic length and content of the clip in the preview, placing it in the assembly, and fine-tuning the trimming there. It is important to understand that the trimming of a clip is not final. You can always change the in/out points to adjust the length or the content.

When you have every clip trimmed and placed in order, you have assembled the program (or program sequence).

Insert Editing

Once you have selected, trimmed, and placed clips on your timeline, you will generally fine-tune the results, using various forms of insert editing.

In computer-based digital editing, insert editing covers any and all procedures used to modify the original assembly of shots on the timeline.

Depending on your software, you may have many different ways to adjust clips, insert new material, change timing, etc. But all systems require an understanding of several different types of edits, including *three-point, four-point, rolling, ripple, slip,* and *slide.*

Three-point edits

As we have seen, every clip has an in point, where it begins, and an out point, where it ends. Think of these as points one and two. When you are inserting a clip into the middle of an existing program, you also have a *program in* point, where the inserted clip will start and a *program out* point, where it will end. Think of the program points as points three and four.

Figure 21-8.
A three-point edit.

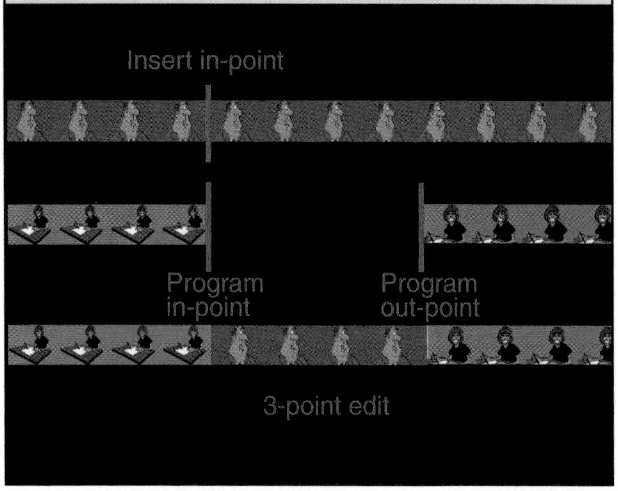

It often happens that you need to replace an exact range of material in the original program with a new clip, in order to maintain precisely the same length. To specify that range, you indicate both the program in and out points. Then you indicate the in point of the replacement clip. When you drop this clip into the original material, your software will start it at the clip in point, run it until it comes to the program out point, and then set the clip out point automatically. This is called a ***three-point edit***, **Figure 21-8**. Most frame-for-frame replacements are three-point edits.

Four-point edits

In some cases, you may have to fit a new clip that is longer or shorter than the program section that it replaces. In this situation, you specify a ***four-point edit***—that is, you select the in- and out-points of both the source clip and the program, **Figure 21-9**. When you make a four-point insert edit, you may choose either of two ways to match the clip length to the program spot. You can change the speed of the source clip to fit. If the difference between source and destination is not great, the resulting slow or fast motion may be undetectable. But if you must stretch, say, a three second source to fill a six-second slot, the speed difference will be obvious.

Alternatively, you can change the length of the source clip to fit the target slot. When you do this, you essentially revert to a three-point edit, letting the software insert the source clip until it runs out of space and then set its out point automatically.

Three- and four-point edits are useful when you cannot change the length of the program or the running time of other shots in it. However, it is more common to insert shots while allowing the rest of the program to adjust to them.

Rolling edits

One way to do this is with a ***rolling edit***. When you make a rolling edit, you retain the original program length. However, the shot following the insert adjusts to compensate. If the replacement is *shorter* than the original, the in point of the following clip moves up to fill the gap. But if the replacement shot is *longer*, the in point of the following clip moves back, to open up the necessary space, **Figure 21-10**. To make a rolling edit with a shorter replacement clip, the next clip must have previously trimmed material at its head. This is a good reason for digitiz-

Figure 21-9.
A four-point edit.

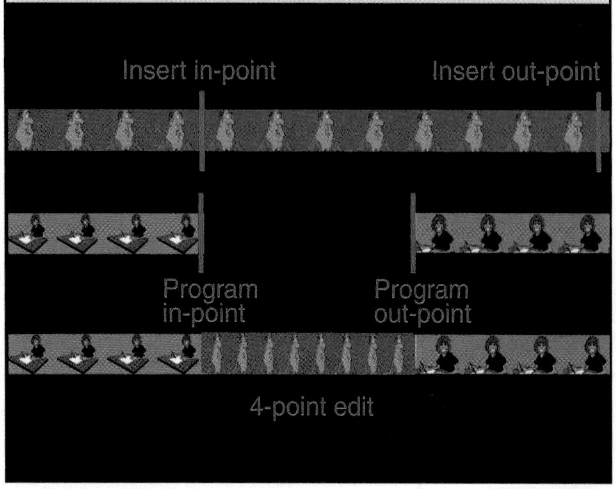

Figure 21-10.
A rolling edit.

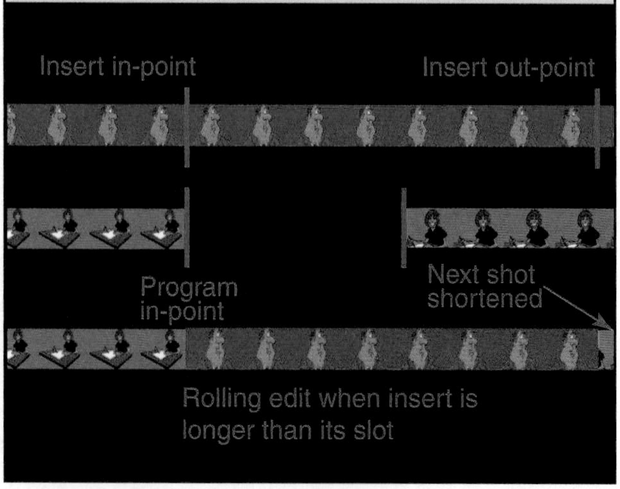

ing and saving at least five extra seconds at the beginning and end of each clip you capture.

Like four-point edits, rolling edits are useful when a program, such as a commercial, must retain a precise overall length. When length is not critical, however, it may be easier to perform a ripple edit instead.

Figure 21-12.
A slip edit.

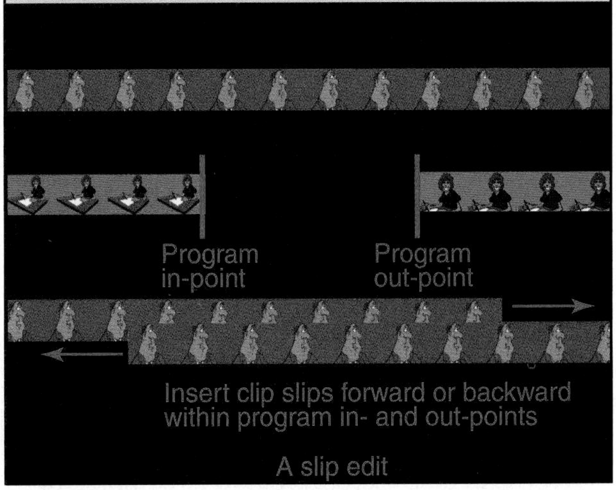

Ripple edits

A ***ripple edit*** is a three-point edit in which all the material following the insert adjusts automatically to fit, **Figure 21-11**. If the insert is *shorter*, all the following material closes the gap by moving back to the end of the new shot. But if the insert is *longer*, all the following material makes room for it by moving over. In practice, you may find that ripple edits are the most common inserts that you make.

Slip and slide edits

If you wish to adjust the program on both sides of the new material, you can select either a slip edit or a slide edit. A ***slip edit*** preserves the in and out points of the preceding and following clips by shifting the in and out points of the source clip, **Figure 21-12**. This allows you to fine-tune the part of the new clip inserted into the program.

A ***slide edit*** preserves the in and out points of the *new* clip, instead, by shortening the clips on both sides. Generally, you can slide the new clip back and forth to decide how much of the previous and/or following shots to shorten, **Figure 21-13**. This allows you to change the length of the source clip without affecting the length of the program.

Editing tools and options

Three-point, four-point, rolling, rippling, slipping, sliding: these options can seem confusing at first; and professional editing software offers many more as well, including other forms of edits, different tools for making the actual edits, different windows in which to adjust clips, and a choice of keyboard or mouse-based commands for many editing functions. In practice, most editors gradually evolve personal working methods to suit their own needs, utilizing the features they prefer and ignoring the others. In other words, you need not master and use every option in a sophisticated postproduction application. Instead, use new features as you need them and then decide whether to continue with them or not.

Figure 21-11.
A ripple edit.

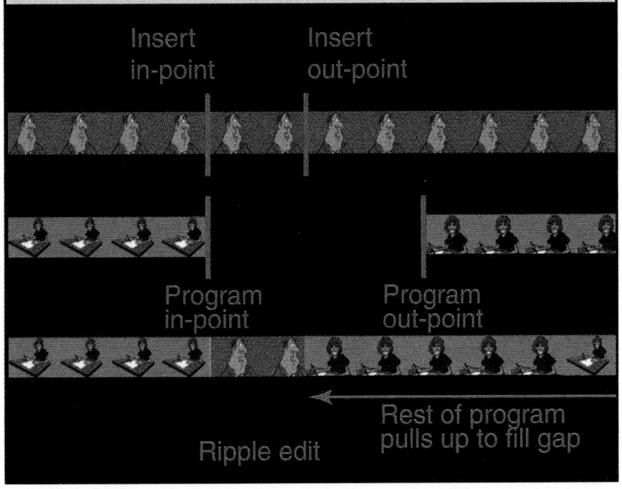

Figure 21-13.
A slide edit.

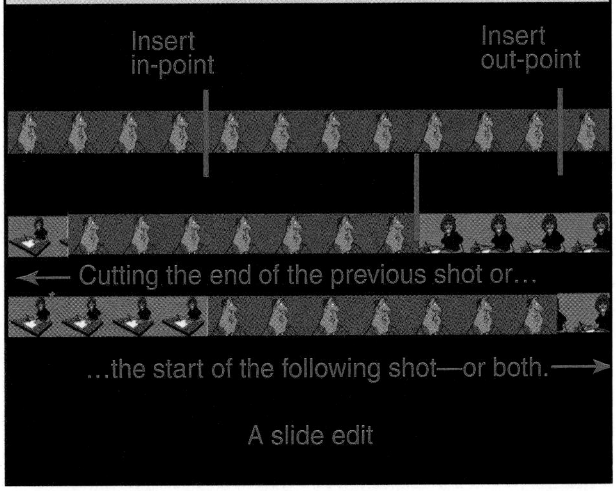

Adding Effects

Program material in digital form is easy to adjust, modify, or change into something quite different. There are hundreds, perhaps thousands of different ways to transform video originals. Each editing application comes with its own proprietary set of effects. So-called "plug-ins" can be added to the original package to extend its capabilities.

Plug-ins are programs that become part of your editing software. They do not have to be opened separately because their commands and other features are automatically displayed in the menus of your application.

You can also export footage to completely separate software applications, work on it, and then import it back into your project.

With so many options available, it would be impractical to discuss every one; so this section covers the general types of image effects that you are likely to encounter.

Transitions

The most common effects are transitions from one sequence to another. Traditional transitions — such as fades, dissolves, and wipes — are supplemented by almost numberless two- and three-dimensional transitions made possible by digital video effects (DVEs). See **Figures 21-14** and **21-15**.

Strictly speaking, fades, dissolves, and wipes are also digital effects, but the term DVE is typically used for the more exotic transitions and other effects.

Superimpositions

Superimpositions are combinations of two or more images displayed together. The most common superimposition is the double exposure, in which two full-screen images are visible at once, **Figure 21-16**. In digital postproduction, however, most superimposition involves replacing parts of one image with parts of another.

Compositing

Compositing is the process of placing part of one image (usually the subject or other foreground object) into a different background. This is commonly done by recording the foreground against a background of a very specific and very uniform

Figure 21-14.
A DVE is dragged between two shots. (Ulead)

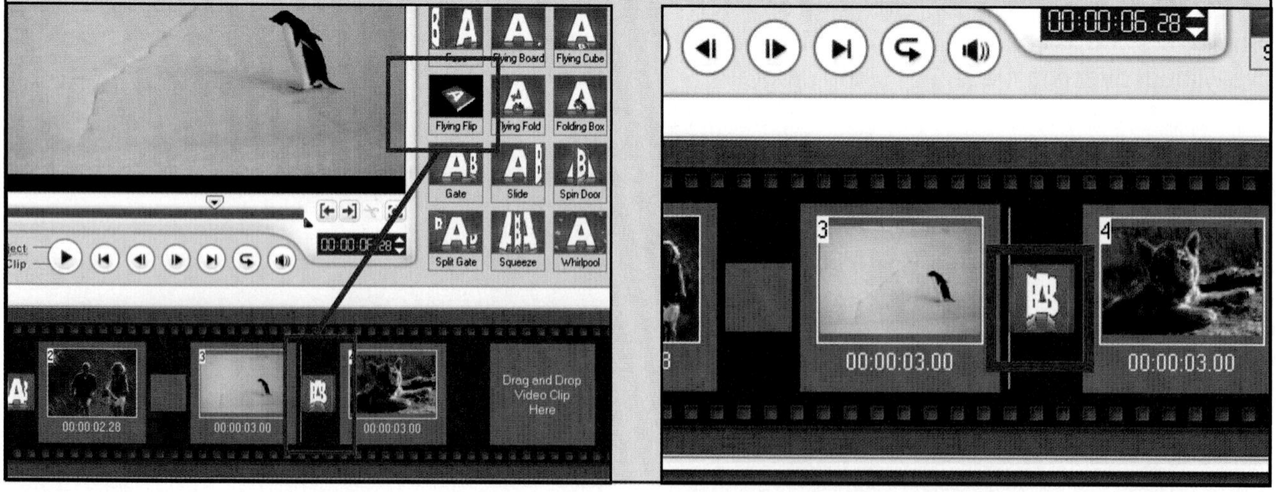

Figure 21-15.
In this "barn door" transition, the first image swings open in two halves to reveal the second. (Ulead)

Figure 21-16.
A superimposition.

Figure 21-18.
A title over the background animation. (Ulead)

solid color. Then the software replaces every pixel of this key color with the corresponding pixel from the background image, **Figure 21-17**.

Titles and Graphics

All editing software includes titling capability, **Figure 21-18**. For more elaborate titles and effects, you can also export material to a separate application such as a paint or photo program. There, you work one frame at a time, adding and animating elements to your original shot. When you have finished, you import the result back into your project. Since full-motion video requires 30 frames per second, working frame-by-frame can be very time-consuming; however, many photo/paint applications can automate parts of the process, copying the changes made in one frame to the next frame, where they can be further modified as the sequence progresses.

Stand-alone programs for compositing and 3D animation software are discussed in Chapter 23.

Image Processing

You can also change the basic nature of your visuals to suit the requirements of your project.

Speed. You can create slow-motion and fast-motion shots.

Direction. You can reverse the direction of the action by "flopping" shots left to right. If, for instance, a left-to-right shot violates a general right-to-left screen direction, you can correct it by flopping it digitally.

In computer graphics applications, the term is "flipping," rather than "flopping."

Composition. You can zoom in to fill the screen with just part of the original composition. This is useful to add emphasis, redesign shot

Figure 21-17.
Compositing. A—The subject in front of a green screen. B—A background animation. C—The two composited together. (Ulead)

Figure 21-19.
Original shot. Shot processed with a "duotone" video filter. (Ulead)

compositions, or eliminate unwanted elements like microphones.

Photographic characteristics. You can change the brightness, contrast, and color balance of your original footage, in order to correct errors in shooting, to match surrounding shots, or to create special effects like a sunset or a moonlit night. You can also turn color footage to black and white or give it an old-fashioned sepia tone, **Figure 21-19**.

Special effects. Most applications include special effects. Though their number is almost limitless, an artificial lens flare is a representative example.

Designing the Audio

When assembling a program, you typically place the production sound for each shot with its video. But that is only the start of audio editing, which includes production sound, background and effects, narration, and music. You are able to manage all these elements only because you have the capabilities of digital editing, **Figure 21-20**.

Digital Audio

Digital audio editing offers so much greater potential than the traditional analog approach that its major advantages are worth considering in detail.

Multiple tracks

To begin with, you can have as many audio tracks as you want—dozens of them, if necessary. These tracks can be mono, stereo, or a mixture, depending on the sound sources used. With nearly unlimited tracks to work with you can create audio as complex as that of most Hollywood films, right on your desktop.

Timeline control

Working with a timeline, you can see all your audio elements in front of you, where you can select them, audition them, move them, and modify them with mouse or keyboard commands.

"Visible" sound

You can see each track as a waveform: a graphic representation of the sound's variations

Figure 21-20.
An audio editing screen. The speaker icons on the right represent sound clips ready for use. (Ulead)

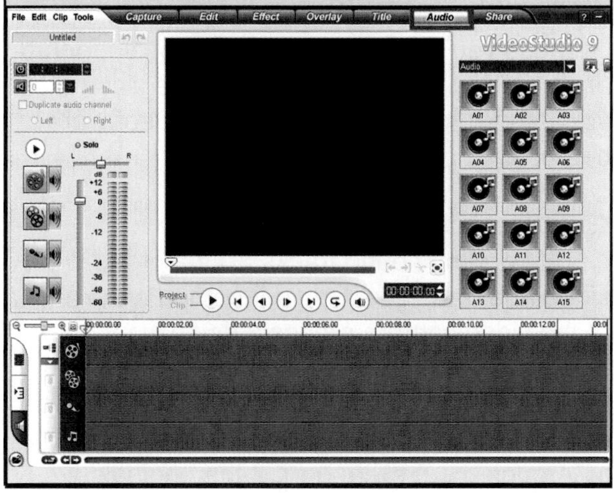

Figure 21-21.
A waveform helps make precise edits.

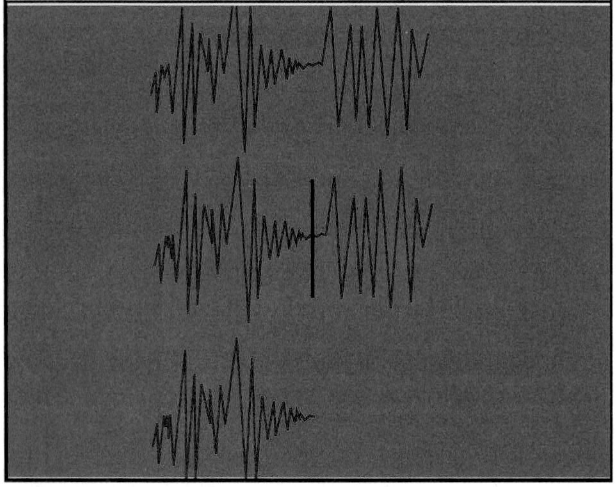

in loudness, **Figure 21-21**. With this tool, you can make extremely precise edits and time music passages exactly.

Synchronization

You can unlock a sound track from its video and move it freely. You can synchronize two different video and audio tracks and then lock them together so that they become a single editing clip. In short, digital audio editing is so powerful that when you are working on a modest desktop computer, you have more creative control than you could find in even a sophisticated analog sound studio. Editing software applications vary in their audio capabilities, but they all include the basic functions of sound capture, editing, and processing.

Capture

Digital sound editing can be so flexible because you work from audio files on your hard drive. So the first step is to get those files onto your drive in a form that your editing software can use.

External sources

Production sound tracks are recorded with video and transferred to your computer as part of the video capture process. If you want only the audio part, you can save it separately and discard the video half of the capture. If the production originals are digital, the audio comes in through the firewire port. If the originals are analog, the audio may by captured by the video card, or by the computer's general purpose sound card. Other external sources, such as microphones and music CD players, are generally captured by the main sound card.

A Simple Audio Production Design

Although editing software may offer close to 100 separate audio tracks, you can achieve very professional results with 12 tracks or less.

To demonstrate, here is a simple audio design for a dramatic program. (For simplicity, this a monophonic program. A stereo version would need twice as many tracks.)

Track	Content
1	Production dialog Track A
2	Production dialog Track B
3	Background A
4	Background B
5	Effects A
6	Effects B
7	Effects C
8	Effects D
9	Music A
10	Music B
11	Misc A
12	Misc B

As you study this layout, notice that all components have at least two tracks (A and B), so that sounds can be cross-faded between them. An extra pair of effects tracks allows the editor to layer up multiple sound effects. A pair of miscellaneous tracks can be used occasionally for dialogue, effects, or music, at places where the regular tracks already have sound in them.

Internal sources

Internal sound sources include files previously recorded or downloaded, and music CD or DVD discs, **Figure 21-22**. Be aware that you usually must use sound card or operating system software to copy files and convert them to formats readable by your editing application.

Audio Editing

With your sound components on your hard drive, you are ready to build a sound track.

Figure 21-22.
Audio tracks can be imported from internally stored files.

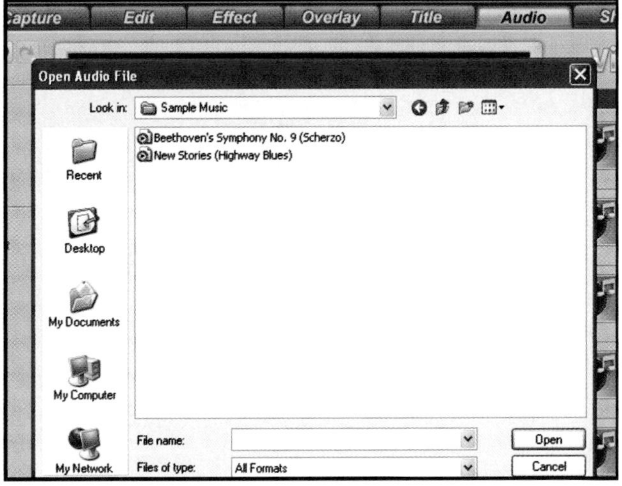

Production sound

The first step in sound editing is to assemble all the production audio. This happens automatically as you select, place, and trim video clips. With the basic production tracks in place, you can modify every one to suit your needs: delete it, replace it with another take (usually for better sound quality), or substitute audio from other shots. If there is a separate production track of ambient (background) sounds, you can lay that under the scene, adjusting its volume to compensate for shot-to-shot differences created by different miking positions.

Sound effects

With the production tracks complete, the next step is to add sound effects. These may have been recorded as part of the production, copied from sound effects discs, or recorded in a Foley (sound effects recording) studio.

Music

Generally, music is the last set of tracks to be laid down. Blended together, the production, background, and effects tracks will sound like a single "live" sound track. Music, however, is perceived by the audience as a separate set of sounds. For this reason, it is often helpful to create the composite live sound track and then balance it with the music.

Transitions

To make transitions from one sequence to another, or simply from one sound to the next, you can blend audio elements together with fades, cross-fades, and split edits. Most editing software allows you to fade sound volume up or down by dragging a volume bar on the time line, **Figure 21-23**.

Fade. A fade-in or fade-out makes a gradual transition. A fast fade-in will prevent a sound from seeming sudden and abrupt. A very long, slow fade-in can bring up a sound so unobtrusively that the audience is not aware of it. Fade-outs are especially useful for dispensing with music that does not have a clean ending. A three-second to eight-second fade-out will generally deliver a graceful exit.

Cross-fade. A cross-fade overlaps the fadeout of one sound with the fade-in of another sound. Cross-fades are often employed to match dissolves on the video track, but you can also use them to blend from one sound to another so that the audience does not notice it.

Split edit. A split edit is a transition in which picture and sound do not change at the same time. When *video leads*, the sound from the old shot continues over the video of the new shot. When *audio leads*, the audio from the new shot begins before the picture, giving the audience a brief foretaste of what is to come. Split edits are especially easy with timeline editing because you can see the overlaps exactly.

Sometimes, a split edit will include a cross-fade at the transition point. However, straight cut split edits can make powerfully dramatic punctuations.

Figure 21-23.
Click on the volume line to establish an edit node ("handle").

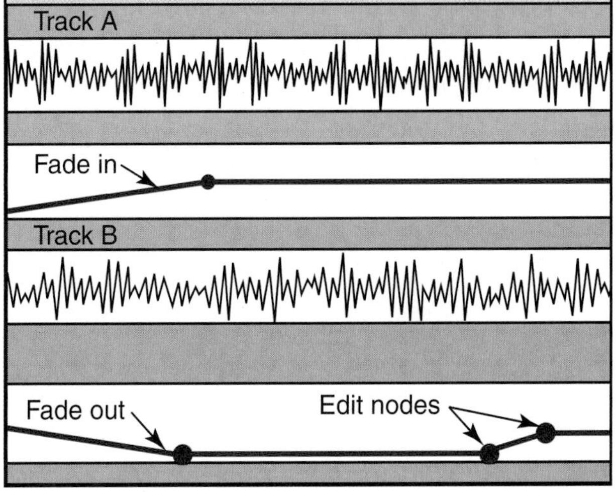

Audio Processing

Fading volume is only the simplest of many ways to manipulate the quality of individual audio tracks. Every editing application has its own set of tools, and plug-in products are available as well. In addition, you can process audio with the software tools provided with computer sound cards or sold separately as computer-based recording studios. Here are just a few of the many ways to process audio tracks.

Stereo imaging

If you are working in stereo, you can place a sound anywhere on a line from left to right, or have it move from one side of the sound image to the other, **Figure 21-24**.

Figure 21-24.
Stereo imaging is adjusted by a slider control. (Ulead)

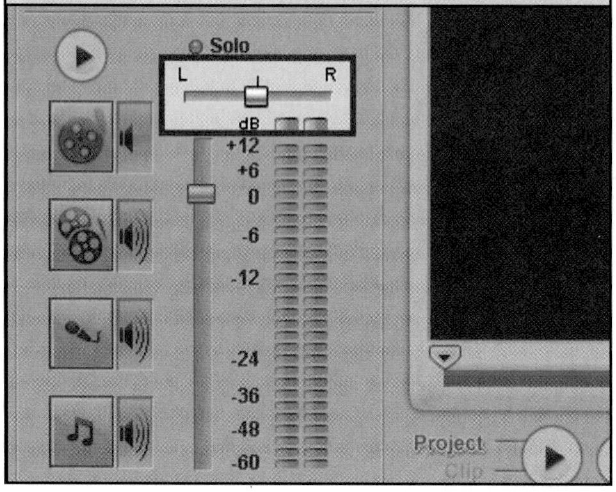

Sweetening

You can clean up audio by adjusting treble and bass, suppressing some sounds, compressing or limiting range of loudness, and eliminating hiss or hum.

Distorting

You can change the character of sound by adding reverberation, changing pitch, or applying one or more of the many sound filters supplied with your editor or available separately.

Storing and Distributing

When you have completely finished your program, you need to clear it out of the computer to make room for the next project and preserve it

Figure 21-25.
Ready for storing and distributing. (Ulead)

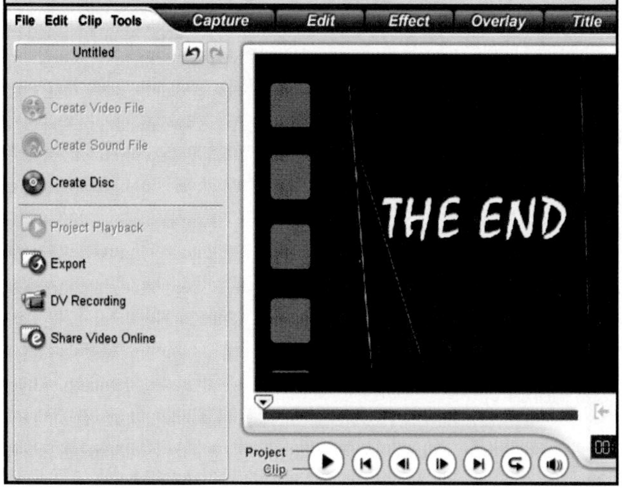

permanently. Also, you need to copy it in a format that can be displayed on a monitor or streamed on the internet. That is, you need to *store* your program and *distribute* it, **Figure 21-25**. These two procedures are not the same.

Storing (archiving)

Storing your program means archiving it in the best quality format available. For example, if you have shot your footage and edited your program in mini-DV format, you will probably want to archive it that way. This stored program is the master from which other copies are made for showing.

Distributing

Distributing your program involves converting it into the format appropriate to the intended display medium and, often, creating support materials for presenting it. For instance, to show your program on TV or a computer monitor, you would probably produce a DVD version of it. This would mean converting the original mini-DV format to the MPEG II format used on DVDs, and then adding menus and perhaps additional program materials.

Storing Finished Programs

Whenever you save a word processing or spreadsheet document, the file you store is a full version of the work you are creating. It is important to understand, however, that this is not

Figure 21-26.
Video Studio Projects are stored as ".vsp" files. (Ulead)

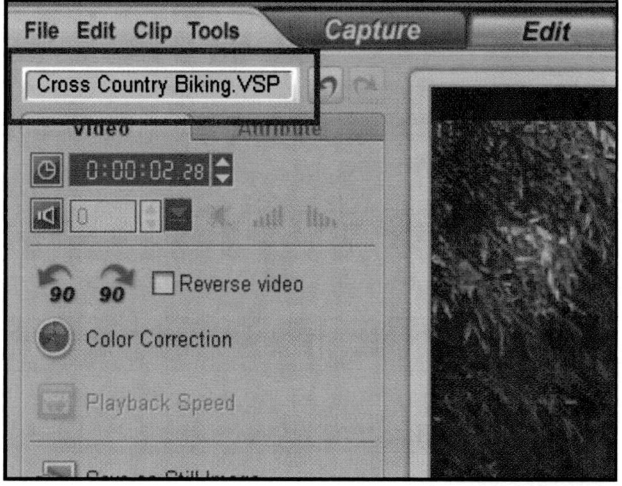

always true of video programs. When you save your work as you edit, you are saving a project, not a program, **Figure 21-26.**

- A *project* is a record of everything you have done to create the program. A project is like a complete set of blueprints for a complex building, but it is not the building itself.
- A *program* is a video that can be played back at full quality, on some form of standard display. It is the building described in the blueprints.

Project files

Project files sometimes do not contain full-quality versions of the shots themselves, but only references to those shots. It is a good idea to save your projects frequently, for protection.

If you plan a major revision of your program, you can save the project, then save it under a different filename and revise this copy. That way, you have a complete backup copy of the original version. This procedure is practical because project files take up relatively little disc space.

Project files are not easily transported from one computer to another because all the shots and other materials for the finished program are in other files that remain behind.

Editing software uses project files to locate program materials on hard drives as it needs them.

Program files

Program files are finished videos. They consume large amounts of disc space because, unlike project files, they contain every frame of video and audio in full-quality versions.

Rendering program files

Your editing software creates a program file by automatically assembling it, following the frame-by-frame instructions in the project file. This assembly process is called rendering. Powerful editing programs running on high-speed computers may ***render*** in real time. That is, they create and recreate the program at full quality as you work on the project. This means that you can see full-quality results at any time, you can catch any defects in the rendering process and correct them on the spot, and you do not need to wait while the software builds your final program.

More modest setups may require you to render your program by batch processing, when you have finished editing. With this approach, you instruct the system to assemble the final program and then wait a considerable time while it does the job.

Most editing software can render trial versions of transitions, effects, titles, etc., so that you can review them as you work.

Aside from the inconvenience of long rendering times, the other drawback of batch rendering is quality control. If the final program contains dropped frames, artifacts, or other imperfections, you will have to repeat the whole process.

Rendering versus archiving

Rendering gives you a final program file stored on a hard drive. But in most cases, you

Figure 21-27.
Rendering and archiving.

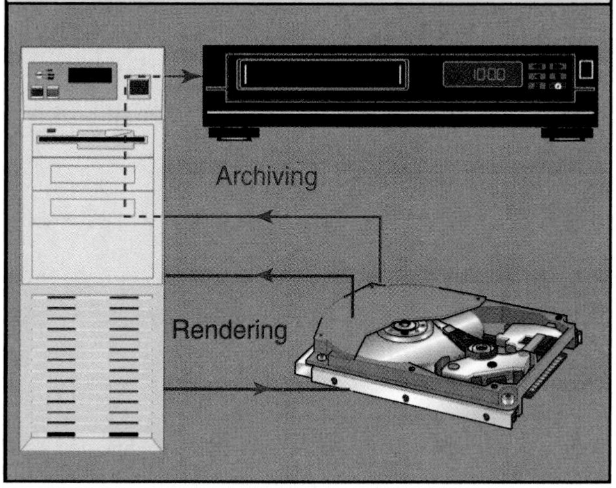

do not want to leave it there, tying up many gigabytes of drive space. Instead you want to *archive* it by storing it permanently in a place where you can preserve it, make copies of it, or show it. Typically, you archive a program by exporting it to a disc recorder or to tape in a camcorder or VCR, **Figure 21-27**.

Formats for archiving

Whether you are using disc or tape, a digital recording format is better for archiving because copies can be made without significant quality loss. If the program may be distributed in multiple formats (or if it might be revised or re-edited later), it is best to keep it in its original digital format, typically the mini-DV (.avi) format of the original camera footage.

On the other hand, if the program is complete, and if it is to be shown only (for example) on DVD, then it can be archived by creating the master DVD, using the DVD's MPEG II format.

Distribution and Display

With the growth of broadband internet connections, it has become possible to download and play video programs on personal computers. This means that these programs must be converted into digital formats suitable for downloading.

For TV display, DVDs have largely replaced other forms of disc recording. VHS tapes, though increasingly becoming obsolete, are still used.

Many computer systems can output video and audio directly in analog format onto conventional videotapes.

With the variety of formats and display systems now in use, preparing programs for distribution has become a whole separate operation.

Chapter 23 explains how to create DVDs.

Varying the Task Order

For clarity, we have discussed each editing operation in order, as if you always captured all of your footage, then assembled all of your shots, added all of your effects, etc. However, one great advantage of digital postproduction lies in your freedom to work on any part of your program at any time, and to perform operations on shots, sequences, or the entire program—or in any combination that you find convenient.

Sequence-based Editing

Many editors prefer to build one sequence at a time—usually, action that happens in one location at one point in the program continuity. Using this approach, you would

- Capture the footage for this sequence.
- Catalog and organize shots and clips for use.
- Assemble the shots in the sequence, trimming each to approximate length.
- Fine-tune the trimming of all the shots in the sequence together.
- Replace and/or add clips by inserting them into the sequence.
- Process the production audio.
- Lay background and sound effects audio.
- Correct shot-to-shot luminance, contrast, and color balance.
- Apply special overlays.

In effect, you complete almost all the editing work on each sequence before going on to the next.

Multiple-pass Operations

Some operations are better performed on entire programs, or at least on segments that are several sequences long. Scene-to-scene transitions, for example, may be easier to keep consistent when you add them all at once.

Music may be easier to evaluate when heard under several sequences together. Musical choices that seemed fine individually may not work as a group, or may feel excessive when heard consecutively.

Successive Approximation

The biggest single advance in digital postproduction is the ability to revisit any part of a program and improve it. In old-style film editing, even changing in- and out-points was tedious and time-consuming, and you could never see the result of adjusting scene transitions because they had to be made later in a laboratory.

In 16mm film editing, removing two frames to adjust a cut point required the editor to file and keep track of a piece of film less than one inch long!

In analog video, you cannot change anything anywhere in the edited tape without re-editing everything that follows it.

In digital postproduction, you are not working with your video and audio originals, but only with copies of them. And even when you instruct your software to render the finished program, the original materials remain untouched. To take full advantage of this major improvement in postproduction procedures, be prepared to re-work parts of your program until you are fully satisfied.

Configuring Digital Projects

Before you can begin digital editing, you must tell the computer the precise electronic characteristics of the program you will work on. That is, you must *configure* the project.

As noted previously, although configuring is the first editing operation, it has been saved for last here because it is rather technical and because it is not always necessary.

Most editing software is supplied with default values that match the most common program settings; so it is possible to begin working with these applications without configuring projects. To take full charge of your programs, however, you need to understand the variables involved and how to set them, **Figure 21-28.**

Figure 21-28.
Project configuration screens. (Ulead)

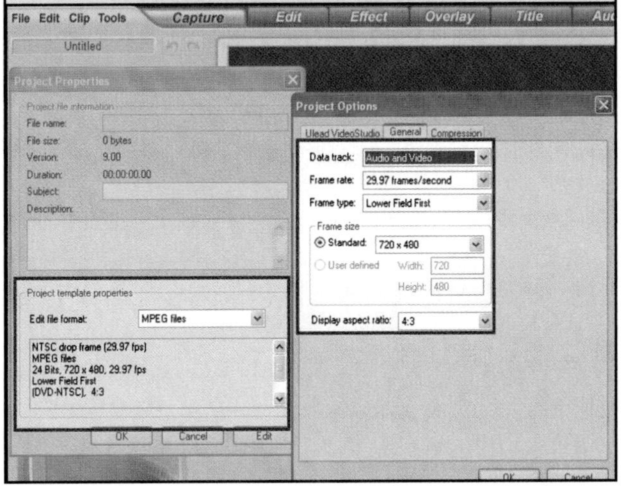

The following sections provide general descriptions and explanations of the many factors involved in configuring a project, starting with the program's basic format(s).

Video Formats

The format of a video is the set of electronic protocols used to record and play it.

In analog video editing, you do not configure the program format because it is already "hard-wired" into all the equipment, as part of the TV standard (such as NTSC) in which you are working.

Computer-based video is format-independent. That means you may start with source material in one format, process it in your computer in a second format, and export it for storage or viewing in yet a third format. In many setups, the format is the same for both source and destination. For instance, you may import mini-DV footage from your camcorder, process it in that format on your computer, and export it back to your camcorder to store your edited program on tape.

In other cases, you may be working with as many as three different formats. For example, if you are editing NTSC original footage for streaming to the Internet, your source will be analog NTSC, your software will work in its native digital format, and your destination will require yet another digital protocol.

Even within basic formats there are choices to be made: full-screen image or partial? Smooth motion or economical 15-frame-per-second movement? The list goes on through many more such decisions. For this reason, you need to understand the major project parameters and set them systematically. The rest of this section discusses each of these parameters.

Video Standards

The NTSC, PAL, and mini-DV formats are all examples of TV "standards," **Figure 21-29.** Video recording standards fall into two broad categories, digital and analog.

Analog standards

Historically, almost all broadcast and cable television has been recorded and displayed in

Figure 21-29.
Mini-DV is the most popular digital format.

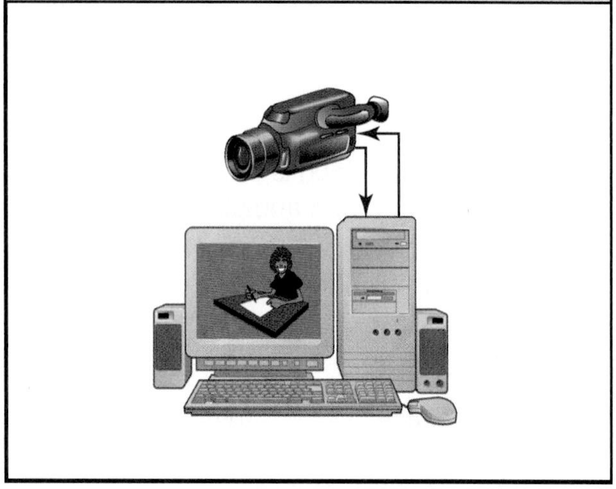

one of three analog standards, PAL, SECAM, or NTSC. The NTSC format is used mainly in North America and Japan. These standards are completely incompatible. If you play a PAL videotape in an NTSC VCR, you will hear only garbled audio and see no picture at all.

Although you can select PAL output in your editing software, you cannot record it successfully on an NTSC machine.

To edit analog video with a computer, you must import it through a video capture device that converts the signal to a digital format that is compatible with your editing software,

Figure 21-31.
A 1394 I/O card does not have to digitize the signal.

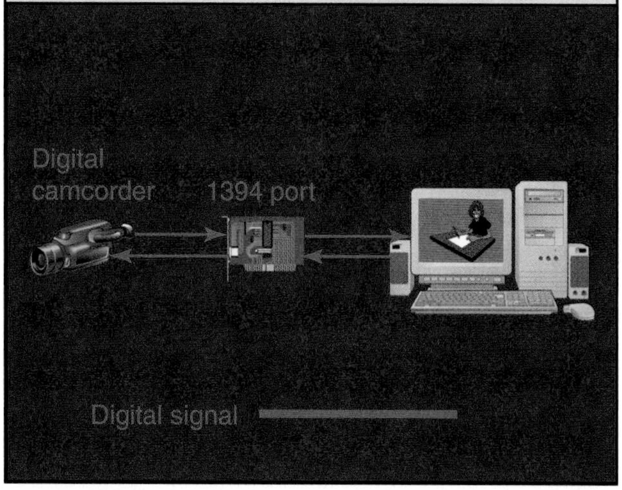

Figure 21-30. When you have completed editing, you can choose to output the digital program through the same device, which will reconvert the signal to analog.

Digital formats

There are many different digital formats to suit the requirements of professional production, consumer video, satellite broadcast, and high definition TV. Since the camcorder records the signal digitally to begin with, it does not have to be converted as it is imported into your computer, **Figure 21-31**. The importing device in your system does need to be

Figure 21-30.
A video capture card converts analog to digital and back again.

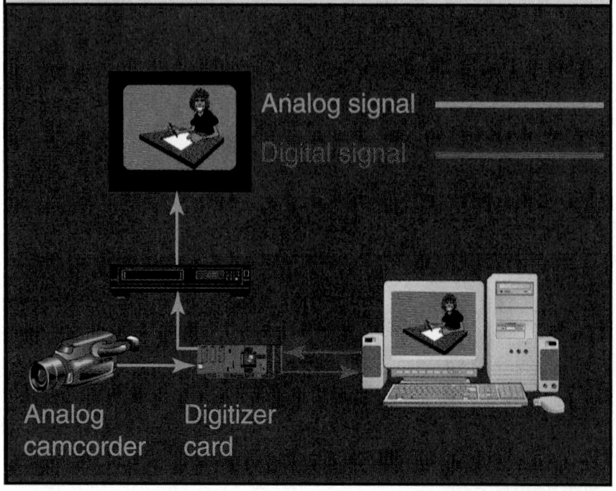

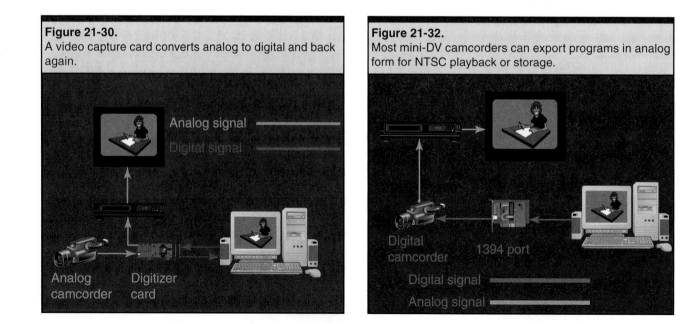

Figure 21-32.
Most mini-DV camcorders can export programs in analog form for NTSC playback or storage.

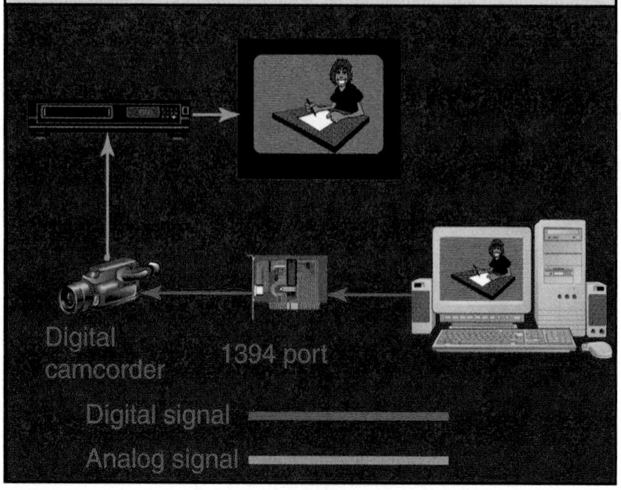

compatible with the digital format of the source material, however.

When exporting your finished program, you must use the format of your storage and display equipment. For example, you may wish to output your program once to mini-DV tape for permanence and quality and again to a VHS VCR for simple TV playback, **Figure 21-32**.

When you have selected your basic input, processing, and output standard(s), computer based editing then requires you to set several parameters within the formats you have chosen. These include time and frame measurement, video scanning system, picture quality, and audio quality.

Time and Frame Measurement

Video (and film) systems do not record continuously. Instead, they divide each second of time into several equal parts and record a picture of each part. Displayed in real time, these separate pictures give the illusion of a continuous, moving image. There are several common ways to record video images, and in every one, the first recording parameter is frame rate.

Frame rate

The ***frame rate*** is the number of individual images (frames) recorded each and every second. It is important to know the frame rate of both your source material and your eventual storage medium, so that your software can process the source material and prepare a compatible final project.

Here are some of the principal frame rates:

- NTSC television: 30 fps (frames per second). The actual frame rate of NTSC video is 29.97 fps, but it is informally referred to as "30 fps."
- PAL and SECAM television: 25 fps.
- Computer full motion: 30 fps.
- Computer limited motion: 15 fps (10 fps is a less-common alternative).
- Sound film: 24 fps.

Timebase

The word ***"timebase"*** refers to the way in which your system calculates time. In most formats, the timebase is identical to the frame rate; the NTSC format used mainly in North America and Japan is an exception. Though we speak of 30 fps for convenience, the true NTSC frame rate is actually 29.97 fps.

The change from a true 30 fps to 29.97 was made for technical reasons when color television broadcasting was developed.

If your computer uses a true 30 fps as the timebase, even the tiny difference of 0.03 frames per second will gradually make the frame count grow increasingly inaccurate. For this reason, it is preferable to set your editing software to a 29.97 fps timebase when you are editing projects intended for TV display.

Time code

The way in which frames are labeled is called time code. ***Time code*** provides a unique address for every frame of video in your program. The most common time code system is SMPTE, which labels each frame of picture and/or sound with a unique identifier. This address consists of the number of hours, minutes, seconds, and frames elapsed since the start of the time code. So, the time code address 00:37:22:09 labels the frame that is 37 minutes, 22 seconds, and nine frames from zero, **Figure 21-33**. The leading "00" indicates that no hours have elapsed.

Figure 21-33.
The parts of a time code address.

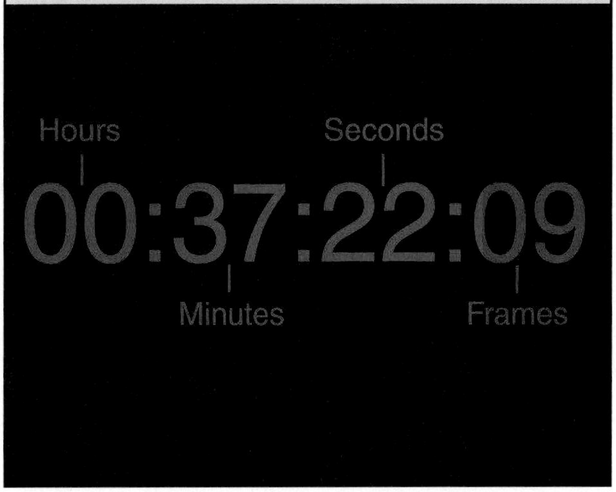

SMPTE stands for the Society of Motion Picture and Television Engineers and is pronounced "SIMP-tee."

Understanding Time Code

An Unchangeable Address?

The statement that each frame is assigned a permanent time code address applies only to a single recording of that frame (and the other frames around it). This means that when you copy a shot you can sometimes copy its time code address with it; but in other cases, your system may assign the material a new set of time codes.

In working with a project, you may be dealing with three separate sets of time code:

- *Source time code*: each clip you import will have time code addresses created in the camcorder.
- *Clip code*: in some editing systems, each separate piece of captured material has its own time code.
- *Project time code*: Your software will time code your project with a single set of continuous addresses.

Though it is not the editor's responsibility to manage time code on the original camera footage, pre-striping every camera tape with time code before using it is often desirable because it ensures that some sort of code is on every second of the tape.

When some older camcorders sense the absence of time code, they reset to zero and start over. If that happens, the editor may find multiple shots on one tape that have the same addresses. This can make locating and importing clips difficult. (Most camcorders sold today do not have this problem.)

Drop-frame and Nondrop-frame Time Code

Time code cannot label an interval shorter than a whole frame, so NTSC time code counts frames as if they were displayed at 30 fps. rather than the actual rate of 29.97 fps. If timecode assigned an address to every frame, the count would very slowly grow inaccurate. After ten minutes, the difference between the time code readout and the actual elapsed time would be over half a second—a huge difference where synchronizing picture and sound is concerned.

To solve the problem, so-called "drop frame" time code skips (drops) two frame numbers at the beginning of each new minute. (Since this method overcompensates just slightly, the two frame numbers are not dropped after every tenth minute.)

"Drop frame" time code does not really drop any frames, only address numbers.

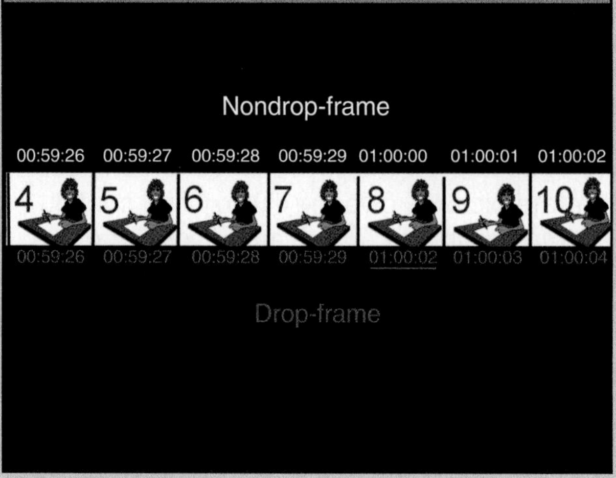

It is important to understand that drop frame time code does not actually eliminate any frames. It simply drops numbers. In other words, between time code addresses ending in "00:29" and "01:02" there are numbers missing, but not pictures.

To continue the address analogy, if two houses side-by-side on Elm Street are 305 Elm and 309 Elm, it does not mean that a house between them has been removed, but only that the number 307 has been skipped in assigning house addresses.

Time code is essential for accurate editing because it allows you to specify the exact frame on which you want every operation to start or end.

Keeping them straight

To understand the differences between frame rate, timebase, and time code, remember:

- *Frame rate* is the actual number of frames per second recorded by a camcorder or created in the editing process.
- *Timebase* is number of units per second with which editing software calculates all its operations. It may or may not be the same as frame rate.

Resolution and Frame Size

Analog television and digital video are different in the way they display images on screens. In conventional television, all monitors are set to match a single standard (such as NTSC) so images in that standard always fill the entire screen.

Digital displays such as computer screens, however, can be set to different resolutions, such as 1280 by 960, 800 by 600, 640 by 480, etc. Consequently, the apparent size of a displayed digital image depends on the relationship between the screen's resolution and the resolution of the image. On a 640 by 480 (VGA) screen, for example,

- a 320 by 240 (VHS quality) picture will fill one quarter of the screen.
- a 640 by 480 (mini-DV) picture will fill the screen exactly.
- a 1280 by 960 picture will fit only one fourth of its area on the screen.

Three different resolutions on a 640×480 screen.

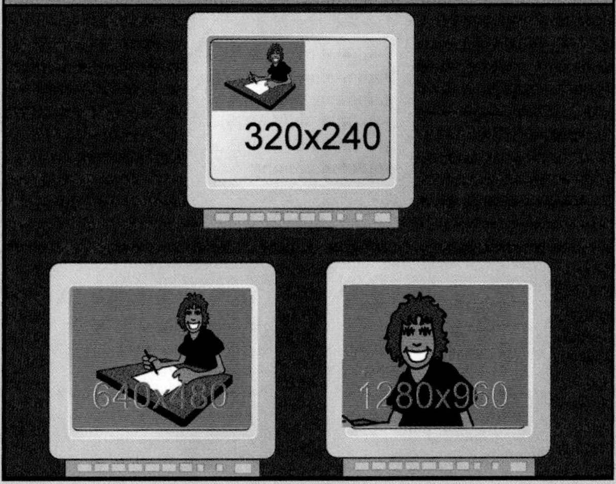

Video editing programs, however, re-size the original images to fill the working window(s).

- *Time code* is the numbering system used to assign a unique address label to every frame. ("Drop frame" time code is used to prevent errors in applying 30 fps time code to a 29.97 fps frame rate.)

Video Scanning System

Each video image (frame) is divided into several hundred horizontal lines, which are "painted" on the display, one at a time. This process is called scanning. There are two approaches to video scanning: interlaced and progressive.

Interlaced video

In this system, every frame of video is made up of two *fields*, one showing only the odd-numbered lines and the other only the even-numbered lines. Since each of the two fields is scanned in 1/60 second, the complete frame is displayed for 1/30 second in the NTSC standard, with the two fields interlaced (mixed together). See **Figure 21-34**.

Interlaced video was developed in the early days of television to work around technical

Figure 21-34.
Interlaced (2-field) video.

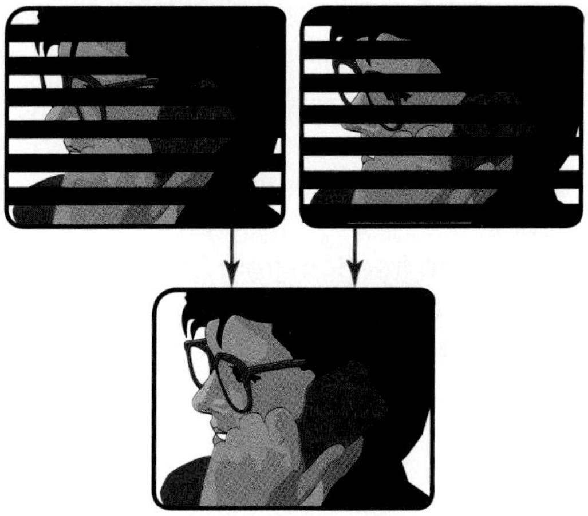

limitations that have since been overcome, but it remains the standard for conventional TV.

When you pause a tape, the image quality degrades because the still picture consists of only one of the two fields. DVDs produce better stills by displaying both fields together.

Progressive scan video

In this system, every odd and even frame line is shown sequentially in one pass. There are no separate fields in progressive scan video, **Figure 21-35**.

Many digital video systems use progressive scanning. In digital postproduction, you may need to convert between interlaced and progressive scan video. An interlaced frame must be "deinterlaced" into a single scan in order to create a full-quality still image. A progressive scan frame must be broken down into two interlaced fields for export to a conventional video display standard such as NTSC.

Figure 21-35.
Progressive scan video.

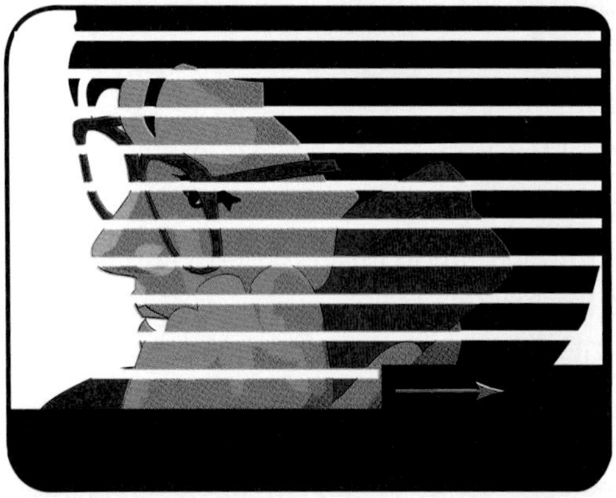

Picture Quality

The apparent quality of a displayed image depends on three factors:

- *Resolution*: The perceived sharpness and amount of detail.
- *Color fidelity*: The accuracy and subtlety of the colors.
- *Integrity*: The freedom from "artifacts" such as spots or areas with a blocky, mosaic appearance.

Of these factors, resolution is perhaps the most important.

Resolution

Resolution is the ability of a video system to resolve (distinguish) fine detail.

Many factors combine to determine resolution, as the sidebar, *Resolution and Frame Size*, explains.

In digital editing, you can set the resolution of each new project. If you are working with digital source material, it is often best to set the project resolution to match. So if you are using mini-DV footage with a 640 by 480 resolution, the project resolution will also be 640 by 480.

Analog video measures resolution differently, so you can only match the quality approximately. If your source is plain VHS, 320 by 240 will preserve the details in the original.

To compensate for small losses during the editing process, it may be advisable to set the digital project resolution one level higher than the minimum needed to capture detail in any analog originals.

Color fidelity

Color fidelity means the accuracy of the colors in the picture, and this quality depends heavily on bit depth. ***Bit depth*** refers to the length of the string of code numbers representing each picture pixel. The more bits recorded per pixel, greater the amount of information about it. Typical color depths are 8-bit, 16-bit, and 24-bit. Consumer digital camcorders can record at least 16-bit color, and most editing software works in at least 24-bit.

Image integrity

Because of the tremendous amounts of data involved, digital video is often compressed. Digital compression can be complicated because there are at least three variables to consider:

- *Compression type*: Some systems, like ***MJPEG***, reduce the data for each individual frame. Others, such as ***MPEG***, record all the information for a "base frame;" but on the frames that follow it, they record only the pixels that have changed since the base frame.

Some standards have grown into whole families. There are now more than half a dozen varieties of MPEG protocols.

- *Compression algorithm*: Each type of compression is available in more than one variety, with designations like "4-1-1" and "4-2-2."
- *Compression level*: Compression systems allow you to set the severity of the compression from slight to vigorous. The milder the compression the better the picture quality—and the bigger the file size as well.

Usually, the compression type and particular algorithm are determined by the codecs in your camcorder and computer. However, you can often set the level of compression for each project.

Codec is the name coined for the program that codes the video digitally and decodes it for analog storage and playback. Some codecs are built into hardware; others are software only.

Aspect Ratio

After frame rate, scanning system, and image quality, the last picture parameter to set is aspect ratio, the relationship of the frame's width to its height. The traditional aspect ratio of video is 4 to 3; four units of width to three units of height, **Figure 21-36**.

Figure 21-36.
A traditional 4 to 3 monitor. (JVC)

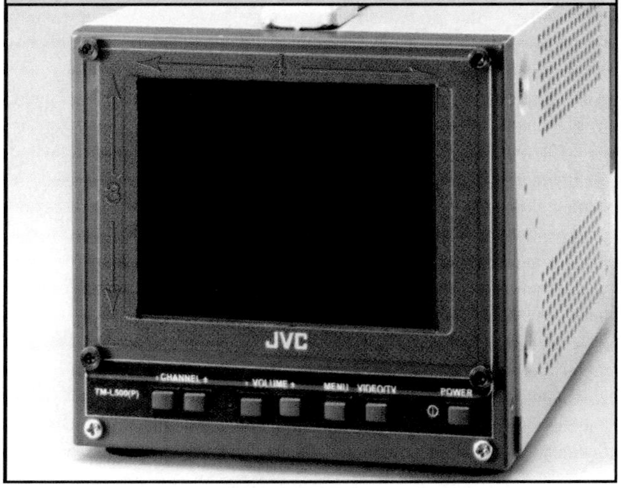

Figure 21-37.
A monitor with a 16 to 9 aspect ratio. (JVC)

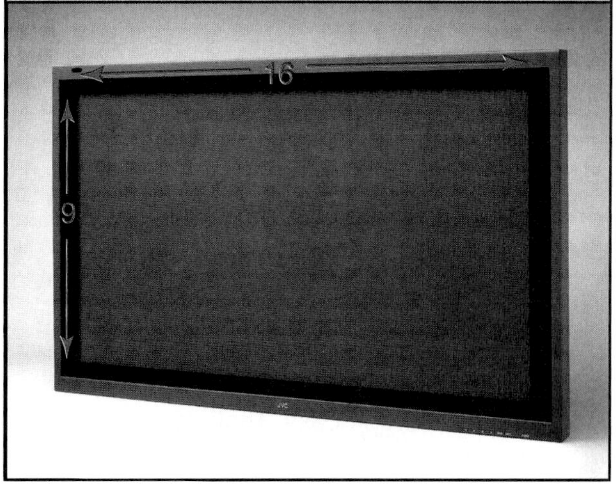

The original 4 to 3 matches the ratio of all movies at the time broadcast TV was developed. Since then, films have been shot in many different aspect ratios, such as 2.2 to 1 and even wider.

Nowadays, however, TV is also displayed in an aspect ratio of 16 to 9, **Figure 21-37**.

This proportion is close to the shape of most movies produced for theatrical release. Many digital camcorders can be set to shoot 16 to 9, and some editing software can modify traditional 4 to 3 video to the wider aspect ratio. As high-definition camcorders and TV sets become more widespread, the 16 to 9 aspect ratio will gradually supplant the older screen shape.

Audio Quality

In addition to setting project parameters for video, you can also select them for your sound tracks.

Sample rate

All digital recording is created by sampling the signal and recording the result thousands of times per second. CD audio is sampled at a rate of 44.1kHz. Audio for the Internet or CD ROM may use a 22kHz ***sample rate***, and there are other rates between the two. (A rate lower than 22kHz is not generally recommended.)

Bit depth

As with video, this is the number of digits (bits) used to record each sample of the audio. Sixteen-bit audio is CD quality. Selecting a lower sample rate and bit depth will reduce the size of your audio files, but the result will be somewhat lower quality.

Matching video

If you set your video at 30 fps for editing, but then play it back at 29.97 frames for NTSC recording, your audio may grow slowly out of synchronization because it was captured to match the slightly faster 30 fps. For this reason, it is best to standardize all video operations at 29.97 fps throughout the capture/edit/export process. That way, your audio will remain in sync with the picture.

It is possible to re-sync 30 fps audio with 29.97 fps video by slowing the audio playback sample rate from 44.1kHz to 44.056kHz.

Working for NTSC playback, you can, as a general rule, safely set a video frame rate of 29.97 and an audio sample rate of 44.1kHz at 16 bits.

Unlike video, which is recorded in separate frames, audio is recorded continuously. For convenience, however, time code addresses are used for audio as well.

Though sizeable, audio files are smaller than video files, so you may not significantly reduce overall project size by skimping on audio quality.

A Lot of Standards

At this point, you may feel that all these technical standards are a confusing nuisance. Keep in mind, however, that they must be set only when changing program specifications. If you work entirely in one format—such as making programs from mini-DV footage for DVD distribution—then you can establish your standards once and leave them that way.

Chapter Review

Answer the following questions on a separate piece of paper. Do not write in this book.

1. Digital editing involves use of video and audio stored in digital form and processed on a ______.
2. An ______ is a small blemish on an image.
3. When a program must retain a precise length, a ______ edit should be used.
   A. slip
   B. rolling
   C. three-point
   D. slide
4. *True or False?* The frame rate of NTSC video is usually referred to as 30 fps, but is actually 29.97 fps.
5. Aspect ratio is the relationship of the frame's ______ to its ______. The traditional aspect ratio of video is 4 to 3.

Technical Terms

Analog: A recording that imitates the original. The level of an analog video signal varies continuously to record the image or sound.

Artifact: A spot or other blemish on a digital image.

Bit depth: The number of digits (bits) used to encode each piece of digital data. Eight-bit, 16-bit, and 24-bit are common bit depths.

Capturing: Bringing program material into a computer from an external source.

Clip: A unit of program material in a digital editing project. A clip may contain anything from part of one shot to several shots together.

Codec: A program that *codes* a signal in digital form and *decodes* it into analog form.

Device control: The ability of software to operate external hardware (camera or VCR) by remote control.

Digital: A recording that repeatedly samples the original continuous signal and records the numerical values of the samples, instead of the signal itself.

EDL: (Edit Decision List) a computerized, frame-accurate record of every component in a program. It can be used to automatically recreate the program from the original sources.

Four-point edit: An edit that specifies the in- and out-points of both the source clip and the place in the program at which the clip will be inserted.

Frame rate: The actual speed with which video frames are displayed. The frame rate of NTSC video is 29.97 frames per second.

MJPEG: A video compression system that reduces the data required to record each frame.

MPEG: A video compression system that records all the data of a "base frame." For subsequent frames, the system records only data that differs from that of the base frame.

Post house: Short for "postproduction house," a professional facility that rents editing setups and personnel for high-quality video editing.

Render: To execute any digital processing of an image. A dissolve between shots, for example, is rendered, frame-by-frame, by the computer.

Ripple edit: An insert edit in which all the material following the inserted clip is moved up or back to accommodate it.

Rolling edit: An insert edit in which the shot following the inserted material is shortened or lengthened to accommodate it.

Sample rate: The number of times per second at which an analog signal level is inspected and digitally recorded. Expressed in kilohertz (kHz). The sample rate of CD-quality audio is 44.1 kHz, meaning that the signal is sampled over 44,000 times per second.

Slide edit: An insert edit in which the out-point of the preceding shot and the in-point of the following shot are adjusted to accommodate the inserted material.

Slip edit: An insert edit in which the in- and out-points of the inserted material are adjusted so that the preceding and following shots remain unchanged.

Three-point edit: An insert edit in which both the in- and out-points of the new material are specified, but only the in-point of the program into which it is inserted.

Timebase: The basis on which time code is assigned to frames. The timebase of NTSC video is 30 frames per second.

Time code: A unique identifier, expressed in hours, minutes, seconds, and frames, assigned to each frame of video.

Window dub: A copy of the original footage in which the time code is visible in a small window at the bottom of the screen.

A nonlinear editing system for high-definition video. (Cinéwave/Pinnacle)

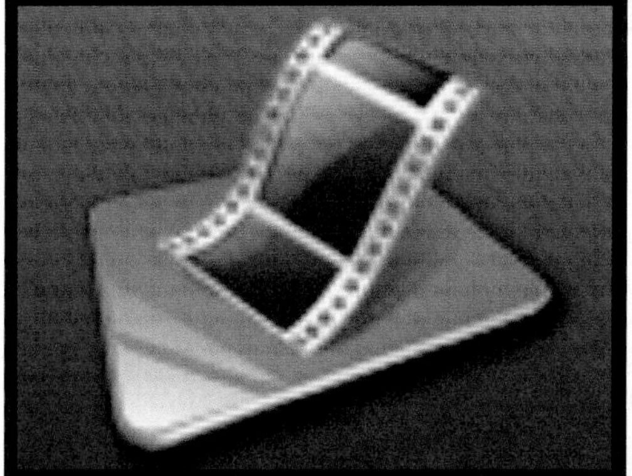

22

Mastering Digital Software

Objectives

After studying this chapter, you will be able to:

- Optimize your computer for digital editing.
- Identify sources of help with your program.
- Familiarize yourself with program operations.
- Develop personal working methods.
- Identify supplementary software programs.
- Explore more advanced functions and programs.

(Sue Stinson)

About Mastering Digital Software

Chapter 21, Digital Editing, covered the basic procedures common to all desktop postproduction systems. The present chapter continues by showing how to learn, use, and master those procedures in the particular editing program that you have chosen. Although there are many such programs available, the discussion is illustrated with Ulead VideoStudio; so if you will be using different software, you may wonder how this chapter can benefit you.

The answer is that this is *not* a chapter on how to use Ulead VideoStudio, but on how to approach *any* new editing software, using the Ulead program as a typical example. The goal is to communicate strategies for mastering editing software in general.

The discussion does not require access to Ulead VideoStudio. As you read, however, you may wish to explore suggested procedures on whichever editing program you are using.

Digital Editing Software

Learning any powerful software demands an investment in time, effort, and patience. Video editing software is particularly powerful because of the volume of data it handles and the variety of tasks it performs. No matter what the product package promises, you cannot expect to install the program and start editing immediately. So no matter how eager you are to begin a project, you will avoid frustration if you budget the time and effort to learn your program first, **Figure 22-1**.

Figure 22-1.
It is natural to want to start immediately. (Ulead)

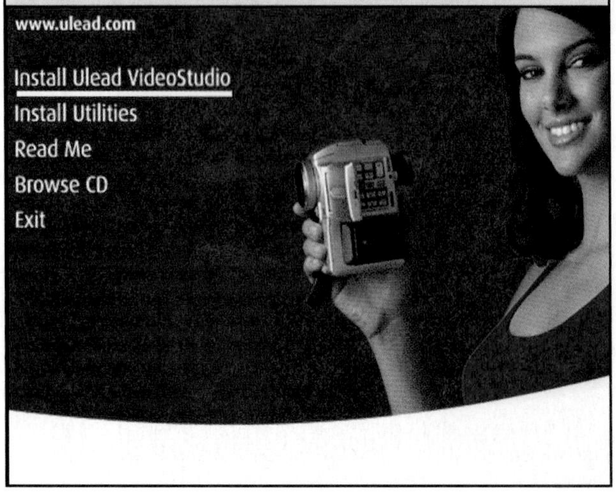

Some programs include ultra-simple options for beginners, as the sidebar explains.

Different Learning Styles

How you actually learn your software depends on your personal style, and no particular approach is superior to the others. You may read the whole manual first and complete all the tutorials (if provided). You may start with simple projects and work toward more ambitious programs. Or you may just fool around until you get the basic idea and then figure out procedures if and when you need them. If that works for you, fine. But if you want to grow comfortable with your editing software quickly, you will probably benefit from a more systematic approach to learning it.

Preparing Your System

Today's computers have powerful processors; large memory and storage capacities; and fast system speeds, communication ports, and hard drives.

When desktop digital editing became available in the 1990s, a fast computer ran at 166 megahertz and a big hard drive held five gigabytes of data.

But even with all this power, video editing places severe demands on computers, especially when capturing footage. When the computer cannot keep pace with the flow of data, video frames are dropped and ***artifacts*** (visual blemishes) mar the images. To prevent or at least minimize these problems, you will want to optimize your computer for editing.

Recommended Hardware

First, you need two fast hard drives. A second drive lets you install the editing software on the system drive and use the other drive for storing the files needed for editing. That way, one drive can handle editing operations while the other manages data in- and output. Dividing the tasks speeds up the process and reduces problems, **Figure 22-2**.

Figure 22-2.
Two hard drives improve capturing by dividing the workload.

Even the best systems drop occasional frames during capture, so it is prudent to screen captured footage immediately and recapture any shots with problems.

If you are capturing analog footage from VHS or hi-8 camcorders, or from VCR tapes or TV, you will need extra hardware to convert the ***analog video*** to a digital format for editing. Video ***capture cards*** can be installed in your computer, but they usually must be coordinated with the computer's sound card (or sound circuits on the motherboard)—often a tricky and time-consuming process. External capture hardware plugs into a computer port and is sometimes easier to use.

Video editing screens are packed with data, so the bigger monitor the better. If your video card allows it, a second monitor can let you spread out your editing tools, **Figure 22-3**.

This works best with software that features "floating" windows, toolbars, preview screens, etc. You can also display other programs like a file management database (which is discussed later in this chapter).

System Tweaks

Today's do-everything operating systems can be set up in many different ways; and different editing programs can benefit from different adjustments. For example, many programs work better if you make (or verify) these settings:

- *DMA enabled.* IDE hard drives work faster when **D**irect **M**emory **A**ccess is enabled.
- *Write-behind caching disabled.* Though this feature improves the performance of some operations, it is not suitable for capturing video.
- *Paging file set at twice RAM size.* If you have, say, one gigabyte of main memory, set the paging ("swap") file size to two gigabytes.

These settings are detailed in user guides, so check your software manual for recommendations.

Other Programs

File capture and program rendering place heavy requirements on a system, so all other

Figure 22-3.
Many graphics programs allow you to "float" elements to different positions and even place them on a different monitor. (Corel)

Figure 22-4.
Shareware programs can help you manage startup applications. (WinPatrol)

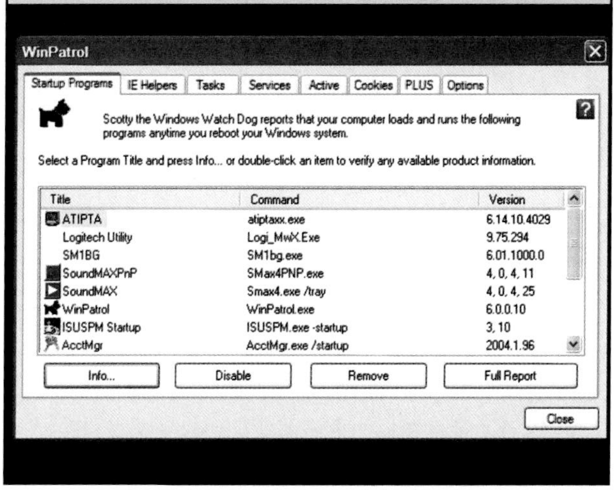

Editing on a Laptop Computer?

Laptop computers are widely advertised as suitable for editing video; and in fact they can be quite competent at this work. But before settling on a laptop as your primary editing system, you may want to consider these points:

- The most powerful laptops are generally not as fast as top-line desktop systems.
- Where equal to desktops, feature-for-feature, laptops are usually more expensive.
- Laptops rarely have two internal hard drives, and external add-on drives reduce the portability (and hence the advantage) of the system.
- Laptop screens are usually not larger than 15 inches; and bigger displays add significantly to system weight and costs.

In short, you can obtain a laptop system that edits video very well, but at a substantial penalty in size, weight, and cost.

programs should be closed during these operations.

If you are comfortable with computer maintenance operations, you can also temporarily disable many of the programs that are run automatically at startup, Figure 22-4.

However, the editing process itself is not as demanding, so you may want to keep one or more support programs open as you work. Common examples inclu de the paint, file management, 3D animation, and other programs discussed later in this chapter.

Reviewing the Basics

No matter what your learning style, you will be well rewarded by an initial review of the materials supplied with your software. Even the best editing programs are not perfectly intuitive, and settings changed during hunt-and-peck explorations can have unexpected effects on later projects.

Getting Started

If there is an installation guide, use it by all means. So-called "quick start" guides, on the other hand, are usually compromises addressed to users who otherwise would not read anything at all. In adjusting your system, installing software, and configuring your program, it is better to follow detailed, step-by-step instructions.

Software Manuals

Program user's guides are useful for reference, but they offer only limited help in getting started. Their drawbacks are often inherent in the form itself. For example, the VideoStudio User's Guide is complete, well-organized, and clearly presented. But its small page size, still smaller black and white pictures, and incomplete index limit its overall effectiveness. The programs themselves generally come with at least a user's guide and a quick reference card. Most of them also include a second, digital copy of the manual that you can access through the help menu in the program. Because these virtual manuals are indexed electronically, information is often easier to find there than in the printed version, **Figure 22-5**.

Tutorials

Tutorials supplied with the software can offer an orientation, but they fall short in delivering much real training. Though designed to provide interactive, hands-on activities they are hampered by their very nature.

For one thing, it is difficult to keep the tutorial software and the editing program open at the same time unless you are working with two monitors. Shifting back and forth on a single screen is inconvenient and interferes with the learning process.

Figure 22-5.
The on-screen version of the VideoStudio user's guide is accessed through the help menu. (Ulead)

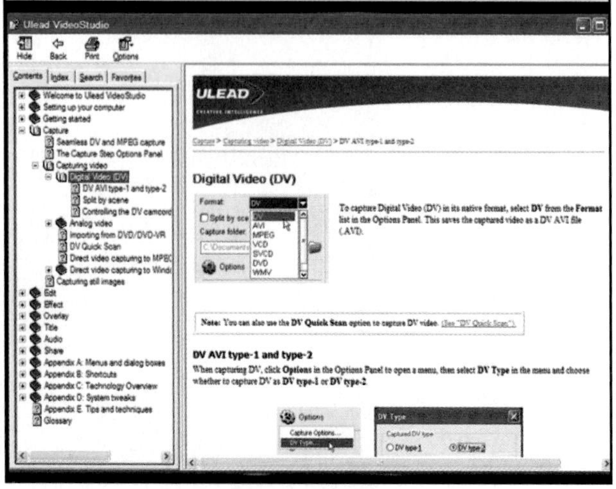

Figure 22-6.
VideoStudio's tutorials are actually demonstrations. (Ulead)

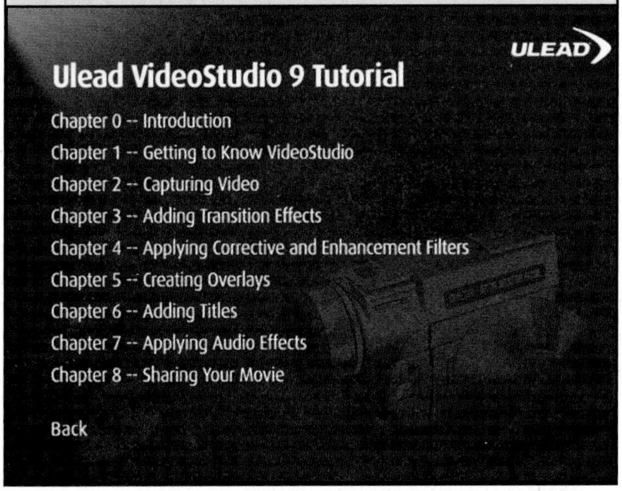

Also, tutorials that instruct you to perform specific actions at exactly predetermined points eventually lose synchronization with what is actually happening on your screen. Once your situation differs from what is expected in the tutorial, the steps prescribed start producing unplanned results.

The VideoStudio support material avoids this problem by compromising. Their "tutorial," is actually not a truly interactive training program, but a guided tour of the software. This in-depth demonstration always works perfectly, but on the other hand, it offers no hands-on training, **Figure 22-6**.

All these orientation materials, then, suffer from various deficiencies; but there are many other sources of information and training, as the sidebar suggests.

Information Resources

Many well-established software applications are supported by a wide variety of training and reference materials.

Many programs can link directly to corporate "mother ship" websites that offer user information, tutorials, and technical support.

Independent websites can be rich mines of information about video editing. Although the Videoguys company sells video products, their site is full of excellent tutorials and other resources.

A tutorial page on a commercial website (Videoguys)

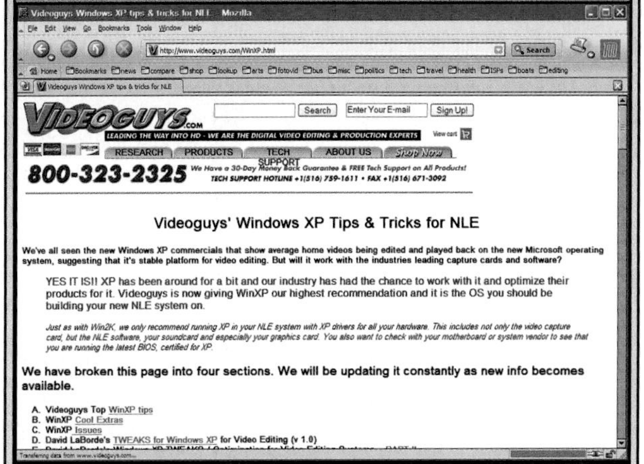

Most large bookstores carry third-party manuals for popular software programs. Review them carefully before selecting one, as they vary widely in quality and relevance to your needs.

Third-party tutorials are often available, usually on DVDs; but their quality may be doubtful and they are difficult to evaluate before purchasing.

Finally, use a search service such as Google to find independent websites and user groups devoted to your software. These sites are often rich sources of information, advice, and support.

Along with a website, many software companies also publish e-mail newsletters and other materials about their programs.

Learning the Interface

Like all desktop software, editing programs present you with a "graphical user interface (GUI)"—a set of pictorial elements that you manipulate to execute commands. Because of the variety of postproduction tasks, editing program ***interfaces*** can look dismayingly complicated, compared to the simpler screens of, say, a word processor.

To get past this initial barrier, you need to learn your program's interface. If it has been well thought out, you will soon see the logic of each component, and moving around the screen will become intuitive.

Again, Ulead's VideoStudio is a good representative example. Its interface, **Figure 22-7**, has seven main parts:

- The step panel.
- The menu bar.
- The options panel.
- The preview window.
- The navigation panel.
- The library.
- The timeline.

Figure 22-7.
The VideoStudio interface. 1—The step panel. 2—The menu bar. 3—The options panel. 4—The preview window. 5—The navigation panel. 6—The library. 7—The timeline. (Ulead)

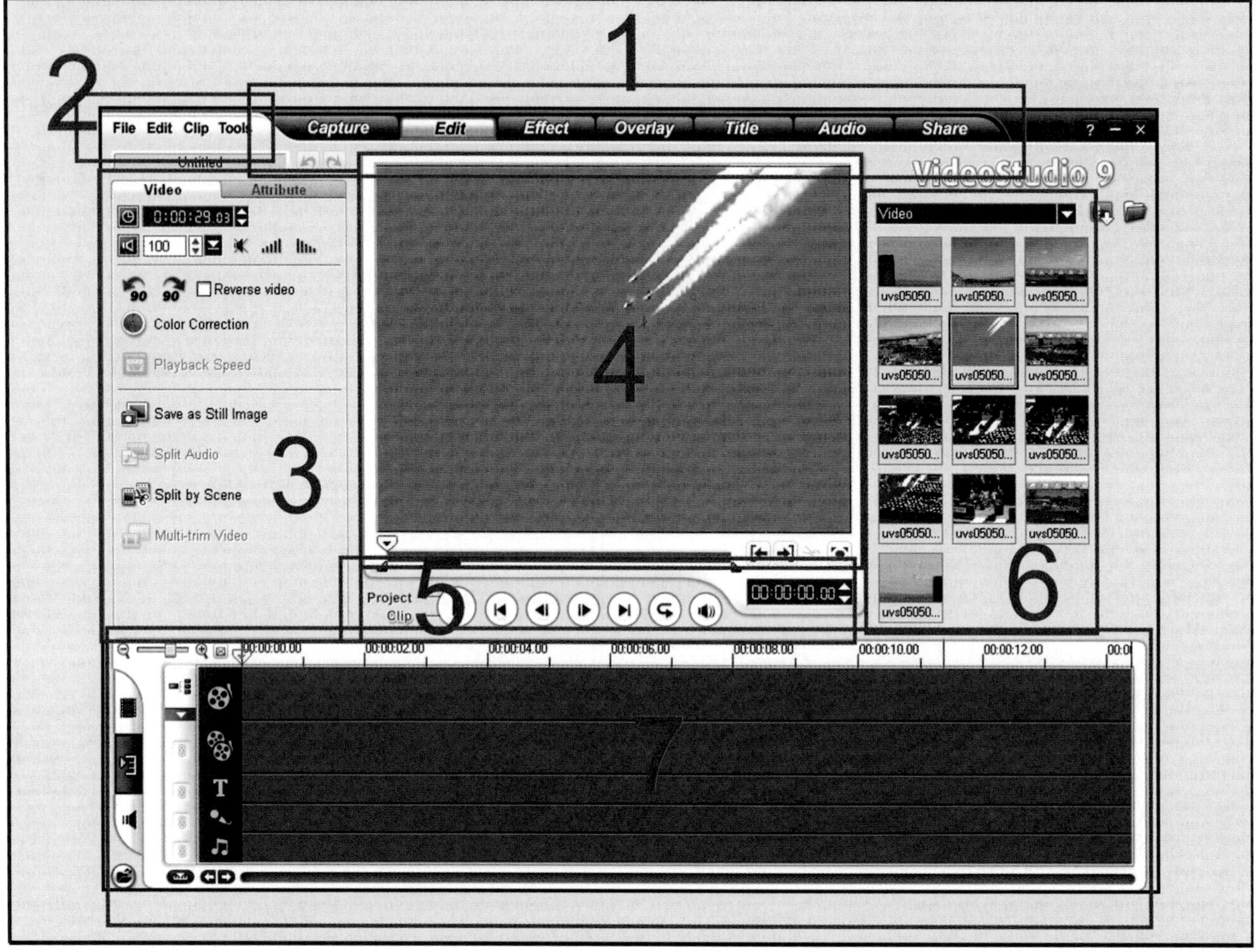

Figure 22-8.
The step panel. (Ulead)

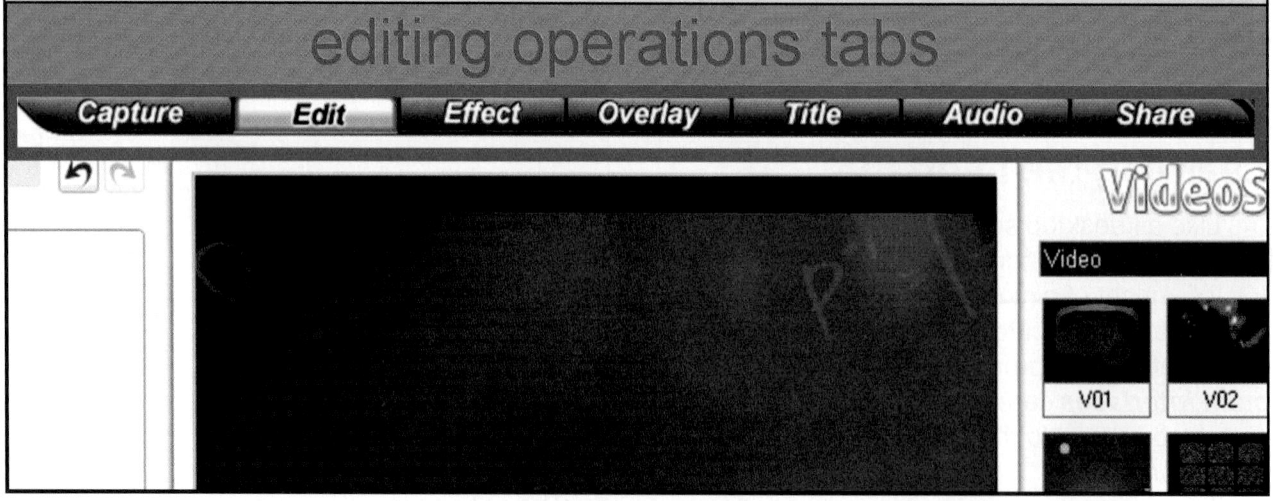

The Step Panel

The program organizes postproduction into several different steps, which you perform more or less (though not perfectly) sequentially. Each tab on the step panel changes the interface to present the tools needed for the operations performed there, **Figure 22-8**.

This organization by steps, one of VideoStudio's more distinctive features, is not found in all editing software.

The program is organized so that the same types of materials and controls show up in the same screen areas, regardless of which step (screen) you are on. To avoid confusing similar-looking screens, get in the habit of frequently checking the step tabs to remind yourself of where you are at the moment.

The Menu Bar

The standard Windows drop-down menu bar provides access to commands and operations that you may need at different times during the editing process, **Figure 22-9**.

Figure 22-9.
The menu bar. (Ulead)

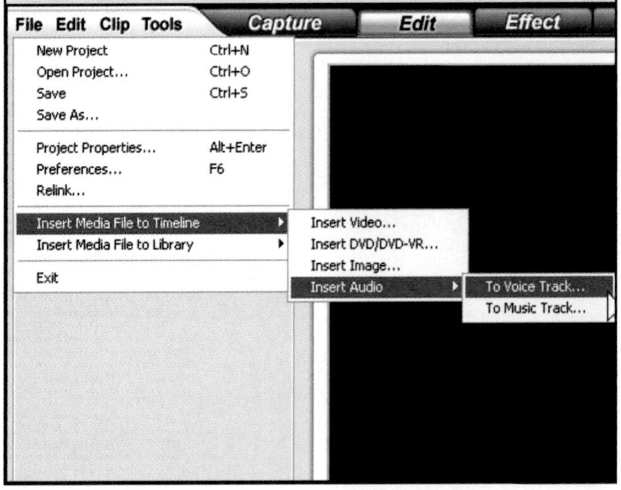

The contents of the step panel and menu bar remain unchanged, no matter which screen you are working on.

The Options Panel

The options panel controls information, settings, and controls appropriate to the operations you are performing, **Figure 22-10**.

Figure 22-10.
The options panel. (Ulead)

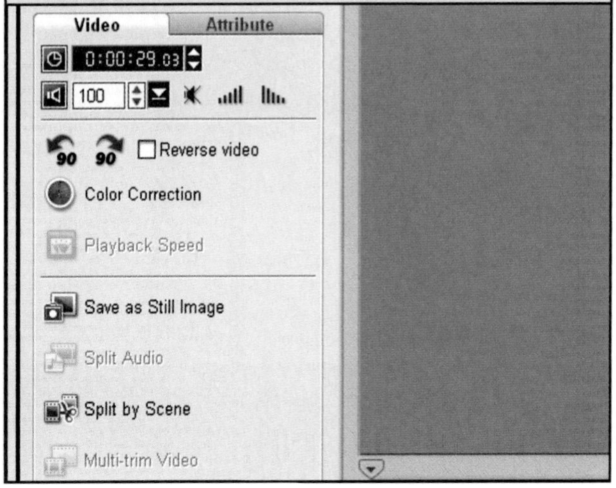

Though it always stays in the same screen location, the options panel information changes markedly from one step screen to the next, **Figure 22-11**.

The Preview Window

The ***preview window*** always shows the ***graphic*** element you are working with, whether it is a clip (shot), an ***effect***, a filter, or a ***title***, **Figure 22-12**.

The preview window can also show you the visuals of the entire project. You ***toggle*** back and forth between individual graphic elements and the whole program by clicking a button to the left of the navigation bar: "***clip***" for the individual element, or "project" for the whole program, **Figure 22-13**. (The selected button turns a brighter blue.)

Most programs show individual elements and whole programs in two or more separate windows. The VideoStudio's single preview window adds work steps but simplifies the screen.

The Navigation Panel

The navigation panel has three different functions:

- It allows you to move (navigate) through the clip or program in the preview window.
- It lets you ***trim*** the length of a clip by setting very precise in- and out-points.
- During the capture step, it controls the operation of the source hardware, when it is a

Figure 22-11.
Information on the options panel changes from one step screen to the next. (Ulead)

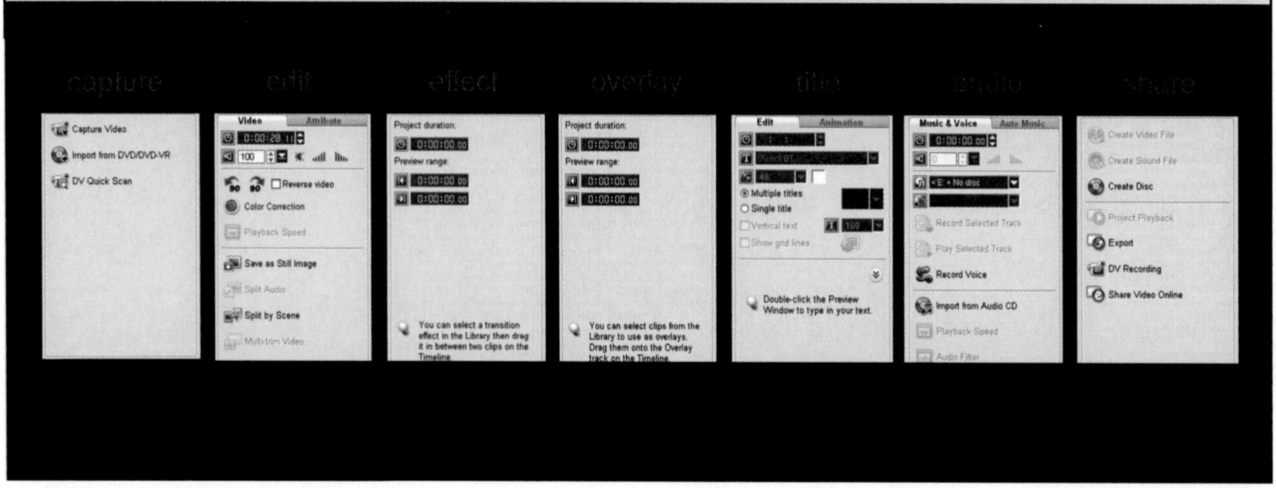

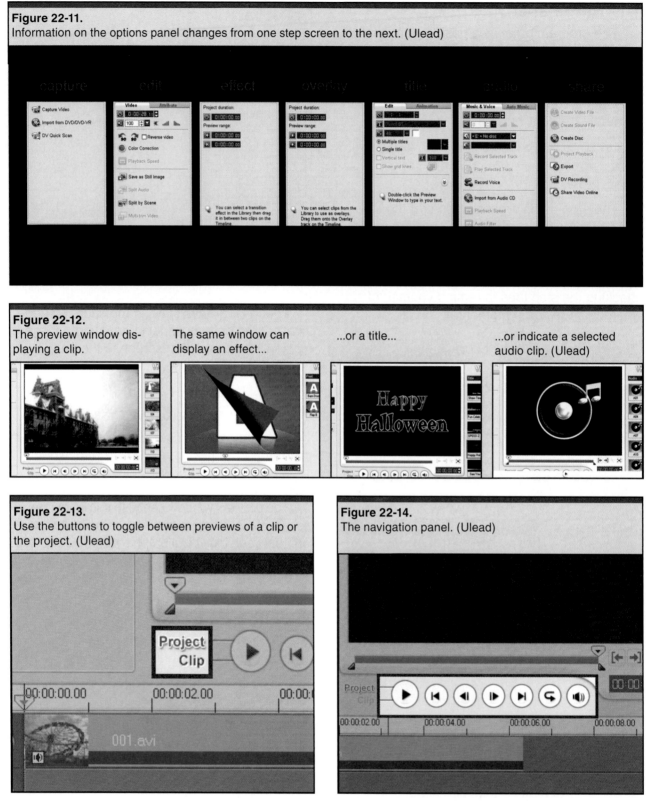

Figure 22-12.
The preview window displaying a clip.

The same window can display an effect...

...or a title...

...or indicate a selected audio clip. (Ulead)

Figure 22-13.
Use the buttons to toggle between previews of a clip or the project. (Ulead)

Figure 22-14.
The navigation panel. (Ulead)

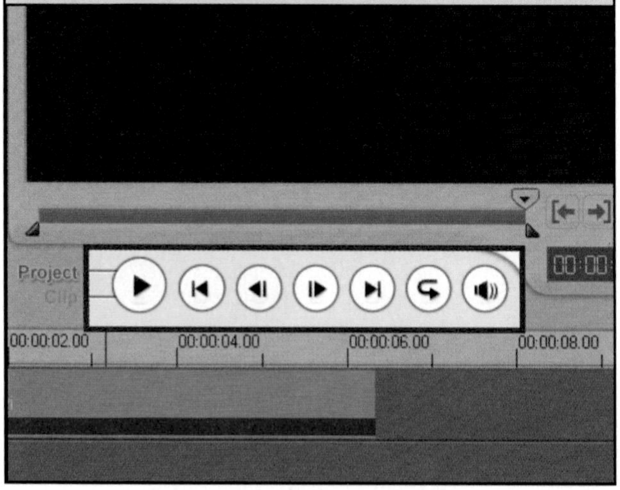

digital camcorder connected to the computer through a 1394 (firewire) port, **Figure 22-14**.

Like the step panel and menu bar, the navigation panel has the same information and functions, no matter which step you are in (that is, which screen you are using).

The library

"Library" is VideoStudio's word for the collection of resources (video clips, still graphics, titles, effects, audio, overlays) that you access from time to time as you work on your project, **Figure 22-15**.

Figure 22-15.
The library. (Ulead)

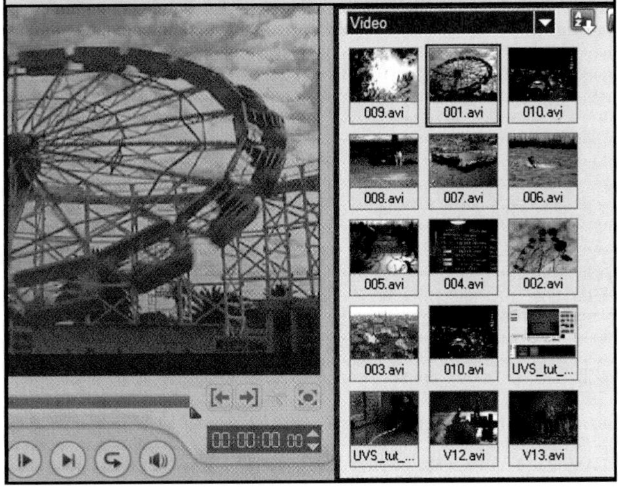

Other programs may call these holding areas "bins" or give them other names.

Like a physical library, this virtual one is organized in sections by subject, and each subject/section contains many individual "books." These sections and individual items are accessed by a tree of menus, **Figure 22-16**.

Some items, such as video clips, still graphics, and audio segments, are placed into the library by the editor. Others, like effects and title templates, are part of the program. Like the menu bar, the library always contains the same items, available from any work screen.

That screen may change automatically, however, if you select a menu item that belongs to a project step tab that is different from the one you are working in.

Figure 22-16.
A menu in the library. (Ulead)

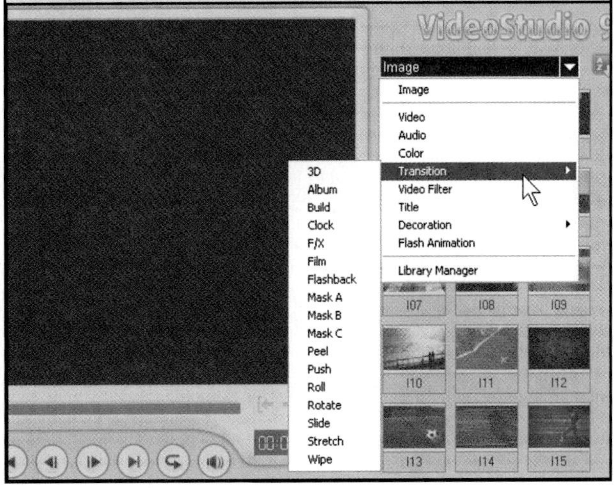

The Timeline

The timeline provides a graphic representation of your program, with the first element on the extreme left and each succeeding element to the right, in order. As you play — or "scrub" (scroll) —through your project, timeline elements move off-screen to the left and new ones appear on the right.

Numbers above the timeline display ***timecode*** locations at every point in the program, in hours/minutes/seconds/frames, **Figure 22-17**.

Simple editing programs use a "***storyboard***" metaphor for the timeline. To use the storyboard, you drag shots into a preferred order like slides on a light table. Professional programs use a "***timeline***" metaphor, in which each shot or other element is represented by a band whose length (matching the timecode addresses above it) represents the length of the shot. As an intermediate program, VideoStudio lets you use either. A third option allows you to display audio tracks in greater detail, **Figure 22-18**.

Ulead uses the term "timeline" for both the display area and one of the display metaphors used in it.

Controls to the left of the timeline allow you to select storyboard, timeline, or audio display formats, **Figure 22-19**.

For the more popular editing programs, you can obtain special keyboards with the keystrokes already labeled and color-coded.

The names and appearances of work areas in other editing programs will be different from

Figure 22-17.
The timeline. (Ulead)

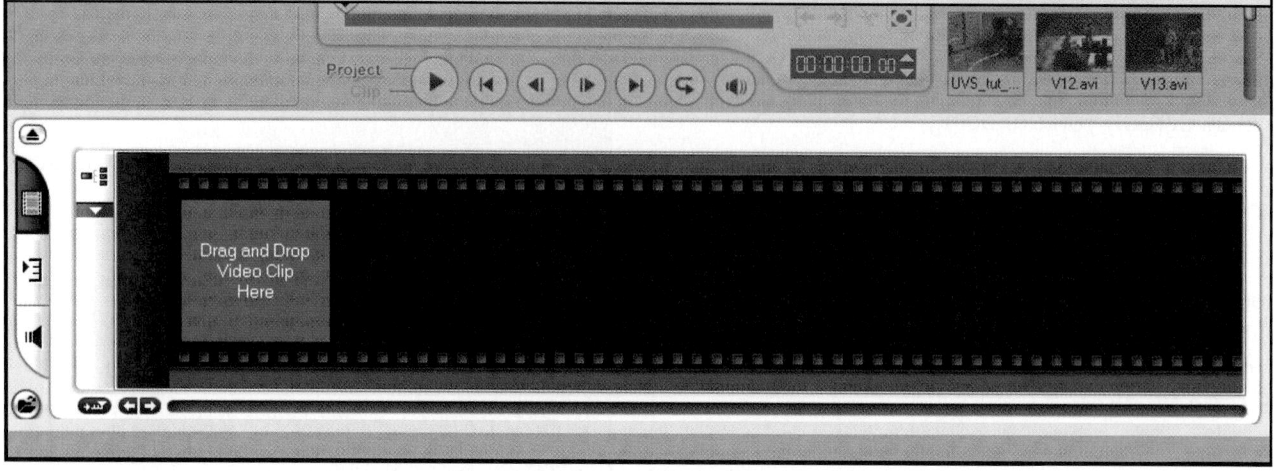

Figure 22-18.
The timeline. A—In storyboard mode. B—In "timeline" mode. C—In audio mode. (Ulead)

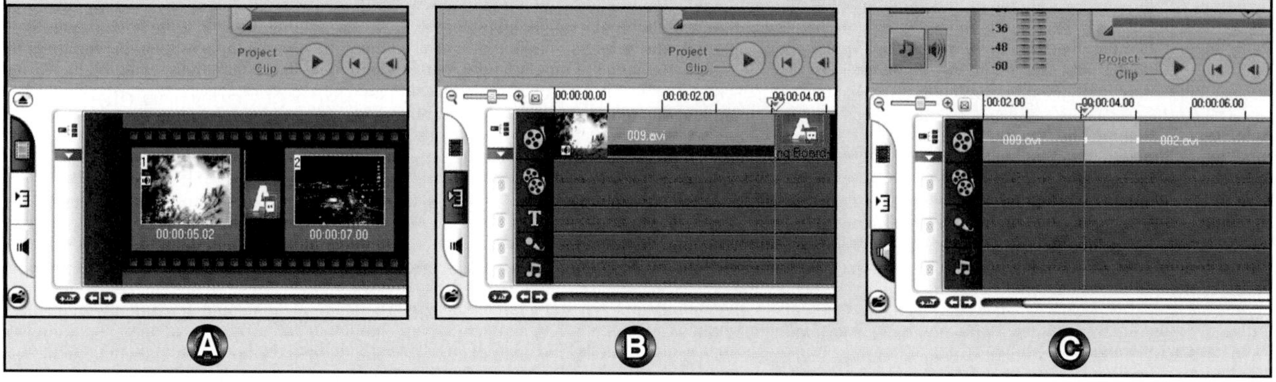

Figure 22-19.
The timeline display selector buttons. (Ulead)

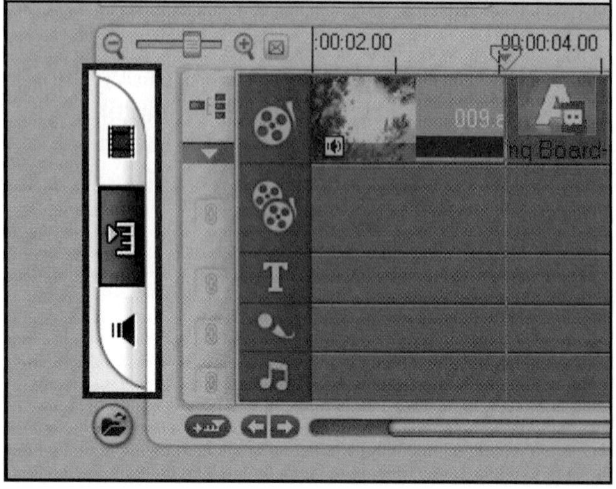

those in VideoStudio, but their functions will be the same or very similar.

Mastering Your Program

Postproduction programs differ, not only in the way their procedures are accessed, but also in the way they are performed. For instance, the editing workflow in VideoStudio does not match the outline presented in the previous chapter of this book, as the sidebar *Different Programs, Different Procedures* shows. Why the difference, and which one is correct?

The answer is that the first list is general, while the second one is specific to VideoStudio; and there lies the big difficulty with digital post-production software. No two programs work quite alike. In word processing or spreadsheet analysis, you can move easily from one program to another, because almost all the tasks have the

Small but Important Features

Editing programs, like other powerful applications, are well-supplied with unobtrusive controls that add versatility and/or convenience. A good program offers dozens of these small features, such as these typical examples from VideoStudio.

Full-screen View

Many times, you will want a closer view of a clip than the small preview window can provide. To enlarge the view, click the button below the lower-right corner of the screen.

The normal preview window.

The full-screen preview window. (Ulead)

Timeline Scale Control

When you are making ***frame-accurate*** edits, you may want the timeline expanded to show you every individual frame. At other times—say, when laying a piece of music under several shots—you may wish to condense the timeline to show you all of them. By dragging the slider, you can make the timeline as tight or as loose as you prefer at the moment.

The timeline can be condensed to show a full minute per unit...

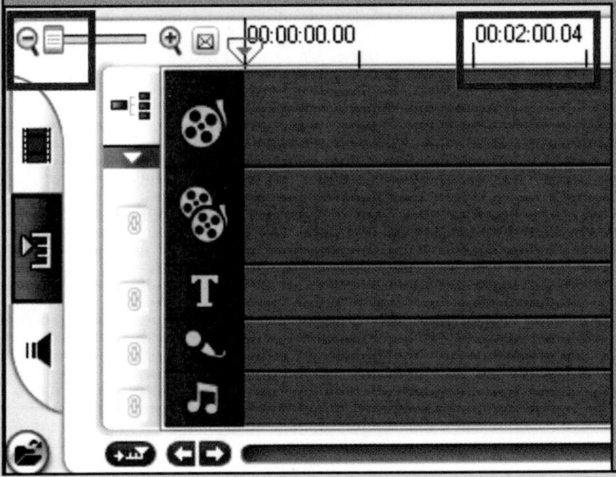

or expanded to show a single frame per unit. (Ulead)

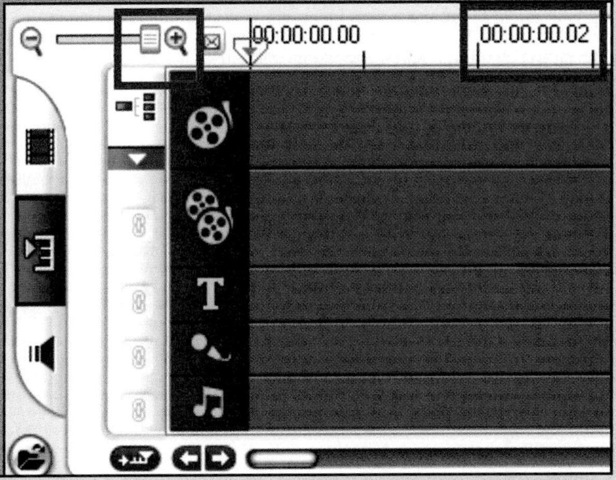

And by clicking the small button that resembles an envelope, you can condense the timeline until the entire project, no matter how long, fits into the window.

Other Helps and Conveniences

You can discover these and many other clever helps by simply moving the mouse cursor over a button or other graphic element and pausing there. Momentarily, a small box will open that labels and explains the feature.

When you become truly proficient in using your software, you may find that keystrokes are faster than mouse operations. Every editing program includes many ***keyboard shortcuts***.

Different Programs, Different Procedures

Chap. 21: Editing Principles	VideoStudio Step Panel Tabs
1. Configuring Digital Projects	1. Capture
2. Capturing Materials	2. Edit
3. Building the Program	3. Effect
4. Adding Effects	4. Overlay
5. Designing the Audio	5. Title
6. Storing and Distributing	6. Audio
	7. Share

same labels and are performed in the same general way. In video editing, that is not the case.

To overcome this problem, you need to understand the basic operations *behind* the windows, labels, and buttons of any one particular program. That way, you can readily move up to more complex software (or sideways to a competitor's product).

What follows, then, is the way in which the basic operations that are common to all editing programs are organized and presented in VideoStudio.

Configuring the Project

VideoStudio begins a new editing project at the main menu, where one choice is "Project Properties," **Figure 22-20**.

Figure 22-20.
The project properties menu. (Ulead)

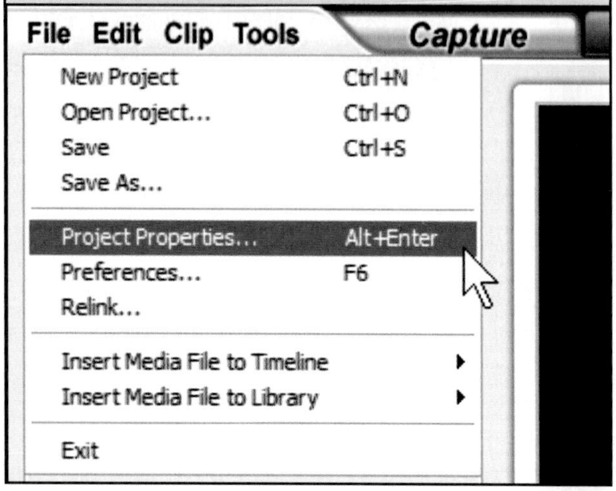

Figure 22-21.
Technical parameters. (Ulead)

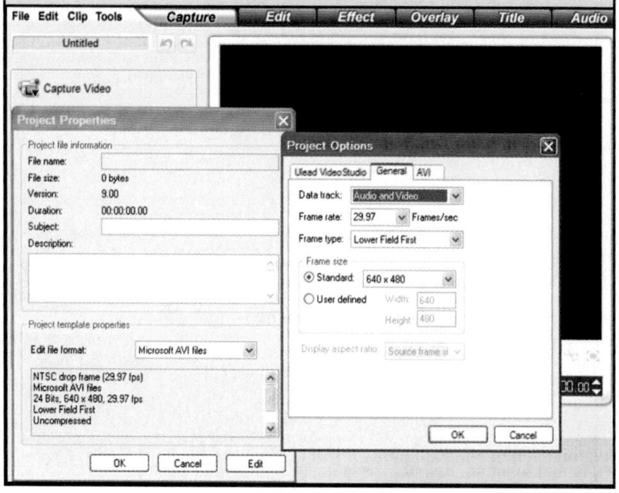

Selecting this item opens a nest of screens on which you set the technical parameters of your project, **Figure 22-21**.

Capturing Materials

Capturing raw footage is the first editing step in any project, so VideoStudio devotes a screen to it, **Figure 22-22**.

Figure 22-22.
The capture screen. (Ulead)

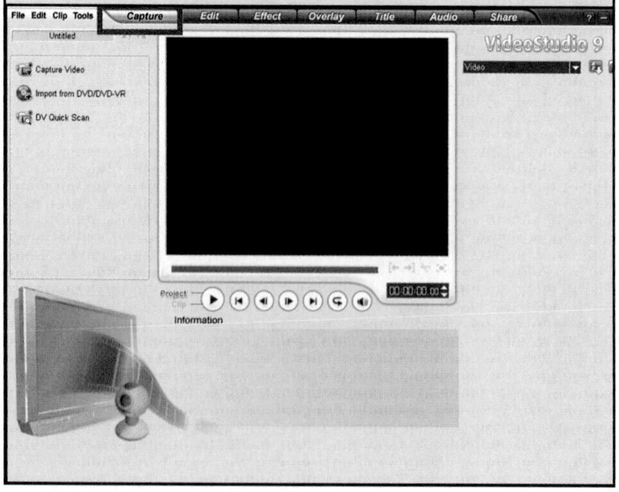

If you check the pull-down menu, you will notice that many selections are unavailable (grayed-out) because they do not apply to the capture process.

Building the Program

The edit step screen in VideoStudio contains the work area for assembling the program, **Figure 22-23**.

Figure 22-23.
The edit screen. (Ulead)

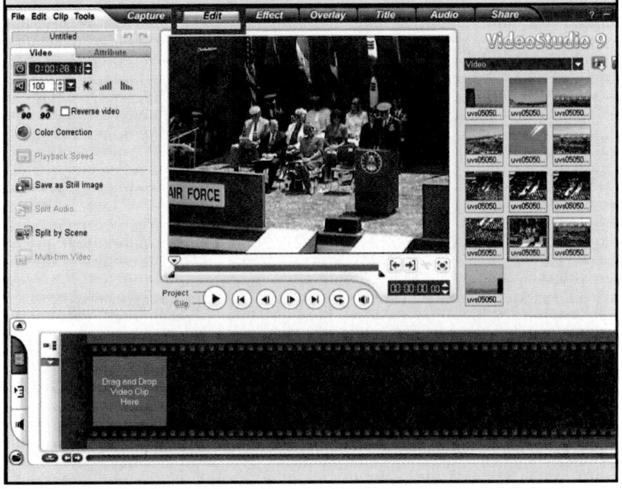

After selecting shots and placing them in order, you can trim them to exact length.

Adding Effects

The "Effect" screen in VideoStudio is somewhat misleading, because this screen handles only one type of computer graphic effect.

Transitions

By "effect," this program really means "***transitions,***" and the library offers dozens of ways to signal a move from one sequence to another, **Figure 22-24**.

Figure 22-24.
Transitions are created on the effects screen. (Ulead)

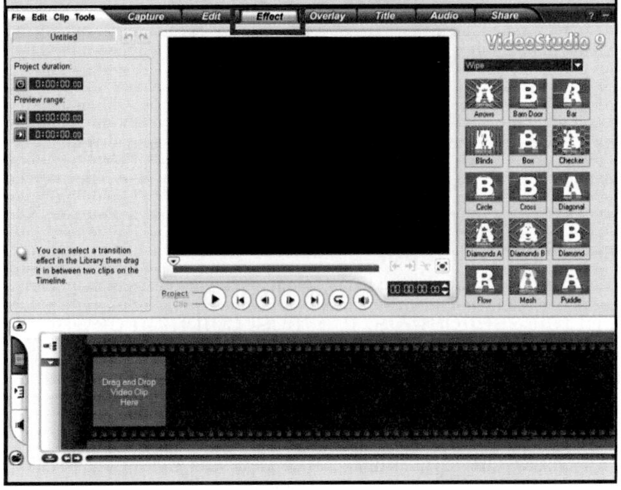

Filters

Filters (and color overlays too) are not considered effects in VideoStudio. Selecting "filters" from the library menu moves you from the effect screen back to the edit screen, **Figure 22-25**.

Figure 22-25.
The filters library on the edit screen. (Ulead)

Overlays

Overlays, however, are given a screen of their own, **Figure 22-26**.

Figure 22-26.
The overlay library on the overlay screen. (Ulead)

Creating Titles

Titles, too, have their own work screen, **Figure 22-27**.

Designing the Audio

Most of the audio operations are performed on the audio screen, **Figure 22-28**.

But some important tasks, like ***splitting*** audio from video (to edit it separately) are performed on the edit screen, **Figure 22-29**.

Figure 22-27.
The titles library on the titles screen. (Ulead)

Figure 22-28.
The audio screen.

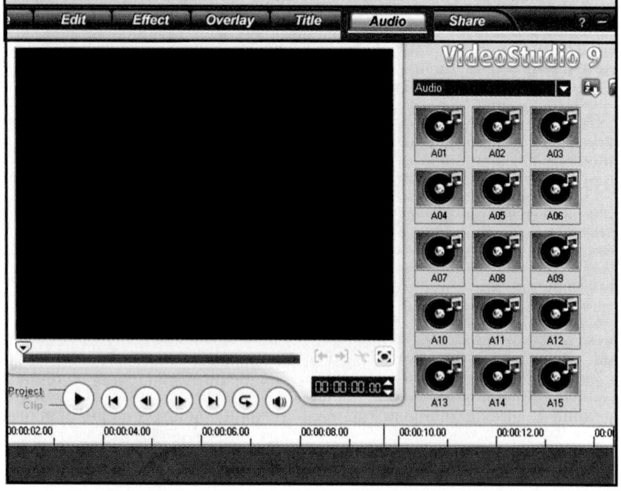

Figure 22-29.
Splitting audio from video. A—The split audio command. (The speaker icon on the timeline shot indicates audio is present). B—When the audio has been split, it shows up separately on an audio track. (Ulead)

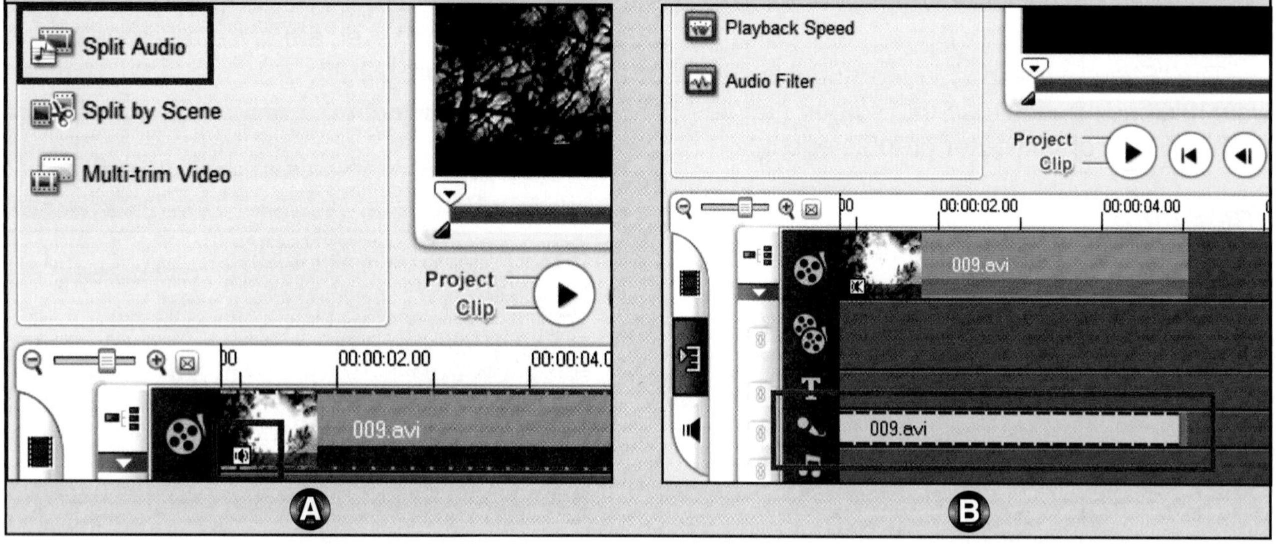

Storing and Distributing

The last phase of a project involves storing it as a finished program and then recording it on an external medium like mini-DV tape or ***DVD disc***. VideoStudio performs these tasks from the share screen.

"Share" is not a very apt word, but it would be difficult to find an equally short term that labels the process more accurately.

Rendering

First, the project must be ***rendered***: re-created as a full-quality program from the blueprint provided by the project file. VideoStudio does not use the term "rendering" for this step, to avoid confusion with other rendering functions in the program. Instead, they say, "create video file," **Figure 22-30**.

That file can be customized for several different release formats, the most common of which is DVD.

DVD authoring

VideoStudio allows the design and creation of relatively sophisticated DVDs, still working from the share screen, **Figure 22-31**.

As you can see, the differences between the general project work flow described in this book and the specific steps in a particular

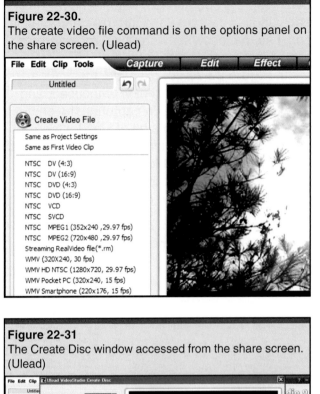

Figure 22-30.
The create video file command is on the options panel on the share screen. (Ulead)

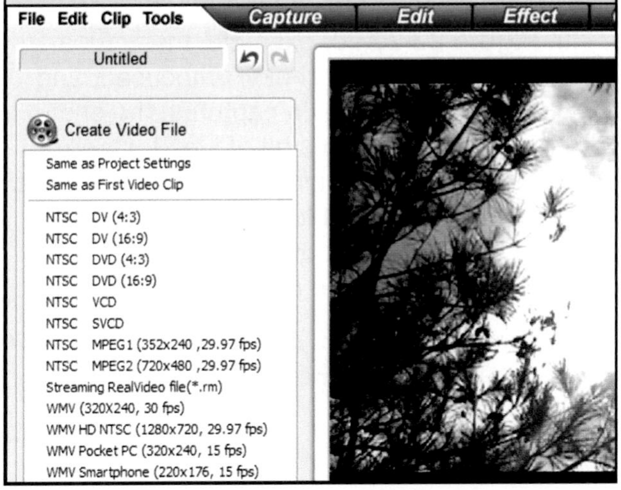

Figure 22-31
The Create Disc window accessed from the share screen. (Ulead)

editing program are easy to reconcile, as long as you understand the principles that lie behind the procedures.

Developing Working Methods

Short projects are easy to edit, and require little in the way of organization. You simply capture all the footage, select and sequence shots, trim clips as needed, garnish with music, effects, and titles, and burn to DVD. But as your programs grow longer and more complex, you can bog down in the size of the job and the wealth of materials.

To avoid that, it helps to develop systematic ways of approaching and managing projects. Of the many things you can do to build a system of postproduction, there are four, in particular, that stand out:

- Plan the project.
- Keep track of materials.
- Focus on sequences.
- Preview the results.

Plan the Project

First, plan the project to suit the video. In a fully scripted program, the shots should be slated by production, scene, shot, and take, **Figure 22-32.** If they are not, catalog the captured footage by referencing it to the script (which you will, of course, use to guide you in editing).

In preparing a documentary, you may not have a script or know where materials will fit in the finished program. Here, you can identify captured footage by date, location, and subject. The same holds true for an anthology program, such as a video school yearbook. In this case, you should have designated sequences (like sections in a printed yearbook). That way, you can complete whole segments of the program while waiting for others.

Figure 22-32.
A fully scripted show should carry identifying slates for every take.

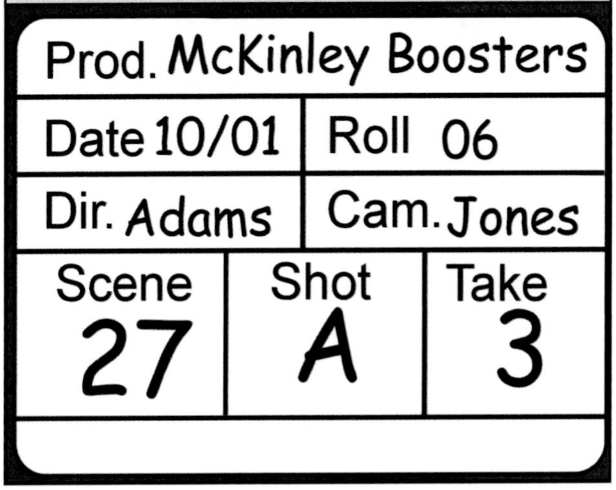
Prod. McKinley Boosters
Date 10/01
Roll 06
Dir. Adams
Cam. Jones
Scene
27
Shot
A
Take
3

For example, you could have the football retrospective edited long before the baseball footage even arrives.

Whether transcribing a camera slate or supplying your own identifications, you will want to rename each clip in your project. To do this, it will be easier to clean up the clips created by

Figure 22-33.
A clip captured, trimmed and renamed. (Sue Stinson)

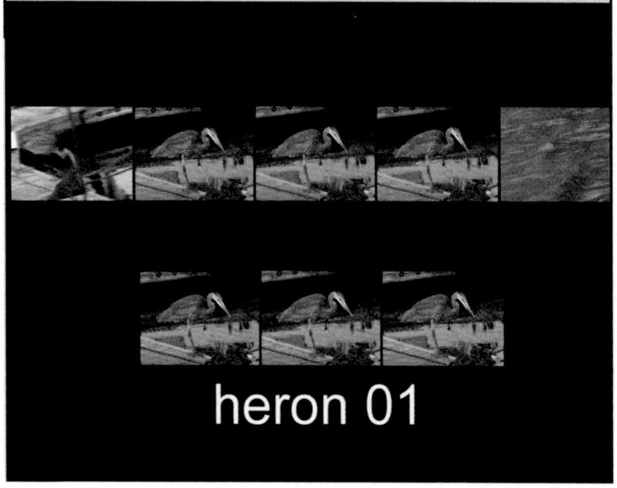

batch capture operations. That means reviewing each shot as captured and trimming and/or splitting it so that each software "clip" represents a camera "take" (one attempt to record a shot).

In **Figure 22-33**, for example, the first part of the heron shot is too shaky, unfocused, and badly composed to use. In capturing the shot, the computer picked up a bit of junk footage (water off the dock) that followed the heron shot. Trimming off the unusable head and foot of the shot saves file space. The last step before storing the trimmed clip is to rename it.

By convention, timelines show only one, or a few, images, no matter how many frames are in the clip.

Example: A Wedding at Stratford

After Sam and Jessica exchanged wedding vows in a 500-year-old house in Stratford-upon-Avon, England, the wedding party floated down the River Avon to another Tudor home, where they dined on an American-style barbecue.

(All wedding photos in this chapter by Sue Stinson)

A video of the event was assembled from footage shot by two different wedding guests.

First, the captured footage was analyzed, revealing that the shots could be assigned to fourteen sequences. A file folder was established for the project and a subfolder for each sequence.

The project file structure.

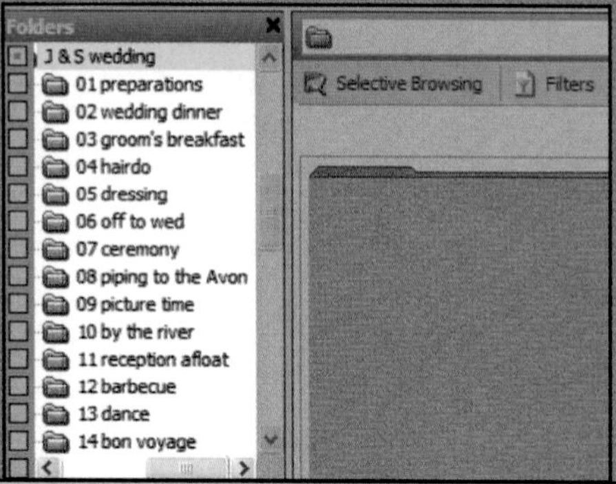

Some of the sequence folders, as displayed by the database.

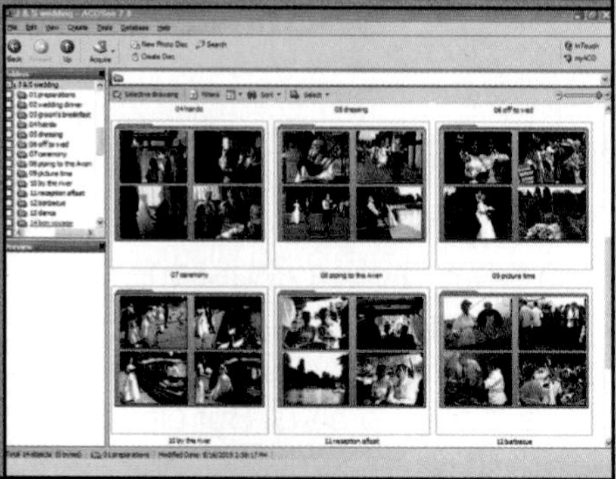

Once the sequence folders were ready, the shots were distributed among them.

The first sequence selected for editing was "Piping to the Avon."

Keep Track of Materials

Once you have prepared your clips, you will need places to put them. All editing programs include the ability to create custom folders for program materials, and many offer database features that allow you to sort by different criteria. To make the most of their abilities,

- Create a master folder for each separate project.
- Add a subfolder for each major segment of the program.
- Establish separate folders for resources like music and sound effects.
- Annotate individual clips.

In renaming captured clips, use slate numbers if available, or a code developed for the program. Avoid names like "Father's closeup," that may be quite clear when you catalog the clip, but meaningless later when you are searching for shots.

Though most editing software has some database capabilities, you may wish to invest in the extra features and power of a stand-alone graphic management program, as explained in the discussion that follows.

Focus on Sequences

To organize your raw footage, you will probably want to capture all of it at once (or as much of it as you have). Once you are ready to begin the project, however, you may find it easier to work one ***sequence*** at a time. It is often better to assemble and trim a relatively small group

After the ceremony, the father of the bride piped the wedding procession down to the River Avon (with bagpipes he had made himself).

Before editing began, each shot in the sequence was renamed to include the sequence and the shot number.

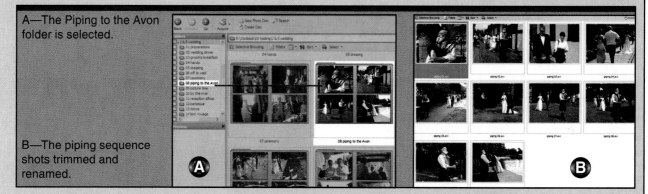

A—The Piping to the Avon folder is selected.

B—The piping sequence shots trimmed and renamed.

And each shot received a brief description in the database, including key words, by which the shot could be searched for at any later time.

The database window, opened by right-clicking the image and choosing "properties."

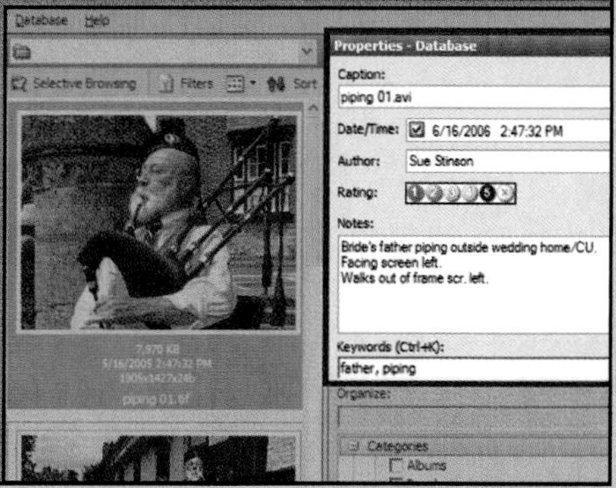

At this point, there were nearly 300 shots catalogued and stored. During editing, clips created from the shots were also stored under related filenames (such as "piping 01 end trim"). With the addition of files for music, background, and effects tracks, the number of files grew to over 500.

Figure 22-34.
It is usually better to edit one sequence at a time. (Ulead/Sue Stinson)

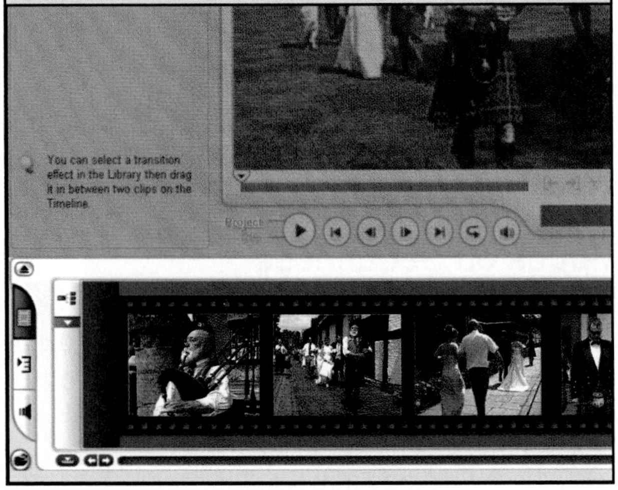

of shots. You can then adjust the ***production audio***, add a ***background track***, and lay individual sound effects, **Figure 22-34**.

Though VideoStudio accommodates only three audio tracks, you can use work-arounds to add more.

It is often better to complete all the sequences in the project before adding transitions, narration and music, which are easier to evaluate when you can judge them all together.

Preview the Results

You should preview your work frequently as you build your sequences. Most programs include a rapid rendering function that quickly produces a lower-quality version of the finished material, complete with transitions, special effects, and graphics. The quality is quite good enough for previewing matters such as shot timing, sequence length, and the like.

However, where extensive image processing and/or effects are involved, you may wish to take the time to create a fully rendered trial version, in order to check for quality.

Finally, never hesitate to return to earlier sequences for changes and touch-ups. Random-access editing is one of the greatest benefits of digital postproduction, and you should make full use of it.

Beware, however, of over-editing. Develop an instinct that tells you when a sequence is finished, and further attempts to improve it may have just the opposite effect.

Supplementing the Software

Today's beginning and intermediate editing programs are complete in the sense that they handle everything from capturing footage to ***burning*** DVDs. But all of them are thin in some areas, and lack other features entirely. At the fully professional level, many programs concentrate on the actual editing process, leaving the more sophisticated aspects of graphics, animation, audio, and effects to other software.

If you are a hobbyist, you may find that a program like VideoStudio will meet your needs indefinitely; but if you grow toward professional-level postproduction, you will want more capability. There are several ways to obtain it.

Program Plug-ins

Some companies offer ***plug-ins***—so-called because they install themselves inside the editing software, where they appear as additions to the regular menus and work windows. Typically, plug-ins are installed to increase the number of transitions and special effects, provide more sophisticated animation capabilities, improve compositing, or perform even more specialized tasks.

Postproduction Suites

Some software vendors, notably Adobe, make the editing software the core of a suite of applications, apparently on the theory that you can upgrade just the operations you choose by mixing and matching components of the suite.

Like Adobe, Ulead Systems coordinates its editing packages with related software, such as PhotoImpact for graphics and DVD Factory for more DVD authoring options.

Upgraded Functionality

In fact, Ulead has three different levels of stand-alone DVD ***authoring*** software: one for advanced amateurs, one for corporate and educational users, and one for creators of DVDs intended for commercial release, in competition with the highly sophisticated discs released by movie distributors.

Sometimes, you find you want to increase the sophistication of your entire editing operation. To accommodate upward mobility, companies like Avid, Adobe, and Ulead offer professional programs with seemingly endless features (and comparably steep learning curves). Some, like Adobe, keep similar interfaces from program to program (Adobe Premiere Elements is actually based on the full-featured Adobe Premiere). Others design separate user interfaces, each optimized for the level of the program (and the user).

Graphic Management Programs

We have noted that third-party graphics management programs can be more satisfactory than the database functions built into some editing packages. Though there are several such programs, *ACDSee*, from ACD Systems is representative.

Figure 22-35 shows the ACDSee work screen, including ***thumbnails*** from a subdirectory of video clips. As this screen indicates,

Figure 22-35.
The ACDSee work screen. 1—File management functions. 2—Clip information. 3—Thumbnail section. 4—Preview. (ACD Systems)

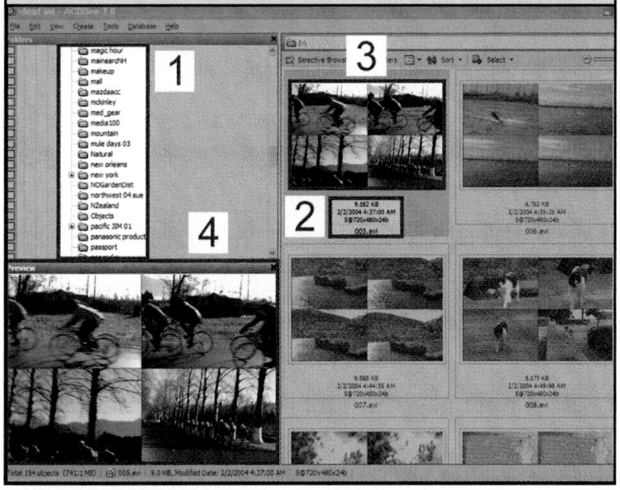

Figure 22-36.
The ACDSee playback window. (ACD Systems)

- All file management functions are available for adding, deleting, moving, and renaming folders and files.
- Considerable information about each clip is displayed.
- The thumbnail contains four stills from its file, sampled from the first frame to the last, to indicate the flow of action in the clip.
- A larger preview of the thumbnail permits closer inspection.

Double-clicking on a thumbnail brings up a full-screen window on which you can play it before deciding to import it into your project, **Figure 22-36**.

Right-clicking the thumbnail opens the menu with the *properties* item. Clicking it, in turn, displays database information about the clip, **Figure 22-37**.

In the sidebar on the wedding, the ACDSee screens were simplified for clarity.

These are some capabilities to look for in choosing a stand-alone graphics management program.

Figure 22-37.
Database information from the properties item. (ACD Systems)

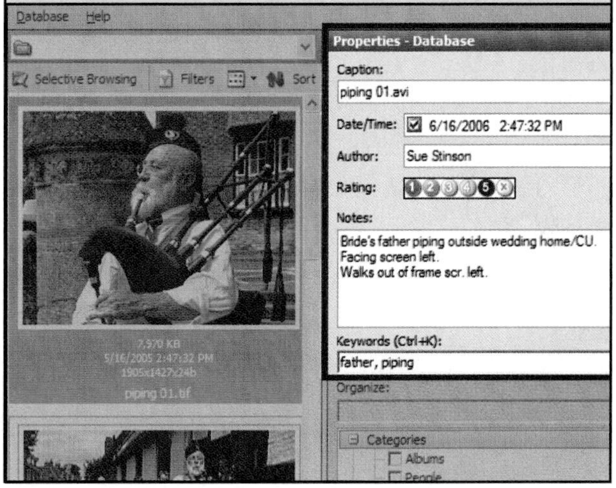

Moving On

This chapter may not have addressed the particular editing software that you are interested in, because to do so for all readers would have required a dozen chapters, eleven of which would have been useless to *you*. Instead, we have tried to cover universal editing procedures by demonstrating them with a single, typical program.

The next chapter follows the same method, using a representative program to explore the additional postproduction operation of DVD authoring.

Chapter Review

Answer the following questions on a separate piece of paper. Do not write in this book.

1. *True or False?* Only entry-level digital editing systems drop occasional frames during capture.
2. Software manuals are useful for ______, but offer only limited help in learning the program.
3. In the ______ metaphor used by professional editing systems, each shot is represented as a band with a length matching the length of the shot. Time code information is displayed above the band.
4. *True or False?* To work efficiently, it is vital to develop systematic ways of approaching and managing projects.
5. As you edit sequences, you should ______ your work frequently.

Technical Terms

Analog video: Video that is recorded as a continuous electrical signal, rather than a series of codes.
Artifacts: Small visual blemishes that mar the quality of an image.
Authoring: In video postproduction, designing and creating a DVD.
Background track: An audio track of ambient sound, usually laid under the production audio track.
Burning: Recording onto disc.
Capture: To import footage from a tape to a computer, digitizing it if the original is analog video.
Capture card: An internal computer accessory card for capturing video.
Clip: A unit of video, audio, graphics, or titles, as used in digital editing.
Drop frames: One or more frames lost during capture, usually because of insufficient computer speed.
DVD disc: A high-density recording medium capable of holding two hours or more of video.
Effect: Short for digital video effect (DVE); technically, any digital manipulation of the video, but commonly used to mean a scene-to-scene transition.
Filter: A digital effect that changes the character — such as color — of the clip(s) it is applied to.
Frame-accurate: Edits made at exactly the single frame desired.
Graphic: Technically, any nonlive-action visual element, but not commonly applied to titles.
Interface: The method by which a user communicates with a computer.
Keyboard shortcut: A keystroke or keystroke combination that performs an action otherwise performed by a mouse.
Plug-in: A program that has no interface of its own, but is operated by commands it has inserted into another program.
Preview window: A window that displays what a piece of visual material looks (or will look) like.
Production audio: The audio track recorded with the video.
Rendered: A final, high-quality version of a finished program created from the lower-quality "blueprint" of the project file.
Sequence: A series of closely related video shots, usually a few minutes long or less.
Splitting: In editing, separating the audio and video tracks of a clip so that each can be processed individually.
Storyboard: In digital editing, a metaphor that represents the project as a succession of slides laid out in order.
Thumbnail: A very small image representing a clip or shot.
Timecode: The address (location) of a clip, expressed in hours, minutes, seconds, and frames.
Timeline: In digital editing, a metaphor that represents a project as a stack of long, narrow strips, one strip for each program element.
Title: Any lettering that appears in a video.
Toggle: To switch back and forth between two states, such as off and on.
Transition: A visual effect signaling the end of one program part and the beginning of the next.
Trim: To indicate the in- and out-points of a clip in order to specify its content and set its exact length.

23 Authoring DVDs

Objectives

After studying this chapter, you will be able to:

- Specify a DVD disc format.
- Select and include video programs.
- Add special sound track and text features.
- Develop a disc organization.
- Design the "look and sound" of a disc.
- Create menus for accessing disc features.
- Prepare a disc project for replication.

(Sue Stinson)

About Authoring DVDs

Authoring is the name given to the process of creating ***DVD*** video discs for playback by computers or DVD players. This process is not a simple mechanical transfer like making a videotape, but a multistep activity that requires planning, craftsmanship, and creativity. In that sense, a video editor is the *author* of a DVD much as a writer is the author of a book.

Aids to Navigation

A DVD such as a feature movie release contains not one program but several—each one recorded at a different location on the disc. (See the sidebar for a typical example.) DVDs provide immediate access to any program on the disc; within each program, they allow jumping to predetermined points (called *chapters*). Often, they also force viewers to watch certain obligatory programs, such as the standard FBI warning screen, **Figure 23-1**.

Figure 23-1.
Though the FBI warning is only a single program shot, it is technically one of the programs on the DVD.

When playing commercial DVDs, notice that some elements will not let you exit before they have finished.

To provide random access to programs and chapters, DVDs are equipped with screens containing ***menus*** much like the user interfaces in computer programs. By selecting and clicking menu items, viewers can navigate through multiple displays to find and access the programs, chapters, and special features they want. Creating this navigation system is a major part of DVD authoring.

In this respect, DVD authoring is similar to Internet Web design, though usually simpler.

Design Elements

Most DVDs are meant for entertainment; and even instructional and promotional products are intended to be as lively as possible. To make them interesting, their menus can include moving images and musical backgrounds. Programs and graphics can be designed to carry out the theme of the disc, **Figure 23-2**. Even sound effects (such as bells or camera shutter clicks) are sometimes added to indicate that selections have been made. Designing the graphics, video elements, titles, music, and sound effects for the DVD can be as creative as editing the programs on it.

Technical Matters

DVD authoring also has a technical side. Since programs may be played on different video displays using different TV standards, the

Figure 23-2.
Most DVD authoring programs include menu templates. (Ulead)

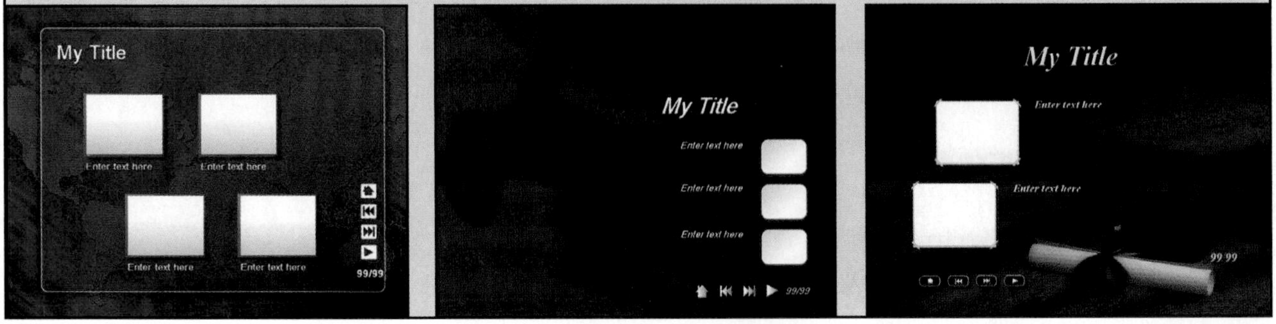

discs must be set up to reflect these differences. Also, they may be played by people speaking different languages, so alternate soundtracks may be selectable by viewers.

Additional soundtracks may also carry commentaries by the director or others connected with the program.

Different sets of titles are often provided for foreign language translation or closed captioning. Each set can be selected and turned on or off by the viewer.

Security Issues

Because DVDs are digital, they are easily duplicated. To hinder the production and sale of pirated copies, DVDs can be equipped with two types of safeguards. To discourage sales in unauthorized markets, they can be regionally encoded, so that discs will work only in players sold in their own region, **Figure 23-3**.

The United States and Canada make up Region 1.

To prevent *all* unauthorized duplication, discs can be encoded with "***copy protection***," which either makes duplicates unwatchable or prevents copying altogether.

Regional encoding and copy protection are optional; but setting the TV standard is not. The various TV systems, notably NTSC (in North America), PAL, and SECAM, are so different that discs made for one standard will not work at all with the others.

Preparing to Publish

Even making the actual discs is not perfectly simple. For one thing, DVDs use a special encoding system that is different from the systems generally used for editing. Programs intended for DVD release must be translated into the "MPEG" encoding system.

Some editing programs can, if desired, translate raw DV footage into MPEG format as they capture it.

From there, individual copies can be made on computer drives that "burn" (record) DVDs. However, commercial discs made in large quantities are manufactured by a completely different process. For these discs, "DLT" (**D**igital **L**inear **T**ape) masters usually must be produced and sent to the commercial duplicating company.

DVD Software

Like video editing software, DVD authoring programs span a wide range of sophistication and complexity. The Ulead products cited in this book are typical. On the one hand, simple DVD authoring is included in *VideoStudio* (the editing program featured in Chapter 22), so discs for home use can be created without additional software. On the other hand, the DVD authoring program, Ulead *DVD Workshop*, **Figure 23-4**, includes capture operations for raw footage, a procedure usually performed with editing software.

Figure 23-3.
DVD regions. (Ulead)

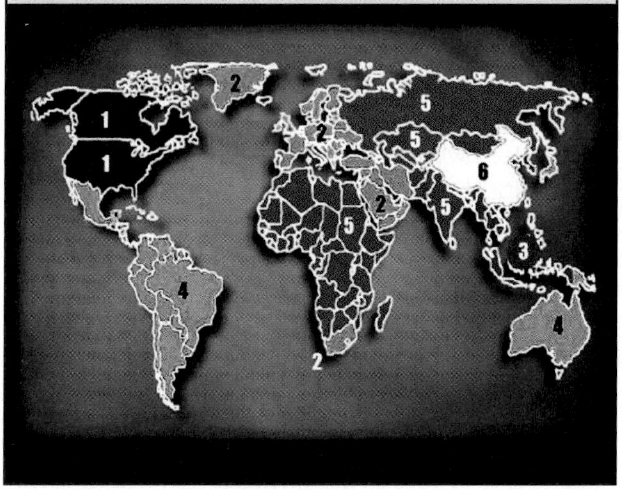

Figure 23-4.
DVD Workshop is intended for advanced amateur and corporate users. (Ulead)

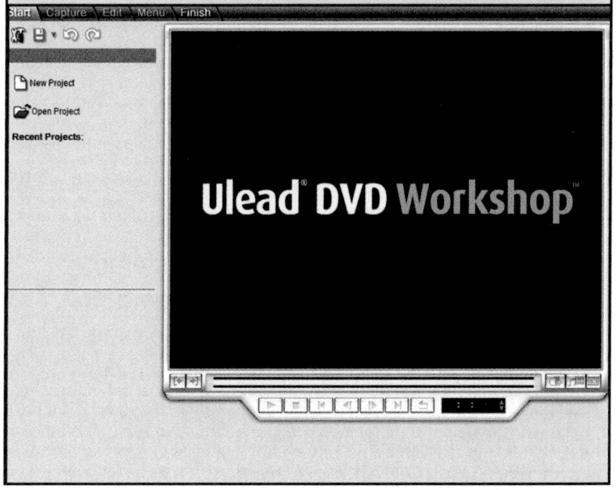

Digital Formats: DV vs. MPEG

Digital camcorders record footage in a particular format (***mini-DV*** for most amateur and entry-level professional models); and video editing software stores captured footage in a format identical or very similar (typically AVI format in Windows-based computers).

DVDs, however, use a different storage format called ***MPEG*** (named for the Motion Picture Experts Group, which devised it) because this format allows great data compression without notable loss in picture quality. (Video program files must be compressed because they contain enormous amounts of data.)

A series of MPEG frames. (Symbolic representation: actual "P" and "B" frames do not look like this.)

There are several versions of MPEG data compression, but their differences need not concern us here.

For this reason, authoring software must convert finished video programs to MPEG format for inclusion on DVDs.

Why not convert captured footage and edit in MPEG to begin with? Because frame-accurate cutting is difficult with MPEG since it does not store every video frame. In each second of video, MPEG designates some individual images as ***I-frames*** (short for "intra-frames") and stores all the picture information within each one. The frames between I-frames, however, are not complete images. All that is stored is the difference between that frame and the previous I-frame.

These frames are labeled P or B; and a complete sequence if I, P, and B frames is called a GOP (Group Of Pictures). Some editing programs can work in MPEG by filling in data missing from a P or B frame with information from the previous I-frame.

By contrast, mini-DV and similar formats use ***intra-frame compression*** for every frame, so you have a complete image wherever you decide to make an edit. Intra-frame compression is not used on DVDs because the resulting data files are larger than comparable files in MPEG format.

In stand-alone DVD applications, Ulead offers three different programs of increasing sophistication. Within even the top-level program, however, selectable "wizards," templates, and automated procedures can simplify the authoring process (at the expense, of course, of sophistication and originality).

This chapter assumes that readers will create completely original, fully professional DVDs; so simplified, semi-automated procedures receive only minimal coverage here.

DVD Workshop is used here for demonstration purposes. Though easy to learn, this program is sophisticated enough to create commercial-grade DVDs, complete with multiple language tracks, different sets of subtitles, regional encoding, and copy protection. In addition, its user interface is similar to that of VideoStudio, so readers of the previous chapter will soon feel comfortable with it.

Authoring a DVD

In summary, creating a DVD involves these essential processes:

- Selecting and adding the programs that viewers will play.
- Designing menus that reference both the programs and the commands to set them up for viewing (audio track, titles, screen shape, etc.).
- Placing "buttons" on the menus that allow viewers to navigate to other menus, set program parameters, and play programs.
- Linking those buttons to the screens, programs, or features that they are designed to access.

But before any of these processes can be undertaken, each new DVD must be set up.

Configuring a DVD Project

The opening screen in DVD Workshop resembles **Figure 23-5**.

Figure 23-5.
The opening screen of DVD Workshop. (Ulead)

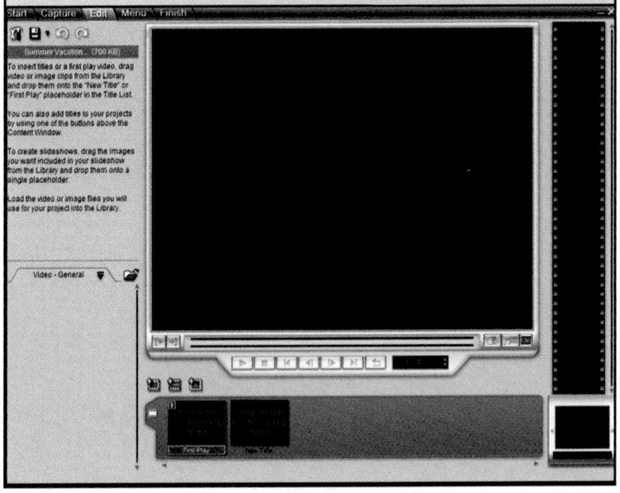

A DVD being authored is called a ***project***, since "program" (or "title") refers to an individual video and there may be several on one disc. To begin a project, select **New Project**, which brings up the new project box, **Figure 23-6**.

Figure 23-6.
The new project box. (Ulead)

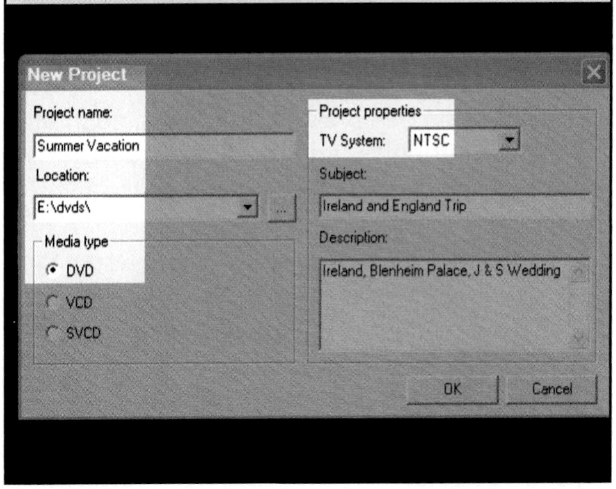

Like editing projects, DVD projects have "project files," which are blueprints for the finished product.

Project Information

Four pieces of information are essential. The *project name* and *location* enable your computer to store the project file.

The *media type* specifies which of several types of discs will contain the finished program and the *TV system* sets the type of TV.

The remaining fields are optional, for your convenience. Under *description*, it often helps to list the names of the programs that will be on the DVD.

A Demonstration DVD Project

The wedding video introduced in the previous chapter was the third of three programs made from footage shot during a summer vacation in Ireland and England. The first was a travelogue on Ireland and the second was a tour of an English palace.

These three programs were brought together to create a single DVD about the vacation. This DVD provides all the examples used in this chapter.

Ireland ...

Blenheim...

The wedding. (Sue Stinson).

Changing Information

It is important to take care in supplying information for a new project. The project name and location can be changed if needed, but it is a good idea to establish a directory for it and keep all components there. Some materials, such as still photos, graphics, and music selections, may come from other sources. By copying them into the project's directory and importing them from there into your DVD program, you can keep all the working elements together.

The media type and TV system cannot be changed. Projects encoded for DVD cannot be burned to other disc types, and vice versa. The TV system can be an issue if the finished program(s) will be shown outside your area. In that case, you will need to select New Project, set TV System to the alternate type, and rebuild the DVD for the different area. (Of course, the programs themselves must match the TV system selected for the DVD. You cannot, for example, place NTSC format programs in a PAL DVD project and expect them to play.)

DVD Workshop defaults to DVD disc format and NTSC system type, the most common settings for North America.

Figure 23-7.
Three programs have been imported from the computer hard drive to the options panel.

The programs are added to the title list panel in the desired order. (Ulead)

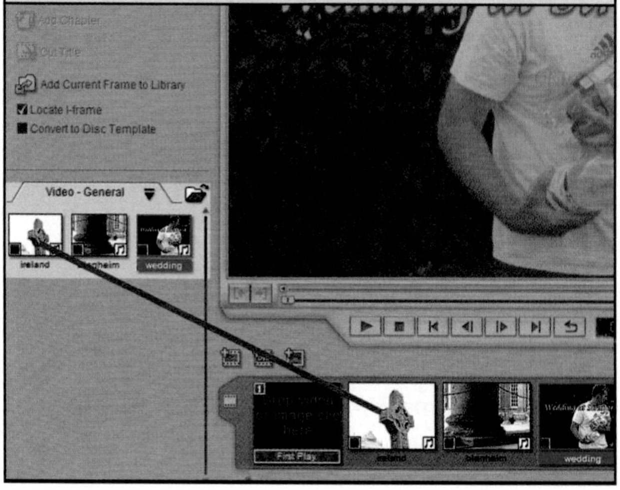

Placing Programs

The first step in authoring a DVD is to select the programs that will be included on it, and the order in which they will be recorded.

Program order is important. Although viewers can play programs in any order, the PLAY or PLAY ALL command buttons show the programs in the order recorded on the DVD.

Ulead DVD Workshop and other authoring applications use the word "title" instead of "program." "Program" is preferred in this discussion to prevent confusion with the textual *titles* that are also components of a DVD.

Selecting Programs

Programs can be captured from camera footage or imported from computer files or DVDs. Here, three programs previously created in Ulead VideoStudio and filed on the computer, are imported to the options panel and then placed in order on the title list, **Figure 23-7**. The title list closely resembles the storyboard in the editing program and functions very similarly.

Creating a "First Play" Clip

Note the empty square on the DVD storyboard labeled ***First Play.*** The program you place here will begin playing as soon as the DVD is started. Here, a short, moody segment on an Irish moor introduces the DVD, **Figure 23-8**. A still from this shot will also form the background for several menus to be constructed later.

Figure 23-8.
Moor shot. The Irish segment will play first. (Ulead)

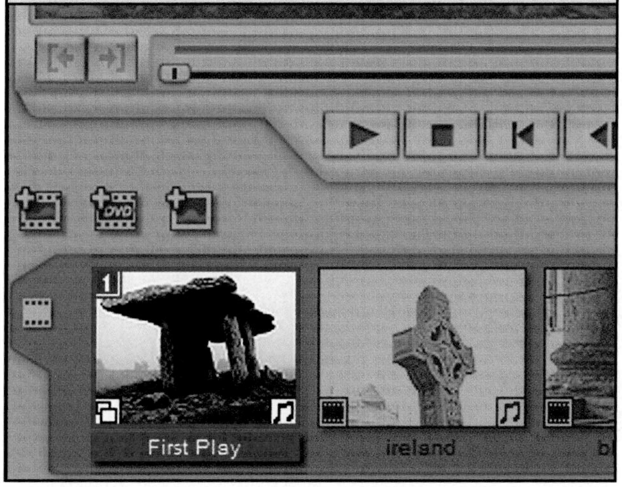

Figure 23-10.
The force first play option. (Ulead)

If you like, you can add a musical background to the opening, either by replacing an existing track or by adding a new track from a music library. As you can see from **Figure 23-9**, you can fine tune the music by adjusting its length to fit the video (or altering the video to fit the length of the music), adjusting its volume level, and fading it in and/or out.

In some cases, you may want to prevent viewers from skipping over this opening segment. In that case, checking the "force first play" box will ensure this, **Figure 23-10**.

Movie DVDs sometimes use this technique to make you watch their previews of other movies.

Figure 23-9.
You can create a custom music (or other audio) track for your opening. (Ulead)

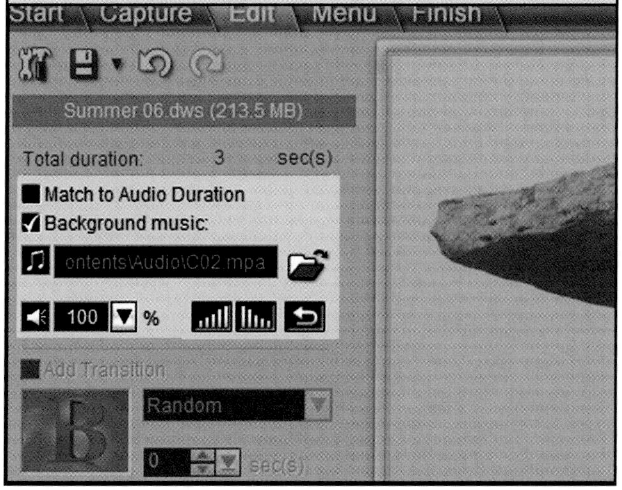

Adding Features

So far, organizing a DVD has not been much different from planning a release videotape. Now, however, you begin to add the selectable options and random access qualities that make the DVD so versatile. The first step is to divide each program into chapters.

Creating Chapters

A ***chapter*** is any piece in a program that can be selected directly from a menu. It begins with a frame of video designated as the chapter start and ends, by default, with the start frame of the following chapter.

Later, in the menu creation process, chapters are named so that viewers know what they contain, but at this stage it is enough to designate them.

Designating chapters

There are several ways to mark the first frames of new chapters. Perhaps the easiest is to ***scrub*** through the program until you reach the first frame of a new sequence (or any other division point you prefer) and freeze the program on that frame. Since the process has several steps, refer to **Figure 23-11** as you read.

First, check the "Locate I-frame" box (1). Since the chapter identifier is an image from the program, a frame other than an I-frame would give you an incomplete image (as explained in the sidebar *Digital Formats: DV vs. MPEG*).

Figure 23-11.
An overview of the screen, showing the steps required in chapter creation. 1—Check the "Locate I-frame" box. 2—Find the first frame of the new chapter. 3—Click "add chapter." 4—A thumbnail identifying the new chapter appears in the chapter list. (Sue Stinson/Ulead).

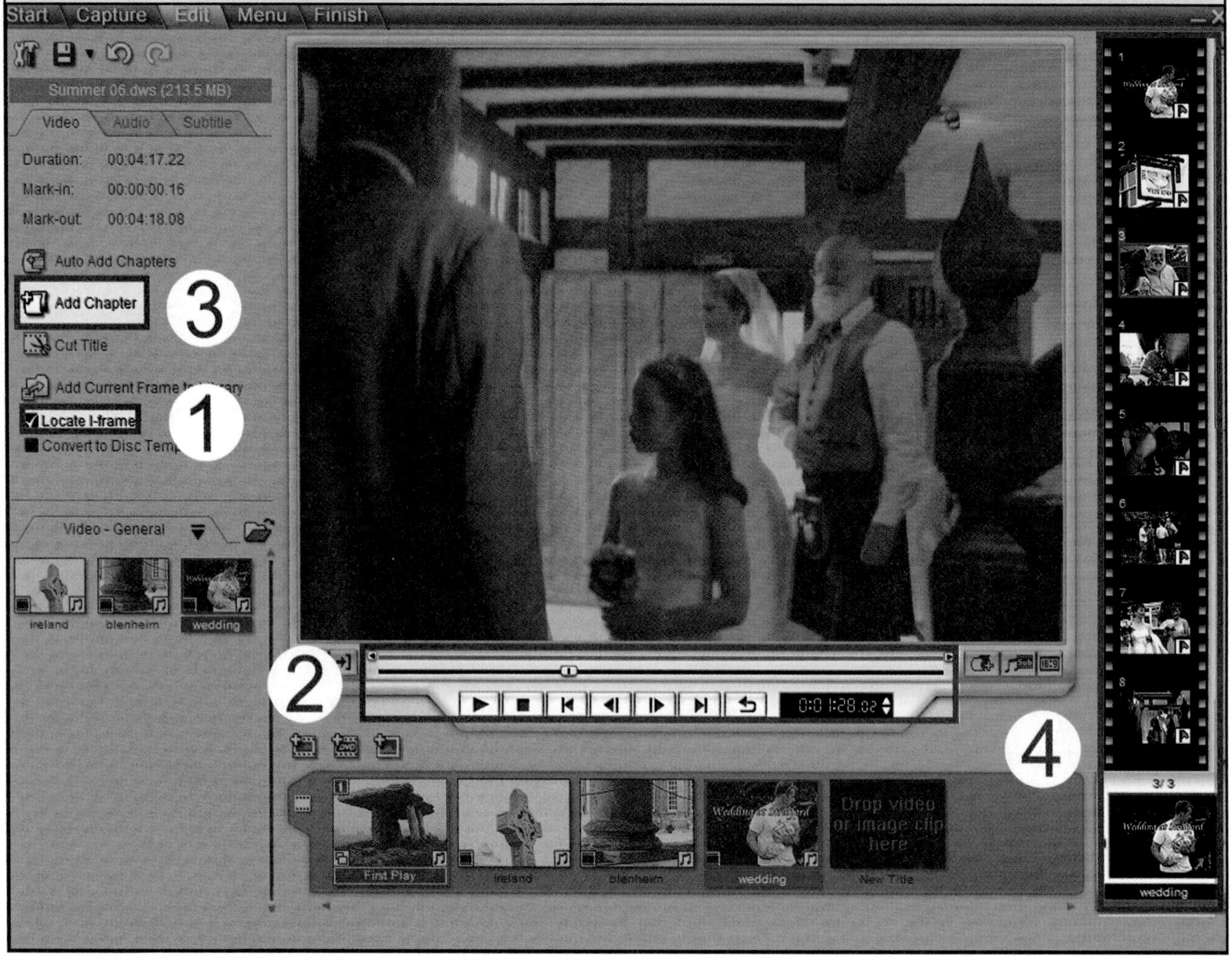

Next, use the slider to find the first frame of the new chapter (2).

Then click "Add Chapter" (3). The thumbnail identifying the new chapter will appear in the chapter list (4).

Changing chapter thumbnails

Sometimes, the first frame of a chapter does not represent the chapter content adequately, or perhaps is it is simply not a high-quality image. For example, the first frame of the sequence, "Dressing for the Ceremony," **Figure 23-12**, is not a very good image. To change it:

- Scrub through the sequence to find a better image and pause there.
- Right-click on the chapter thumbnail to open a menu. Select "Set Chapter Thumbnail."

The thumbnail will change to match the image on the preview screen.

It is important to understand that the chapter will still begin on the first frame of the sequence, as originally selected, even though its identifying thumbnail has been changed.

Adding Alternate Audio Tracks

It is common practice to supply DVDs with alternate sound tracks selectable by viewers. These are often foreign languages or director comment tracks. These tracks are added from computer files, **Figure 23-13**.

Later in the authoring process, you will provide viewers with information about these sound tracks and the means to select and enable or disable any

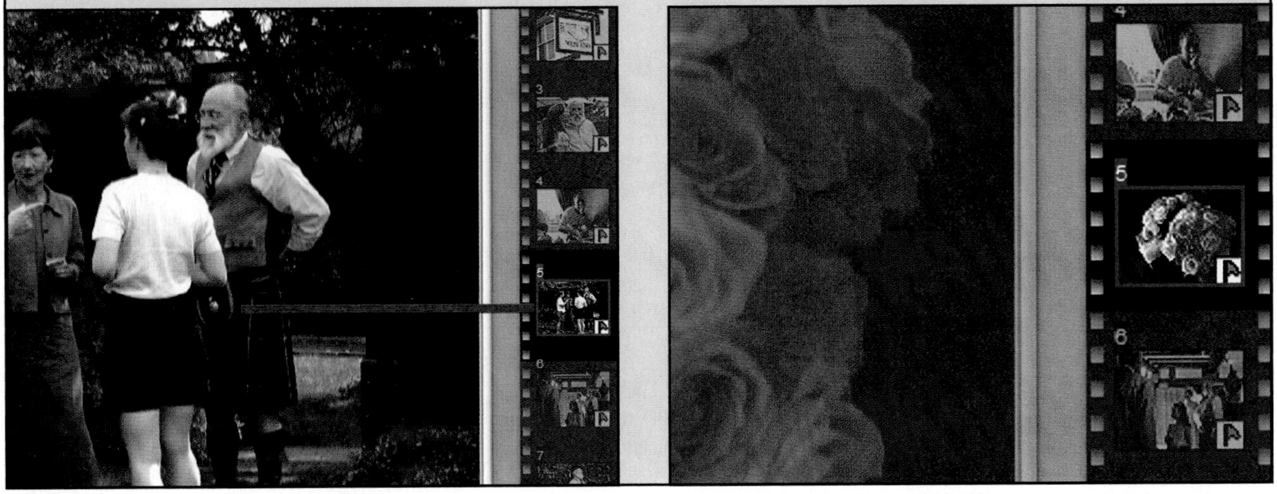

Figure 23-12.
The first frame of the chapter is not particularly striking. Another frame from the sequence makes a better chapter thumbnail. (Sue Stinson/Ulead)

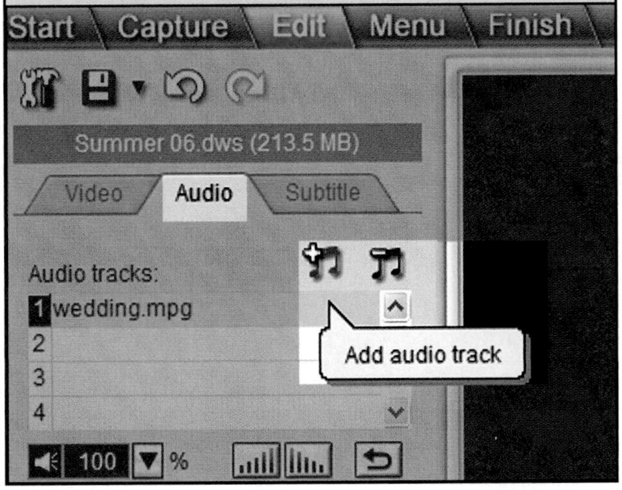

Figure 23-13.
Adding or deleting an audio track. Selectors are note icons with + or - signs. (Ulead)

Figure 23-14.
The language settings box is accessed from the options panel. (Ulead)

one of them. Since the software does not know whether a particular track is a director's commentary or a Greek translation, the next task is to provide this information**, Figure 23-14**.

Adding Titles

You may also want to add one or more sets of subtitles. These are usually foreign language translations or captions for the hearing-impaired. The process is essentially the same as adding audio tracks. After selecting the subtitle tab in the options box, you select "Add/Edit Subtitles" to open the subtitle work screen, **Figure 23-15**.

Adding subtitles can be a lengthy process. Sometimes it is better to write them all as a text file and then import them piece-by-piece to the appropriate place in the program. In **Figure 23-16**, the ***title*** has been typed directly into the subtitle field on the work screen.

Note that the options on the right side of the work screen provide complete control over subtitle formatting.

As with audio tracks, the subtitle tracks must be identified. Selecting "Specify language settings" opens a work screen for doing this, **Figure 23-17**.

Figure 23-15.
The add/edit subtitles command. (Ulead)

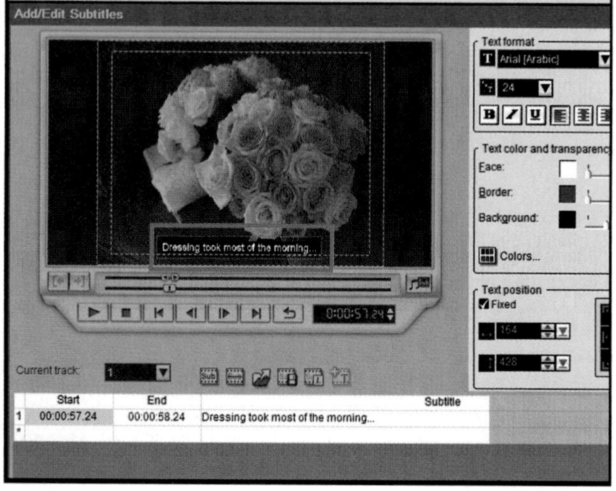

Figure 23-16.
The subtitle work screen. (Ulead)

Altogether

At this point, all the major content components of the DVD are in place:

- The opening.
- The programs.
- The program chapter divisions.
- The audio tracks.
- The subtitles.

But as yet, there is no way for the viewer to navigate among these features and select the components desired for viewing the programs.

Creating *navigation tools* is the next step in authoring the DVD.

Figure 23-17.
The specify language settings selection is an icon that must be learned. (Ulead)

Creating Menus

Navigation is accomplished with menus. Organizing and linking these menus can be a simple matter of following on-screen instructions, or a complex process for a custom design. (See the sidebar *A DVD Menu Tree* for an example of a complex menu structure.)

Templates and Wizards

If you are using a ***template*** supplied with your DVD authoring software, creating a navigation system can be quite simple. The DVD menu in **Figure 23-18** was completed in less than ten minutes by following on-screen prompts provided by a menu-creation ***wizard***.

It was customized by replacing the template text with the program titles and dragging each program icon into a button on the menu.

The small icons on the left allow viewers to play the previous program, the next program, or all programs in order.

Designing the Menu

Templates hide the organization of the DVD behind pre-structured menus.

They can also make features like alternate audio tracks and subtitles difficult to access.

A DVD Menu Tree

Here is a menu outline for the DVD of a typical feature film.

- **FBI Warning**
- **Studio Logo**
- **Preview One**
- **Preview Two**
- **Main Menu**
  - **Play Movie**
  - **Chapters**
    - **1-12**
      - **1 2 3...**
    - **13-24**
      - **13 14 15...**
    - **25-36**
      - **25 26 27...**
  - **Setup**
    - **Screen**
      - **Full screen**
      - **Letterbox**
    - **Sound**
      - **Dolby 5.1**
      - **Stereo**
    - **Language**
      - **English**
      - **French**
    - **Titles**
      - **English**
      - **French**
      - **Off**
  - **Extras**
    - **Commentaries**
      - **The director**
      - **The two stars**
    - **Documentaries**
      - **"The Making of..."**
      - **"Special Effects"**
    - **Trailers**
      - **Teaser**
      - **Theatrical trailer**
    - **Music Video**
    - **Deleted Scenes**
      - **Scene 1**
      - **Scene 2**
      - **Scene 3**
    - **Cast Biographies**
      - **Star 1**
      - **Star 2**
      - **Star 3**
      - **Director**

In addition, each screen must have buttons allowing the viewer to move back up the tree to higher level menus.

Figure 23-18.
A blank menu template.

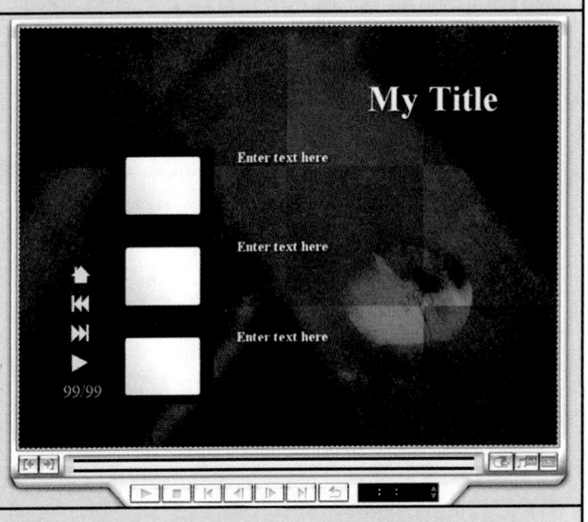

The demonstration vacation DVD menu created from the template. (Ulead)

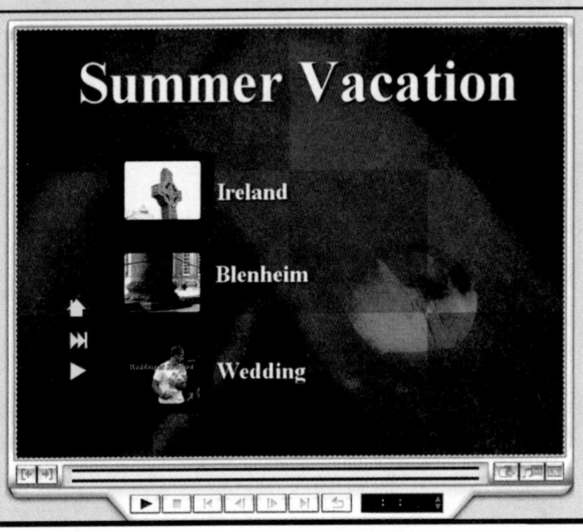

If the DVD is at all complex, it will need custom menus, all designed from scratch and linked in a hierarchy.

The Hierarchy

The components of a DVD are organized in a hierarchy much like the table of organization in a business. If the organization of the vacation DVD were expressed that way, it would resemble **Figure 23-19**.

However, the structure of a DVD is created as a hierarchy of linked menus; so the vacation DVD organization actually looks like **Figure 23-20**.

For clarity, Figure 23-20 includes only the wedding "tree" from top to bottom. Figure 23-21 shows them in more detail.

Figure 23-19.
A typical table of organization.

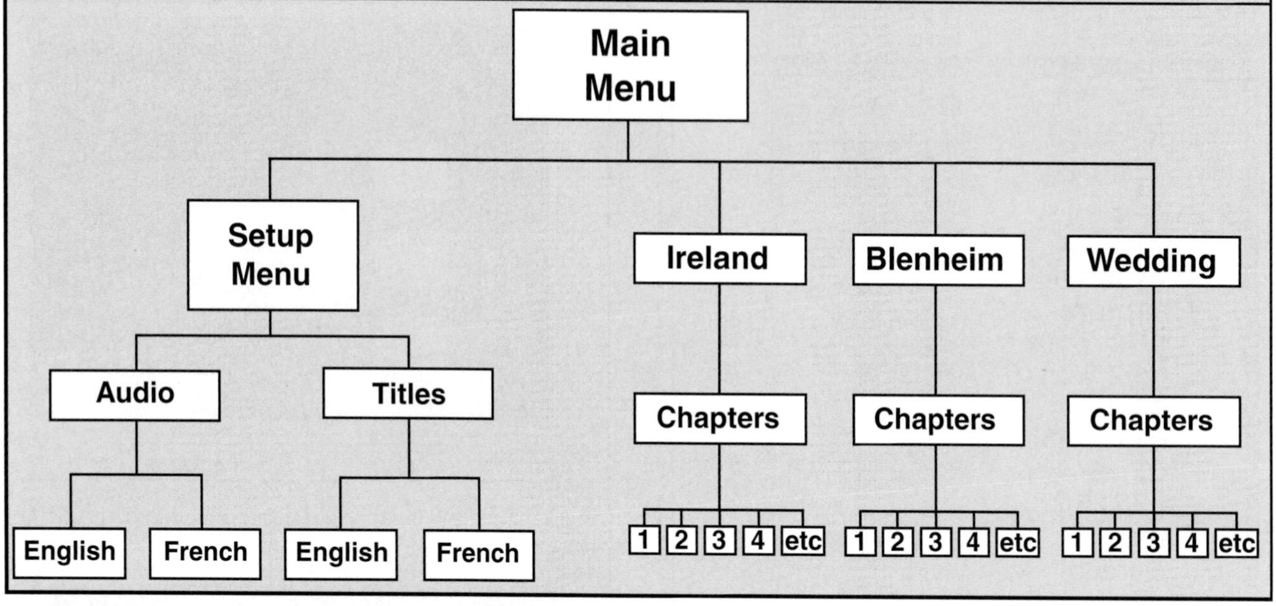

Figure 23-20.
The vacation menu structure.

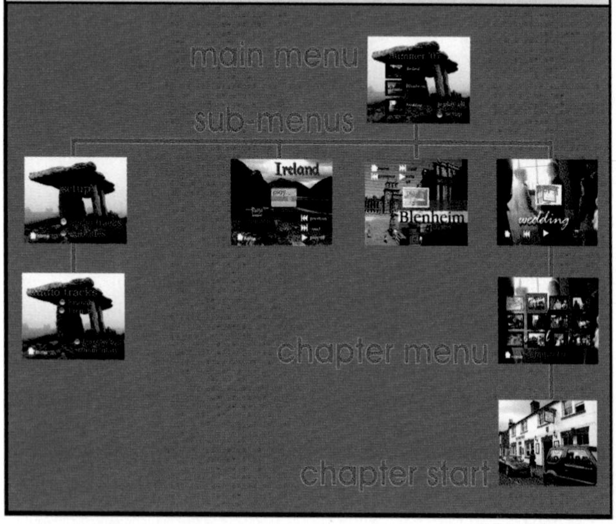

The links

As shown in Figure 23-21B, the wedding menu (like the others) has certain features:

- *Play*, to start the wedding program.
- *Home*, to return to the main menu.
- *Play previous*, to start the previous program on Blenheim palace.
- *Play all*, to start with the Ireland program and play all three programs.
- *Chapters*, to access the chapters menu.

Unlike other menus, this one lacks a *play next* button because the wedding is the final program on the disc.

Each of these graphic elements is called a ***button*** (whether button-shaped or not) and each button activates a *playlist*. Other, purely decorative graphic elements, are called *objects*.

Menu Components

Objects, buttons, and playlists are the building blocks of DVD menus.

Objects

Objects are images (still photos or graphics), frames copied from video, or text (whether program titles or menu commands). Objects are inert—that is, clicking on them has no effect.

A special type of object is called a ***placeholder.*** A placeholder is a spot on a menu template that will become an object when you drag a suitable element onto it.

Although placeholders are not used in original menus, they are covered below, for completeness.

Buttons

Buttons are graphic elements that are linked to programs, commands, or other menus, **Figure 23-22**. They are like switches: when activated, they "turn on" the items they are linked to.

Buttons can take almost any form, from icons imported from a library, to custom graphics, to stills from videos—or even tiny screens on which the linked videos actually play.

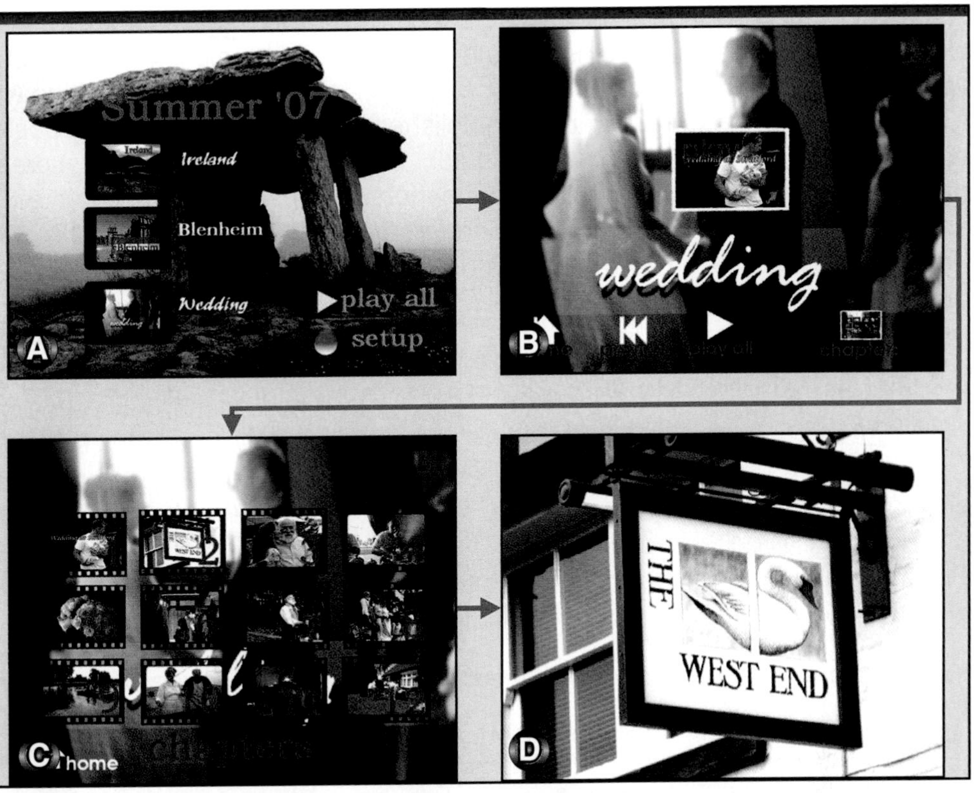

Figure 23-21.
Linked menus. A—The main menu. B—The wedding video menu. C—The wedding chapter menu. D—The start of a chapter. (Sue Stinson)

Figure 23-22.
All the items except the background image are buttons. (Ulead)

Figure 23-23.
The Ireland button is linked to the Ireland menu through the playlist. (Ulead)

Playlists

Playlists, as their name reveals, are lists containing one or more names identifying programs, menus, or commands in the DVD. Each button has its own playlist. When a viewer activates the object, the item(s) on its list will play.

To see how this works, study **Figure 23-23**. "Ireland" is a button on the main menu. Its playlist contains one item: "MENU-01." As you can see from the inset detail, MENU-01 is the menu screen for the program on Ireland. So, activating the Ireland button causes the link in its playlist to open the Ireland program menu.

If a playlist contains more than one item, each will play in order.

Building the Menus

Menus can be extremely sophisticated, with audio (usually music) that loops as long as the menu is on-screen. The buttons can be small screens in which the videos they link to are actually playing. Objects and buttons can move in a variety of CG effects.

Each DVD authoring program accomplishes these tasks in slightly different ways, so our demonstration DVD project will stick to the more generic operations. At this point, the menu tree has been designed, and now each menu will be built, starting with the main one.

In actually building menus, you will probably go through the sequence of placeholder-object-playlist-text for each item.

Background

First of all, we need a background. The "First Play" video includes a moody shot of a prehistoric tomb, so we capture a still from it as an object and scale it to fill the frame, **Figure 23-24**.

If we were working with a template, placeholders would be in position to receive objects. Though they are not present in our custom design, we will add them for demonstration purposes, **Figure 23-25**.

Buttons

Now we are ready to place the buttons that will link to the individual program menus. (Assume that these menus have already been created and so are available to be linked.) See **Figure 23-26**.

We still need buttons for two commands: *play all* and *setup*. *Play all* is a standard button, automatically linked to the command to play all the programs on the DVD in order. *Setup* is not a

Figure 23-24.
The background for the main menu. (Sue Stinson)

Figure 23-25.
Placeholders added to the main menu screen.

Figure 23-26.
Buttons added.

Figure 23-27.
Utility buttons added.

standard button, so we pick a button shape, open its playlist, and link it to the setup menu. (The buttons are in the button library. Two selections from that library are shown in **Figure 23-27.**)

Text

So far, *you* know what the buttons activate but viewers do not (the titles on the program buttons are too small for easy reading). So the last step is to add ***text***, using formatting commands very similar to those in most graphics and word processing programs, **Figure 23-28**.

Checking Your Work

You will repeat this process for every other menu in the project. Before leaving each one, you should verify two important things:

- that menu components–especially buttons –do not overlap, and
- that every button has an appropriate link in its playlist.

To do this, click the buttons highlighted in **Figure 23-29**. Green borders will appear to show the boundaries of objects and buttons, and each button will display the item in its playlist.

Figure 23-28.
Adding text completes the menu.

Completing the Project

With the menu system designed and the individual menus created, only two steps remain: previewing the DVD and preserving it permanently by archiving it.

Figure 23-29.
The controls for the borders and links are below the screen. (Ulead)

Previewing

Because DVD menu systems can be complex, it is prudent to preview your project before locking it up. Most authoring software includes a "virtual DVD player" that simulates all the functions of an actual player. To check your work, make sure that every button is linked to the appropriate element (program, menu, or command). Play every selection on the DVD, looking for small problems with audio tracks, graphic elements, and the like, **Figure 23-30**.

Archiving

There are three ways to store your DVD project permanently. Most authoring programs offer all three methods.

DVD burning

The first is to simply burn a disc, that is, create an actual DVD of your program. This option is fine if you plan to keep the project on your computer hard drive, or else make a few discs and discard the original materials. Though you can load the DVD files back onto your computer, you can only play them as you would in a DVD player. You can no longer work with them.

Creating a disc image file

A ***disc image file*** is a file that contains all the materials used in your DVD project. With it, you can reopen the project for additions or revisions and prepare a new DVD. Since you can

Figure 23-30.
The preview screen has a virtual DVD remote control. (Ulead)

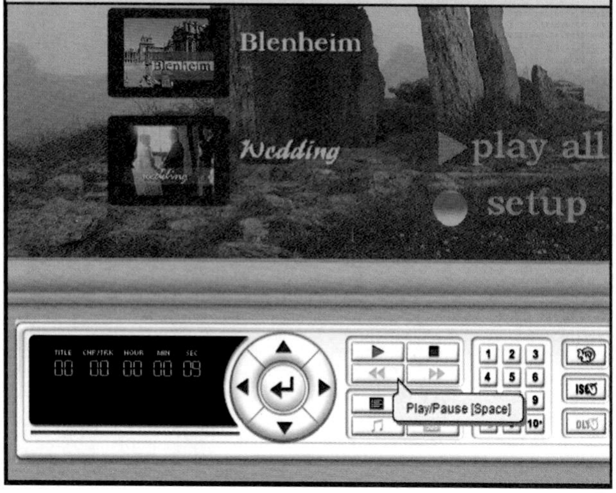

record a disc image file on a DVD as a data file, you can upload it whenever you wish to use it. That way, you can clear your drive of the very large files generated by DVD projects.

Creating a digital linear tape

Digital linear tape ("DLT") is the standard format for supplying DVDs to commercial duplicators for mass production. Although you probably will not have a DLT machine, you can instruct your program to store the DVD in DLT format, which can then be copied to disc as a data file and delivered to a facility with DLT capabilities.

Change Is the Only Constant

It is wise to archive your DVD in a format that can be reopened and worked on in the future, because recording and playback systems are always changing. Today's popular format may be unplayable in a few years, like the once-universal 78 rpm shellac audio discs (they were called "records"), **Figure 23-31**.

As long as your software can read your working files (and your operating system can run your software) you will be able to reformat your DVD for whatever playback medium comes along.

Figure 23-31.
The 78 rpm record was once the standard audio medium.

Chapter Review

Answer the following questions on a separate piece of paper. Do not write in this book.

1. To discourage piracy, DVDs are often equipped with two types of security measures: ______ and copy protection.
2. *True or False?* To prevent viewers from skipping over the opening segment, the "force first play" option is selected when creating the DVD.
3. To permit the viewer to navigate the DVD easily, you must create ______.
4. Buttons are graphic elements that are linked to programs—when activated, they ______ the items they are linked to.
5. Your project can be archived by burning a CD, creating a ______ file, or creating a DLT (digital linear tape).

Technical Terms

Authoring: The process of designing and assembling DVDs.

Button: An element on a DVD menu that, when activated, links to a program, a menu, or a function.

Chapter: A designated section of one program on a DVD.

Copy protection: A system of encoding designed to prevent copying DVDs.

Disc image file: A computer file containing all the components of a DVD project.

Digital linear tape (DLT): A DVD storage file format for recording on a special half-inch tape that will be the master for mass-producing DVDs.

DVD: A computer disc format originally designed to hold movies, but now used for all-purpose data storage, too.

First play: A menu or program that plays automatically when a viewer starts the DVD.

I-frames: "Intra-frames" that contain all the image data. (See *MPEG*).

Intra-frame compression: A digital video recording system that saves storage space by condensing and recording the data for every frame.

Menus: In DVD authoring, screens displaying buttons for navigating to other menus, programs, or functions.

Mini-DV: The most popular recording format for small camcorders. Uses intra-frame compression.

MPEG: A recording format that records full information only for selected frames (called "I-frames"), plus partial information for other frames.

NTSC: The TV standard used in North America and Japan. Other major standards include PAL and SECAM.

Object: An item on a menu that does not link to anything.

Placeholder: On a menu template, an indicator of where an item should be placed to make it a button.

Playlist: A list, for each button, containing the names of one or more menus, programs, or functions activated by that button.

Project: The process of authoring a DVD.

Regional encoding: Code on a DVD that permits it to be played in only one designated world region.

Scrub: To move quickly through a program by dragging a handle along a track below a preview window.

Template: A pre-designed menu that is customized by substituting appropriate buttons and text.

Text: Anything written on a DVD menu.

Title: In DVD authoring, any program or menu. (A common term, but not used in this sense in this book.)

TV standard: A broadcast system, such as NTSC, SECAM, or any of several varieties of PAL.

Wizards: Mini-programs that guide users through the menu creation process step by step.

As a chapter thumbnail, this image may represent sequences captured at the end of the wedding day. (Sue Stinson)

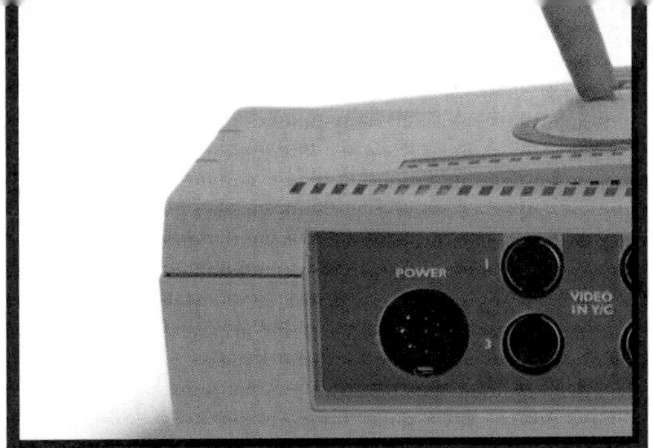

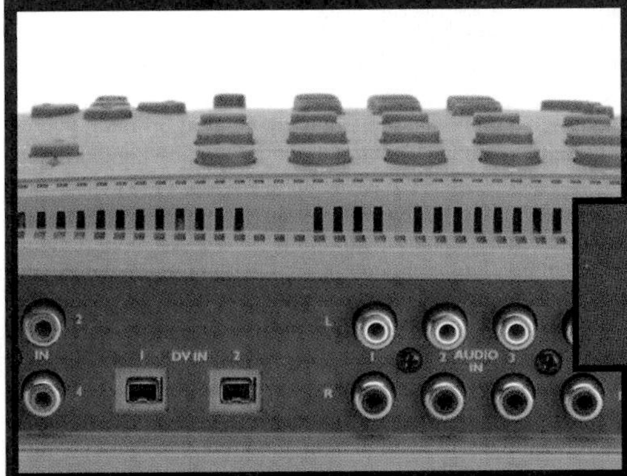

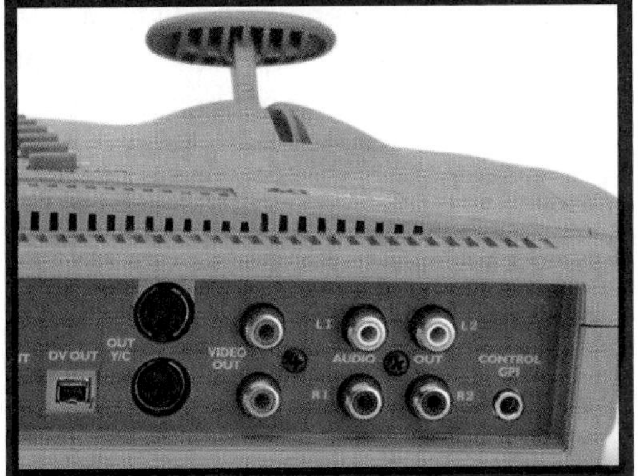

24

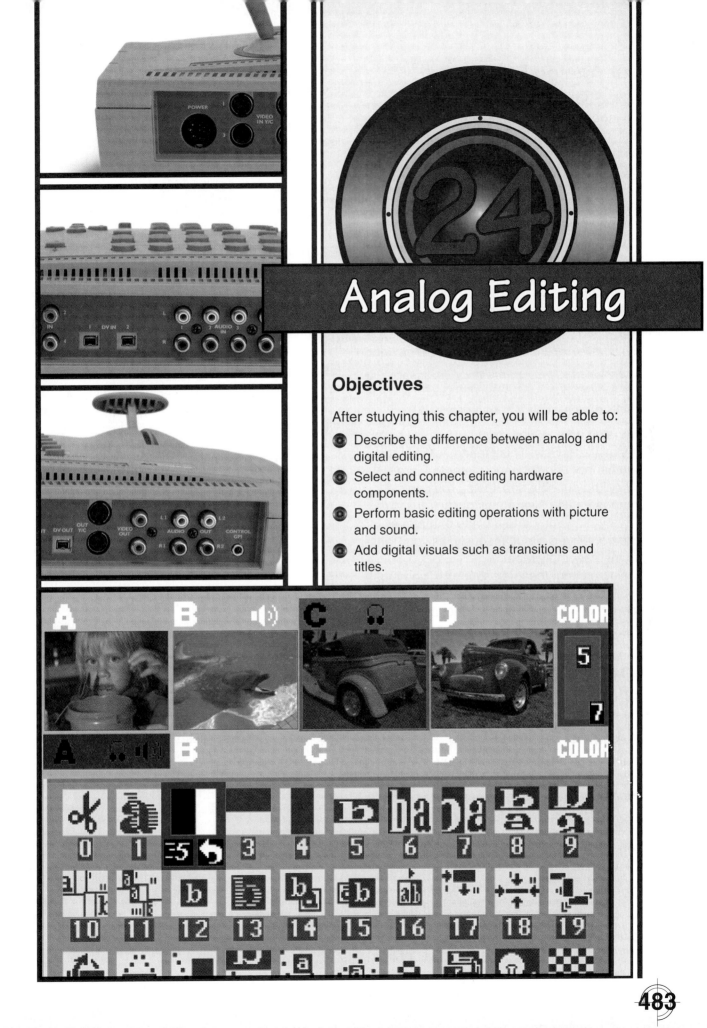

Analog Editing

Objectives

After studying this chapter, you will be able to:

- Describe the difference between analog and digital editing.
- Select and connect editing hardware components.
- Perform basic editing operations with picture and sound.
- Add digital visuals such as transitions and titles.

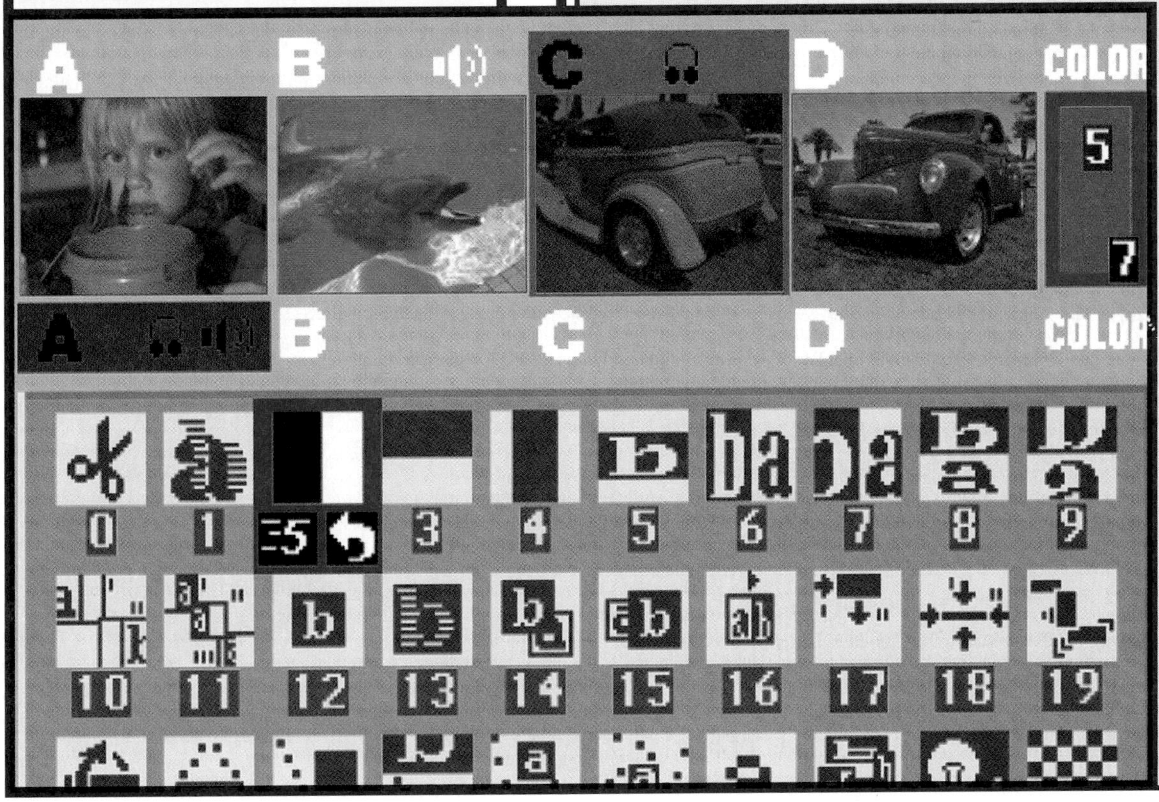

About Analog Editing

Analog editing is the name given to traditional tape-based video postproduction, to contrast it with digital editing, which uses materials stored in a computer. Tape-based editing is *linear* because you cannot jump instantly to any point in a videotape. To reach a desired spot from any starting place, you must roll through every shot between those two points. On complex projects with multiple source tapes, finding raw footage can be time-consuming.

Like the source tapes, the assembly tape containing the project being edited is also linear. To create it, you must begin by recording the opening shot in the program and then record all the rest of the shots in chronological order.

Two exceptions to this rule are the video insert and audio dub functions, which can add new material in the middle of previously edited footage.

In analog editing, you must perform all operations simultaneously. This means, for example, that you cannot add a dissolve later to a previously edited cut between shots. You cannot superimpose a title over a previously transferred shot. Everything you do to a piece of material must be performed simultaneously, as you transfer it to the assembly tape.

Analog editing, then, is *sequential*: you must assemble each shot in strict chronological order, and it is *concurrent*: you must create all the parts of the finished shot at the same time.

You can, however, premix certain kinds of audio tracks before feeding the result to the assembly track.

Why Analog Editing?

In almost all postproduction today, computer-based digital editing has replaced the traditional tape-based analog approach because of digital editing's speed, versatility, and output quality. For simpler projects and casual video shooters, however, tape-based editing still has certain advantages.

For one thing, it requires no special equipment. Although you could buy elaborate analog hardware, you can perform simple editing tasks very effectively with just a camcorder and a consumer VCR. This makes it cost-effective for the person who edits only occasionally. Also, analog editing is relatively simple, so you can master the basics in half an hour. Finally, analog editing lets you experiment with postproduction before deciding whether or not to invest money and learning time in a full-fledged digital system.

An Analog Editing Setup

Before you can edit videos, you must assemble and connect the components of an editing system. We will start with a very simple setup and then explain how additional pieces can improve convenience and enlarge your editing capabilities.

Essential Editing Components

The most basic editing setup consists of three pieces: a camcorder, a VCR, and a TV set, **Figure 24-1**. The camcorder functions as the source deck, in which you play back the original footage. The VCR is the record deck on which you assemble your edited program. You watch what you are doing on the TV set.

The VCR on which you are building your finished program is usually called the assembly deck.

Adding an Editing VCR

If you decide to add dedicated editing equipment to a basic setup, the first investment should probably be an editing-quality VCR. Edit quality

Figure 24-1.
A simple editing setup.

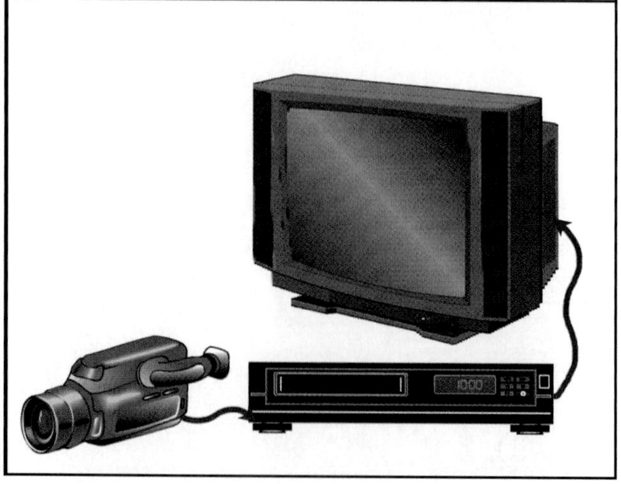

VCRs have flying erase heads, which allow perfect cuts from shot-to-shot without the rainbow-like patterns that sometimes mar edits on simple VCRs. Flying erase heads also permit you to replace previously transferred video, audio, or both with other material. This function is essential for more advanced editing, as we will see.

Adding an Editing Monitor

You can edit quite capably with just one monitor (or TV set), due to the way in which the assembly deck VCR displays footage. The signal sent to the TV set depends on the mode in which the assembly VCR is operating, **Figure 24-2**:

- STOP, RECORD, or RECORD/PAUSE modes display the signal passed through from the source VCR (or camcorder), so you can see your original footage.
- PLAY or PAUSE modes display the signal already recorded on the assembly tape.

Nevertheless, you may want to provide your source deck with a monitor of its own. When working with only one monitor, it can be hard to remember quickly whether the image on the screen is from the source deck or the assembly deck. For this reason, a second monitor connected directly to the source camera is a convenience, **Figure 24-3**.

Figure 24-3.
A source monitor always displays the source image.

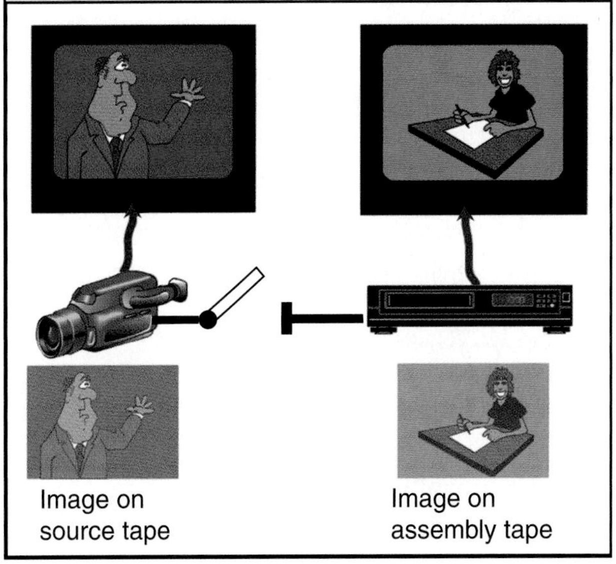

In summary then, here are the simplest, least expensive setups for analog editing:

- Camera to home VCR, with TV set.
- Camera to editing VCR, with TV set.
- Camera to editing VCR, with TV set and separate source video monitor.

Adding a Switcher/DVE Generator

A digital switcher, **Figure 24-4**, accepts and switches video and audio input from two VCRs,

Figure 24-2.
Image display with different VCR modes.

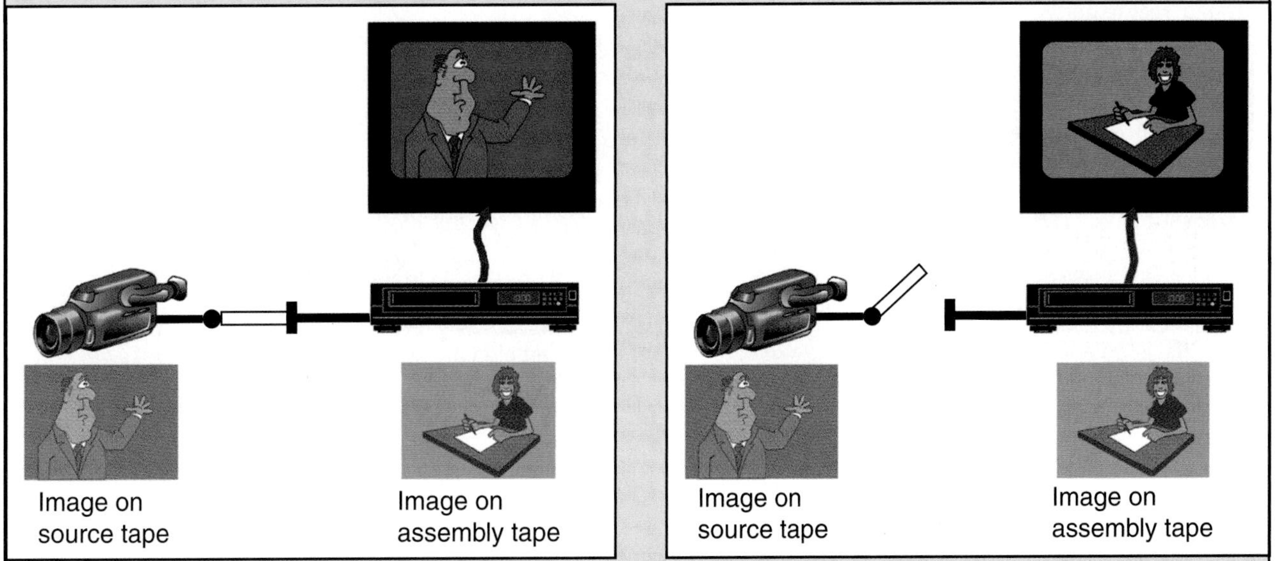

A—STOP, RECORD, or RECORD/PAUSE modes. B—PLAY or PAUSE modes.

Analog vs. Digital; Linear vs. Nonlinear

Analog vs. digital

The terms analog and digital refer to the way in which the video and audio signals are encoded.

Analog signals are recorded as subtle variations in the voltage of the electrical signal. Graphed, they resemble the accompanying diagram. Analog recording closely reproduces the continuous voltage fluctuations created by the imaging circuits and the microphone.

The varying voltage of an analog signal.

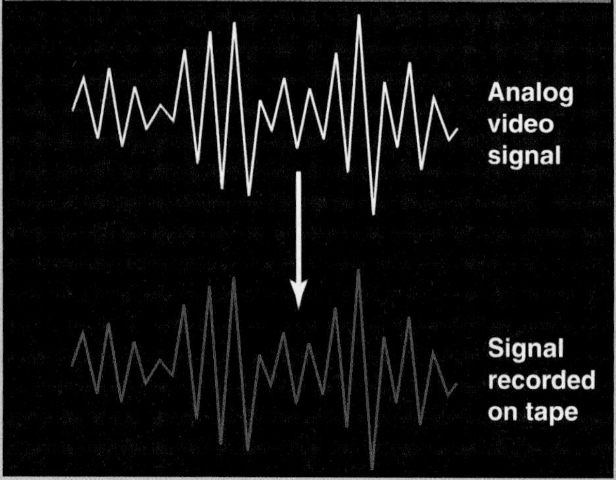

Digital recording does not copy the fluctuating voltage of the signal. Instead, it inspects (samples) that signal at intervals, notes its strength at each point, and expresses both values in the ones and zeros of binary notation. It is these graph coordinates that constitute digital recording, rather than the original signal. Because the samples are taken so frequently (many thousands of times each second), the resulting numerical values accurately represent the original voltage patterns.

In summary, an analog recording copies the original signal directly, while a digital recording represents it as a series of numbers instead. In general, digital recording is preferred to analog because the resulting files can be copied endlessly (like all computer files) with 100 percent accuracy. By contrast, copies of copies of analog recordings (like Xerographic copies) deteriorate in quality from one ***generation*** to the next.

Linear vs. nonlinear

The terms linear and nonlinear refer to the way in which you can gain access to your materials.

Linear materials are recorded on tape, so you cannot jump instantly to any part of the record. Instead, you must roll through the tape to the desired point. Linear recording systems include:

- Analog audio tape.
- Digital audiotape (DAT).
- Analog video tape (VHS, 8mm, etc.).
- Digital video tape (Mini-DV, Digital 8, etc.).

Nonlinear materials are recorded so that you can quickly access your material, wherever it is stored. Nonlinear recording systems include:

- Print materials (books, magazines, newspapers).
- Computer files (on hard drives, DVDs, and CD-ROMs).

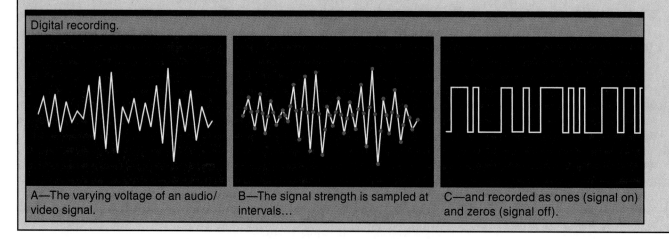

Digital recording.

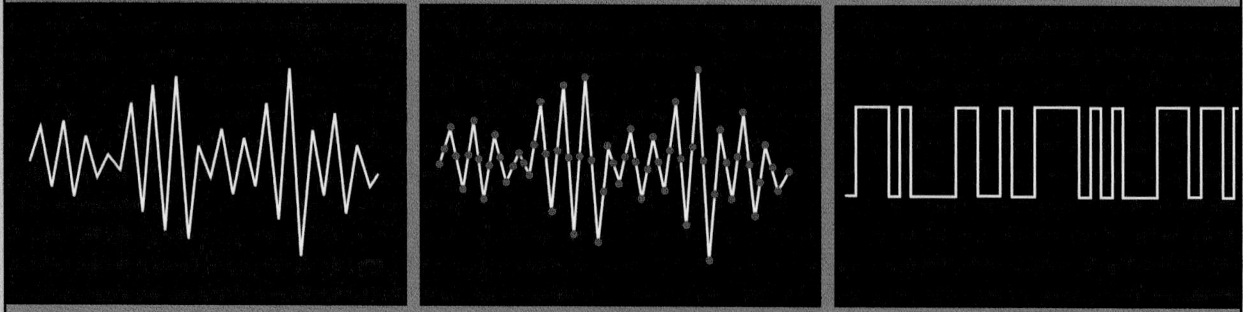

A—The varying voltage of an audio/video signal.

B—The signal strength is sampled at intervals...

C—and recorded as ones (signal on) and zeros (signal off).

- CD audio disks.
- Analog video disks (laser disks).
- Digital video disks (DVD).

Notice that both analog and digital recording systems exist in both linear and nonlinear formats:

	Analog recording	Digital recording
Linear access	VHS, 8mm	Mini-DV tape
Nonlinear access	Laser disk, vinyl LP	DVD, CD, hard drive

Noncomputerized vs. computerized

It is inaccurate to say that nonlinear video is computerized while linear video is not. Computers have long played important roles in traditional linear editing. Edit controllers and switcher/DVE generators contain small, specialized computers. In professional postproduction, computers have been used for decades to create and store edit decision lists (EDLs) to help automate final program assembly.

Temporary vs. permanent

Finally, it is not true that digital video is permanent, while traditional analog video decays and fades away. The fact is that any signal recorded on magnetic tape or disk will deteriorate over time, regardless of whether it is analog or digital.

Unlike analog recordings, however, digital recordings can be copied and recopied without appreciable quality loss (***generation loss***). In practical terms, this means that digital video is potentially permanent, as long as fresh copies are made at suitable intervals.

Optically scanned disks, whether analog or digital, are not subject to signal decay because they are not magnetic recording media. However, the disks themselves are not truly permanent. They are vulnerable to scratches and other physical injuries, as well as decay of the recording media.

and usually from additional sources like microphones and CD players.

A switcher is essential for making transitions like dissolves between shots, and for mixing audio sources. All digital switchers include digital video effects (DVEs) for transitions like wipes, flips, and fly-ins, and for special visual effects like mosaic, strobe, and compositing.

Figure 24-4.
A digital switcher. (Videonics)

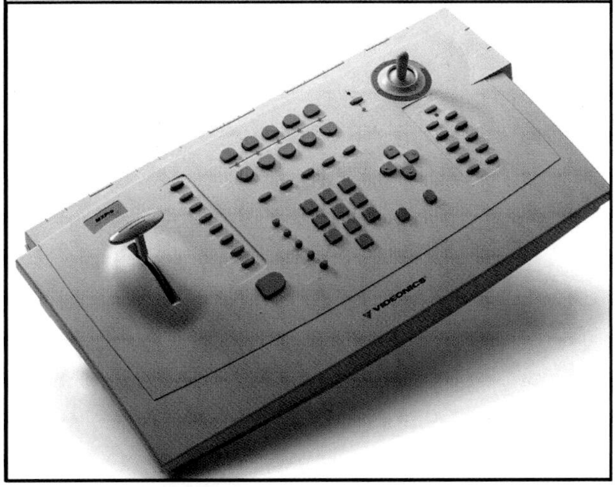

Setting up an Editing System

If your editing system is nothing more than your camcorder, home VCR, and TV set, there is practically no setup. Simply cable from the camcorder's video and audio outputs to the comparable inputs on your VCR, **Figure 24-5**. The VCR is already cabled to display audio and video on the TV set.

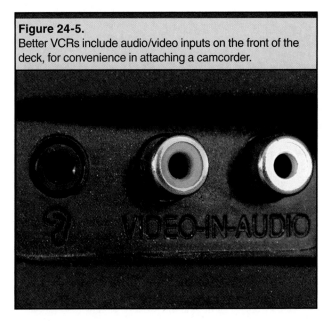

Figure 24-5.
Better VCRs include audio/video inputs on the front of the deck, for convenience in attaching a camcorder.

Flying Heads

Heads are tiny electromagnets that record, play, or erase video signals. ***Fixed heads*** remain still while the tape rolls past them. ***Flying heads*** are mounted on a rapidly spinning ("flying") drum. All video record heads revolve on this drum, but some audio record and video/audio erase heads are fixed.

The flying heads on a VCR drum.

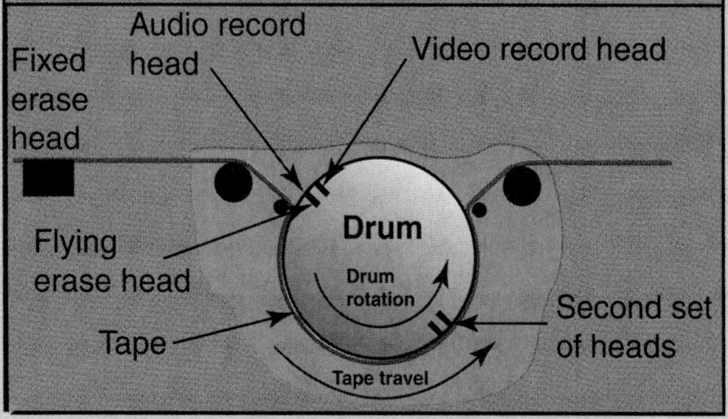

Because the tape moves past the spinning drum at an angle, the flying heads record the signal in a series of diagonal bars. This is called "helical scan" recording.

The diagonal stripes represent the video track.

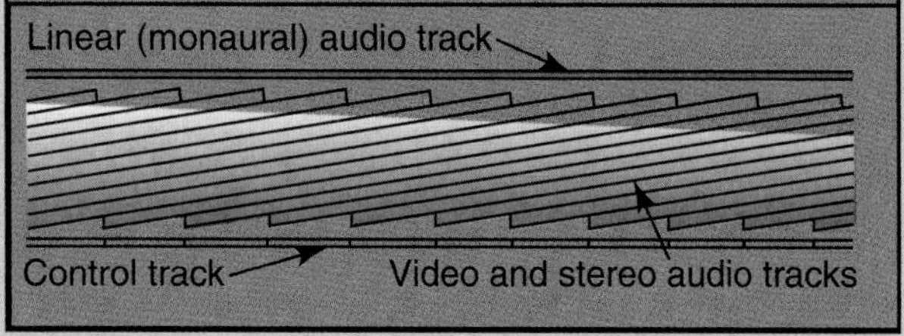

Fixed heads, however, record (or erase) in a continuous linear pattern.

Most home VCRs are equipped only with fixed erase heads. Every time you press the record button on your VCR, you also activate the fixed erase head, which cleans any previously recorded signal off the tape before it reaches the record heads.

Because the fixed erase head does not operate in a helical pattern, it always leaves partially erased track bars.

A fixed erase head destroys parts of several track passes.

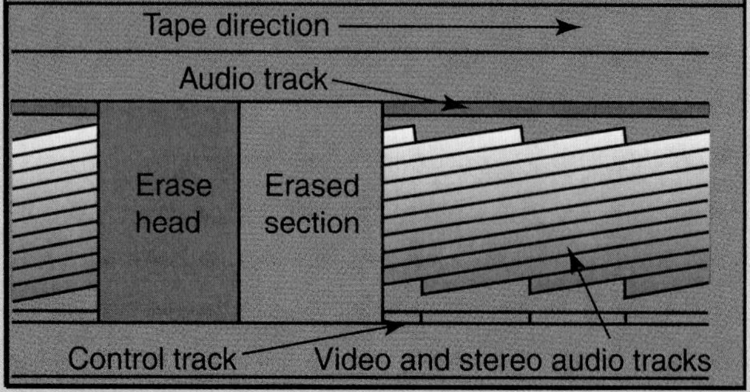

These partially erased signals create wavy lines and rainbow patterns between shots. Also, because the fixed erase head is in front of the record head, it erases material that is not replaced when the re-recording stops.

Flying erase heads solve this problem because they erase in the same helical pattern as the record heads. Editing decks are equipped with both fixed and flying erase heads. When you engage the VIDEO INSERT or AUDIO DUB controls on an editing deck, the flying erase heads operate instead of the fixed erase head, and you can drop replacement video and/or audio into the middle of previously recorded material.

Flying erase heads conform to the recording pattern.

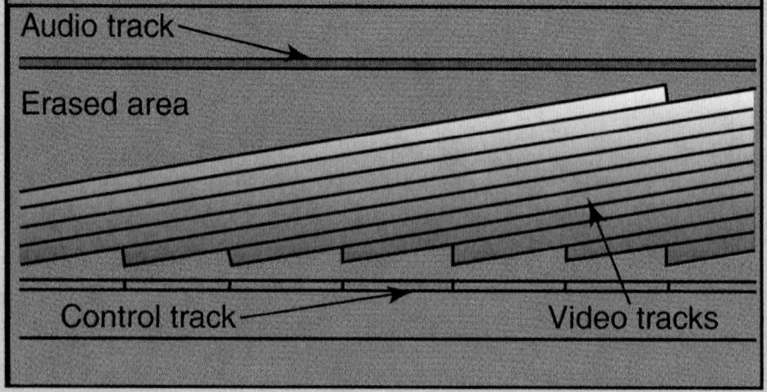

For all but the simplest projects, however, it will pay to create a setup specifically for editing. Doing this means arranging the hardware and cabling components together.

Arranging the hardware

Since people in most western cultures read and work from left to right, you may want to position your components this way, with the camcorder to the left and the assembly VCR to the right. If you are working with a TV set, it is probably not worth moving, but if you have small monitors, place each one directly on top of its VCR. If the VCR is a camcorder, position its monitor behind it instead.

Inexpensive Monitors

Monitors are television display units that lack channel tuners and RF (antenna lead) inputs. Typically, they produce higher-quality pictures than all but the most expensive TV sets, and they include sophisticated setup controls for adjusting their displays. All professional editing installations use monitors.

Small (13-inch) TV sets are often a good alternative to professional monitors because they typically cost far less. And though their picture is not studio quality, it is quite good enough for evaluating raw footage on the source VCR. Also, many home VCRs have only one set of composite (and sometimes (S-connector) outputs. On the other hand, all of them include RF outputs, so you can connect them to the RF input on a TV set without tying up any editing connections. (RF jacks and plugs are the standard connectors used with TV antenna or cable leads.)

RF connections allow you to connect a TV set to your source VCR or camcorder.

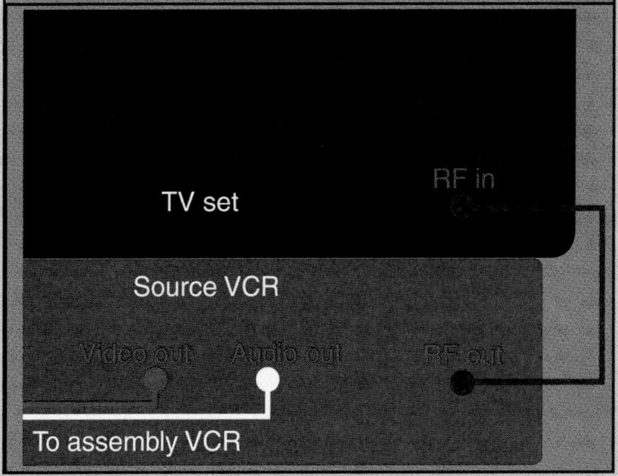

Recycling old hardware

Because of the RF connection, any surplus TV set, black and white or color, will work adequately as a source deck monitor. You can even use old computer monitors, if they have the "RGB" inputs that were common in the early generations of Apple and IBM desktop models. (RGB computer monitors lack RF inputs, however, so they require a second line-level output on the VCR.)

The key point is that a source monitor does not require quality high enough to evaluate image sharpness or color, because the same image is displayed on the assembly deck monitor.

Cabling the components

Linking equipment is relatively simple if you follow a few standard procedures:

- Always cable from a video or audio "out" plug on the source to an "in" plug on the destination. Remember that all analog plugs work in one direction only. The most common reasons for not transmitting picture or sound are failure to plug from "out" to "in" and inserting jacks in the wrong plugs.

In this discussion, a port on a piece of hardware is called a "jack" and a pin on the end of a cable is called a "plug."

- Always use Y/C (S-video) connectors, **Figure 24-6**, for video when available, for better picture quality.
- RCA plugs and jacks are color coded: yellow for composite video, white for left stereo audio, and red for right stereo audio. If you are connecting to a mono assembly deck, use the yellow jack for composite video and the white jack for audio (it carries both the left stereo and the monaural tracks).

Figure 24-6.
Most hardware offers both composite video jacks (yellow) and Y/C video, also called S-video. (Videonics)

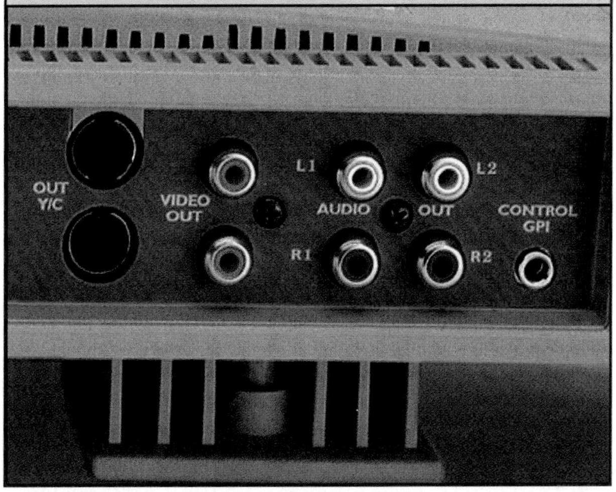

Color coding on hardware jacks is meaningful (for example, a yellow jack *always* means composite video) but color codes on cable plugs are purely for convenience. Any plug will work in any jack of the same type. To help trace connections, though, it is good procedure to match plug colors to those of the jack you insert them into. If you are

Figure 24-7.
Connecting hardware properly.

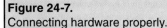

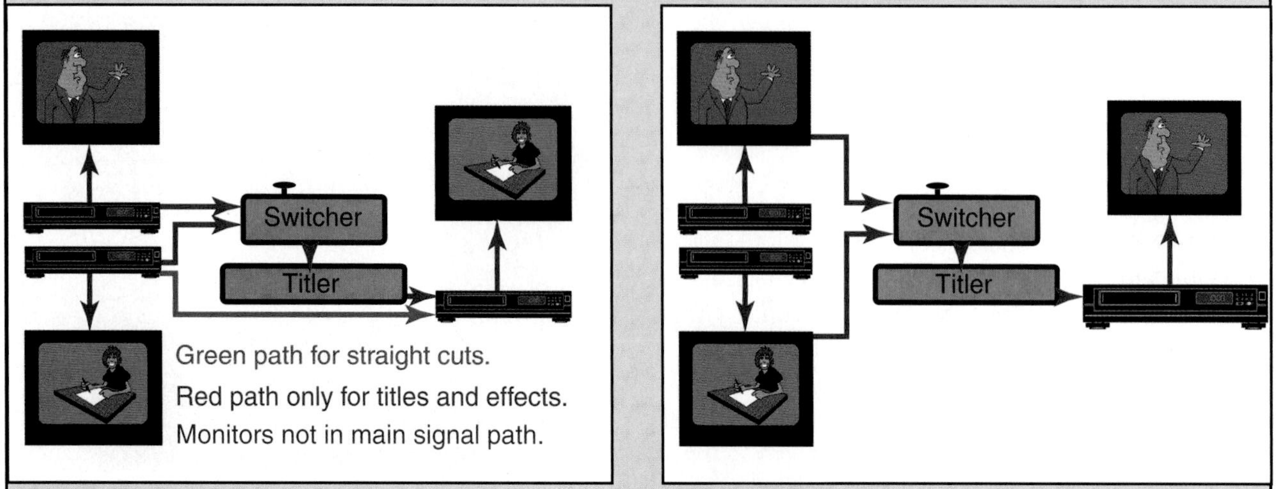

A—*Correct:* Monitors are cabled separately; other processors are hooked up only when needed.

B—*Incorrect:* Monitors and additional equipment are in the signal path.

using TV sets, connect them to their decks via RF connectors. If your connections are not working properly, start troubleshooting at the last component and check plugging and switch settings, working upstream toward the signal source.

Connecting other hardware

If you have a switcher/DVE generator, it is placed downstream of the source deck(s). Often this setup will involve a second source deck and monitor, to permit transitions between shots (A/B roll editing).

Since additional editing components degrade the signal slightly as it passes through them, they should be inserted into the editing path only when required, **Figure 24-7**.

Editing Picture and Track

With your equipment set up and cabled, you are ready to begin editing. The simplest method is called *assembly editing*. If you have a deck with flying erase heads, you can also do insert editing; if you have a second source deck and a switcher, you can perform A/B roll editing. We will look at each method, in turn.

Assembly Editing

Assembly editing means building a program by transferring one shot after another, in strict chronological order, from the first one to the last.

Assembly editing is sometimes called "crash cutting."

Every assemble edit requires the same five steps:

1. Position the ***assembly*** (record) ***tape***.
2. Find the new shot on the ***source*** (camera) ***tape***.
3. Rewind the source tape.
4. Roll the source tape.
5. Enable RECORD on the assembly deck.

In this discussion, VCR functions are printed in all-capital letters, like RECORD, and we use the generic term "enable" to mean activate them—cause them to start operating.

Here is each step in greater detail.

1. **Position assembly tape.** Using slow motion, if possible, find the last frame of the previously recorded shot that you wish to keep and PAUSE the assembly deck on the next frame after it. Then, enable RECORD to place the assembly deck in RECORD/PAUSE mode.
2. **Find new shot start frame.** Using the same slow motion technique, pause the source tape on the first frame of the new shot that you wish to copy to the assembly tape.
3. **Set up source tape preroll.** Note the numbers on the source deck time counter. Then rewind the source tape about five seconds or a bit more and pause it again.
4. **Roll source tape.** Enable PLAY on the source deck. When you reach the "runup point" on its time counter...

5. **Enable RECORD on assembly tape.** Copy the new shot until several seconds *past* the point where you think you want it to end (out-point), **Figure 24-8**.

Key terms are explained in the sidebar, *Runup and Preroll*. Some VCRs shift from RECORD/PAUSE to RECORD when you press PAUSE a second time. With others, you press the PLAY control instead.

In discussing source shots, the first frame transferred to the assembly tape is called the ***in-point*** and the last frame is the ***out-point***, **Figure 24-9**.

In illustrations such as Figure 24-9, the individual pictures are not single frames, but merely indicators of the type used in computer editing timeline displays.

When you repeat this process with the next edit, you will back up the assembly deck to the desired out-point of this newly copied shot and go into RECORD/PAUSE. That way, the next shot copied will replace the unwanted tail end of this one. With practice, you will find that you can copy source material to your assembly tape with frame-accurate precision.

Assembly editing permits you to make professional edits with a minimum of equipment. But it does not allow you to fix mistakes or add new material to previously edited shots. For that, you need to perform *insert editing* instead.

Insert (and Dub) Editing

Insert editing means replacing a piece of picture, sound, or both with new material.

Figure 24-8.
Always record past your probable out-point.

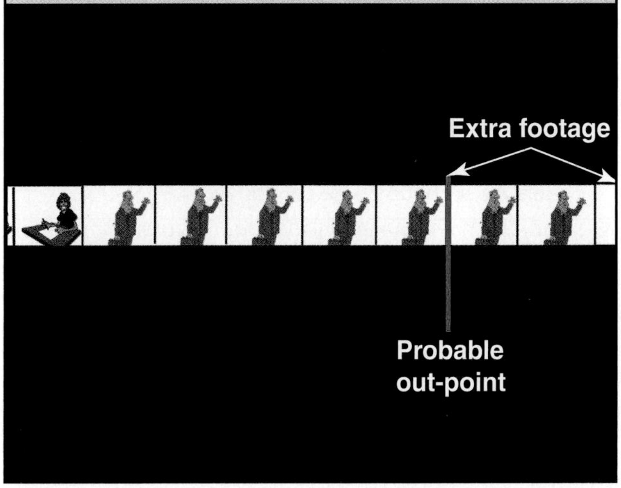

Figure 24-9.
In- and out-points.

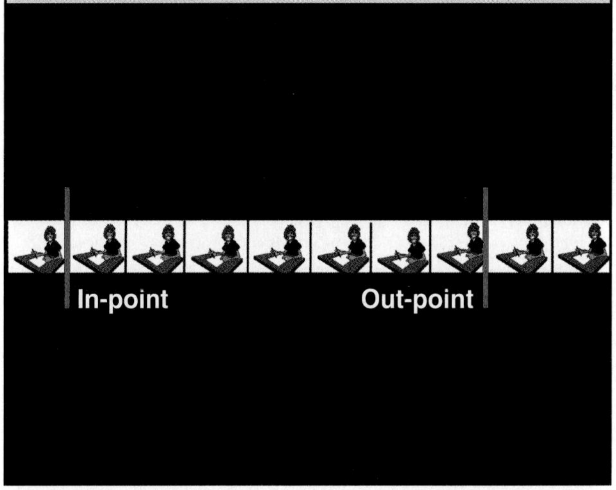

As noted earlier, inserts and dubs require an assembly deck with flying erase heads.

The terminology is confusing because the entire procedure is called "insert editing." However, replacing a picture is called "inserting" but replacing audio track is called "dubbing." Most editing VCR controls are labeled INSERT for picture and DUB (or AUDIO DUB) for sound.

Using video insert

A video insert allows you to replace previously transferred video while leaving the original sound under the new picture. For example, suppose you made a regular transfer of the woman's shot, **Figure 24-10**. Using the INSERT command, you can replace the middle of her shot with the man's shot, while her audio continues over it.

Figure 24-10.
A video insert.

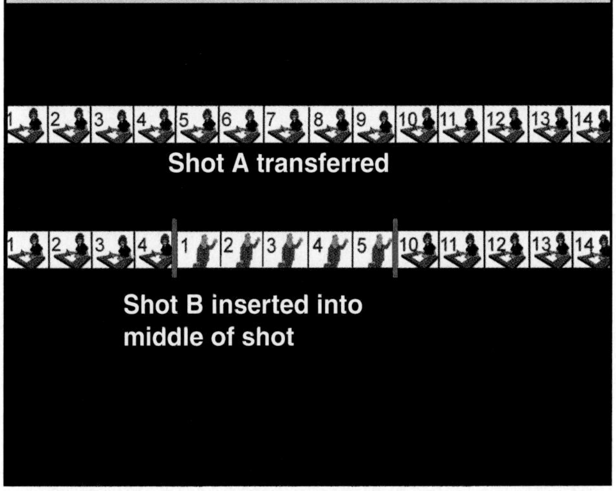

Runup and Preroll

It is important to understand that your assembly (record) VCR will start recording exactly where you pause the tape, but it will not start transferring the instant you enable RECORD. It takes the deck between one and two seconds before it actually begins laying a signal down on tape.

Enabling RECORD at the in-point loses 1-2 seconds.

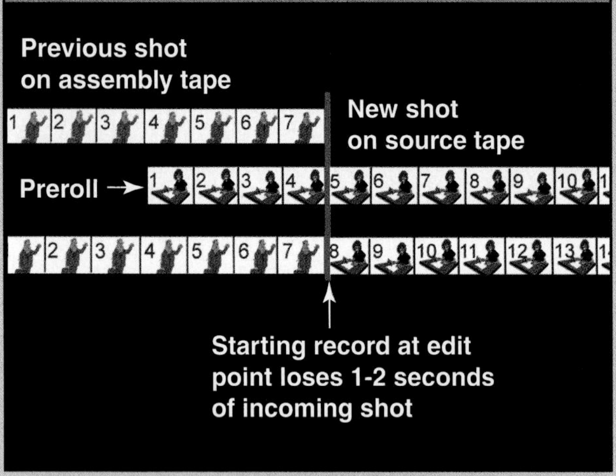

This means that you cannot copy a shot accurately by starting the source and assembly decks together or by playing the assembly tape and enabling RECORD when you see the in-point go by on the source tape. Either method will produce edits that miss the first few seconds of the shot.

That is why you need to calculate runup and use preroll.

Runup

Runup is the exact amount of time between enabling RECORD and starting to actually copy a shot. If you know the runup time of your record deck, you can make frame-accurate copies by enabling RECORD just that much earlier.

For example, suppose you wish to begin a new shot at 01:20:10 on your source deck counter. (That means "1 hour, 20 minutes, and 10 seconds.") Suppose that your VCR takes exactly two seconds to start recording, which means that two seconds is its runup time.

To start copying at 01:20:10, then, you would need to enable RECORD two seconds early, when the source deck counter reads 01:20:08.

Preroll

Like RECORD on the assembly deck, PLAY on the source deck does not start as soon as you enable it. You must give the source deck time to get up to speed and begin sending a high-quality picture. This extra time is called ***preroll***. You do not need to determine preroll time precisely; it is generally safe to allow

Enabling RECORD at the runup point captures the first frame of the incoming shot.

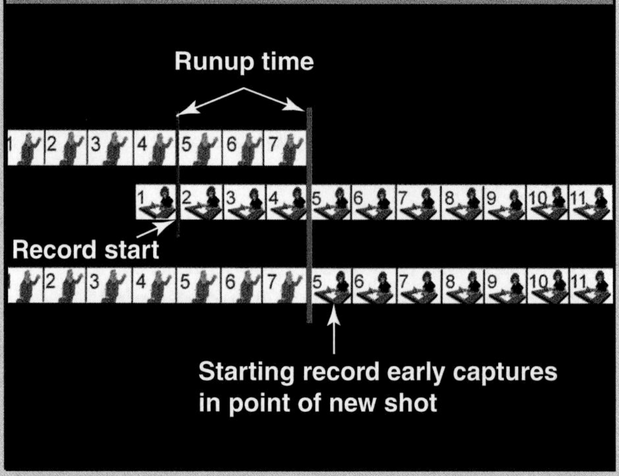

it about five seconds. So if your in-point is 01:20:10, then back up the source tape five seconds for preroll, to 01:20:05.

Counting down

In our example, the transfer sequence would resemble this (using the numbers on the source deck counter):

01:20:05: enable source PLAY, to allow for preroll.

01:20:06

01:20:07

01:20:08: enable assembly RECORD, to allow for runup.

01:20:09

01:20:10: transfer begins at actual source shot in-point.

Calculating runup

VCR runup times vary by brand and model, and even by individual machine. You can learn the runup time of your particular assembly deck by trial and error. But if you have a digital stopwatch (or stopwatch function on a wristwatch) you can determine it precisely. Here is the procedure:

- Begin videotaping a closeup of the stopwatch set to 00:00:00:00 (most stopwatches are accurate to one-one hundredth of a second), and then start it running. Stop taping after the watch has counted perhaps 20 seconds.
- Cue up the stopwatch footage in your source VCR at the head of the shot, before the watch is activated.
- Cue up a second tape in your assembly VCR and place the deck in RECORD/PAUSE.
- Play the stopwatch footage. At a selected point, such as 5 seconds exactly (00:00:05:00) enable RECORD on the assembly deck and copy the running stopwatch for a few seconds.

- Play back the shot on the assembly tape, noting the stopwatch time on the very first frame recorded (use slow motion or single frame advance, if possible). Suppose, for instance, that the reading is 00:00:06:50.
- Subtract the start reading from the first frame reading:
  6:50 (first frame actually recorded)
  –5:00 (frame at which RECORD was enabled)
  1:50 (runup time)

Due to lags in reaction time, your results will probably resemble those shown in the photos, where the nominal five-second mark is actually 00:05:12 and the first recorded frame is at 00:06:53. The runup time, then, is 1 and 41/100 seconds. For editing purposes, figuring 1.4 seconds is accurate enough.

For frame-accurate cutting, you would always enable RECORD 1.5 seconds before the in-point of the shot you wish to transfer. (Remember that this is only an example, and the runup time on your VCR will probably be different.)

Calculating runup time. A—Stopwatch reading when RECORD is enabled. B—First recorded frame of stopwatch.

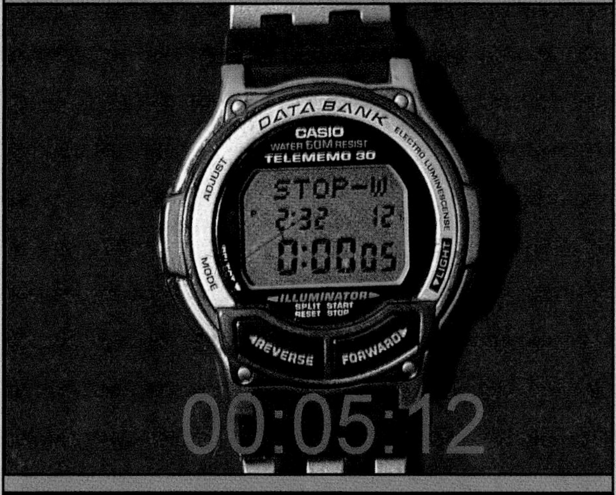

Using audio dub

Audio dubbing replaces sound instead of picture. It is useful for adding narration, sound effects, or music.

You cannot use the dub function to combine (mix) audio tracks. The dub function completely replaces one sound track with another.

Suppose, for example, that you want to dramatize a car crash without wrecking an actual car. To do this, you let the car exit the frame and then use the AUDIO DUB co mmand to replace the remaining production track with a car crash effect. Audio dub can also be used for dialogue replacement, Foley sound effect recording, and narration, but its most common application is probably for laying music under picture.

Procedures for insert and dub

Insert and dub use exactly the same steps, except for two very important differences. Instead of preparing your assembly deck by cuing the tape and entering RECORD/PAUSE, you enable INSERT/PAUSE or DUB/PAUSE. This difference is critical because if you mistakenly use RECORD/PAUSE, you will replace both picture and track and the fixed erase head will ruin the material following the transfer.

There is no way to fix this mistake. To recover, you will have to go back to the beginning of the original shot, re-transfer it, and continue editing from there.

And, instead of transferring past your intended out-point, stop the insert or dub *precisely* at the out-point or you will record over material that you wish to save.

Two notes about *insert* and *dub:* some assembly decks perform runup automatically when you use insert or dub. Experiment to see how your VCR works. Also, some decks allow you to use both insert and dub at the same time. This is useful for revising earlier work by laying in both the picture and sound of replacement footage.

For relatively simple dubs and inserts, these procedures work well in analog editing systems. If, however, you mix and match audio and video in more complex combinations, consider computerized nonlinear editing instead.

Where Did the Audio Dub Go?

VCRs record stereo audio by "multiplexing" (combining) the audio and video signals on the diagonal recording tracks.

When making audio dub edits, it is important to understand exactly what happens to the audio tracks on the tape:

- The original sound remains on the stereo tracks.
- The newly dubbed sound replaces the original audio on the linear, monaural track.

Most VCRs automatically default to stereo playback mode, so when you audition an audio dub, you may hear only the original track instead. To prevent this confusion, set your VCR to play back in mono mode.

The audio dub is audible only when played back in mono.

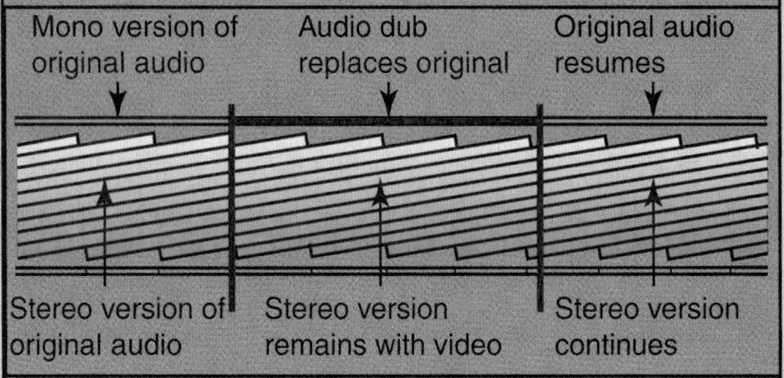

A/B Roll Editing

A/B roll editing involves combining picture and/or track from two sources (called A and B) in real time. An A/B roll editing setup is illustrated in **Figure 24-11**.

Typically, both program sources are VCRs, but the B source can also be a CD player, a microphone, or any other video or audio input.

In a typical application such as a dissolve, Deck A plays the shot being copied to the assembly deck, while deck B plays the tape containing the next shot. At the transition point, the switcher cross-fades (dissolves) from Shot A to Shot B and continues the transfer.

This discussion makes all switches from the A deck to the B deck. In actual editing, you will often switch from B to A, too.

Figure 24-11.
A simple A/B roll setup.

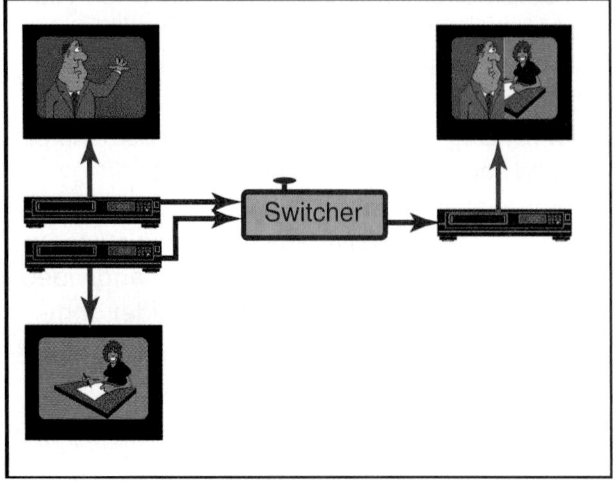

Essentially the same procedure is used to create other effects, to make composite shots, and to superimpose titles over video. A/B roll editing is also useful for combining the video of one source with the audio of another, without using the audio dub function.

A/X Roll Editing

Some camcorders have built-in digital effects, especially wipes and dissolves. If you are using one of these camcorders as your assembly deck, you can make so-called "A/X" transitions with your camera. (They are called A/X because a B roll is not involved.)

Basically, when you enable the camera's effect function, it captures the last frame of Shot A as a still and holds it in a memory called a frame buffer. Then, when you record Shot B, the camera automatically creates a transition between the still in its memory and the incoming signal.

Here is the general procedure for creating an A/X effect; in this example, a wipe (the specific steps depend on the design of the camcorder):

- Before transferring the outgoing (A) shot, enable the WIPE function on the camcorder.
- Transfer Shot A, ending it at the frame you want frozen.
- Transfer the incoming shot (B) in the usual way, **Figure 24-12**.

In planning an A/X transition, avoid cutting the A shot during obvious on-screen movement because the suddenly frozen frame will reveal the trick.

Figure 24-12.
Making an A/X transition: At the end of shot A... ...stop recording after the subject has left the frame. Transfer shot B in the normal way. The camcorder will create the transition automatically.

Transitions, Effects, and Titles

Transitions (like dissolves), visual effects (like superimpositions) and titles all involve two or more images on-screen at once. For this reason, most transitions and effects are achieved through the Digital Video Effects (DVE) functions of A/B switchers.

Stand-alone titling hardware can overlay titles on pictures without going through a switcher.

Working with these program enhancements requires care and planning because in analog editing, you must do everything simultaneously. You cannot return to previously edited shots and add titles or transitions. You cannot superimpose an image on visuals that you have already copied to the assembly tape.

Transitions

Transitions are visual effects that signal changes in program material. Traditional transitions like fades and dissolves indicate changes in time and place. Transitions that wipe, flip, fly in, tumble, turn like pages, or otherwise transform image A into image B are used to provide visual variety and excitement. Suggestions for selecting and using transitions appear elsewhere in this book. Here, we are concerned with the techniques used to create them in a analog editing setup.

From here on, we will use the acronym for Digital Video Effects: DVE.

Programming transitions

You preset a transition (or any other effect) on your digital switcher. That way, at the transition point in your program, you need only enable the DVE, which creates the effect automatically. Programming presets include:

- *Speed*: You can make the effect last anywhere from a fraction of a second to several seconds.
- *Background*: If you are not using A/B roll editing, the effect will bring the new shot in on top of a background, whose color you can select.
- *Edge*: Some transitions, especially wipes, include edges between the old and new images. Depending on the features of your switcher, you may be able to make the edge soft, hard, or highlighted with a colored line. As you can see from **Figure 24-13**, each type creates a somewhat different visual effect.

Figure 24-13.
Edge transitions. A—Soft edge wipe. B—Hard edge wipe. C—Colored edge wipe.

Graphic Effects

All switchers (and some camcorders, which can be used as assembly edit decks) include popular graphic effects such as:

- **Horizontal or vertical flop.** You can often correct an error in screen direction by reversing a shot left-to-right, **Figure 24-14**, or turning it upside down.
- **Mosaic.** Mosaic breaks the image up into small squares (you can program their exact size on the switcher).
- **Monochrome.** Monochrome lets you turn your image into black and white.
- **Paint.** This effect breaks the image into pieces and colors them with coarser hues.
- **Negative.** This effect can be useful for copying the negatives of still photographs as direct positives.

Figure 24-14.
Although flopping can be a useful effect, beware of image elements, like reversed words or letters, that can give away the flopped shot. (Western Recreational Vehicles)

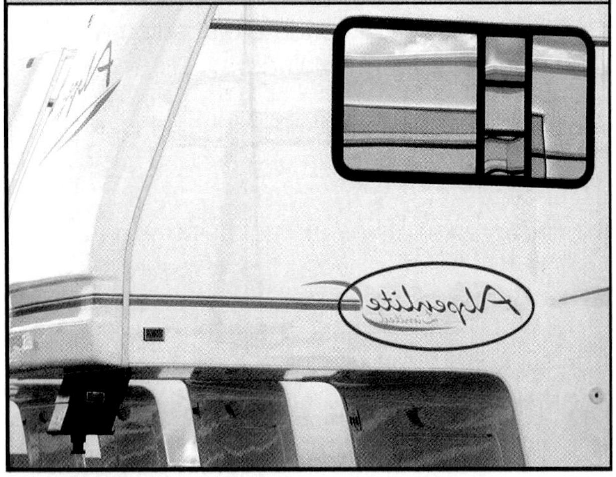

Motion Effects

Alterations of image movement can have many practical uses:

- **Strobe.** This effect creates jerky movement by repeating frames. Though exaggerated strobe effects have limited usefulness, a slight strobe can simulate the jerkiness of silent movies.
- **Still.** Still is a useful DVE effect because, unlike the PAUSE control on a VCR, it produces a perfectly motionless and undistorted picture. However, both methods yield an inferior quality image because they display only one of the two fields that make up a full frame.
- **Slow motion.** Slow motion is created not in a DVE but in the source deck. To achieve it, simply operate the VCR in slow motion as you copy the shot.
- **Fast motion.** Use exactly the same technique for fast motion.

Multiple Images

There are three types of screen display that contain more than a single video image.

Superimpositions

Superimpositions are pairs of images in which one picture is underneath the other and is partly visible through it. They are easily created by running one image on the A roll, the other on the B roll, and manually keying a partial dissolve between them.

Split screen

Create a split screen with a different image on each part by manually executing a partial wipe and then stopping part-way.

Picture-in-picture

This effect opens up a small second image within the main image, often in one corner of the screen. Picture-in-picture is useful when an on-screen person is required to comment on the main image—whether a teacher explaining a shot of a science experiment or a person signing for hearing-impaired viewers.

Compositing

Compositing means replacing one part of an image with part of another image. Compositing is used to insert someone or something into a scene, or to add an image or even a setting behind the subjects.

Perhaps the most widespread use of compositing is in TV news, where maps are composited behind the weather reporters.

Titles

In analog editing, titles must be inserted in chronological order; if superimposed over picture, they must be done as each shot is transferred to the assembly tape. There are three kinds of titlers: camera titlers, stand-alone titlers, and computer software.

Camera titlers

Some consumer camcorders allow you to create titles in the camera and superimpose them over shots as you tape them. Normally,

Figure 24-15.
Titling. A—A stand-alone titler. B—Titlers can create sophisticated composite effects. (Videonics)

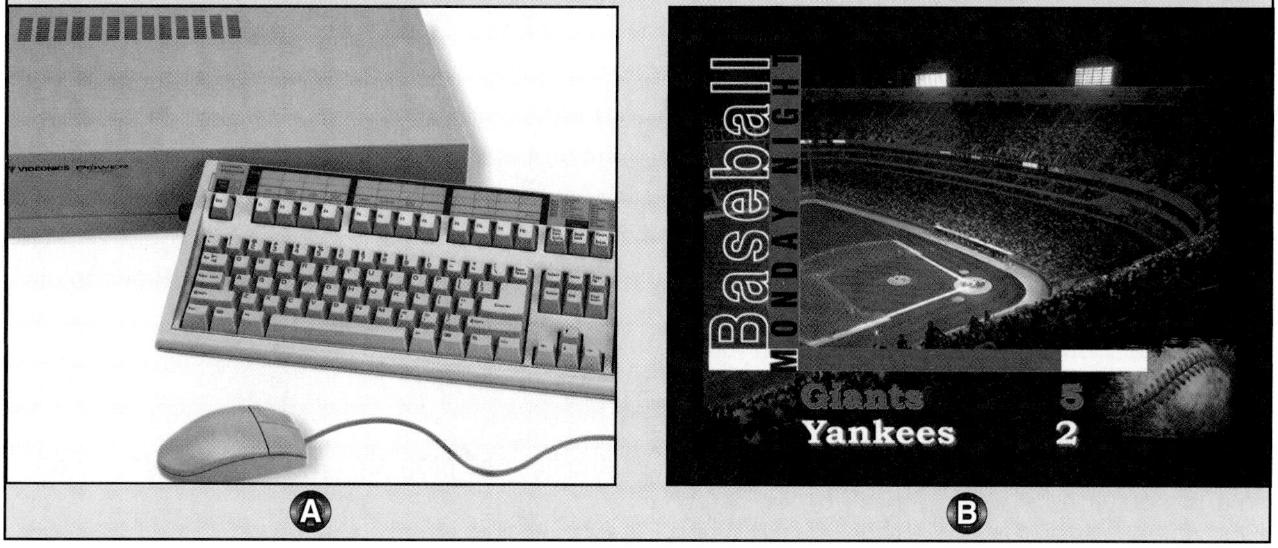

that would force you to make all titling decisions while shooting; but if you are using your camcorder as an assembly deck, you can add titles during postproduction, too. Unfortunately, most cameras produce crude-looking letters with very little choice of type, color, or size.

Stand-alone titlers

Video titlers are available as separate pieces of hardware, **Figure 24-15**. In addition to creating titles in a wide variety of colors, fonts, and sizes, they can also composite titles over live-action footage.

Like switchers, titlers are now hard to find, but they do appear on Internet auction and want-ad Web sites.

Computer-generated titles

With a computer presentation program like Corel Presentations or Microsoft PowerPoint, you can create very sophisticated video titles.

You will also need a computer card or accessory box to translate computer video output into NTSC television signals.

Create your titles on a composite color background (usually blue or green). Because the computer-generated color is absolutely uniform, you can key titles beautifully over live action, through your DVE. For best results, always evaluate the titles on an NTSC monitor as you create them. A computer screen does not show them as they will appear on TV. For complex designs, record the titles on a work tape and key them from the B roll source deck, for more precise control.

From Analog to Nonlinear

Analog editing offers a simple, inexpensive entry into the world of video postproduction. Beyond this entry level, however, the traditional approach has been largely replaced by digital, nonlinear editing. Computer-based postproduction offers all the power and flexibility of film editing, but at a fraction of the cost. Nonlinear editing was explored in detail in Chapters 21, 22, and 23.

Chapter Review

Answer the following questions on a separate piece of paper. Do not write in this book.

1. In analog editing, you must perform all operations ______.
2. RCA plugs and jacks are color-coded ______ for composite video.
3. *True or False?* In analog editing, the process of inserting video (replacing picture material) is called dubbing.
4. ______ are visual effects that signal changes in program material.
5. Superimpositions, ______, and picture-in-picture are three types of screen display that contain more than a single video image.

Technical Terms

Analog editing: Working with both source materials and finished program on videotape, rather than in a computer.

Assembly editing: Constructing a video program by transferring shots in sequence to a master tape.

Assembly tape: The tape in the record VCR, on which the program is created.

Audio dubbing: Replacing a segment of previously recorded audio without disturbing the accompanying video.

Compositing: Combining foreground elements from one image with the background of another.

Fixed head: A record, play, or erase head that operates in a continuous line on the tape passing across it.

Flying head: A record, play, or erase head that operates on the tape in short discrete diagonal segments.

Generation: A level of tape copy. An original camera tape is first generation, a copy made from it is second-generation, a copy of the copy is third generation.

Generation loss: The decline in video and audio quality from one analog generation to the next.

Head: A tiny electromagnet that encodes signals by magnetizing particles on videotape (record), decodes previously recorded signals by reading their magnetic patterns (play), or destroys previously recorded signals by giving all tape particles an identical magnetic alignment (erase).

In-point: The first frame of a new piece of material.

Insert editing: Replacing part of a previously edited video and/or audio segment with new material. The opposite of assembly editing.

Out-point: The last frame of a new piece of material.

Preroll: Starting the source deck playback prior to the in-point of the shot being transferred.

Runup: The difference between the point in time when the RECORD function is enabled and the first frame of recorded material.

Source tape: The tape providing the raw material for editing; usually, but not always, an original camera tape.

Glossary of Technical Terms

A

Act: A major division of a program, typically ten or more minutes long, and (on TV) divided from other acts by fade-out/fade-in transitions.

Action line: An imaginary line separating camera and subject. Keeping the camera on its side of the line maintains screen direction.

Additive editing: Creating a program from raw footage by starting with nothing and adding selected components.

Amp (ampere): In lighting, the amount of power being drawn by a light.

Analog: A recording that imitates the original. The level of an analog video signal varies continuously to record the image or sound.

Analog editing: Working with both source materials and finished program on videotape, rather than in a computer.

Analog video: Video that is recorded as a continuous electrical signal, rather than a series of codes.

Angle of view: The breadth of a lens' field of coverage, expressed as an arc of a circle, such as "10°."

Aperture: The opening in the lens that admits light. The iris diaphragm can vary the aperture from fully open to almost or completely closed.

Artifacts: Small visual blemishes that mar the quality of an image.

Assembly editing: In analog editing, constructing a video program by transferring shots in sequence to a master tape.

Assembly tape: The tape in the record VCR, on which the program is created.

Associative continuity: A sequence of shots related by similarity of content.

Asymmetrical balance: A composition in which dissimilar elements have equal "visual weight."

Audio: Collectively, the sound components of an audiovisual program.

Audio dubbing: In analog editing, replacing a segment of previously recorded audio without disturbing the accompanying video.

Authoring: In video postproduction, designing and creating a DVD.

Autoexposure: The camera system that delivers the correct amount of illumination to the recording mechanism, regardless of the light level of the shooting environment.

Autofocus: The camera system that ensures that the subject in the image appears clear and sharp.

Available light: The natural and/or artificial light that already exists at a location.

B

Background light: A light splashed on a wall or other backing to lighten it and add visual interest.

Background track: An audio track of the characteristic sounds of an environment, such as ocean surf, city traffic, or restaurant noises, usually laid under the production audio track.

Balanced line: A three-wire microphone cable designed to minimize electrical interference.

Barn doors: Metal flaps in sets of two or four, attached to the front of a spotlight to control the edges of the beam.

Beat: A short unit of action in a program, often, though not always, corresponding to a scene.

Bit depth: The number of digits (bits) used to encode each piece of digital data. Eight-bit, 16-bit, and 24-bit are common bit depths.

Blacking the tape: Prerecording an entire tape with a pure-black picture, no sound, and time code (or in analog formats, a control track).

Blocking: A performer's movement within a shot. It must be rehearsed and memorized so that the actor hits the correct spots with respect to camera and lighting, and so that the movements can be repeated exactly for other angles on the same action.

Boom: A studio microphone support consisting of a rolling pedestal and a horizontal arm.

Booming: Moving the entire camera up or down through a vertical arc (also *craning*).

Breaking news: News that is covered as it is happening (or soon afterward).

Brightness: The position of a pictorial element on a scale from black to white.

Broad: A small, rectangular, open light used mainly for fill and background lighting.

Budgeting: Predicting the costs of every aspect of a production and allocating funds to cover it.

Buffer shot: A shot placed between two others to conceal differences between them.

Buildup: A sequence of title cards, each one adding a new line of information.

Burning: Recording onto disc.

Business: Activities performed during a shot, such as writing a letter or filling a vase with flowers.

Button: An element on a DVD menu that, when activated, links to a program, a menu, or a function.

C

Camcorder: An appliance that both captures moving images (***cam***era) and stores them on tape (re***corder***).

Camera angle: The position from which a shot is taken, described by horizontal angle, vertical angle, and subject size.

Camera dolly: A rolling camera support. A cart dolly resembles a wagon; a pedestal dolly mounts the camera on a central post.

Camera light: A small light mounted on the camera to provide foreground fill.

Capture: To import footage from a tape to a computer, digitizing it if the original is analog video.

Capture card: An internal computer accessory card for capturing video.

Cardioid: A spatial pattern of microphone sensitivity, named for its resemblance to a Valentine heart.

CCD (Charge-Coupled Device): One type of camcorder imaging chip that converts optical images into electronic signals.

Century stand (C-stand): A telescoping floor stand fitted with a clamp and usually an adjustable arm, for supporting lights and accessories.

Chapter: A designated section of one program on a DVD.

Cheat: To move a subject from its original place (to facilitate another shot) in a way that is undetectable to the viewer.

Cinema verité: A type of documentary attempting perfect fidelity to actual events, with a minimum of intervention by its makers (often referred to as simply "*verité*").

Clip: A unit of video, audio, graphics, or titles, as used in digital editing.

Codec: A program that *codes* a signal in digital form and *decodes* it into analog form.

Color shot: A view of the scene, or a detail of it, not directly part of the action.

Color temperature: The overall color cast of nominally "white" light, expressed in degrees on the Kelvin scale. Sunlight color temperature (5200K) is cooler (more bluish) than halogen light color temperature (3200K).

Color temperature meter: A light meter that measures the relative blueness or redness of nominally "white" light.

Commercial: A very short program intended to sell a product, a person, or an idea.

Compositing: A digital editing process in which elements of one image replace elements of another image to create a combination of both.

Composition: The purposeful arrangement of visual elements in a frame.

Concept: The organizing principle behind an effective program. Often called an angle, perspective, or slant.

Conflict: The struggle between opposing forces that creates dramatic action.

Continuity: The organization of video material into a coherent presentation.

Contrast: The difference between the lightest and darkest parts of an image, expressed as a ratio ("e.g., four-to-one").

Cookie: A sheet cut into a pattern and placed in the beam of a light to throw distinctive shadows such as leaves or window blinds.

Copy protection: A system of encoding designed to prevent copying DVDs.

Correcting: Making small, continuous framing adjustments to maintain a good composition.

Cover: Additional angles of the main subject, or shots like inserts and cutaways, recorded to provide material needed for smooth editing.

Crew: Production staff members who work behind the camera. In larger professional productions, the producer, director, and management staff are not considered "crew."

Cross cutting: Also called "intercutting," cross cutting is showing two actions at once by alternating back and forth between them. It can also be used to show three or even more parallel actions.

Cut together: A term with two meanings: 1) to follow one shot with another; 2) to select edit points on outgoing and incoming shots so that an edit is not apparent.

Cutaway: A shot other than, but related to, the main action.

D

Day-for-night: A method of shooting daylight footage so that it appears to have been taken at night.

Decoding: Identifying and understanding the elements in a composition.

Default: A setting that is selected automatically unless the user changes it manually.

Delivery system: The method (such as Web site, TV monitor, or kiosk) by which a program will be presented, and the situation (alone at a desk, in a training room, in a crowded store) in which it will be watched.

Depth of field: The distance range, near-to-far, within which subjects appear sharp in the image.

Device control: The ability of software to operate external hardware (camera or VCR) by remote control.

Dialogue: Speech by performers on-screen.

Diffusion: White spun glass or plastic sheeting placed in the light path to soften and disperse it.

Digital: A recording that repeatedly samples the original continuous signal and records the numerical values of the samples, instead of the signal itself.

Digital linear tape (DLT): A DVD storage file format for recording on a special half-inch tape that will be the master for mass-producing DVDs.

Digital video effect (DVE): Informally, any digitally-created transitional device other than a fade or dissolve. (Technically, fades and dissolves are digital, too).

Digital zooming: Increasing the subject size by filling the frame with only the central part of the image.

Digitize: To record images and sounds as numerical data, either directly in a camcorder or in the process of importing them to a computer.

Disc image file: A computer file containing all the components of a DVD project.

Dissolve: A fade-in that coincides with a fade-out, so that the incoming shot gradually replaces the outgoing shot. Typically used as a transition between sequences that are fairly closely related.

Documentary: A type of nonfiction program purporting to communicate information about a real-world topic.

Dollying: Moving the entire camera horizontally (also *trucking* and *tracking*).

Dramatic structure: The organization of a story to build interest and excitement.

Dress: To add decorative items to a set or location. Such items are called "set dressing."

Dropped frames: One or more frames lost during capture, usually because of insufficient computer speed.

Dutch: Referring to an off-level camera. The "put dutch on a shot" or "dutch the shot" is to purposely tilt the composition.

DVD disc: A high-density recording medium capable of holding two hours or more of video.

E

Editing: Creating a video program from production footage and other raw materials.

Editorial documentary: A documentary that attempts to win viewers over to its position or point of view.

EDL (Edit Decision List): A computerized, frame-accurate record of every component in a program. It can be used to automatically recreate the program from the original sources.

Emphasis: The process of calling attention to a pictorial element.

Equalization: The adjustment of the volume levels of various sound frequencies to balance the overall mixture of sounds.

Equalizer: A device for adjusting the relative strengths of different audio frequencies.

Expressionism: A lighting style that adds a heightened emotional effect, without regard for lighting motivation.

F

Fade-in: A transition in which the image begins as pure black and gradually lightens to full brightness. Used to signal the start of a major section such as an act or an entire program.

Fade-out: A transition in which the image begins at full brightness and gradually darkens to pure black. Used to signal the end of a major section such as an act or an entire program.

Filename: The identification of an editing element as it is stored in a computer. Filenames may or may not be identical to timecode addresses or slate numbers.

Fill light: The light that lightens shadows created by the main (key) light.

Film: An audiovisual medium that records images on transparent plastic strips by means of photosensitive chemicals.

Filter: (1) A digital effect that changes the character—such as color—of the clip(s) it is applied to. (2) In lighting, a sheet of colored or gray-tinted plastic placed over lights or windows to modify their light.

First play: A menu or program that plays automatically when a viewer starts the DVD.

Fishpole: A location microphone support consisting of a hand-held telescoping arm.

Fixed head: A record, play, or erase head that operates in a continuous line on the tape passing across it.

Flag: A flat piece of opaque metal, wood, or foam board placed to mask off part of a light beam.

Flashback: A sequence that takes place earlier in the story than the sequence that precedes it.

Floodlight: A large-source instrument that typically lacks a lens; used for lighting wide areas.

Fluorescent lamp: A lamp that emits light from the electrically charged gasses it contains.

Flying head: A record, play, or erase head that operates on the tape in short, discrete, diagonal segments.

Focal length: Technically, one design parameter of a lens, expressed in millimeters (4mm, 40mm). Informally, the name of any particular lens, such as, "a 40mm lens."

Focus: Photographically, the part of the image (measured from near to far) that appears sharp and clear; also, generally, the object of a viewer's attention.

Foley studio: An area set up for recording real-time sound effects synchronously with video playback.

Four-point edit: An edit that specifies the in- and out-points of both the source clip and the place in the program at which the clip will be inserted.

Frame: The border around the image. Also, to frame something is to include it in the image by placing it inside the frame.

Frame off: To frame off is to exclude something from an image by placing it outside the frame.

Frame rate: The actual speed with which video frames are displayed. The frame rate of NTSC video is 29.97 frames per second.

Frame-accurate: Edits made at exactly the single frame desired.

f-stop: A particular aperture. Most lenses are designed with preset f-stops of f/1.4, f/2, f/2.8, f/4, f/5.6, f/8, f/11, f/16, and f/22.

G

Gaffer: The chief lighting technician on a shoot.

Gag: Any effect, trick, or stunt in a movie.

Gain: The electronic amplification of the signal made from an image, in order to increase its brightness.

Generation: A level of analog tape copy. An original camera tape is first generation, a copy made from it is second generation, a copy of the copy is third generation.

Generation loss: The decline in video and audio quality from one analog generation to the next.

Glamorous lighting: Lighting that emphasizes a subject's attractive aspects and de-emphasizes defects.

Graphic: Technically, any nonlive-action visual element, but not commonly applied to titles.

H

Halogen lamp: A lamp with a filament and halogen gas enclosed in an envelope of transparent quartz.

Head: A tiny electromagnet that encodes signals by magnetizing particles on videotape (record), decodes previously recorded signals by reading their magnetic patterns (play), or destroys previously recorded signals by giving all tape particles an identical magnetic alignment (erase).

Head room: The distance between the top of a subject's head and the upper edge of the frame.

High key: Lighting in which much of the image is light, with darker accents.

I

I-frames: "Intra-frames" that contain all the image data. (See *MPEG*.)

Image: A single unit of visual information. An image may last for many frames, until the subject, the camera, or both, create a new image by moving. Most shots contain several identifiable images.

Image stabilization: Compensation to minimize the effects of camera shake. Electronic stabilization shifts the image on the chip to counter movement; optical stabilization shifts parts of the lens instead.

Incandescent lamp: A lamp with a filament enclosed, in a near-vacuum, in a glass envelope ("bulb").

Incident meter: A light meter that measures illumination as it comes from the light sources.

Infomercial: A program-length commercial masquerading as a regular program.

Infotainment: Documentaries intended to entertain as much as to inform.

In-point: The first frame of a new piece of material.

Insert: A close shot of a detail of the action, often shot after the wider angles, for later *insertion* by the editor.

Insert editing: Replacing part of a previously edited video and/or audio segment with new material. The opposite of assembly editing.

Instrument: A unit of lighting hardware such as a spotlight or floodlight.

Interface: The method by which a user communicates with a computer.

Intra-frame compression: A digital video recording system that saves storage space by condensing and recording the data for every frame.

Iris diaphragm (iris): A mechanism inside a lens (usually a ring of overlapping blades) that varies the size of the lens opening (aperture).

J

Jump cut: An edit in which the incoming shot is visually too similar to or too different from the outgoing shot.

K

Key light: The principal light on a subject.

Keyboard shortcut: A keystroke or keystroke combination that performs an action otherwise performed by a mouse.

L

Lamp: The actual bulb in a lighting instrument.

Large-source: A lighting instrument, such as a scoop or other floodlight, with a big front area from which light is emitted. Light from large-source instruments is relatively soft and diffuse.

Lavaliere: A very small microphone clipped to the subject's clothing, close to the mouth.

Lead room: The distance between the subject and the edge of the frame toward which it is moving.

Leading lines: Lines in a composition that emphasize an element by pointing to it.

Leading the eye: Using compositional techniques to direct the viewer's attention.

LED (Light Emitting Diode): An electronic light source for special applications.

Letterboxed image: A wide screen image displayed in the center of a regular TV screen, with black bands above and below it filling the frame.

Library footage: Film or video collected, organized, and maintained to be rented for use in documentary programs (also called *stock footage*).

Line reading: A vocal interpretation of a line that includes its speed, emphases, and intonations. In simple terms, "You came back!" is one reading of a line, and "You came *back*?" is a different reading.

Linear continuity: A sequence of shots organized chronologically and/or logically.

Lines: Scripted speech to be spoken by performers.

Live: Being presented as it is transmitted from the recording video cameras. "Live on tape" means a recorded, but largely unedited presentation of a live program.

Look room: The distance between the subject and the edge of the frame toward which it is looking.

Looping: Replacing dialogue in real-time by recording it synchronously with video playback.

Low key: Lighting in which much of the image is dark, with lighter accents.

M

Magic hour: The period, of up to two hours before sunset, characterized by long shadows, clear air, and warm light.

Magic realism: A lighting style that creates a dreamy or unearthly effect, often enhanced digitally in postproduction.

Magnification: The apparent increase or decrease in subject size of an image, compared to the same subject as seen by the human eye. Telephoto lenses magnify subjects; wide angle lenses reduce them.

Marks: Places within the shot where the performer is to pause, stop, turn, etc. These spots are identified by marks made of tape or chalk lines.

Match point: The places, in two shots, where they can be cut together to make the action appear continuous.

Media opportunities: Events created specifically for the purpose of being covered by news organizations.

Medium key: Lighting an image so that neither light nor dark tones dominate the image (a term used only in this book).

Menus: In DVD authoring, screens displaying buttons for navigating to other menus, programs, or functions.

Microphone (mike): A device that converts sound waves into electrical modulations, for recording.

Mini-DV: The most popular recording format for small camcorders. Uses intra-frame compression.

Mixer: A device that balances the input strengths of signals from two or more sources, especially microphones.

Mixing: The blending together of separate audio tracks, either in a computer or through a sound mixing board.

MJPEG: A video compression system that reduces the data required to record each frame.

Montage: A brief, multilayer passage of video and audio elements designed to present a large amount of information in highly condensed form.

Motivated lighting: Lighting that imitates real-world light sources at the location.

MPEG: A recording format that records full information only for selected base frames (called "I-frames"). For subsequent frames, the system records only data that differs from that of the base frame.

N

Narration: Spoken commentary on the sound track.

Naturalism: A lighting style that imitates real-world lighting so closely that it is invisible to most viewers.

Neutral density filter: In lighting, a gray sheet filter placed over windows to reduce the intensity of the light coming through them. May also be a glass filter placed over the lens for the same purpose.

NTSC: The TV standard used in North America and Japan. Other major standards include PAL and SECAM.

O

Object: An item on a DVD menu that does not link to anything.

Objective insert: A detail of the action presented from a neutral point of view.

On-camera light: A small light mounted on the camera to provide foreground fill.

Optical zoom: Changing a lens' angle of view (wide angle to telephoto) continuously by moving internal parts of the lens.

Out-point: The last frame of a new piece of material.

P

Pan: (1) To pivot the camcorder horizontally on its support. (2) In lighting, a large, flat instrument fitted with fluorescent or other long tube lamps.

Panning: Pivoting the camera horizontally in place.

Parallel cutting: Presenting two or more sequences at once by showing parts of one, then another, then the first again (or a third) and so on.

Parallel time streams: Two or more lines of action presented together, either by alternating pieces of the various streams, or by splitting the screen for simultaneous presentation.

Pedestaling: Moving the camera straight up or down on its central support.

Perspective: The simulation of depth in a two-dimensional image.

Pickup: A shot obtained later to record action that was either missed or inadequately covered previously.

Pickup pattern: The directions (in three dimensions) in which a microphone is most sensitive to sounds.

Pictorial realism: A lighting style in which lighting, though motivated, is exaggerated for a somewhat theatrical effect.

Picture plane: The actual two-dimensional image.

Placeholder: On a DVD menu template, an indicator of where an item should be placed to make it a button.

Playback: (1) Previously recorded video and/or audio reproduced so that actors or technicians can add to or replace parts of it synchronously, in real-time. (2) Studio-quality music recording reproduced so that performers can synchronize lip movements with it while videotaping.

Playlist: A list for each DVD button containing the names of one or more menus, programs, or functions activated by that button.

Plug-in: A program that has no interface of its own, but is operated by commands it has inserted into another program.

Point of view (POV): A vantage point from which the camera records a shot. Unlike a camera *angle*, a point of view is not described by subject distance ("closeup," etc.) and unlike a *setup*, a point of view is not concerned with production equipment.

Post house: Short for "postproduction house," a professional facility that rents editing setups and personnel for high-quality video editing.

Practical: A lighting instrument that is included in shots and may be operated by the actors.

Preroll: Starting the source deck playback prior to the in-point of the shot being transferred.

Presenter: An on-camera narrator who speaks directly to the viewer.

Preview window: A window that displays what a piece of visual material looks (or will look) like.

Production audio: The audio track recorded with the video.

Production track: The "live" sound recorded with the video.

Program: Any complete video presentation, from a five-second commercial to a movie two or more hours long.

Project: The process of authoring a DVD.

Protection shot: A shot taken to help fix potential problems with other shots.

Pulling focus (following focus, racking focus): Changing the lens focus during a shot to keep a moving subject sharp.

R

Realism: A lighting style that looks like real-world lighting, though it is slightly enhanced for pictorial effect.

Reconstructed footage: Re-enactments of past events so that they can be videotaped.

Reflective meter: A light meter that measures illumination as it bounces off the subjects and into the camera lens.

Reflector: A large silver, white, or colored surface used to bounce light onto a subject or scene.

Regional encoding: Code on a DVD that permits it to be played in only one designated world region.

Rendering: Creating a full-quality version of material previously edited in a lower-quality form.

Rim light: A light placed high and behind a subject to create a rim of light on head and shoulders, to help separate subject and background.

Ripple edit: An insert edit in which all the material following the inserted clip is moved up or back to accommodate it.

Roll to raw stock: To advance a videotape through previously recorded sections to blank tape, in preparation for additional recording.

Rolling edit: An insert edit in which the shot following the inserted material is shortened or lengthened to accommodate it.

Rugged lighting: Lighting that emphasizes three dimensional qualities and surface characteristics of a subject.

Rule of thirds: An aid to composition in the form of an imaginary tic-tac-toe grid superimposed on the image. Important picture components are aligned with the lines and intersections of the grid.

Runup: The difference between the point in time when the RECORD function is enabled and the first frame of recorded material.

S

Sample rate: The number of times per second at which an analog signal level is inspected and digitally recorded. Expressed in kiloHertz (kHz). The sample rate of CD-quality audio is 44.1 kHz, meaning that the signal is sampled over 44,000 times per second.

Scale: The perception of an object's size by comparison to another object.

Scene: A short segment of program content, usually made up of several related shots.

Scoop: A type of floodlight used mainly in TV studios.

Screen: A mesh material that reduces light intensity without markedly changing its character.

Screen direction: The orientation of on-screen movement with respect to the left and right edges of the frame.

Screen time: The length of real-world time during which a sequence is displayed on the screen (in contrast to the length of video world time that apparently passes during that sequence).

Script: A full written documentation of a program, including scenes, dialogue, narration, stage directions, and effects, that is formatted like a play script.

Scrub: To move quickly through a program by dragging a handle along a track below a preview window.

Sell: To add details to increase the believability of a screen illusion.

Sequence: A segment of a program, usually a few minutes long, consisting of related, organized material.

Setting focus: Adjusting the lens to make the subject appear clear and sharp.

Setup: An arrangement of production equipment (typically camera, microphone, and lighting) placed to record shots from a certain point of view.

Shoot: To record film or video; also, "a shoot" is an informal term for the production phase of a film or video project.

Shot: An uninterrupted recording by a video camera.

Shutter: The electronic circuitry that determines how long each frame of picture will accumulate on the imaging chips before processing. The standard shutter speed is 1/60 per second for NTSC format video.

Silk: A fabric material that both reduces light intensity and directionality, producing a soft, directionless illumination.

Slate: A written video and/or spoken audio identification of a component such as a shot, a line of narration, or a sound effect.

Slide edit: An insert edit in which the out-point of the preceding shot and the in-point of the following shot are adjusted to accommodate the inserted material.

Slip edit: An insert edit in which the in- and out-points of the inserted material are adjusted so that the preceding and following shots remain unchanged.

Small-source: Describing a lighting instrument such as a spotlight, with a small front area from which light is emitted. Light from small-source instruments is generally hard-edged and tightly focused.

Softlight: A lamp or small light enclosed in a large fabric box, which greatly diffuses the light.

Sound: The noises recorded as audio.

Sound effects: Specific noises added to a sound track.

Source tape: The tape providing the raw material for editing; usually, but not always, an original camera tape.

Specular reflections: Hard, bright reflections from surfaces such as water, glass, metal, and automobile paint that create points of light on the image. Often controllable by a polarizer.

Speed: The light-gathering ability of a lens, expressed as its maximum aperture. Thus, an f/1.4 lens is "two stops faster" (more light sensitive) than an f/2.8 lens.

Spin: The management of information for somebody's benefit.

Split edit: An edit in which the audio and video of the new shot do not begin simultaneously. When "video leads," the sound from the preceding shot continues over the visual of the new shot. When "audio leads," the sound from the new shot begins over the end of the preceding visual.

Splitting: In editing, separating the audio and video tracks of a clip so that each can be processed individually.

Spotlight: A small-source lighting instrument that produces a narrow, hard-edged light pattern.

Staging in depth: Positioning subjects and camcorder to exploit perspective in the image.

Standup: A report presented on camera, usually by a reporter.

Stock shot: A shot purchased from a library of pre-recorded footage for use in a program; collectively called "stock footage."

Storyboard: (1) Program documentation in graphic panels, like a comic book, with or without dialogue, narration, stage directions, and effects. (2) In digital editing, a metaphor that represents the project as a succession of slides laid out in order.

Straight cut: An edit in which audio and video change simultaneously; also, an edit that does not include an effect such as a fade or dissolve.

Subjective continuity: A sequence organized according to the feelings of the director/editor, rather than by chronological or other objective criteria.

Subjective insert: A detail of the action presented from a character's point of view.

Subtractive editing: Creating a program by removing redundant or poor-quality material from the original footage and leaving the remainder essentially as it was shot.

Symmetrical balance: A composition in which visual elements are evenly placed and opposed.

T

Tabletop: Videography of small subjects and activities on a table or counter.

Take: A single attempt to record a shot.

Talent: Every production member who performs for the camera.

Tally light: A small light on a camera that glows to show that the unit is recording.

Telephoto: A lens or a setting on a zoom lens that magnifies subjects and minimizes apparent depth by filling the frame with a narrow angle of view.

Teleprompter: A machine that displays text progressively as a performer reads it on-camera.

Television: Studio-based, multicamera video that is often produced and transmitted "live."

Template: A pre-designed DVD menu that is customized by substituting appropriate buttons and text.

Tent: A white fabric draped all around a subject to diffuse lighting completely for a completely shadowless effect.

Text: Anything written on a DVD menu.

Three-point edit: An insert edit in which both the in- and out-points of the new material are specified, but only the in-point of the program into which it is inserted.

Three-point lighting: So-called "classic" subject lighting, consisting of key, fill, rim, and background lights.

Throw: The distance between a light or camera and the subject lit; also, with lights, the maximum useful distance between those points.

Thumbnail: A very small image representing a clip or shot.

Tilting: Pivoting the camera vertically in place.

Timebase: The basis on which time code is assigned to frames. The timebase of NTSC video is 30 frames per second.

Timecode address: The unique identifying code number assigned to each frame (image) of video. Timecode is expressed in hours, minutes, seconds, and frames, counted from the point at which timecode recording is started.

Timeline: In digital editing, a metaphor that represents a project as a stack of long, narrow strips, one strip for each program element.

Title: Any lettering that appears in a video.

To frame: To feature a subject in a composition.

To frame off: To exclude an object completely from the screen.

Toggle: To switch back and forth between two states, such as off and on.

Transducer: The component of a microphone that converts changing air pressure ("sound") into an electrical signal ("audio").

Transition: A visual effect signaling the end of one program part and the beginning of the next.

Treatment: A written summary of a program, formatted as narrative prose, that may be as short as one paragraph or as long as a scene-by-scene description.

Trim: To indicate the in- and out- points of a clip in order to specify its content and set its exact length.

Trimming: Removing unwanted material from the beginning and/or end of a shot.

Tripod: A three-legged camera support that permits leveling and turning the camera.

TV standard: A broadcast system, such as NTSC, SECAM, or any of several varieties of PAL.

U

Umbrella: A silver- or fabric-covered umbrella frame used to reflect light onto subjects from its concave side, or to filter light through its convex side.

Unbalanced line: A two-wire microphone cable subject to electrical interference, but less bulky and expensive, for use in amateur applications.

V

Video: An audiovisual medium that records on magnetic tape or other storage device by electronic means; also, single-camera taped program creation in the manner of film production rather than studio television.

Visual literacy: The ability to evaluate the content of visual media through an understanding of the way in which it is recorded and presented.

Voice-over: Spoken commentary on the sound track from someone who is not in the image.

Voltage: The electrical potential or "pressure" in a system, nominally 110 volts in North America.

W

Wallpaper: Footage intended to take up screen time while the audio presents material that cannot be shown.

Wattage: In lighting, the power rating of a lighting instrument. 500, 750, and 1,000 watt lamps are common.

White balance: The camera setting selected to compensate for the color temperature of the light source that is illuminating the subject.

Wide-angle: A lens or a setting on a zoom lens that minimizes subjects and magnifies apparent depth by filling the frame with a wide angle of view.

Widescreen video: Video using a screen proportioned 16 to 9, in contrast to the traditional TV screen's 12 to 9 proportion.

Window dub: In analog editing, a copy of the original footage in which the time code is visible in a small window at the bottom of the screen.

Wipe: A transition between sequences in which a line moves across the screen, progressively covering the outgoing shot ahead of it with the incoming shot behind it.

Wizards: Mini-programs that guide users through computer operations, step by step.

Z

Zoom: To magnify or reduce the size of a subject by changing the focal length of the lens.

Zooming: Changing the lens' angle of view to fill the frame with a smaller area (zooming in) or a larger area (zooming out). The effect is to reduce or expand the size of subjects in the frame.

Index

A

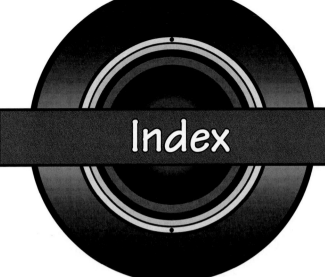

B

C

D

E

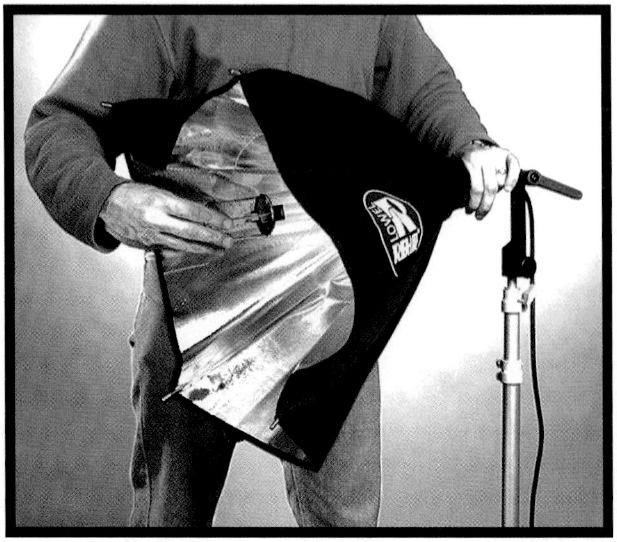

F

G

H

I

J

K

L

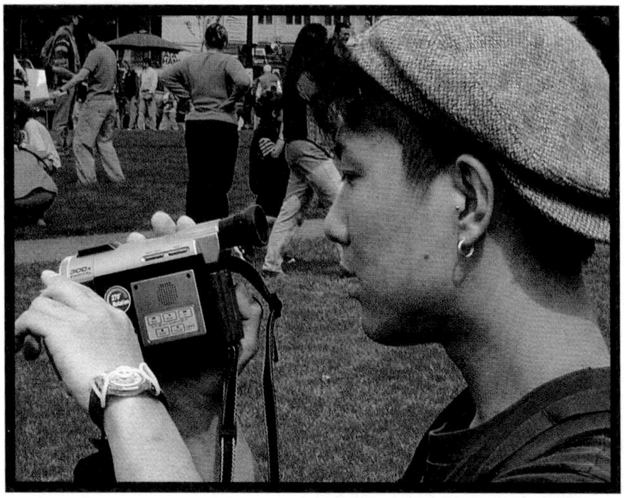

M

Mix
000.07:13
knights
knights2
marina
car
beach
beach2
sunset
music1
weplay
Silence
Timeline
Range
Add
Remove
Create
Fade
Volume
0 dB
Correction
0 dB
Silence
000.06:01

N

O

P

R

S

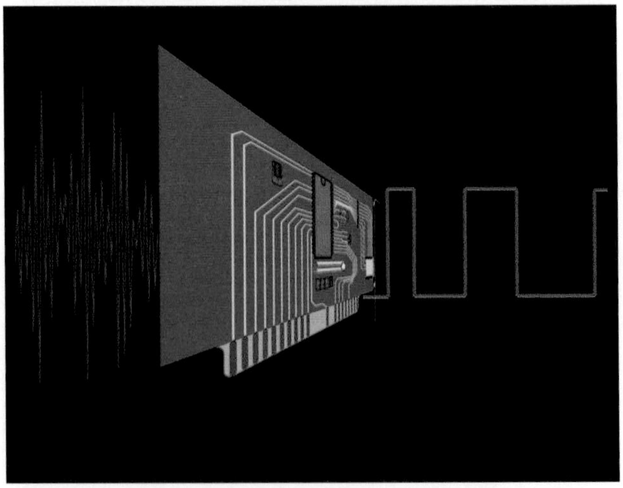

T

U

V

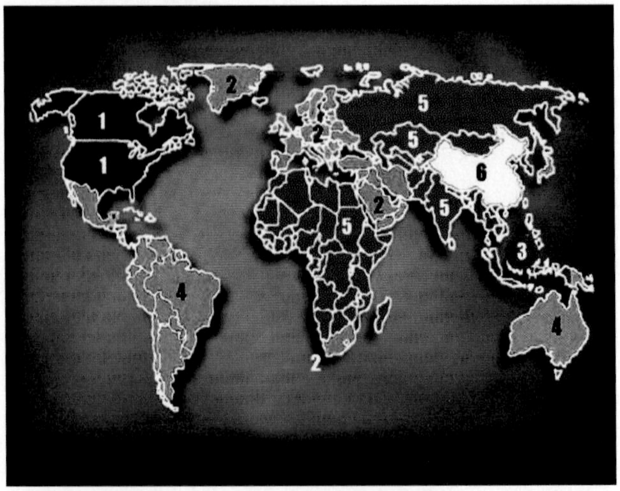

W

Z

OFF ZEBRA
ON